Environmental Hazards

The much expanded sixth edition of *Environmental Hazards* provides a fully up-to-date overview of all the extreme events that threaten people and what they value in the twenty-first century. It integrates cutting-edge material from the physical and social sciences to illustrate how natural and human systems interact to place communities of all sizes, and at all stages of economic development, at risk. It also explains in detail the various measures available to reduce the ongoing losses to life and property. Part I of this established textbook defines basic concepts of hazard, risk, vulnerability and disaster. Attention is given to the evolution of theory, to the scales and patterns of disaster impact and to the optimum management strategies needed to minimize the future impact of damaging events. Part II employs a consistent chapter structure to demonstrate how individual hazards, such as earthquakes, severe storms, floods and droughts, plus biophysical and technological processes, create distinctive impacts and challenges throughout the world. The ways in which different societies can make positive responses to these threats are placed firmly in the context of sustainable development and global environmental change.

This extensively revised edition includes:

- A new concluding chapter that summarizes the globalization of hazard and critically examines the latest perspectives on climate-related disasters.
- Fresh views on the reliability of disaster data, disaster risk reduction, severe storms, droughts and technological hazards.
- More boxed sections with a focus on both generic issues and the lessons to be learned from a carefully selected range of recent extreme events.
- An annotated list of key resources, including further reading and relevant websites, for all chapters.
- 183 diagrams, now in full colour, and available to download on: www.routledge.com/9780415681063/.
- Over 30 colour photographs and more than 1,000 references to some of the most significant and recent published material.

Environmental Hazards is a clearly written, authoritative account of the causes and consequences of the extreme natural and technological processes that cause death and destruction across the globe. It draws on the latest research findings to guide the reader from common problems, theories and policies to explore practical, real-world situations and solutions. This carefully structured and balanced book captures the complexity and dynamism of environmental hazards and has become essential reading for students of every kind seeking to understand this most important contemporary issue.

Keith Smith is Emeritus Professor of Environmental Science and former Dean of Natural Sciences at the University of Stirling. He is a Fellow of the Royal Society of Edinburgh.

'The latest edition of *Environmental Hazards* provides a reliable guide to the ever changing field of natural hazards and disasters. The sixth edition covers a remarkable range of interdisciplinary topics in an accessible manner. The text is a unique resource for anyone wanting to understand how human society on planet Earth often finds itself in peril, and what we can do about it'.

Roger Pielke, *Professor of Environmental Studies,*
University of Colorado at Boulder, USA.

'*Environmental Hazards* has become the indispensable text for hazards students and scholars. The new edition brings together a wealth of updated and new case studies and examples. The common structure adopted for the chapters in Part II enables useful comparisons between hazard types and the varied risks and adaptation opportunities they present. This is a detailed and thorough treatment of the complex approaches to and challenges of hazard management'.

Dr Maureen Fordham, *Enterprise Fellow*
Principal Lecturer in Disaster Management,
University of Northumbria, UK.

Environmental Hazards

Assessing risk and reducing disaster

Sixth edition

Keith Smith

Routledge
Taylor & Francis Group

LONDON AND NEW YORK

First published 1991
Second edition 1996
Third edition 2001
Fourth edition 2004
Fifth edition 2009

Sixth edition 2013
by Routledge
2 Park Square, Milton Park, Abingdon, Oxon OX14 4RN

Simultaneously published in the USA and Canada
by Routledge
711 Third Avenue, New York, NY 10017

Routledge is an imprint of the Taylor & Francis Group, an informa business

British Library Cataloguing in Publication Data
A catalogue record for this book is available from the British Library

Library of Congress Cataloging in Publication Data
A catalog record has been requested for this book

ISBN: 978–0–415–68105–9 (hbk)
ISBN: 978–0–415–68106–3 (pbk)
ISBN: 978–0–203–80530–5 (ebk)

Typeset in Minion and Univers
by Florence Production Ltd, Stoodleigh, Devon, United Kingdom

Printed and bound in India by Replika Press Pvt. Ltd.

Contents

List of figures   viii
List of plates   xiv
List of tables   xvi
List of boxes   xviii
Preface to the sixth edition   xxi
Preface to the first edition   xxii
Acknowledgements   xxiv

PART I
THE NATURE OF HAZARD   **1**

1   Hazard in the environment   **3**

   A   Introduction   3
   B   What are environmental hazards?   4
   C   Hazard, risk and disaster   11
   D   Earlier perspectives   14
   E   Current views: the complexity
       paradigm   18
   F   The organizational context   20

2   Dimensions of disaster   **23**

   A   Introduction   23
   B   Defining disaster   24
   C   Measuring disaster: archives   28
   D   Explaining disaster: time trends   31
   E   Explaining disaster: spatial patterns   37
   F   Managing disaster   42

**3   Complexity, sustainability and
vulnerability**   **46**

   A   Introduction   46
   B   Complexity science   46
   C   Complexity and disasters   47
   D   An example: the Bam earthquake   48
   E   Sustainability and disasters   51
   F   Vulnerability and resilience   52
   G   Drivers of vulnerability and
       disaster   61

**4   Risk assessment and
management**   **71**

   A   The nature of risk   71
   B   Risk assessment   75
   C   Risk perception and
       communication   81
   D   Risk perception in practice   84
   E   Risk management   86
   F   The role of information technology   91

**5   Reducing the impacts of
disaster**   **96**

   A   Scoping the task   96
   B   Protection: hazard resistance   99
   C   Mitigation: disaster aid   106
   D   Mitigation: insurance   114

E Adaptation: preparedness   118
F Adaptation: predictions, forecasts
   and warnings   124
G Adaptation: land use planning   127

PART II
THE EXPERIENCE AND
REDUCTION OF HAZARD   137

6 Tectonic hazards: earthquakes
and tsunamis   139
A Earthquake hazards   139
B Earthquake behaviour   142
C Primary earthquake hazards   146
D Secondary earthquake hazards   148
E Protection   153
F Mitigation   160
G Adaptation   164

7 Tectonic hazards: volcanoes   176
A Volcanic hazards   176
B The nature of volcanoes   177
C Primary volcanic hazards   179
D Secondary volcanic hazards   185
E Protection   189
F Mitigation   193
G Adaptation   193

8 Mass movement hazards   205
A Landslide and avalanche hazards   205
B Landslides   208
C Landslides: cause and triggers   213
D Snow avalanches   218
E Protection   220
F Mitigation   224
G Adaptation   226

9 Severe storm hazards   235
A Atmospheric hazards   235
B The nature of tropical cyclones   236
C How tropical cyclones develop   238
D Tropical cyclone hazards   241
E Severe summer storms   247
F Severe winter storms   251

G Protection   255
H Mitigation   259
I Adaptation   261

10 Weather extremes, disease
epidemics and wildfires   268
A Introduction   268
B Extreme temperature hazards   270
C The nature of disease epidemics   273
D Infectious diseases and climate   278
E Disease hazard reduction   283
F Wildfire hazards   286
G The nature of wildfires   288
H Wildfire hazard reduction   294

11 Hydrological hazards: floods   299
A Flood hazards   299
B Flood-prone environments   302
C The nature of floods   309
D Protection   318
E Mitigation   323
F Adaptation   329

12 Hydrological hazards:
droughts   337
A Drought hazards   337
B Types of drought   339
C Causes of drought hazards   351
D Protection   358
E Mitigation   361
F Adaptation   363

13 Technological hazards   371
A Introduction   371
B The scale and nature of the
   hazard   373
C An outline of theory   378
D Technological hazards in
   practice   379
E Perception: the transport and
   nuclear industries   385
F Protection   391
G Mitigation   391
H Adaptation   393

**14 Environmental hazards in a
    changing world   402**
    A Introduction   402
    B The globalization of hazard   403
    C Environmental change   405
    D Air pollution and climate
       change   407

    E Geophysical paths to disaster   410
    F Climate change and environmental
       hazards   419

Bibliography   435
Index   471

Figures

1.1	Environmental hazards at the interface between the natural events system and the human use system	4
1.2	Extreme geophysical events and severe system failures within a framework of global change and sustainability issues	8
1.3	A generalized spectrum of environmental hazards from physical to human causes	8
1.4	A matrix showing possible combinations of physical exposure to hazard and human vulnerability in relation to risk and security	9
1.5	Sensitivity to environmental hazard expressed as a function of annual rainfall and societal tolerance	10
1.6	Relationships between the severity of hazard, probability and risk	12
1.7	Schematic evolution of a drought disaster	13
1.8	The track of two Category-5 hurricanes across Central America	18
2.1	Possible losses and gains in disaster	25
2.2	A disaster-impact pyramid	26
2.3	Annual number of natural and technological disasters 1975–2009	31
2.4	Annual number of Great Natural Catastrophes according to type of event recorded 1950–2009	32
2.5	Annual number of people reported killed by natural and technological disasters 1975–2009	33
2.6	Annual total of reported economic losses in natural disasters 1980–2011	34
2.7	Annual total of overall and insured losses from Great Natural Catastrophes recorded 1950–2009	34
2.8	Annual number of volcanic eruptions and eruptions of size 0.1 km^3 or greater 1790–1990	35
2.9	Annual number of Great Weather Catastrophes in meteorological, hydrological and climatological categories recorded 1950–2009	37

2.10 Economic damages reported by country and disaster type for the 10 costliest
     natural disasters recorded 1991–2005   40
2.11 Global pattern of the UN Human Development Index, 2010   41
2.12 Global pattern of the UN Disaster Risk Index (DRI)   42
2.13 The reduction of risk through pre-disaster protection and post-disaster recovery   43
3.1 The DNA model applied to complexity in disaster causation   47
3.2 Location map of the city of Bam, Iran   48
3.3 The Swiss Cheese model of disaster   51
3.4 Percentage of the national population living on less than US$1 per day in 2007–8
     throughout the world   62
3.5 Socio-economic factors and fatality rates in flash floods during July 1993 in Nepal   64
3.6 Observed and predicted future growth in the urban population globally and by region   68
4.1 Risk plotted relative to benefit for various voluntary and involuntary activities   73
4.2 A probabilistic event tree for a hypothetical gas pipeline accident   75
4.3 Generalized statistical relationships between the magnitude and the frequency and
     return period of damaging natural events   77
4.4 The probability of occurrence of floods of various magnitudes during a period
     of 30 years   78
4.5 Annual maximum wind gusts at Tiree, western Scotland, 1927 to 1985   79
4.6 The effects of a change to increased variability on the occurrence of extreme events   80
4.7 The effects of a change to an increased mean value on the distribution of extreme
     events   80
4.8 Changes in human sensitivity to hazard due to variations in physical events and
     changes in societal tolerance   81
4.9 Sequential approach to natural hazard risk management in Switzerland   87
4.10 The ALARP approach to risk management   89
5.1 Energy release on a logarithmic scale for selected hazardous geophysical events   96
5.2 Simplified world map of selected natural hazards   97
5.3 Three categories of disaster reduction strategy   98
5.4 Flood defence along the coastline of Belgium and the Netherlands   101
5.5 The effectiveness of deflecting dams in steering snow avalanches   102
5.6 A theoretical illustration of the resistance of an engineered building to wind stress   103
5.7 Daily number of disaster victims attending hospitals in Guatemala City in relation
     to the arrival of medical supplies and emergency hospitals after the 1976 earthquake   107
5.8 Overview of the aid players involved in humanitarian emergencies   109
5.9 Annual number of Presidential Disaster Declarations in the USA 1953–2011   110
5.10 Annual total of humanitarian aid 1990–2008   112
5.11 Cumulative donor response to appeals for aid in the period following four major
     disasters   113
5.12 The accumulation of insured losses after the Northridge earthquake, 1994   115
5.13 A typical set of stakeholder groups involved in hazard reduction planning   119
5.14 Evacuation map for Galle City, Sri Lanka   120
5.15 Map showing the expected location of displaced households and available disaster
     shelters in the greater Memphis, Tennessee, area   122
5.16 The Village Disaster Risk Management Training model (VDRMT)   123

5.17	Generic model of a well-developed hazard forecasting and warning system	126
5.18	Map showing seismic shaking hazards from earthquakes in California	129
5.19	Proposed regulation map for volcanic risk reduction around Mount Pelée, Martinique	130
5.20	Flood Insurance Rate Map and coastal high-hazard area of Lee County, Florida	132
5.21	A matrix of the Swiss hazard zoning system	133
5.22	Debris flow hazard map of the alluvial fan at Llorts, Andorra	133
5.23	Portion of an earthquake fault zone map in California	135
6.1	World map of major tectonic plates and the distribution of active earthquakes and volcanoes	142
6.2	Map of the damage following the 1995 earthquake in Kobe, Japan	144
6.3	Schematic illustration of the four main types of earthquake waves	146
6.4	Map of the Mount Huascaran rock avalanche disasters in the Peruvian Andes	149
6.5	Changes in water level in the Pacific Ocean that created the tsunami of 11 March 2011	152
6.6	Typical evolution of a tsunami wave	153
6.7	Progress of a tsunami wave across the Pacific Ocean	154
6.8	Relationships between earthquake intensity and building damage in the 1995 Kobe earthquake	155
6.9	Schematic illustration of the effects of ground shaking on various types of buildings	157
6.10	Depiction of tsunami protection works	160
6.11	Structural measures and insurance adopted by residents of California against earthquakes	165
6.12	Earthquake prediction in New Zealand	167
6.13	The Pacific Tsunami Warning System	169
6.14	Hypothetical pattern of earthquake shaking in firm rock	171
6.15	Illustration of variations in ground shaking due to surface geology	172
6.16	Earthquake hazard planning in Ano Liossia, Athens, Greece	173
6.17	Typical example of coastal land planning for tsunami hazards	174
6.18	Part of the tsunami hazard map of the city of Hilo, Hawaii	174
7.1	Section through a composite volcanic cone	178
7.2	The influence of distance on hazardous volcanic phenomena	180
7.3	Hazard zone map for the Nevado del Ruiz volcano, Colombia	187
7.4	The distribution of lahar deposits on the slopes of Merapi volcano, Java	188
7.5	Simplified map of the fishing port of Vestmannaeyjar, Heimay, Iceland in 1973	190
7.6	Diagrammatic section of the tunnel system at Kelut volcano, Java	191
7.7	Organizational flow chart for a volcanic emergency plan	195
7.8	The stages of a generic volcanic-earthquake-swarm model	198
7.9	Proposed destinations for evacuees from a major eruption at Mount Vesuvius, Italy	200
7.10	The island of Hawaii zoned according to the risk from lava flows	201
7.11	Map of volcanic hazards at Galeras volcano, Colombia	202
7.12	Volcanic hazards around Mount St Helens, USA	203
8.1	Annual number of landslide publications 1945–2008	206

8.2   Annual number of avalanche fatalities in the USA 1950/51 to 2009/10
       winter seasons   207
8.3   Down-cutting by rivers can cause landslides   212
8.4   The characteristic profile of a rotational landslide   212
8.5   A map showing the area of land disturbed in the Vaiont landslide of 1963 215
8.6   Landslide activity in relation to rainfall in the tropics   216
8.7   The two most common types of snow-slope failure   219
8.8   Idealized slope section showing avalanche hazard reduction measures   222
8.9   Survival after an avalanche   226
8.10   Map of the area of the Tessina landslide in Northern Italy   228
8.11   Avalanche hazard management in the western USA   229
8.12   Distribution of landslides on Tonoas Island, Federated States of Micronesia in 2002 230
8.13   Reduction of landslide incidence by hazard management in Hong Kong, 1948–96 232
9.1   Some effects of hurricane 'Katrina' on New Orleans in August 2005   237
9.2   Destructive energy of hurricane wind speeds as compared to a tropical storm 238
9.3   The nature of the storm surge hazard   239
9.4   World map of the location and frequency of tropical cyclones   240
9.5   A model of the structure of a tropical cyclone   243
9.6   Population changes in coastline counties in the USA affected by hurricanes since
       1960   246
9.7   Annual hurricane damage during the twentieth century in the United States 246
9.8   Number of hailstorms for counties in England and Wales 1930–2004   251
9.9   Generalized tracks of severe wind-storms crossing western Europe 1999–2010 252
9.10   Insured losses suffered in European wind-storms in 1990   253
9.11   Hurricane losses to residential structures in the south-eastern USA   258
9.12   Percentage of households in Florida prepared for hurricanes   261
9.13   Average annual accuracy of Atlantic hurricane forecasts   263
10.1   Map of locust activity in the State of Victoria, Australia   269
10.2   Frequency of heat-related deaths by age in Philadelphia, USA during July 1993 272
10.3   Number of excess deaths recorded in France during the 2003 heat-wave   272
10.4   Spatial distribution of *P. falciparum* malaria endemicity in 2007   280
10.5   World map of areas reporting cases and outbreaks of cholera 2007–2009 282
10.6   Deaths and damages caused by bushfires in Australia   289
10.7   Sources of wildfire ignition in two different regions   290
10.8   Seasonal patterns of bushfire activity in Australia   291
10.9   The 'Ash Wednesday' bushfires of 16 February 1983 in Australia   293
10.10 Residence in Victoria, Australia and ownership of fire-fighting equipment 296
11.1   Flood hazard thresholds as a function of depth and velocity of water flow 300
11.2   Areas of potential flooding in England and Wales   301
11.3   Types of flooding in Bangladesh   305
11.4   Human vulnerability to flooding in Vietnam   307
11.5   Causes of floods in relation to other environmental hazards   310
11.6   Map showing extensive flooding over Pakistan in August 2010   310
11.7   Influence of urbanization on the hydrological cycle   312
11.8   Idealized flood hydrographs from rural and urban areas   313

11.9	Height of the storm surge in the North Sea on 31 January 1953	314
11.10	Annual losses from river floods in the USA 1904–2010	315
11.11	The number of planning applications for development on floodplain land in England 1996/97 to 2001/2	317
11.12	Engineered measures to protect land and development against river and marine floods	319
11.13	Flood stages of the Mississippi river, USA, during July 1993	320
11.14	Idealized flood hydrographs for water inflowing and discharging from a reservoir	320
11.15	Simulated flood discharges on the upper Mississippi river, USA during July 1993	320
11.16	Schematic of flood-proofed residential buildings on a river floodplain	322
11.17	Humanitarian funding levels at 22 July 2011 after the Pakistan floods emergency 2010	324
11.18	Schematic representation of the river flood hazard	328
11.19	Floodplain map for the Avon River at Northam, Western Australia	332
11.20	Adjustment to the flood hazard at Soldiers Grove, Wisconsin, USA	333
11.21	Draft reconstruction and relocation plan for the town of Grantham, Queensland, Australia	335
12.1	The development of a drought regime	338
12.2	A classification of drought types	340
12.3	An idealized flow duration curve for a river	342
12.4	Examples of droughts in Australia during the second half of the twentieth century	344
12.5	Percentage area of the United States in severe and extreme drought from January 1895 to August 2009	346
12.6	Annual corn yields in the USA 1960–1989	346
12.7	Rainfall patterns over Ethiopia	350
12.8	Countries of the Sahel region prone to drought	352
12.9	Sahelian rainfall during the rainy season as a percentage of the 1961–90 mean	353
12.10	Rainfall in eastern Australia during September–February in relation to the Southern Oscillation Index	354
12.11	Time-series of rainfall anomalies during June–September 1877–2006 over the core Indian monsoon region	355
12.12	Reservoir storage and flow regulation on the river Blithe, England	359
12.13	Idealized emergence of a water supply drought	359
12.14	Changes in water storage in reservoirs along the upper river Tone, Japan	360
12.15	Drought response and food security in sub-Saharan Africa	367
12.16	The use of check dams across intermittent water courses	369
13.1	Annual number of deaths 1900–84 from industrial accidents	376
13.2	Inverse relationship between the failure rate for all dams and the number of dams constructed between 1850 and 1950	383
13.3	Safety challenges and organizational responses required in the process industry	385
13.4	Disaster preparedness for people in the lowest income quartile, compared to the rest of the population in Alabama	395
13.5	Map of the pre-incident layout of the Buncefield fuel depot site	398
13.6	Idealized risk contours within the Consultation Distance around a high-hazard chemical site in the UK	400

13.7   Example of risk acceptance and land planning around major hazard establishments   400
14.1   A map of the KOF Index of Globalization in 2011   404
14.2   The increase in atmospheric carbon dioxide 1960–2010 at Mauna Loa, Hawaii   409
14.3   The global land–ocean temperature index 1880–2010   410
14.4   Idealized depiction of the two phases of the Walker circulation   412
14.5   Relationships between El Niño events and epidemics of Ross River virus in south-east Australia   414
14.6   Average track of winter wind-storms across the North Atlantic Ocean   415
14.7   Conceptual view of the oceanic 'conveyor belt'   416
14.8   Diagrammatic view of the interactions between extreme events, societal vulnerability and disaster reduction measures   422
14.9   Potential spread of malaria (*P. falciparum*) due to climate change   426
14.10 Annual maximum flood series for the Mississippi river at St Paul 1893–2002   427
14.11 Annual totals of winter precipitation in the Mediterranean region 1902–2010   428
14.12 Annual variations in the NAO Index 1870–2010   430
14.13 The progressive rise in global mean sea level   431

Plates

1.1   Timanfaya National Park on Lanzarote, Canary Islands   10

1.2   Multispectral colourized satellite image of hurricane 'Mitch' at 20.28 UTC on 26 October 1998   19

2.1   A woman cooks food outside a slum dwelling on stilts above a polluted waterway in Manila, Philippines   38

2.2   People queue for food aid distributed through the UN World Food Programme in Port-au-Prince, Haiti in January 2010   44

3.1   The ancient citadel of Bam, Iran   49

3.2   An old man made homeless by the $M_W = 7.6$ magnitude earthquake of October 2005 in Pakistan-administered Kashmir   53

3.3   An area of deforestation in Para State, Brazil during 2005   66

4.1   Emergency response to a 2009 road traffic accident on Highway 2/E30 leading to Berlin near Hanover, Germany   83

4.2   A scientist from the United States Geological Survey Volcano Observatory on Hawaii Island uses a portable GPS receiver to track changes in the surface topography   91

5.1   Part of the 17th Street Canal levee in New Orleans in August 2008 recently reconstructed after damage by flood-waters from hurricane 'Katrina'   100

5.2   Office buildings in Concepcion, Chile designed and purpose-built to withstand earthquake and tsunami stresses   104

5.3   Rescue workers recover a body from rubble created by the $M_W = 8.0$ magnitude earthquake at Hangwan, Sichuan Province, China, in 2008   107

6.1   A tsunami wave generated by the $M_W = 9.0$ magnitude Great Tōhoku earthquake on 11 March 2011   150

6.2   Examples of structural collapse in different types of buildings   158

7.1   Eruption of the Soufrière Hills volcano on the island of Montserrat in August 1997   181

7.2   The Eyjafjallajökull volcano in southern Iceland erupts on 16 April 2010   183
7.3   The spire of a church protrudes from ash, mud and rock deposited by pyroclastic
       flows that destroyed Plymouth, Montserrat   191
8.1   The Ferguson rockslide above Highway 140 in the Merced River canyon, California   210
8.2   Head scarp of the Las Colinas landslide, Santa Tecla city, El Salvador   217
8.3   A combination of snow fences and mature forest cover protect valley development
       from avalanches in the Austrian Alps   223
9.1   Residents in Tegucigalpa, Honduras, dig themselves out of flood debris in
       November 1998   245
9.2   A developing tornado threatens Caddo County, Oklahoma   248
9.3   Friends and neighbours help residents to assess damage after a tornado in
       Columbia, North Carolina, 2011   249
9.4   Highway congestion on route 37 heading north-west to San Antonio, Texas
       as residents evacuate to escape a hurricane, 1999   265
10.1   Children wade across a stream flowing through the Kroo Bay slum area of
       Freetown, Sierra Leone   283
10.2   A man uses a 'fogging' technique to spray insecticide against mosquitoes
       in Thailand   284
10.3   A fire truck moves away from an advancing bushfire in the Bunyip State Forest
       near Melbourne, Australia   292
11.1   A woman carries a water jug whilst wading through floods in Shyamnagar
       Upzila, Satkhira District, Bangladesh   306
11.2   Flooding along the Phra Pinkalao road, Bangkok, in November 2011   316
11.3   Flooding in the centre of the city of York, England during autumn 2000   324
12.1   The carcass of a dead animal lies near the UNHCR Dadaab refugee camp, in
       the North Eastern Province Kenya during August 2011   348
12.2   Pastoralists collecting water during March 2006 from a hole dug in a dry river
       bed in Ethiopia   357
12.3   A mother from Somalia nurses her 18-month-old malnourished son   362
13.1   Attempts to extinguish fires on the Deepwater Horizon oil rig on 21 April 2010   374
13.2   The Atal Ayub colony at Bhopal, Madhya Pradesh, India in 1992   382
13.3   A view of the fire at the Buncefield oil storage terminal near Hemel Hempstead,
       England on 11 December 2005   399
14.1   Firemen attempt to control a forest fire started in 1997 to clear land for
       development in East Kalimantan, Indonesia   408
14.2   Male residents of Fogafale, Tuvalu in the Pacific Ocean reinforce their small
       protective wall   433

Tables

1.1	Major categories of environmental hazard	5
1.2	Disaster-related deaths in high- and low-income countries 1980–2004	14
1.3	The evolution of environmental hazard paradigms	15
2.1	Disasters responsible for at least 100,000 deaths recorded since AD 1000	24
2.2	Number of disasters, deaths and economic damage in three different databases	28
2.3	List of disaster types and sub-types recorded in EM-DAT	30
2.4	Recorded deaths and economic losses in the five deadliest and the five costliest disasters 2000–10	39
3.1	Variables tested for use in the UNDP Disaster Risk Index	55
3.2	Variations in vulnerability at the household and family level	56
3.3	Socio-economic aspects of vulnerability in disaster	57
3.4	Restoration of power supplies in Los Angeles following the Northridge earthquake 1994	58
3.5	Community resilience indicators	60
3.6	Economic and social impact of disasters in Pacific Island nations 1950–2004	69
4.1	Basic elements of quantitative risk analysis	76
4.2	Differences between risk assessment and risk perception	82
4.3	Twelve factors influencing public risk perception	86
4.4	A risk matrix for industrial accidents	90
4.5	The use of remotely sensed imagery in hazard management	92
5.1	The world's 10 costliest natural disasters	117
5.2	Proportion of residents in the San Francisco Bay area, California taking loss-reducing actions	121
6.1	The 10 largest earthquakes in the world since 1900 and the 10 deadliest on record	140
6.2	The proportion of agricultural assets destroyed by the 1993 Mahrashtra earthquake	140
6.3	Annual frequency of occurrence of earthquakes of different magnitudes	144

6.4	Worldwide recorded fatalities from tsunamis 1995–2011	151
6.5	The number of people surviving after rescue following the Kobe earthquake	161
6.6	Earthquake safety checklist	165
6.7	Loma Prieta earthquake losses by hazard type	172
7.1	Human impacts of volcanic hazards in the twentieth century	177
7.2	Selected criteria for the Volcanic Explosivity Index	179
7.3	Effects of large volcanic eruptions on weather and climate	182
7.4	Precursory phenomena observed before a volcanic eruption	196
8.1	Classification of landslides	201
8.2	Impact pressure and the damage from snow avalanches	220
8.3	Vegetation characteristics and avalanche frequency	233
8.4	The Swiss avalanche zoning system	233
9.1	Severe storms as compound hazards	236
9.2	The Saffir/Simpson hurricane scale	237
9.3	The world's 10 deadliest tropical cyclones in the twentieth century	241
9.4	Fujita scale of tornado intensity	248
9.5	The 10 deadliest tornadoes recorded in the USA	249
9.6	The five most severe winter storms affecting the north-east USA	255
9.7	Hurricane 'Katrina' estimated insured losses	260
9.8	Numbers killed and evacuated in cyclone emergencies in Bangladesh	261
10.1	Selected infectious diseases with epidemic potential	279
11.1	Five flood disasters in Europe 2000–10	300
11.2	Reduction in flood losses on the Mississippi and Missouri rivers	321
11.3	Aid requirements and beneficiaries following the 2010 floods in Pakistan	323
11.4	Numbers of flood victims rescued by air and boat in the Mozambique floods of 2000	329
11.5	Fact-file on the post-1993 flood relocation of Valmeyer, Illinois	334
12.1	Drought severity ranked by return period	341
12.2	Major droughts and their impact in Australia	343
12.3	Adoption of adjustments to drought in Bangladesh	364
12.4	Global monitoring and warning for drought and food shortages	366
13.1	Some examples of early technological accidents	372
13.2	Annual death toll 1970–1985 from natural and man-made disasters	375
13.3	Technological disasters 1900–2011 by continent and disaster type	375
13.4	The 10 deadliest transport, industry and miscellaneous accidents 1900–2011	377
13.5	Simplified criteria for notification of a 'major accident' to the European Commission	384
13.6	Deaths per 10^9 kilometres travelled in the UK	387
13.7	The International Nuclear Event Scale	388
13.8	Major technological accidents causing UK deaths during the twentieth century	393
14.1	Global change paths tending to increase environmental hazards	406
14.2	Known impact craters ranked by age	418
14.3	Energy release, environmental effects and fatality rates of extra-terrestrial impacts	419
14.4	Changes to selected climatic extremes since 1950 and projected to 2100	423
14.5	The world regions most vulnerable to coastal flooding due to future sea-level rise	432

Boxes

1.1   Tōhoku 2011: a major na-tech disaster   6

1.2   The twentieth-century paradigm debate   16

2.1   Types of disaster impact   25

3.1   The Swiss Cheese model   51

4.1   Quantitative risk assessment   76

4.2   The ALARP principle   89

5.1   Avalanche-deflecting dams in Iceland   102

5.2   Advantages and disadvantages of commercial insurance   116

5.3   Earthquake hazard zoning in California   134

6.1   The 2010 Haiti earthquake: a long-term disaster   141

6.2   The Modified Mercalli earthquake intensity scale   145

6.3   Ground shaking in earthquakes   147

6.4   Earthquake safety and buildings   156

6.5   Problems of aid delivery in the aftermath of a major earthquake   162

7.1   Airborne volcanic ash and aviation   183

7.2   Emergency response in Montserrat 1995   190

7.3   Crater-lake lahars in the wet tropics   193

8.1   The Vaiont landslide   215

8.2   How snow avalanches start   219

8.3   The Vargas landslides   225

8.4   The Tessina landslide warning system   228

9.1   Hurricane 'Katrina': lessons for levees and for lives   242

9.2   The dream of severe storm suppression   256

9.3   Improving hurricane evacuation in the United States   265

10.1   Diseases and disasters   275

10.2   The emerging flaviviruses   277

11.1	Flood hazards on the Yangtze River, China	303
11.2	Floods in England: the summer of 2007	326
12.1	Drought in Australia	342
12.2	Drought and famine in the Horn of Africa	349
13.1	The 1984 gas disaster at Bhopal, India	281
13.2	The 1986 nuclear disaster at Chernobyl, Belarus	389
13.3	The 2005 explosions and fires at the Buncefield fuel site, UK	398

Preface to the sixth edition

It is over 20 years since the first edition of *Environmental Hazards* was published. Since then, our understanding of the environment and its hazards has improved. The theoretical base is stronger and more sophisticated tools for hazard monitoring and risk mitigation have become available. The whole field of study has matured from a relatively small sub-discipline into a mainstream, policy-driven area of active and relevant research. Positive outcomes have not always followed. The financial resources and the political will required for effective disaster reduction are often lacking. Surprise remains a common reaction when the Indian Ocean tsunami (2004), hurricane 'Katrina' (2005) and the Japan earthquake (2011) inflict death and destruction in these widely separated places. Environmental hazards pose important – even growing – threats which are rarely capable of simple solutions. Complex on-going processes – globalization, climate change, population growth, resource depletion, increasing material wealth – influence the death and destruction that disaster brings. This applies to all nations, although it is the poorest countries, and the most disadvantaged people, who suffer most.

Environmental Hazards strives to explain the drivers of hazard and outline the measures that can reduce the disaster losses. From the outset, an account limited to 'natural' forces was insufficient and technological hazards, for example, have always been included. The scope of the book has widened further as fresh material has claimed its rightful place within a dynamic framework of emerging research and its applications. This new edition provides an up-to-date and balanced overview by drawing on multi-disciplinary sources. Although the structure of the book will be familiar to existing users, the content has been substantially re-written and expanded. There are more case studies, now supported by full-colour diagrams and photographs to illustrate real world situations, backed up by a comprehensive updated bibliography.

Over the years, the information highway leading to hazards and disasters has become increasingly congested. It is hoped that this book will continue to provide the reader with a useful road map that includes signposts along the way that encourage exploration of some of the minor routes that lie beyond the confines of this book.

Keith Smith
Braco, Perthshire
April 2012

Preface to the first edition

This book has been written primarily to provide an introductory text on environmental hazards for university and college students of geography, environmental science and related disciplines. It springs from my own experience in teaching such a course over several years and my specific inability to find a review of the field which matches my own priorities and prejudices. I hope, therefore, that this survey will prove useful as a basic source for appropriate intermediate to advanced undergraduate classes in British, North American and Antipodean institutions of higher education. If it encourages some students to pursue more advanced studies, or provides a means whereby other readers become more informed about hazardology, either as policy makers or citizens, then I will be well satisfied. Without a wider appreciation of the factors underlying the designation by the United Nations of the 1990s as the International Decade for Natural Disaster Reduction (IDNDR), the important practical aims of the Decade to improve human safety and welfare are unlikely to be achieved.

The term 'environmental hazards' defies precise definition. Not everyone, therefore, will endorse either my choice of material or its treatment in terms of the balance between physical and social science concepts. In this book, the prime focus is on rapid-onset events, from either a natural or a technological origin, which directly threaten human life on a community scale through acute physical or chemical trauma. Such events are often associated with economic losses and some damage to ecosystems. Most disaster impact arises from 'natural' hazards and is mainly suffered by the poorest people in the world. Within this context, my intention, as expressed in the subtitle, has been to assess the threat posed by environmental hazards as a whole and to outline the actions which are needed to reduce the disaster potential.

The structure of the book reflects the need to distinguish between common principles and their application to individual case studies. Part I, 'the nature of hazard', seeks to show that, despite their diverse origins and differential impacts, environmental hazards create similar sorts of risks and disaster-reducing choices for people everywhere. Here the emphasis is on the identification and recognition of hazards, and their impact, together with the range of mitigating adjustments that humans can make. These loss-sharing and loss-reducing adjustments form a recurring theme throughout the book. In Part II, 'The experience and reduction of hazard', individual environmental threats are considered under five main generic headings (seismic hazards, mass movement hazards, atmospheric hazards, hydrologic hazards and technologic hazards). In this section the concern is for the assessment of specific hazards and the contribution which particular mitigation strategies either have made, or may make, to reducing the losses of life and property from that hazard.

Keith Smith
Braco, Perthshire
July 1990

Acknowledgements

This book could not have been completed without generous assistance from many sources.

The previous (fifth) edition was co-authored by David Petley, University of Durham, who made important revisions to seven chapters. His valuable contributions are gratefully acknowledges both in general here and also *in situ* within the text. This edition benefits from the input of seven anonymous reviewers who made numerous insightful and supportive suggestions, the vast majority of which I have been pleased to incorporate. A special date is owed to Nick Scarle, Senior Cartographer, School of Environment and Development, University of Manchester, who prepared the diagrams with great care and skill. As always, Routledge HQ has exercised a highly professional blend of advice and encouragement. I wish to thank Andrew Mould and his team for their support, this time recognizing the practical say-to-day help provided by Faye Leerink and Casey Mein. Finally, I would like to thank the University of Stirling for continuing support over many years and for granting full access to library facilities well into my period of retirement.

The raw material has come from an ever-widening group of sources. Some have been especially fruitful; notably the disaster database maintained by the Centre for Research on the Epidemiology of Disasters (CRED) at the University of Louvain, the annual *World Disasters Reports* published by the International Federation of Red Cross and Red Crescent Societies (IFRCRCS) in Geneva and various organizations in the USA, such as the United States Geological Survey (USGS) and the Federal Emergency Management Agency (FEMA), that place a wealth of information in the public domain. The authors and the publisher would like to thank the following learned societies, editors, publishers, organizations and individuals for permission to reprint, or reproduce in modified form, copyright material in various figures and tables as indicated below. Every effort has been made to identify, and make an appropriate citation to, the original sources. If there have been any accidental errors, or omissions, we apologize to those concerned.

LEARNED SOCIETIES

American Association for the Advancement of Science for Figure 4.1 from *Science* by C. Starr.

American Geophysical Union for Table 6.7 from *EOS* by T.L. Holzer, Table 7.2 from *Journal of Geophysical Research* by C.G. Newhall and S. Self and Table 7.3 from *Reviews of Geophysics* by A. Robock.

American Meteorological Society for Figures 9.2 and 9.7 from *Weather and Forecasting* by R.A. Pielke Jr and C.W. Landsea.

American Planning Association for Figure 11.20 from the *Journal* by E. David and J. Meyer.

Institute of Foresters of Australia for Figure 10.9 from *Australian Forestry* by A. Keeves and D. R. Douglas.

International Glaciological Society for Figure 5.5 from *Annals of Glaciology* by T. Jóhannesson.

Oceanography Society for Figure 14.7 from *Oceanography* by W.S. Broecker.

The Geological Society Publishing House for Table 14.2 from *Meteorites: Flux with Time and Impact Effects* by R.A.F. Grieve.

The Royal Society of London for Table 13.6 from *Risk: Analysis, Perception and Management* by D. Cox *et al.*

PUBLISHERS

Academic Press, Orlando, for Figure 7.10a from *Volcanic Activity and Human Ecology* by P.D. Sheets and D.K. Grayson (eds).

Australian Government Publishing Company, Canberra, for Figure 10.8 from *Bushfires in Australia* by R.H. Luke and A.G. McArthur.

Blackwell Publishers for Figure 9.1 in *Geology Today* by T. Waltham, Figure 10.3 in *Risk Analysis* by Poumadere *et al.*, Table 6.2 in *Disasters* by S. Parasuraman.

Cambridge University Press for Figure 1.6 from *The Business of Risk* by P.G. Moore and Figure 14.5 from *Human Frontiers, Environments and Disease; Past Patterns, Uncertain Futures* by T. McMichael.

Commonwealth of Australia, Canberra for Figure 12.4 from Bureau of Meteorology website.

Controller of Her Majesty's Stationery Office, London for Figures 13.5 and 13.6.

Elsevier for Figures 6.2 from *Journal of Hazardous Materials* by S. Menoni, 6.16 from *Engineering Geology* by P. Marinos *et al.*, 5.22 from *Geomorphology* by M. Hürlimann *et al.*, 7.2 from *Environmental Hazards* by D.K. Chester *et al.*, 7.4 from *Journal of Volcanology and Geothermal Research* by F. Lavigne *et al.*, 7.11 from *Journal of Volcanology and Geothermal Research* by A.D.H. Artunduaga *et al.*, 8.11 from *Cold Regions Science and Technology* by R. Rice Jr, 9.11 from *Reliability Engineering and System Safety* by Z. Huang *et al.*, 10.7b from *Global Environmental Change B* by A. Badia *et al.*, 10.10 from *Fire Safety Journal* by J. Beringer, 11.11 from *Applied Geography* by N. Pottier *et al.* and Table 14.5 from *Global Environmental Change* by R.J. Nicholls *et al.*

S. Karger AG, Basel, for Figure 5.7 from *Epidemiology of Natural Disasters* by J. Seaman, S. Leivesley and C. Hogg.

Kluwer Academic Publishers, Dordrecht, for Figure 6.17 from *Tsunamis: Their Science and Engineering* by K. Iida and T. Iwasaki (eds).

Osservatorio Vesuviano in co-operation with the United Nations IDNDR Secretariat for Table 4.5 from *STOP Disasters* by G. Wadge.

Oxford University Press, New York, for Figure 1.1 from *The Environment as Hazard* by I. Burton, R.W. Kates and G.F. White.

Plenum Publishing Company for Table 13.2 from *Risk Analysis* by A.F. Fritzsche.

Springer-Verlag, Berlin for Figures 5.13 and 7.8 from *Monitoring and Mitigation of Volcanic Hazards* by D.W. Peterson and S.R. McNutt.

J. Wiley and Sons, Chichester, for Figure 9.7 from *Hurricanes: Their Nature and Impacts on Society* by R.A. Pielke Jr and R.A. Pielke Sr.

ORGANIZATIONS

California Seismic Safety Commission for Table 6.6 from *California at Risk* by W. Spangle and Associates Inc.

Illinois State Water Survey for Figure 11.13 from *The 1993 Flood on the Mississippi River in Illinois* by N.G. Bhowmik.

Colorado Avalanche Information Center for Figure 8.2.

Federal Emergency Management Agency for Figure 5.9.

Munich Re Insurance Company, Munich, for Figures 2.4, 2.7, 2.9, 5.12 and 9.4.

United Nations for Table 4.5 from *STOP Disasters* by G. Wadge.

United Nations Environment Programme, Nairobi, for Figure 2.8 (upper) from *Environmental Data Report*.

United States Geological Survey, Denver, Colorado for Figure 11.15 from *Effects of Reservoirs on Flood Discharges on the Kansas and Missouri River Basins (Circular No. 1120E)* by C.A. Perry.

United States Geological Survey, Virginia, for Figure 7.12b from *The 1980 Eruptions of Mount St. Helens* by C.D. Miller, D.R. Mullineaux and D.R. Crandell.

United States Geological Survey for Figures 6.6 and 14.10.

University of Toronto Department of Geography for Figure 1.5 from *The Hazardousness of a Place* by K. Hewitt and I. Burton.

University of Toronto, Institute of Environmental Studies, for Tables 4.1 and 4.3 from *Living with Risk: Environmental Risk Management in Canada* by I. Burton, C.D. Fowle and R.J. McCullough (eds).

INDIVIDUALS

D. Atkins, Colorado Avalanche Information Center, for Figure 8.2.

Professor R.G. Barry, University of Colorado, for Figure 9.5.

Professor A. Bernard, Free University of Brussels, for Figure 7.6.

Dr K.R. Berryman, DSIR, Wellington for Figure 6.12.

H. Brammer, Hove, for Figure 11.3.

Dr W.S. Broecker, Columbia University, Palisades, New York, for Figure 14.7.

Dr W. Bryant, California Geological Survey, Sacramento, CA, for Figure 5.23.

Dr D.R. Crandell, US Geological Survey, Denver, for Figure 7.12a.

Dr J. de Vries, University of California, Berkeley, for Figure 4.8.

Dr D.R. Donald, US Department of Agriculture, for Figure 12.6.

Dr N.J. Duijm, Technical University of Denmark, for Figure 13.7.

Dr D.J. Gilvear, Stirling University, for Figure 12.12.

Professor G.W. Housner, California Institute of Technology, for Figure 6.2.

Professor M. Hulme, University of East Anglia, for Figure 12.9.

Professor R.W. Kates, Clark University, for Figure 1.7.

Dr A. Malone, University of Hong Kong, for Figure 8.13.

Professor P.G. Moore, London Graduate School of Business Studies, for Figure 1.6.

T. Omachi, Infrastructure Development Institute, Japan, for Figure 12.14.

Dr D. Ruatti, International Atomic Energy Authority, Geneva, for Table 13.5.

Marjory Roy, formerly of the Meteorological Office, Edinburgh, for Figure 4.5.

Dr W.D. Smith, DSIR, Wellington, for Figure 6.12.

Dr J.C. Villagrán de León, United Nations University for Figure 5.14.

Dr J. Whittow, Reading University, for Figure 6.4.

Part One

THE NATURE
OF HAZARD

'We have met the enemy and it is us.'

Attributed to Walter Kelly

CHAPTER ONE

Hazard in the environment

1

A INTRODUCTION

In the early twenty-first century, the Earth supports a human population that is more numerous and – in general – is healthier and wealthier than ever before. At the same time, there is an unprecedented awareness of the risks that face people and what they value. Some of this concern is associated with the death and destruction caused by 'natural' hazards like earthquakes and floods. Other anxieties focus on the risks that originate in the built environment, such as industrial accidents and failures of technology that are seen as 'man-made'. In addition, there are widespread fears about more elusive, still-emerging dangers, like climate change, sea-level rise and the loss of biodiversity.

A paradox exists between material progress and these feelings of insecurity. This is because economic development and environmental hazards are rooted in the same on-going processes of global change. As the world population grows, so more people are exposed to hazard. As that population becomes more prosperous, so more personal and corporate wealth is placed at risk. As agriculture intensifies and urbanization spreads,

so more complex and expensive infrastructure is exposed to damaging events. These trends towards potential large-scale losses are underpinned by rising levels of human consumption that impose heavy burdens on natural assets such as land, forests and water. Uncertainties are raised about environmental quality, the availability of key resources and their sustainability into the future. Many people in the 'less developed countries' already experience insecure lives and livelihoods because of poverty, weak governance and a dependence on a degraded resource base that makes them especially vulnerable to 'natural' hazards and other threats.

The power of modern communications, including non-stop news coverage and the rise of social networking via cell phones and the internet, permits the rapid dissemination of information about the latest disaster, often in graphic detail. Despite – or perhaps because of – this constant flow of information, it is difficult to place individual disasters in context and make broader assessments of risk. Is the world really becoming a more dangerous place? If so, what are the causes? What are the main environmental threats? What is a disaster? Why do advanced nations still remain vulnerable to some natural processes? Why do disasters kill more people in poor countries than in rich countries? What effects do disasters have on economic development? Why do some disasters create much greater losses than the physical scale of the event suggests? Is climate change an environmental hazard? What are the best means of reducing the impact of hazards and disasters in the future?

Most people accept that it is impossible to live in a totally risk-free environment. We all regularly face some degree of risk, whether it is to life and limb in a road accident, to our possessions from theft or to our personal space from noise or other types of pollution. In some cases, we adopt risks through individual 'life-style' choices, such as smoking cigarettes, overeating or participating in dangerous sports. Some self-imposed risks, like smoking tobacco or driving a car, result in a significant number of premature deaths and other losses over a period of time. But these are familiar, everyday risks with consequences dispersed through the whole population. They do not create large-scale deaths and damages that are so

concentrated in time and space that they disrupt whole communities in what is commonly perceived as a 'disaster'.

B WHAT ARE ENVIRONMENTAL HAZARDS?

This book concentrates on the more extreme, often rapid-onset, events that directly threaten human life, property and other assets by means of acute physical or chemical trauma on a relatively large scale. Such losses follow the sudden release of energy or materials in concentrations greatly in excess of normal background levels. In this book, the term *environmental hazard* is limited to events originating in, and transmitted through, the natural and built environments that lead to human deaths, economic damage and other losses above certain predefined thresholds of loss. In fact, thresholds of loss are used to define *disaster* (see section 1.C, page 12). Hazards and disasters are two sides of the same coin; each merges into the other and neither can be fully understood from the standpoint of either physical science or social science alone. As shown in Figure 1.1, they are linked to wider issues like global environmental change and the many interacting factors that determine the prospects for sustainable development in the future.

Two main types of environmental hazard can be identified (Table 1.1).

1 Natural hazards

The most important group of hazardous events is normally classed as 'natural'. The United Nations International Strategy for Disaster Reduction (UN/ISDR, 2009) defined a *natural hazard* as:

any natural process or phenomenon that may cause loss of life, injury or other health impacts, property damage, loss of livelihoods and services, social and economic disruption or environmental damage.

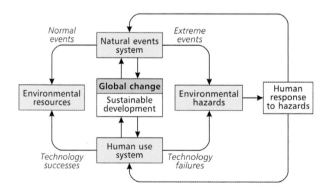

Figure 1.1 Environmental hazards lie at the interface between the natural events system (extreme events) and the human use system (technology failures). They interact with global change and sustainable development and are also influenced by societal responses to reduce disaster. Adapted from Ian Burton *et al.*, *Environment as Hazard* (1978); by permission of Oxford University Press, Inc.

Table 1.1 Major categories of environmental hazard

Natural hazards (extreme geophysical and biological events)
Geologic – earthquakes, volcanic eruptions, landslides, avalanches
Atmospheric – tropical cyclones, tornadoes, hail, ice and snow
Hydrologic – river floods, coastal floods, drought
Biologic – epidemic diseases, wildfires
Technological hazards (major accidents)
Transport accidents – air accidents, train crashes, ship wrecks
Industrial failures – explosions and fires, release of toxic or radioactive materials
Unsafe public buildings and facilities – structural collapse, fire
Hazardous materials – storage, transport, misuse of materials

Notes: Drought is a slow-onset natural hazard. Some hazards are combined na-tech events.

This type of description is well rooted in the literature but fails to provide a scale of loss. It is most suitable for hazards like earthquakes and volcanic eruptions, where the damaging processes are truly 'natural' in origin because they remain unaffected by human actions.

However, the Earth's surface and atmosphere are increasingly subject to anthropogenic change. This suggests that, although all 'natural hazards' are triggered by physical forces, certain events and their outcomes may be influenced by human actions, whether deliberate or unintended. In other words, some types of natural hazard become *quasi-natural hazards*. For example, the disaster impact of a river flood may be inadvertently increased by deforestation in the catchment area or deliberately decreased by the construction of a control dam. Where an increase in either the frequency or severity of hazardous physical events can be attributed to degraded land or over-exploited resources the term *socio-natural hazard* is sometimes used.

One group of hazards has a clear hybrid identity. These are the *natural-technological ('na-tech') hazards* that arise when extreme natural processes lead to the failure of industrial structures and other assets within the built environment. Although triggered by natural forces, the main threat often comes from pollution in the atmos-

phere or surface waters, due to accidental releases of dangerous substances (Showalter and Myers, 1994). Examples include the radioactive pollution arising from damage to the Fukushima nuclear power plant (Japan) in March 2011 by the Tōhoku earthquake and tsunami (Box 1.1). In September 2003 Italy suddenly lost over one-quarter of its electricity supply when storm-force winds in Switzerland damaged the transmission line importing power into the country. Floods and earthquakes destroy river dams and allow the uncontrolled release of stored water to create damage downstream. Sometimes na-tech flood disasters occur in the absence of infrastructural failure. In 1963 a landslide displaced water over the top of the Vaiont dam in Italy and caused many deaths, even though the dam itself remained intact.

2 Technological hazards

The other chief subset of disaster threats comes from the built environment and has been defined by the UN as:

> *hazards originating from technological or industrial conditions, including accidents, dangerous procedures, infrastructure failures or specific human activities, that may cause loss of life,*

Box 1.1 Tōhoku 2011: a major na-tech disaster

The Great Tōhoku earthquake in Japan is a classic example of a na-tech disaster. In this case, the 'knock-on' series of events triggered by an earthquake, quickly followed by tsunami waves, caused extensive damage to a nuclear power facility and released significant amounts of radioactive material into the environment.

Step One   On 11 March 2011, the island of Honshu, north-east Japan was struck by an earthquake of magnitude $M_W = 9.0$. At the time, this was the largest instrumentally recorded earthquake ever to hit Japan. The offshore epicentre was roughly 70 km east of the Oshika Peninsula of Tōhoku, on the Sanriku coast, with a hypocentre 30 km below sea level. The earthquake resulted from thrust faulting at the plate boundary between the Pacific and Indo-Australian–Fiji plates (see Figure 6.1). At this point, the Pacific Plate is moving to the west at a rate of 83 mm yr^{-1}. The boundary is a subduction zone where the Pacific Plate descends beneath Japan at the Japan Trench (USGS, 2011). Since 1973, nine events of magnitude $M_W = 7.0$ or greater have occurred here, although no earthquake during the twentieth century attained a magnitude of $M_W = 8.0$ or more.

Step Two   The earthquake triggered tsunami waves up to almost 40 m high in places that struck the coast several minutes later. Some waves travelled inland for up to 10 km. The Sanriku coast has many deep coastal bays which constrain and amplify the height of approaching tsunami waves, and similar disasters in historic times occurred in 1611, 1854, 1896 and 1933. The coast is protected by extensive sea walls, some 12 m high, but most were easily over-topped.

   The combined death toll from the earthquake and tsunami was at least 20,000, with 14,000 homes destroyed and 100,000 properties damaged. The Japanese Red Cross sent 230 response teams and over 2,000 evacuation centres were set up in north-east Japan. This was the most expensive disaster in Japanese history. Total economic damages were estimated at a record US$366 billion, with insured property losses of US$20–30 billion. About US$4,000 billion was wiped off the Nikkei 225 stock market index, which initially fell by over 6 per cent.

Step Three   The tsunami flooded the coastal Fukushima I nuclear power station. The plant, operated by the Tokyo Electric Power Company, was 40 years old and produced 4,696 MW of electricity. It comprised six boiling water reactors, designed to withstand a $M_W = 8.2$ earthquake and 5.7 m tsunami wave. The earthquake activated an automatic shut-down system and emergency generators started to run the sea-water pumps used for cooling the reactors. But the entire plant, including the diesel generator building, was struck by 14 m high tsunami waves. As a result of generator failure, four reactors began to overheat and three ultimately suffered meltdown. Explosions caused by a build-up of hydrogen gas in the outer containment buildings released radioactive material and levels of 400 millisieverts (mSu) were recorded at No 4 reactor. This compares with the 350 mSu criterion adopted for evacuation at Chernobyl. A 20 km exclusion zone was declared around the Fukushima I nuclear power station, with a 10 km zone for the Fukushima II power plant. In total, around 80,000 residents were evacuated. Initially the incident was rated 5 on the 7-point International Nuclear Event Scale but later it was reassessed at the highest level. This was the first Category 7 nuclear accident since the Chernobyl disaster of 1986, indicating risks to human health and environmental contamination from leakage of cooling water and contamination of coastal waters.

Following the emergency phase, the main priority was to cool the reactors with recirculated water to safe temperatures below 100°C. In the initial response, helicopters were used to drop limited amounts of sea-water and, in the absence of relief generators and with the loss of electrical power on site, radioactive steam had to be released manually. Eventually, large quantities of sea-water were pumped ashore and by December 2011 the plant was declared to be in 'cold shutdown'. However, water cooling will be necessary for several years while the radioactive fuel in the reactors slowly decays. The removal of fuel from the three most-damaged reactors is unlikely to occur within the next 10 years. Full decommissioning of the plant would include the decontamination of an area extending to 2,000 km², the disposal of an estimated 90,000 tonnes of contaminated sea-water and the removal of millions of cubic metres of topsoil. This process could cost up to US$50 billion and take 40 years to accomplish.

injury, illness or other health impacts, property damage, loss of livelihoods and services, social and economic disruption or environmental damage.

Most threats in this category arise from human errors which expose flaws in the design and/or the functioning of built structures or industrial-scale processes. Once again, the definition provides no numerical scale of loss.

The causes of environmental hazards, whether natural or technological, are reasonably well understood. As shown in Chapter 2, extensive databases exist to illustrate the spatial and temporal patterns of disaster. However, there is a growing recognition that apparently localized threats are often linked to wider-scale processes. For example, the slope failure that produces a landslide, or the rain-storm that produces a river flood, may originate respectively through tectonic and ocean–atmosphere mechanisms operating far beyond the mountain range or the river valley where the impact occurs. Many physical processes in the Earth's crust and its atmosphere are driven by forces that operate on hemispheric, or even planetary, scales and deploy vast amounts of energy and materials. A few – like asteroid collisions with the Earth – have the potential to create a global catastrophe not yet experienced in human history. In turn, certain hazardous

processes, on both local and regional scales, are influenced by *global environmental change (GEC)*. Some change is due to natural variations in climate, but human-induced modifications of the Earth–atmosphere system play an increasing role.

Global interactions and environmental changes are not necessarily hazards in themselves. Sometimes they offer opportunities as well as risks, but all too often they amplify the disaster potential of existing natural and technological threats. This situation highlights the need for a better understanding of the worldwide *coupled human–environment system (CHES)*. A wider perspective ensures that hazards and disasters are now seen as complex issues nested within a setting that includes global change and sustainability issues (Figure 1.2).

Hazards and disasters can no longer be viewed as one-off, site-specific events capable of regulation by local responses alone. In some instances the wider links and consequences are fairly clear; for example, global warming drives sea-level rise, which leads to increased risks from coastal floods. Other questions, such as the adverse effects of natural hazards on economic development, or on the future security of food and water supplies, are less self-evident. Unlike traditional 'natural hazards', many of these complex interrelationships have become prominent recently and there

Figure 1.2 Extreme geophysical events and severe system failures within a framework of global change and sustainability issues.

is a lack of historical experience and scientific record to enable understanding. As a result, some key issues – like the impact of climate change on environmental hazards – remain controversial within both scientific and political circles (see Chapter 14).

Most environmental hazards can be placed on an approximate scale of causation, ranging from entirely natural forces to examples subject to considerable human influence (Figure 1.3) As the scale moves from rarer, uncontrolled

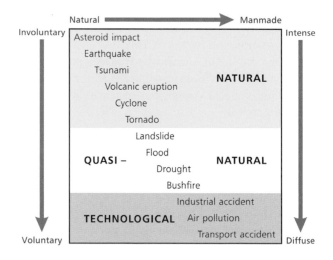

Figure 1.3 A generalized spectrum of environmental hazards from physical to human causes. Risks with a high level of human causation are more readily accepted and more diffuse in terms of disaster impact.

natural events (asteroid impact, earthquake) through more 'quasi-natural' hazards (landslide, bushfire) towards more common technological-type hazards (transport accident, air pollution), so disaster impacts tend to become less concentrated within a particular community. As the extent of human causation grows, the risk spreads and the public tends to be more accepting of any loss. Entirely voluntary life-style hazards, such as cigarette smoking or mountaineering, are excluded from this book because they are wholly man-made, self-inflicted risks. Similarly, hazards of mass violence are excluded because crime, warfare and terrorism are intentional harmful acts on humans by humans. On the other hand, some societal characteristics do influence hazard impacts, either directly or indirectly. For example, certain epidemics of infectious disease are accepted in this book because they are not only a major cause of disaster deaths but are also related to environmental conditions. Factors such as poverty, ill-health and environmental degradation are not hazards in themselves but they have indirect effects by amplifying human vulnerability to hazardous events (see Chapter 3, page 52).

In summary, the chief features of environmental hazards are:

- The origin of the event is clear and produces known threats to human life or well-being (a rain-storm produces a flood that causes death by drowning).
- The warning time is normally short (the events are often rapid-onset).
- Most of the direct losses, whether to life or property, are suffered shortly after the event.
- The human exposure to hazard is largely involuntary, normally due to the location of people in a hazardous area.
- The resulting disaster justifies an emergency response, sometimes on an international humanitarian scale.
- Uncertainty about, and wide variations in, loss from year to year make risk assessment and loss reduction difficult.

Many hazardous events represent the extremes of a statistical distribution that, in a different context, would be regarded as a natural resource (Kates, 1971). For example, normal river flows are a community benefit, providing water power, drinking supplies, amenity, etc., but very high flows bring a flood hazard. Many beneficial uses of water depend on river-control technology, in the form of embankments, bridges and dams. Water under human control in a reservoir is perceived as a resource but, if technology fails and the dam collapses, then a flood disaster may result. It is important to realize that environmental hazards spring neither from a vengeful God nor from a hostile environment. Rather the environment is 'neutral'. It is the human use of the environment, both natural and man-made, which identifies resources and hazards through human perception.

Human sensitivity to environmental hazards is determined by the *physical exposure* of people and their assets to potentially damaging events and by the degree of *human vulnerability* (or *resilience*) to such damaging events. These factors are dynamic, and change through space and time. Generally speaking, exposure and vulnerability are hazard specific in any one location, such as a river floodplain, but broader relationships can be displayed in a simple matrix (Figure 1.4). Industrialized nations have invested heavily to reduce exposure and vulnerability by improving environmental security. For example, Japan is a mountainous country with three-quarters of the population housed on alluvial plains and along coasts below mean sea level. The country is exposed to multiple hazards – earthquakes, tsunamis, tropical cyclones, snow-storms – but many protective structures and safety strategies are in place to curb disaster losses. Conversely, poor countries with high exposures to risk find it difficult to fund hazard protection and to reduce exposure. The resulting high level of vulnerability means that their disaster losses are disproportionately high when compared with the damage inflicted on resource-rich nations. In particular, they suffer the adverse effects of more

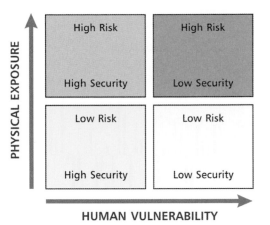

Figure 1.4 A matrix showing possible combinations of physical exposure to hazard and human vulnerability in relation to risk and security.

frequent and less extreme events that would be much less hazardous elsewhere.

All natural and human systems exhibit some degree of variability, but most socio-economic activities are geared to an expectation of 'average' environmental conditions. In Figure 1.5 the central yellow zone represents an acceptable, or tolerable, range of fluctuation for any 'element' vital for human survival or well-being. The element could be a natural process (rainfall) or a technical process (chemical production). Within this zone, the element is perceived as a beneficial *resource*. However, when it fluctuates over a critical threshold beyond the 'normal' band of tolerance, the element becomes a *hazard*. Thus, very high or very low rainfall will be deemed to create a flood or a drought; high releases of gases from a factory will be perceived as air pollution. The *hazard magnitude* can be determined by the peak deviation beyond the threshold on the vertical scale, and the *hazard duration* from the length of time the threshold is exceeded on the horizontal scale. The potential time-scale of hazard duration ranges over at least seven orders of magnitude, from the seconds or minutes of ground-shaking experienced in earthquakes to drought conditions that can persist for decades.

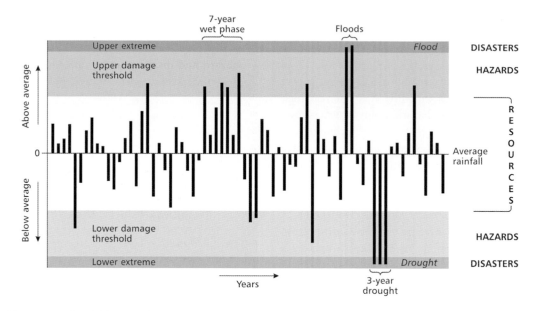

Figure 1.5 Sensitivity to environmental hazard expressed as a function of annual rainfall and societal tolerance. Within the yellow band of tolerance, variations are perceived as resources; beyond the damage thresholds they are perceived as hazards or disasters. Adapted from K. Hewitt and I. Burton, *The Hazardousness of a Place: A Regional Ecology of Damaging Events* (University of Toronto Press, 1971). Reprinted with permission of the publisher.

Plate 1.1 Timanfaya National Park on Lanzarote, Canary Islands, covers 50,000 km² of ecologically important land created by volcanic eruptions between 1730 and 1736. Limited tourist activity is permitted showing that economic benefits can flow from natural hazards. (Photo: Keith Smith)

The ever-changing balance between environmental hazards and resources – between risks and rewards – can be illustrated in many ways. For example, the Mediterranean coastal area has a high dependence on tourist revenue. Visitors are attracted by dry, sunny summers and a warm sea. A few volcanic sites, like Etna, Vesuvius and Santorini, are of special scenic interest. Although these sites pose a risk, limited volcanic activity – expulsion of steam or minor ejections of ash – probably adds to their tourist potential. Such natural assets are termed *ecosystem services*. However, a major volcanic eruption, or even the prediction of one, would lead to the closure of the surrounding area and a collapse in local tourist income. In the longer term, a sustained climatic trend to hotter, more arid summers would lead to water shortages and forest fires, and change the perceived desirability of the area for tourism.

Human populations are especially at risk on the margins of hazard tolerance, where small physical changes create large socio-economic impacts, like the effects of rainfall variability on agriculture in semi-arid areas. Over a long period of time, frequent but unpredictable low-level variability around a critical threshold may have as much significance as the rare occurrence of more extreme events. Sudden change is an integral part of all natural systems, but the very rarest events may not be recognized as threats. It is often only when such changes are observed by humans – and perceived as a threat – that a hazard exists. In other words, hazards are a human interpretation of events because they seem to be extreme or rare within the lifetime experience of individuals. Most people will be aware that floods are a hazard; few will be aware that meteorite strikes on Earth are a hazard, because they have been rare in historic times.

C HAZARD, RISK AND DISASTER

In order of decreasing severity of impact, environmental hazards create the following threats:

- to people – death, injury, disease, mental stress
- to goods – property damage, economic loss
- to environment – loss of flora and fauna, pollution, loss of amenity.

Threats to human life are normally given the highest priority, followed by losses to material assets. Most disasters are characterized by a minimum level of human mortality. Fatalities and economic damages can be assessed directly by counting deaths and monetary losses after a disaster. In general, these two measures are the basis for scaling hazard impacts in disaster. Although the environment is valued by humans, it attracts much less attention in disaster assessment. At present, relatively little human mortality can be explicitly linked either to episodes of severe environmental pollution or to progressive declines in ecosystem quality. It is also more difficult to calculate the value of environmental resources on conventional financial scales.

Hazard and disaster can be ranked according to impact criteria, and the probability of a hazardous event can be placed on a scale from zero to certainty (0 to 1). The relationship between a hazard and its probability can then be used to determine the overall level of risk, as shown in Figure 1.6. Risk is sometimes taken as synonymous with hazard, but risk has the additional implication of the statistical chance of experiencing a particular hazard. Hazard is best viewed as a naturally occurring, or human-induced, process or event with the potential to create loss, i.e. a general source of future danger. Risk is the actual exposure of something of human value to a hazard and is often measured as the product of probability and loss. Thus, we may define *hazard* – the cause – as:

a potential threat to humans and their welfare arising from a dangerous phenomenon or substance that may cause loss of life, injury, property damage and other community losses or damage.

Then *risk* – the likely consequence – becomes:

the combination of the probability of a hazardous event and its negative consequences.

The difference between hazard and risk can be illustrated by two people crossing an ocean, one in a large ship and the other in a rowing boat (Okrent, 1980). The hazard (deep water and large waves) is the same in both cases but the risk (probability of capsize and drowning) is much greater for the person in the rowing boat. This analogy shows that, whilst the type of danger posed by earthquakes – for example – may be similar throughout the world, people in the poorer, less developed countries are often more vulnerable and at greater risk than those in the richer, more developed countries. When large

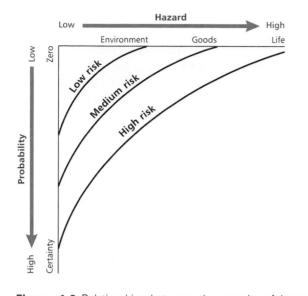

Figure 1.6 Relationships between the severity of hazard, probability and risk. Hazards to human life are rated more highly than damage to economic assets or the environment. After P.G. Moore, *The Business of Risk* (1983). © Cambridge University Press 1983, reproduced with permission.

numbers of people are killed, injured or otherwise adversely affected, the event is termed a *disaster*. Unlike hazard and risk, a disaster is an actual happening, rather than a potential threat. So we may broadly identify *disaster* – the actual consequence – as:

> *a serious disruption of the functioning of a community or a society involving widespread human, material, economic or environmental losses or impacts which exceed the ability of the affected community or society to cope using its own resources.*

> UN/ISDR (2009)

Environmental hazards stem from natural events, but disasters are social phenomena that occur when a community suffers exceptional levels of disruption and loss. Although a hazardous event can occur in an uninhabited region, risk and disaster can exist only in areas where people

and their possessions are located. The sequence of events leading to a disaster can be shown as:

If a community is threatened by an extreme event, the risk may be contained or minimized in some way, but, if people or infrastructure cannot be protected, a disaster is the likely outcome. A disastrous train of events can occur if humans place unrealistic demands on the environment or select a technology that eventually fails, with harmful consequences (Hohenemser *et al.*, 1983). Figure 1.7 illustrates a disaster sequence for drought, with linked causal stages at the top line and possible control stages below. If the control measures fail, famine-related deaths are a possible consequence. In practice, direct cause-and-effect linkages rarely operate, and complex emergencies develop. For example, when fires and explosions occurred in the 1906 San Francisco earthquake, due to the rupture of gas supply pipes, the primary hazard was strong ground shaking, the secondary hazard was soil liquefaction and the tertiary hazard was fire and explosion.

Given this framework, what is the risk of disaster from environmental hazards? In general, the profile of disaster loss portrayed in the media is not matched by the actual incidence of deaths or damages. Headline media reports are, by definition, non-routine and arise infrequently. For example, although Mileti *et al.* (1999) found that natural hazards in the United States killed about 1,250 people per year and injured a further 5,000, only one-quarter of the fatalities and

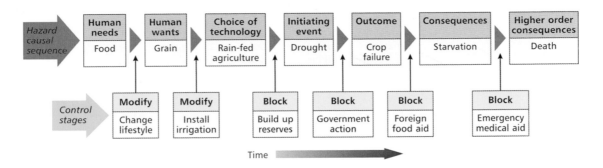

Figure 1.7 Schematic evolution of a drought disaster. The sequence of events is shown at the top of each box and the disaster consequences are shown in the lower segments. Six potential intervention stages, designed to reduce disaster, are linked to pathways between the hazard steps by vertical arrows. After Hohenemser *et al.* (1983). Reproduced from C. Hohenemser *et al.*, The nature of technological hazard. *Science* 220: 378–84 (1983). Reprinted with permission from AAAS.

half the injuries resulted from major disasters. Most deaths stemmed from small, frequent events (lightning strikes, car crashes in fog and local landslides). In Italy, with a landslide risk second only to Japan's amongst the industrialized nations, the death rate from road accidents is over 200 times that from landslides but still remains low for the population as a whole (Guzzetti, 2000). According to Fritzsche (1992), a mere 0.01 per cent of the US population has died from natural disasters. Similarly, although natural hazards in the USA regularly damage public facilities like roads, water systems and buildings, the losses are only 0.5 per cent of the capital value of the nation's infrastructure. Average disaster relief costs are less than 1 per cent of the total federal budget (Burby *et al.*, 1991).

As noted by Sagan (1984), deaths and injuries from disasters are often reported as safety issues, especially in the developed countries, where there is a strong risk-averse culture. These accidental deaths are perceived differently from chronic human illnesses, which are viewed as on-going health issues. In the more developed countries (MDCs), average mortality from all causes is strongly dependent on age. The death rate tends to be high during the first few years of life but soon drops sharply. It then rises steadily until, at

age 70 and beyond, it exceeds infant mortality. This pattern reflects the importance of lifestyle factors and degenerative diseases in the western world, where some 90 per cent of all deaths are due to heart disease, cancers and respiratory ailments. Tobacco consumption is a major factor; worldwide about 3 million people die prematurely each year through smoking.

Accidental deaths from all causes rarely constitute more than 3 per cent of mortality in the MDCs. The situation in less developed countries is rather different, where the per capita risk of a disaster-related death has been estimated to be between 4 and 12 times greater than in the industrialized countries. Strömberg (2007) estimated that the one-third of the world's population living in low-income countries suffers almost two-thirds of all disaster-related deaths. Table 1.2 shows the large disparity in mortality rates due to differences in national wealth and type of government between high- and low-income countries when the countries are standardized for population size and exposure to disaster. Once again, environmental hazards are not the only cause, given the greater presence of risks like disease epidemics and armed conflicts in the LDCs.

In summary, the cumulative losses from 'headline disasters' are relatively low in relation to

Table 1.2 Disaster-related deaths in high- and low-income countries 1980–2004

Country category	Number of disasters	Average population (million)	Exposed population (million)	Killed in disasters	GDP per capita	Democracy index
High-income	1,476	828	440	75,425	23,021	9.5
Low-income	1,533	869	496	907,810	1,345	3.2

Source: D. Strömberg (2007) Natural disasters, economic development and humanitarian aid. *Journal of Economic Perspectives* 21: 199–222. Reproduced with permission.

Notes: Exposed population – the population share in each country that lives in areas in the top three deciles of risk exposure to volcanic activity, earthquake, floods, landslides or drought multiplied by the population in the country and summed over the countries in that income group.

other causes of premature death and damage, especially in the MDCs. Disasters and accidental losses are newsworthy because the impacts are highly concentrated in space and time and often provide striking photographs and television footage. Some media influences on disaster reporting are discussed in Chapter 2.

D EARLIER PERSPECTIVES

Our understanding of hazards and disasters has changed through history. A concern for earthquake and famine began in the earliest times (Covello and Mumpower, 1985). Great catastrophes were seen then as 'acts of God' – a divine punishment for moral misbehaviour – rather than as a consequence of human use of hazard-prone land. This view encouraged an acceptance of disasters as external, inevitable events. Eventually, as in the case of frequently flooded land, communities learned to avoid the most dangerous sites. Later still, organized attempts were made to limit the damaging effects of natural hazards, an approach that led to the development of the four hazard paradigms recognized in Table 1.3.

The *engineering paradigm* originated with the first river dams constructed in the Middle East over 4,000 years ago, whilst attempts to defend buildings against earthquakes date back at least 2,000 years. This approach is based on 'hardening' built structures to withstand most hazard stresses and evacuating people from harm by emergency action. The growth of the earth sciences and civil engineering practices over the following centuries led to increasingly effective structural responses designed to control the damaging effects of certain physical processes. By the end of the nineteenth century new measures, like weather forecasting and severe storm warnings, could also be used. It is largely undertaken with the aid of science-based government agencies and remains a necessary and important strategy today.

Before the mid-twentieth century, there was limited understanding of the interactions between environmental hazards and people. The *behavioural paradigm* originated with an American geographer, Gilbert White (1936, 1945). He saw that natural hazards are not purely geophysical phenomena outside of society but are linked to societal decisions to settle and develop hazard-prone land, often for economic motives. White was critical of the undue reliance placed on engineered structures to control floods and other hazards in the USA and introduced the social perspective of *human ecology*. This interpretation stems from earlier work in the 1920s, notably by Harlan H. Barrows, who applied concepts from ecology – such as interconnectivity, spatial organization and system behaviour – to the functioning of human communities. The basic idea was that the interactive nature of human–environment relations defines the well-being of both. In other words, human ecology links the physical and social sciences to provide a more

Table 1.3 The evolution of environmental hazard paradigms

Period	Paradigm name	Main issues	Main responses
Pre-1950	Engineering	What are the physical causes for the magnitude and frequency of natural hazards at certain sites and how can protection be provided against them?	Scientific weather forecasting and large structures designed and built to defend against natural hazards, especially those of hydro-meteorological origin.
1950–70	Behavioural	Why do natural hazards create deaths and economic damage in the MDCs and how can changes in human behaviour minimize risk?	Improved short-term warning and better longer-term land planning so that humans can adapt and avoid sites prone to natural hazards.
1970–90	Development	Why do people in the LDCs suffer so severely in natural disasters and what are the historical and current socio-economic causes of this situation?	Greater awareness of human vulnerability to disaster and an understanding of how low economic development and dependency contribute to disaster.
1990–	Complexity	How can disaster impacts be reduced in a sustainable way in the future, especially for the poorest people in an unequal and rapidly changing world?	Emphasis on the complicated interactions between natural and human systems, leading to improvement in the long-term management of hazards according to local needs.

balanced approach to resolving the conflicts that arise between human needs and the sustainability of the environment.

Gilbert White was the first person to question whether truly 'natural' hazards exist at all. He proposed that, instead of attempting to control nature's extreme events – like floods – people should adapt their behaviour to the uncertainties raised by the magnitude and frequency of physical processes. Although revolutionary at the time, this view led to a blended approach. Engineers continued to build to standards designed to withstand natural forces and scientists introduced other technocratic advances, for example in hazard monitoring and warning schemes. Simultaneously, social scientists explored how disasters might be reduced through human adjustments, such as insurance and better land planning. This combined hazards-based viewpoint became widely accepted and was summarized in several books from the North American research school (White, 1974; Burton, Kates and White, 1978, rev. 1993).

The *development paradigm* emerged during the 1970s as a more theoretical and radical alternative (Box 1.2). It drew on experience in the less industrialized parts of the world, where natural disasters create more severe impacts, including large losses of life. Answers were sought in the longer-term, root causes of disasters and the research focus shifted from hazards to a disasters-based viewpoint and from the MDCs to the LDCs. The link between under-development and disasters was scrutinized, and it was concluded that economic dependency increased both the frequency and the impact of natural hazards. Human vulnerability – a feature of the poorest and the most disadvantaged people in the world – became an important concept for understanding disaster impacts (Blaikie *et al.*, 1994; Wisner *et al.*, 2004).

In the late twentieth century, these two contrasting views were still evident (Mileti *et al.*, 1995). Physical scientists, including civil engineers and meteorologists, were associated with the agent-specific, hazard-based behavioural paradigm using technical solutions plus some

Box 1.2 The twentieth-century paradigm debate

Environmental hazards are open to many interpretations. In the past, divisions arose between the more technology-based behavioural paradigm, adopted by many government bodies, and the more theoretical development paradigm, favoured by some social scientists.

The behavioural paradigm

Engineering responses to environmental hazards go back a long way but modern approaches began in the USA. Following the 1936 Flood Control Act, the US Army Corps of Engineers constructed major flood-control works (dams and levees) throughout the country. This strategy appeared rational during the 1930s and 1940s, due to growing confidence in the relevant scientific fields (meteorology, hydrology), political demands for greater development of natural resources and the availability of capital for public works.

Gilbert White was at first a lone voice in arguing that flood control works should be integrated with non-structural methods, like land use planning, to produce more comprehensive floodplain management. His view recognized the role played by human behaviour in creating hazards. Urban development of flood-prone land was attributed to 'behavioural' or cultural faults, including a mis-perception – by developers and home-owners alike – of the risk/reward balance that exists when hazardous land is occupied for economic gain. Within the developing countries, other forms of behaviour, such as deforestation or the over-grazing of land, were considered irrational and thought to contribute to disaster. The universal purpose of disaster reduction was to prevent these temporary disruptions to 'normal' life.

Although White's ideas gained some attention, 'technical fix' solutions dominated. It was believed that, in the fullness of time, the transfer of technology from the developed to the developing world, as part of an overall modernization process, would solve its problems too. Many centralized organizations were created because only government-backed bodies had the financial resources and expertise needed to apply science and engineering on the required scale. The United Nations, in particular, sprouted a number of agencies responsible for international disaster mitigation at this time.

According to Hewitt (1983), the behavioural paradigm had three thrusts:

- Despite some acknowledgement of the role of human behaviour in the occupation of hazard-prone land, the prime aim was to contain nature through engineering works, such as flood embankments and earthquake-proofed buildings, allied with land use controls.
- Other measures included field monitoring and the scientific explanation and statistical assessment of geophysical processes. Modelling and prediction of damaging events followed the introduction of advanced technical tools, e.g. remote sensing and telemetry.
- Priority was given to strengthening bureaucracy for disaster planning and emergency responses, mostly operated by the armed forces. The notion that only a military-style organization could function in a disaster area was attractive to governments because it emphasized the authority of the state when re-imposing order.

This paradigm has a cultural emphasis but also contains practical methods for loss reduction. It remains important but has been described as an essentially Western interpretation of disaster. Critics see this approach as materialistic, reflecting undue faith in technology and capitalism leading to 'quick fix' remedies. It has also been faulted for overemphasizing the role of individual choice, against the power of financial bodies and other institutions, in hazard-related decisions; for neglecting environmental quality, for example in the construction of control schemes and the drainage of wetlands as a flood reduction measure, and for down-playing human vulnerability in disasters.

The development paradigm

This philosophy emerged due to slow progress in reducing disaster losses in poorer countries. It originated with social scientists who believed that disasters in the Third World arise principally from the workings of the global economy and the marginalization of disadvantaged people. Extreme natural events were seen as 'triggers' of deeply rooted and long-standing problems, especially poverty. This more radical interpretation of disaster proposes fundamental change in economic, social and political systems. Contrary to the behavioural paradigm, it dwells on the long-term common features of disaster and stresses the limits to individual actions imposed by powerful financial and political interests.

The development paradigm was ably summarized in the work of Wisner *et al.* (2004), who envisage disasters as the outcome of a direct clash between the socio-economic processes that create human vulnerability and the natural processes that create geophysical hazards. There are several key points:

- Disasters are caused largely by human exploitation rather than by natural or technological processes. Macro-scale root causes of vulnerability lie in the economic and political systems that exercise power and influence, both nationally and globally, and result in marginalizing poor people.
- On-going pressures, such as chronic malnutrition, disease and armed conflict, channel the most vulnerable people into unsafe environments, such as flimsy housing, steep slopes or flood-prone areas, either as a rural proletariat (dispossessed of land) or as an urban proletariat (forced into shanty towns). Effective local responses to hazards are limited by a lack of resources at all levels.
- 'Normality' in the Western sense is an illusion. Frequent disaster strikes are characteristic, rather than unusual, and reinforce socio-economic inequalities. Disaster reduction in poor countries depends on fundamental changes and a re-distribution of wealth and power. Modernization – relying on imported technology and 'quick fix' measures – is inappropriate. Instead, self-help using traditional knowledge and locally negotiated responses is seen as a better way forward.

In summary, the development view is based on the theory that disasters spring from under-development arising from political dependency and unequal trading arrangements between rich and poor nations. The poorest sections of society are forced to over-use the land and other resources, so that this behaviour cannot be regarded as 'irrational'. Specifically, rural over-population, landlessness and migration to unplanned, hazard-prone cities are the inevitable outcomes of capitalism, which is the root cause of environmental disaster.

In the immediate future, the political economy of the world is unlikely to change sufficiently in ways favoured by the development lobby. However, the paradigm has been helpful in refining some key concepts, such as the importance of poverty and vulnerability amongst disadvantaged people everywhere. Geophysical processes are not the sole contributor to disaster impact, any more than humanitarian aid is a permanent solution to deep-seated socio-economic problems in poor countries. A better understanding of socio-economic conditions is clearly needed and human vulnerability analysis and mapping is now routinely undertaken alongside geophysical risk assessments when planning for disaster reduction.

adaptation measures derived from human ecology. In contrast, social scientists, such as sociologists and anthropologists, drew on the development paradigm and adopted a cross-hazard, disaster-based view that stressed failings within political and social systems, together with the need to improve the efficiency of human responses to all types of mass emergency (Quarantelli, 1998).

McEntire (2004) argued that, whilst the behavioural and development paradigms enriched the study of hazard and disaster, each approach has crucial shortcomings. He made a case for a broader, more integrated view that sees disaster as the outcome of complicated interactions between many variables – physical, technological, social and institutional. Similarly, Dynes (2004) called for a vision that broadens beyond the Western focus on the rapid-onset hazards threatening largely urban communities to embrace modern-day threats that range from the multi-layered emergencies afflicting the rural poor in the LDCs to the disasters that still occur in the richest mega-cities of the MDCs.

E CURRENT VIEWS: THE COMPLEXITY PARADIGM

The need for a new paradigm can be illustrated by a tale of two hurricanes (Petley, 2009). On 28 October 1998, hurricane 'Mitch' made landfall on the coast of Honduras as a Category 5 storm – the strongest category of tropical cyclone. Over the next three days it crossed Honduras, Nicaragua and Guatemala, creating much destruction in Central America (Figure 1.8a). By 2 November at least 11,000 people had been killed and a similar number were missing. Most deaths resulted from mudslides and flash floods which caused economic damage, estimated at over US $5 billion, in areas that were already poor. Nine years later, on 2 September 2007, hurricane 'Felix', another Category 5 storm, made landfall on the border between Honduras and Nicaragua at almost the same location as 'Mitch' (Figure 1.8b). It also

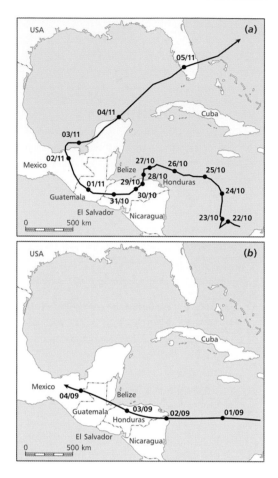

Figure 1.8 The track of two Category-5 hurricanes across Central America. Upper map shows hurricane 'Mitch' in October–November 1998; lower map shows hurricane 'Felix' in September 2007. After Petley (2009).

tracked across Honduras, Nicaragua and Guatemala, bringing strong winds and intense rainfall. But this time, the losses were far fewer. The estimated number of fatalities was 135, less than one per cent of the deaths in hurricane 'Mitch', whilst the economic damage was a fraction of that previously recorded.

The two storms had a similar size and intensity, and followed similar tracks, but had different impacts. Why should this be so? A behaviourally based answer would stress the forces of nature:

Hurricane Mitch
NOAA-14 AVHRR Composite
Multi-spectral False Color Image
October 26, 1998 @ 2028 UTC

GULF OF MEXICO

BELIZE

HONDURAS

PACIFIC OCEAN

NICARAGUA

CARIBBEAN SEA

Multispectral colourized satellite image of hurricane 'Mitch' at 20.28 UTC on 26 October 1998. This Category 5 Atlantic storm, one of the deadliest on record, produced winds of 155 knots around the low pressure centre. (Photo: NCDC/NOAA)

perhaps the intensity and duration of rainfall was much greater for hurricane 'Mitch' than for 'Felix'. This is possible because hurricanes are graded according to maximum wind speeds rather than rainfall, the characteristics of which were responsible for most of the losses in 'Mitch'. A development-based response would stress the vulnerability of the local population. For example, after hurricane 'Mitch', new disaster reduction measures were implemented – relocation of people away from the most dangerous areas, plus improved emergency planning – that would have reduced the impact of 'Felix' to some extent.

Hazard researchers and disaster managers have now merged the natural and social sciences in a more even-handed way. The *complexity paradigm* – termed sustainable hazard mitigation by Mileti and Myers (1997) – looks beyond local, short-term loss reduction in order to mesh disaster reduction with a realistic development agenda that secures a more sustainable future. A new hazard paradigm does not imply a complete rejection of previous ideas but represents a shift in emphasis. Most successful paradigms capture best practice from the past and absorb that experience into a fresh approach. The focus here has shifted from preparedness and emergency response towards mitigation that includes long-term recovery and improvement, as well as societal issues like vulnerability and resilience (Wenger, 2006). But previous strategies remain relevant. It is difficult to envisage a world in which well-designed engineering works, good land planning and effective humanitarian aid play no part in disaster reduction.

The complexity approach embeds hazards and disasters within global issues like climate change and sustainability (see Chapter 3). Humans are not simply the victims of hazards; they themselves contribute to hazardous processes and to disaster outcomes. Human actions over-exploit and degrade natural resources through processes like deforestation and global warming that, in turn, amplify the risk from natural hazards like river floods and sea-level rise. The exact relationships between 'traditional disasters' and 'complex emergencies' – and between these disasters and the forces of global environmental change – are presently unclear. This is because we are only just starting to understand the extent of human domination of the Earth's ecosystems and the extent to which this influences the vulnerability of societies to extreme events (Messerli *et al.*, 2000).

Chronic uncertainty fuels the fear of catastrophic threats of global significance. For example, the terrorist attack of 11 September 2001 in New York City was (until hurricane 'Katrina') the most costly disaster in US history, releasing at least US$20 billion for aid and anti-terror measures. It brought hazards of mass violence to the fore and led to the US concept of 'homeland security'. It also alerted the insurance industry to the threat from other 'super' hazards. Future disasters are expected to be larger in scale than in the past, due to the growing complexity of human society and the concentration of people in urban areas. Mega-disasters, capable of cutting across regional boundaries and existing socio-economic systems, have to be considered. These include threats like disease pandemics and the collision of meteorites

with settled parts of planet Earth. Some threats – like climate change – are already global in scope.

F THE ORGANIZATIONAL CONTEXT

Policy makers have become increasingly aware of the globalization of hazards and disasters. The United Nations is responsible for creating an international framework for disaster reduction. The process began in 1990 with the International Decade for Natural Disaster Reduction (IDNDR), a programme driven by concerns that disaster losses threatened the sustainability of future population growth and wealth creation, especially in the developing countries. This was followed in 1994 by the mid-term World Conference on Natural Disaster Reduction, held in Yokohama, Japan. That conference highlighted some policy failings, including an excessive emphasis on scientific solutions, reliance on the transfer of hazard-mitigating technologies to poorer countries and a relative neglect of the social, economic and political dimensions of disaster. In effect, the Yokohama conference became the first global forum to recognize the significant contribution that human vulnerability makes to disaster losses.

More recently, parallel efforts have reflected concern about climate change. The United Nations Framework Convention on Climate Change (UNFCCC) is an international treaty, established in 1992, designed to stabilize greenhouse gas concentrations in the atmosphere at a level that protects the climate system for present and future generations. In particular the treaty encouraged the MDCs to help poorer countries adapt to the adverse effects of climate change. In 1997 the Kyoto Protocol was adopted. This Protocol came into force in February 2005 and placed a legal requirement on developed countries to reduce greenhouse gas emissions by at least 5 per cent of the combined emissions recorded in 1990 during the period 2008 to 2012. Yet another disaster-related initiative was the Millennium Declaration of September 2000, an ambitious agenda which formalized 18 Millennium Development Goals set out by world leaders for reducing poverty and improving lives. All these initiatives drew attention to the links between poverty and disasters. They highlight the concern that environmental hazards have the capability to undermine the Millennium Development Goals and that disaster risk reduction has a key role to play in encouraging sustainable development for the future.

In 2000, as a policy follow-up to the IDNDR, the UN member states adopted the *International Strategy for Disaster Reduction (ISDR)* as the prime mechanism for raising political efforts to reduce natural and man-made disasters (UN/ISDR, 2004). The mandate of ISDR is to act as the focal point within the UN system for the coordination of disaster reduction and to ensure that disaster reduction becomes integral to all sustainable development, environmental protection and humanitarian policies. In effect, the ISDR consists of a wide array of partnerships comprising governments, intergovernmental and non-governmental organizations, financial institutions, scientific and technical bodies as well as the private sector and civil society. The ISDR secretariat is based in Geneva, with a liaison office in New York. It operates through a network of five regional offices in Bangkok, Cairo, Brussels, Nairobi and Panama, with related functions located in Dushanbe, Suva, Bonn and Kobe.

Further developments took place at the Hyogo World Conference on Disaster Reduction held at Kobe, Japan, during 2005. This meeting produced the *Hyogo Framework for Action (HFA)*, adopted by 168 countries for actions during 2005–15. Once again, the need to build communities more resilient to disaster in order to achieve sustainable development in the future is acknowledged. Specifically, the HFA aims to:

- ensure that disaster risk reduction is a national and local priority supported by strong institutions

- identify, assess and monitor disaster risks and enhance the provision of early warning
- increase capacity, knowledge and innovation to build a culture of safety and hazard resilience at all levels
- integrate all disaster reduction measures – preparedness, mitigation and programmes to lower vulnerability – into sustainable development policies
- add risk reduction into the design and implementation of disaster emergency response, recovery and reconstruction programmes.

The HFA document encourages a collaborative strategy that includes national governments, regional bodies and local communities. Unlike the Kyoto Protocol, it has no legal status. It would be tempting to conclude that disaster risks are different from, and are treated less seriously than, the adverse effects of climate change. This would be wrong. Previously, responses to disaster have been viewed as short-term emergency actions, whilst climate change has been seen as a slow-onset, multi-generational problem. In truth, *Disaster Risk Reduction* and *Climate Change Adaptation* share mutual goals that can often be achieved by similar means. They need solutions now. It is important that this complementarity is recognized so that synergies can be harnessed to reduce environmental risks from all sources in a more comprehensive and sustainable manner for the future.

FURTHER READING

Degg, M.R. and Chester, D.K. (2005) Seismic and volcanic hazards in Peru: changing attitudes to disaster mitigation. *The Geographical Journal* 171; 125–45. An example of how hazard paradigm shifts can be applied to disaster reduction.

McEntire, D.A. (2004) Development, disasters and vulnerability: a discussion of divergent theories and the need for their integration. *Disaster Prevention and Management* 13: 193–8. A thoughtful critique of the behavioural and development paradigms.

Montz, B.E. and Tobin, G.A. (2011) Natural hazards: an evolving tradition in applied geography. *Applied Geography* 31: 1–4. A short update on the geographical contribution.

Strömberg, D. (2007) Natural disasters, economic development and humanitarian aid. *Journal of Economic Perspectives* 21: 199–222. This paper takes a global view of the differential impacts of, and the human responses to, disaster.

UN/ISDR (2004) *Living with Risk: A Global Review of Disaster Reduction Initiatives*. United Nations, Geneva. Sets out the general direction of international action for disaster reduction.

Wisner, B. *et al.* (2004) *At Risk: Natural Hazards, People's Vulnerability and Disasters*. Routledge, London and New York. An overview of hazards and disasters with a focus firmly on human vulnerability.

WEB LINKS

Aon Benfield UCL Hazard Research Centre, London www.abuhc.org

Natural Hazards Research and Applications Information Center, Colorado www.colorado.edu/hazards/

Overseas Development Institute, London www.odi.org.uk

UN International Strategy for Disaster Reduction www.unisdr.org

CHAPTER TWO

Dimensions of disaster

A INTRODUCTION

In the 30 years between 1974 and 2003, more than 2 million people were killed in over 6,350 'natural' disasters worldwide (Guha-Sapir *et al.*, 2004). In addition, 5.1 billion individuals were directly affected by these events, including 182 million people left homeless; economic costs were estimated at US$1.4 trillion. These are significant losses, but all disaster data should be interpreted with care. The records are difficult to assemble. Disaster impacts, trends and patterns are complex and often controversial. This is especially so when policy-related questions are raised about apparent increases in the frequency and/or magnitude of extreme geophysical events and the adverse societal consequences.

Most disaster losses are due to a small number of very high-magnitude events. Taking a long-term view, Table 2.1 lists all recorded disasters since AD 1000 that claimed at least 100,000 lives. Four hazard types were involved – earthquakes, tropical cyclones, floods and droughts – and were responsible for over 20 million deaths. The list is notable for the frequent appearance of Asian countries. Indeed, over 70 per cent of these

disasters were located in Asia, with almost 40 per cent in China alone. This reflects the geographical size of Asia, the high proportion of the world's population living there, the length of written records available for China and – not least – the hazardous nature of the physical environment. Famine is excluded from Table 2.1 although it is often linked with drought. Both drought and famine can last for several years. For example, in 1932–33, 7 million people died from famine in the Soviet Union, and between 1959 and 1962, 29 million people died from famine in China.

It is vital to compile reliable disaster data in order to identify time trends and spatial patterns that will inform policy making. Problems start with a lack of standardized methods for data collection in the field and inconsistencies in defining and assessing key impacts. Even if the organization responsible for archiving the data adopts clear and consistent definitions, and uses transparent methods for processing the information, the original suppliers of the information may not have done so (Guha-Sapir and Below, 2006). An absence of agreed methodologies can lead to misunderstanding and undermine confidence in the information.

Table 2.1 Disasters responsible for at least 100,000 deaths recorded since AD 1000

Year	Country	Type of disaster	Fatalities
1931	China	Flood	3,700,000
1928	China	Drought	3,000,000
1971	Soviet Union	Epidemic	2,500,000
1920	India	Epidemic	2,000,000
1909	China	Epidemic	1,500,000
1942	India	Drought	1,500,000
1921	Soviet Union	Drought	1,200,000
1887	China	Flood	900,000
1556	China	Earthquake	830,000
1918	Bangladesh	Epidemic	393,000
2010	Haiti	Earthquake	316,000
1737	India	Tropical cyclone	300,000
1850	China	Earthquake	300,000
1881	Vietnam	Tropical cyclone	300,000
1970	Bangladesh	Tropical cyclone	300,000
1984	Ethiopia	Drought	300,000
1976	China	Earthquake	290,000
1920	China	Earthquake	235,000
2004	Indian Ocean	Tsunami	230,210
1876	Bangladesh	Tropical cyclone	215,000
1303	China	Earthquake	200,000
1901	Uganda	Epidemic	200,000
1622	China	Earthquake	150,000
1984	Sudan	Drought	150,000
1923	Japan	Earthquake	143,000
1991	Bangladesh	Tropical cyclone	139,000
2008	Myanmar	Tropical cyclone	138,000
1948	Soviet Union	Earthquake	110,000
1290	China	Earthquake	100,000
1786	China	Landslide	100,000
1362	Germany	Flood	100,000
1421	Netherlands	Flood	100,000
1731	China	Earthquake	100,000
1852	China	Flood	100,000
1882	India	Tropical cyclone	100,000
1922	China	Tropical cyclone	100,000
1923	Niger	Epidemic	100,000
1985	Mozambique	Drought	100,000

Note: These figures are approximate. The list is biased towards the recent past because of the unavailability of earlier records. Adapted from Munich Re (1999), CRED database and other sources.

B DEFINING DISASTER

1 Recording disaster

It is easy to see that the events in Table 2.1 were important, but there is no internationally agreed definition of what constitutes a 'disaster'. The most obvious feature is that the losses have to be sufficiently large to disrupt the functioning of a community beyond its ability to cope. But quantifying the precise impact thresholds necessary for such disruption – whether for fatalities, economic damage or other losses – is a challenging task. Loss of human life is usually the prime indicator. This may seem to be a readily available statistic, but it may not be so in countries lacking detailed population data. Information on other human impacts such as personal injury, disease and homelessness is more difficult to obtain, even in well-governed nations. In the absence of standardized methods of survey and assessment, the collection of robust information on property damage and less direct economic losses is also unlikely.

Many disasters are compound events and create instant problems of classification. To avoid double counting of impacts, each loss category, such as deaths or damages, should be recorded once only and allocated to a specific cause, but practical problems arise. For example, when an earthquake triggers a landslide, should deaths be attributed to the earthquake (the trigger event) or the landslide (the direct cause)? Will deaths occurring days or weeks later from injuries sustained in the landslide be recorded? Generally speaking, when recording mass fatalities, the initial trigger is named. This means that the effects of 'secondary' hazards, such as landslides, are under-estimated (see Chapter 8). For long-duration disasters, such as drought, there may be doubts about the exact date of the event. The precise location of a disaster can also be difficult to identify, especially for events (like floods and droughts) that cross administrative boundaries. As a result, Peduzzi *et al.* (2005) linked an

Box 2.1 Types of disaster impact

Few disaster reports include any losses beyond the direct and tangible impacts (Figure 2.1). *Direct effects* are the first-order consequences that occur immediately after an event, such as the deaths and economic loss caused by the throwing down of buildings in an earthquake. *Indirect effects* emerge later and may be more difficult to attribute to the event. These include factors such as mental illness resulting from shock, bereavement and relocation from the area. *Tangible effects* are those for which it is possible to assign monetary values, such as the replacement of damaged property. *Intangible effects*, although real, cannot be properly assessed in monetary terms. For example, many important archaeological sites in Italy are at risk from landslides, floods and soil erosion (Canuti *et al.*, 2000).

- *Direct losses* are the most visible consequence of disasters, due to the immediate damage, such as building collapse. They are comparatively easy to measure, although methodologies are not standardized and surveys are always incomplete. For example, loss estimates for insurance purposes are probably more accurate than some field-based surveys. However, insurance claims can be deliberately inflated and there is a lack of insurance cover in poor countries. Direct losses are not always the most significant outcome of disaster.

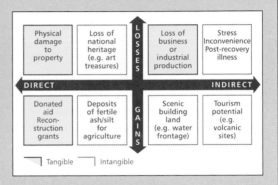

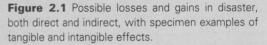

Figure 2.1 Possible losses and gains in disaster, both direct and indirect, with specimen examples of tangible and intangible effects.

- *Direct gains* represent benefits flowing to survivors after a disaster, including various forms of aid. People with skills in the construction trade may obtain well-paid employment in the restoration phase following the event and, occasionally, some longer-term enhancement of the environment may occur. On the Icelandic island of Heimaey, volcanic ash resulting from the 1973 eruption was used as foundation material to extend the airport runway, and geothermal heat has been extracted from the volcanic core.

- *Indirect losses* are the second-order consequences of disaster, like the disruption of economic and social activities. As property values fall, consumers save rather than spend, business becomes less profitable and unemployment rises. Again, data are incomplete. For example, no financial losses are reported for epidemics, although the premature death of active workers inevitably results in a loss of manpower and productivity. Ill-health effects often outlast other losses. Psychological stress affects the victims of disaster directly and has an indirect influence on family members and rescue workers. The symptoms include shock, anxiety, stress or apathy and are expressed through sleep disturbance, belligerence and alcohol abuse. Attitudes of blame, resentment and hostility may also occur.

- *Indirect gains* are less well understood. They are the long-term benefits enjoyed by a community as a result of its hazard-prone location. Little systematic research has been undertaken, for example, into the balance between the on-going advantages of a riverside site (flat building land, good communications, water supply and amenity), as compared with the occasional losses suffered in floods.

automated geographical information system (GIS) to Centre for Research on the Epidemiology of Disasters (CRED) disaster data to create a rapid mapping procedure for floods, earthquakes, cyclones and volcanoes. It was found that over 80 per cent of these events could be successfully geo-referenced by this scheme.

Although damaging extreme events may be classified according to natural and technological *causes*, they must breach certain human *effects* thresholds before they are recorded as disasters. One consequence is the widespread under-reporting of smaller-scale disasters. This is most common in LDCs that lack the reliable, up-to-date demographic and socio-economic statistics needed to provide a precise record of loss. For example, what total should be used when, as often happens, a wide range of deaths are given or when mortality is assessed 'in the thousands'? How can the data include those persons reported missing, those who die later from their injuries or from secondary effects, such as famine or disease? So far, a definition for disaster-related 'injury' has not been agreed.

Despite these limitations, data on disaster mortality are often more accurate than other loss information, including that relating to financial costs. Munich Re (2005) estimated that the percentage of natural catastrophes with good-quality reporting of economic losses in the period 1980 to 1990 was approximately 10 per cent. By 2005 that figure had risen to about 30 per cent, but still left about two-thirds of natural disasters lacking accurate information. Once again, the problem arises from inconsistencies in data collection, especially for indirect economic losses. For example, if a flood damages a bridge and farmers cannot transport their goods to the local market, should the loss in produce sales be included in the calculation? This is preferable, but data are rarely available. Similarly, little account is taken of the on-going financial burden of disaster preparedness, even though these resources are not then available for other community needs.

In summary, most disaster audits are limited to estimates of direct deaths, injuries and immediate damage and capture only part of the impact picture (Box 2.1). Many disaster survivors suffer indirect impacts, including the loss of a relative, malnutrition, physical or mental illness, loss of employment, debt and forced migration. Some consequences can persist for years after the event, but few follow-up surveys are made. The extent of homelessness immediately after a disaster is often the only indirect loss for which reasonably reliable statistics exist.

2 Reporting disaster

The mass media play a major role in raising disaster awareness and as a primary source of information for the agencies that collect disaster data. News organizations are well equipped to collect and transmit information continuously in times of crisis, and knowledge of a disaster spreads rapidly from the hazard zone to a global audience (Figure 2.2). This flow of information can stimulate public interest in the event and may also influence the flow of disaster aid. Eisensee and Stromberg (2007) found that emergency relief decisions by the US federal government were strongly influenced by media reports of natural disasters. More aid was provided when disasters

Figure 2.2 A disaster-impact pyramid. Consequences and awareness of the event spread outwards from a small number of people directly affected in the hazard zone to the global population, via the mass media.

were not crowded out of the headlines by unrelated but more newsworthy items. For example, in order to have the same chance of receiving relief funds, a disaster during an Olympic event required three times the number of casualties of a disaster that occurred when media pressure was at its lowest level.

Media reports during the early phases of an emergency tend to be patchy in cover and unreliable in content, due to uncertainties at the time. Unfortunately, there is rarely any correction of such reports before interest levels decline and other stories take over. As shown by Ploughman (1995), unbalanced reporting of natural disasters is long standing for newspapers in the USA, although local – as opposed to national – news media can provide good in-depth reporting (Rashid, 2011). In a review of media coverage of disaster responses by aid organizations, Ross (2004) noted that journalists lacked knowledge about humanitarian issues and suffered from crisis fatigue. News editors prioritize stories according to death totals and the availability of graphic video material. Disaster reports on commercial TV channels are partly news and partly entertainment. Short-term dramatic disasters (earthquakes or volcanoes) attract a lot of news coverage but chronic issues (like drought) are more likely to be dealt with much later in a current affairs feature. When there is a dependence on advertising revenue, the media tend to concentrate on the prosperous target markets of commercial sponsors. This can lead to under-reporting of disaster impacts on poorer social groups in disadvantaged areas (Rodrigue and Rovai, 1995).

Miller and Goidel (2009) identified the main features of news organizations that lead to reporting bias. First, the coverage of disasters is episodic and creates the impression of a series of unrelated events rather than a coherent narrative. The focus is typically victim orientated. During the 'Ash Wednesday' bushfires in Australia, residents were portrayed as helpless victims, with little mention of more positive aspects like early warning and emergency response (McKay, 1983). The media rarely explore the wider context of disaster and pose deeper questions, such as: why do so many Americans locate close to hurricane-prone coasts, or so many Australians live in suburbs prone to bushfires? Other examples of media reliance on predetermined story lines include exaggerating the physical scale of events and highlighting rapid-onset events at the expense of long-term disasters such as drought. Second, the media employ social stereotypes that may falsely portray affected residents. Such bias existed in the US media after hurricane 'Katrina' struck New Orleans in 2005. Victims were often characterized as poor, black and – sometimes – unworthy of public support. Stories of looting, lawlessness and criminal damage were prominent, even though this was quite rare (Tierney *et al.*, 2006). When African-Americans broke into shops to obtain food, it was 'looting', whereas the same behaviour by others was seen as an act of survival. Just as many white victims featured in TV interviews as black victims, even though blacks were more numerous in the New Orleans population and probably suffered greater loss.

The coverage of disasters on television news channels is much influenced by the visual impact of film reports (Greenberg *et al.*, 1989; Wrathall, 1988). According to Garner and Huff (1997), media reporting shows an excessive concentration on the emergency phase of disaster, especially if striking images of distressed victims are available. There is also an undue concentration on events close to home. Adams (1986) studied the reporting by American television of 35 natural disasters (each causing at least 300 fatalities) in various parts of the world and found that the world was prioritized by geographical location so that the death of one Western European equalled 3 Eastern Europeans or 9 Latin Americans or 11 middle Easterners or 12 Asians. In summary, more coverage of the underlying causes of disaster, rather than the routine focus on panicked residents, helpless victims and looting of evacuated properties, would greatly improve the media balance.

C MEASURING DISASTER: ARCHIVES

Disaster records are held by many bodies, including international organizations, national government agencies, insurance companies and academic institutions. These bodies have different resources and reasons for data collection, so the resulting information varies in content and quality. Guha-Sapir and Below (2002) compared disaster data compiled by two European reinsurance companies with the EM-DAT database developed by a university-based research centre. The survey revealed major inconsistencies (Table 2.2). The EM-DAT record contained many more fatalities, whilst the two reinsurance companies recorded higher levels of economic loss. This reflects the priorities of the organizations concerned; the reinsurance companies were mainly concerned with economic losses, with CRED emphasizing the humanitarian aspects of disaster.

Table 2.2 The total number of disasters, people killed and economic damage as recorded in three different databases for four countries

	CRED	Munich Re	Swiss Re
Honduras			
No. of events	14	34	7
Number killed	15,121	15,184	9,760
Number affected	2,982,107	4,888,806	0
Total damage ($US million)	2,145	3,982	5,560
India			
No. of events	147	23	1,220
Number killed	58,609	877	65,058
Number affected	706,722,177	2,993,281	16,188,723
Total damage ($US million)	17,850	112	68,854
Mozambique			
No. of events	16	23	4
Number killed	106,745	877	233
Number affected	9,952,500	2,993,281	6,500
Total damage ($US million)	27	112	2,085
Vietnam			
No. of events	55	101	36
Number killed	10,350	11,114	9,618
Number affected	36,572,845	20,869,877	2,840,748
Total damage ($US million)	1.915	3,402	2,681
Totals			
Total no. of events	232	387	167
Total number killed	189,825	96,418	84,669
Total number affected	756,139,629	277,490,405	19,035,971
Total damage ($US million)	21,937	29,629	79,180

Source: After D. Guha-Sapir and R. Below, R. *The Quality and Accuracy of Disaster Data: A Comparative Analysis of Three Global Data Sets*. Working Paper prepared for the Disaster Management Facility, World Bank, CRED (2002). Copyright CRED 2009.

Datasets cover varying time periods. Some of these are too short to provide a valid sample and may be restricted in other ways. Some concentrate on certain types of loss, like the reinsurance companies. Several are limited to specific types of hazard either worldwide – e.g. volcanoes (Witham, 2005), landslides (Petley, Durham University database) – or regional – e.g. floods and mass movements in Switzerland (Hilker *et al.*, 2009). In a review of 31 databases covering natural and technological disasters, Tschoegl *et al.* (2006) concluded that a lack of standardization in definitions, differences in disaster classification, inadequate accounts of methodology and variations in the availability of the sources undermined the usefulness of the information.

A quality framework for disaster databases was recommended by Below *et al.* (2010). The key features are:

- *Prerequisites and sustainability.* The archiving body should have enough institutional support and other resources to maintain the dataset over a representative period of time.
- *Data accuracy and reliability.* The information should be as complete as possible, with good geographical coverage; procedures should be in place to test for bias and other faults.
- *Methodology.* The raw information should be processed according to clear concepts and definitions regarding issues like entry criteria, storage and back-up.
- *Credibility.* There should be evidence of the expertise and impartiality of the archiving body that includes assurances on transparency and quality-control procedures.
- *Serviceability.* The information should be useful and convenient, it should be easy to interpret, have perceived relevance and be disseminated in a timely way.
- *Accessibility.* The data should be readily accessible to a wide variety of users and contact details should be available for those seeking further information.

The most complete record of disasters, used by United Nations agencies and adopted in this book, is maintained by the Centre for Research on the Epidemiology of Disasters (CRED) at the University of Louvain, Belgium. The Emergency Events Database (EM-DAT) covers natural and technological disasters, as described in Table 2.3, from 1900 onwards (Sapir and Misson, 1992; Guha-Sapir *et al.*, 2004). Information prior to 1988 was absorbed from the previous records of the US Office of Foreign Disaster Assistance (OFDA) but is less complete. CRED information is updated daily, whilst various checks and revisions occur at three-monthly and annual intervals. Such quality control ensures that losses are properly verified after the emergency disaster phase.

For inclusion in EM-DAT, a disaster must have killed 10 or more persons, or affected at least 100 people, although an appeal for international assistance, or a national government disaster declaration, will take precedence over other criteria. For displaced persons, or drought and famine to register, at least 2,000 people have to be affected. The clear emphasis is on human impacts, deaths and disruption, rather than economic loss or environmental damage. Another important resource is the catalogue of 'Great Natural Catastrophes' (NATCAT) held by the insurance group Munich Re. The NATCAT dataset follows broad UN disaster definitions but has higher thresholds governing the entry of fatalities and other impacts, although these criteria are not always spelled out precisely. One practical consequence of this is that far fewer events are catalogued in the NATCAT lists than in EM-DAT.

The quality of disaster data is under continuous scrutiny, largely to ensure that users understand the limitations of the information. Gall *et al.* (2009) identified several problems:

- *Hazard bias.* User assumption – every hazard is represented; in reality, selective reporting of hazards may occur based on the priorities of the collecting agency.

Table 2.3 List of disaster types and sub-types recorded in EM-DAT

Natural disasters		Technological disasters	
Disaster types	**Disaster sub-types**	**Disaster types**	**Disaster sub-types**
Drought		Industrial accident	Chemical spill
Earthquake			Explosion
Epidemic			Radiation leak
			Collapse
Extreme temperature	Cold wave		Gas leak
	Heat wave		Poisoning
Famine	Crop failure		Fire
	Food shortage		Other
	Conflict		
	Drought	Miscellaneous accident	Explosion
Slide	Avalanche		Collapse
	Landslide		Fire
			Other
Volcano			
Wave/surge	Tsunami	Transport accident	Air
	Tidal wave		Boat
Wildfire	Forest fire		Rail
	Scrub fire		Road
Windstorm	Cyclone		
	Hurricane	Conflict	Intrastate
	Storm		International
	Tornado		
	Tropical storm		
	Typhoon		
	Winter storm		

Source: After CRED at http://www.cred.be (accessed 16 February 2010).

Note: Cyclone, hurricane and typhoon are different names for the same event used in different parts of the world. Some major types of natural disaster (drought, earthquake and flood) are not sub-typed but all types of technological disaster are sub-typed.

* *Temporal bias.* User assumption – losses are comparable over time; in reality, changes exist in recording procedures and due to increases in population and wealth.
* *Threshold bias.* User assumption – all losses are counted; in reality, a filtering process means that many small-scale losses go unreported.
* *Accounting bias.* User assumption – all types of loss (human, economic, direct, insured) are included; in reality, this is a highly variable feature of disaster loss estimation.
* *Geographical bias.* User assumption – losses are comparable across geographical areas and units; in reality, changing political and administrative boundaries and internal variations make this impossible.
* *Systemic bias.* User assumption – losses are computed uniformly; in reality, economic losses may, or may not, be adjusted for price inflation. Where a range of loss estimates exists, an agency may select the highest, lowest or the mean estimate for archiving.

Even with the highest-quality data, it can be difficult to draw totally valid conclusions. The widespread use of *absolute impact* thresholds obscures differences in the *relative impact* of disasters between – and within – individual countries. For example, a US$100 million loss would be caused by a much smaller and higher-frequency event in – say – California than in Bangladesh. Generally, datasets of financial losses, like those produced by the reinsurance industry, tend to emphasize the MDCs, where the exposed assets are higher and more important to the compiler. Conversely, datasets that prioritize fatalities tend to be biased towards the LDCs, where large numbers of vulnerable people live. Global datasets are aggregated from information from individual nation states but national statistics do not capture local impacts, including those on very vulnerable groups such as the poor or ethnic minorities. Small, isolated communities are especially at risk of being ignored. The loss of 10 able-bodied men from a remote fishing village would be far more devastating for the survival of that community than the death of 100 men in a large city.

Ideally, disaster impact should be scaled according to regional or local population numbers, the economic functions and the financial resources available for recovery in both the public and private sectors. This rarely happens, although CRED has attempted to identify so-called '*significant' natural disasters*, where the number of deaths per event is 100 or more, damage amounts to 1 per cent or more of the annual national gross domestic product (GDP) and the number of affected people is 1 per cent or more of the total population. These relative measures for damage and affected people indicate more accurately than absolute national totals the effect of disasters on poor countries with weak economies and small populations.

D EXPLAINING DISASTER: TIME TRENDS

1 Disaster trends

Although the CRED archive goes back to 1900, the systematic recording of disasters on a global scale did not really begin until 1964. Since then there has been a rise in the recorded number of natural disasters from an annual average of less than 50 before 1965 to around 250 per year in the 1990s. Fewer technological disasters occur, but the general rate of increase is similar. In a study of 'Great Natural Catastrophes' for the period 1950–99, Munich Re (2005) found that the number of reported events in the 1990s was 4.5 times the number recorded in the 1950s and overall losses increased from US$48 billion in 1950–59 to US$575 billion in 1990–99 (based on 2005 cost values). Much of this increase was attributed to climate-related disasters.

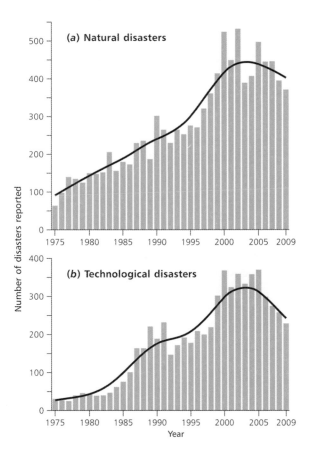

Figure 2.3 (a) Annual number of natural disasters reported 1975–2009, with trend; (b) annual number of technological disasters reported 1975–2009, with trend. After CRED EM-DAT. Copyright CRED 2009.

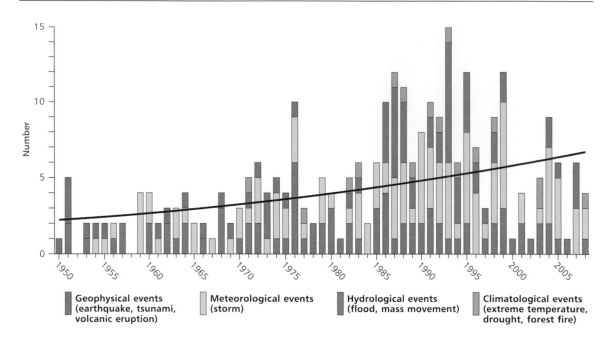

Figure 2.4 Annual number of Great Natural Catastrophes according to type of event recorded 1950–2009, with trend. After Munich Re NATCATSERVICE (2010).

The rise in reported natural disasters continued during the last quarter of the twentieth century and peaked around the year 2000 (Figure 2.3a). Technological disasters followed a similar trend (Figure 2.3b). Due to the higher entry criteria, the smaller number of NATCAT 'Great Natural Catastrophes' recorded during 1950–2009 shows more inter-annual variability and a less well-defined peak arising slightly earlier in the 1980s and 1990s (Figure 2.4). The importance of climate-related events (storms, floods, mass movements, extreme temperatures, droughts and wildfires) is clear. The time trend of fatalities in CRED data is different. For EM-DAT mortality during 1975–2009 there was marked inter-annual variability but a general decline with the lowest values after the year 2000 (Figure 2.5a). Conversely, the graph for technological-related deaths, a much smaller sample size, follows the long upward trend in the number of reported disasters much more closely and a decline is not apparent until the first decade of the twenty-first century (Figure 2.5b).

These differences suggest that deaths from environmental hazards as a whole – and from natural hazards in particular – have been stable or decreasing for some time, perhaps due to the growing success of disaster reduction measures.

Different trends exist for other types of disaster loss. For example, the number of people affected by natural disasters appears to have grown since the mid-1980s, in line with the number of recorded events (Guha-Sapir *et al.*, 2004). More detailed information is available on financial losses. Figure 2.6 shows total annual economic damage from natural disasters from 1980 to 2011 based on CRED data, with the losses benchmarked to 2011 US$ values (CRED, 2012). All the high-cost peaks are associated with exceptional disasters reported during the last 20 years, notably the Kobe, Japan earthquake of 1995, hurricane 'Katrina', USA in 2005 and the 2011 Tōhoku earthquake and tsunami in Japan. NATCAT data provided by Munich Re displays a similar picture, with a more distinct upward trend from 1950 to

2009 in both overall and insured economic losses (Figure 2.7). Although the death toll was relatively small in most of the high-cost years, the growing financial impact has alarmed the insurance industry, particularly with respect to Atlantic hurricanes. 'Katrina' alone cost US$45 billion in 2005 and worldwide insured losses in that year reached about US$100 billion. In several high-loss years the disaster costs exceeded international spending on development aid throughout the world.

Policy makers and others are always interested in time trends of disaster loss when the trend is upwards. There are three main reasons why disaster losses might display a sustained rise: changes in data recording methods (improvements in data collection), changes in the physical environment (an increased frequency and/or intensity of hazard events) and changes in the societal environment (greater human exposure and vulnerability to hazard). These factors are not mutually exclusive and it is important to identify the main drivers of trends, in order to devise the most suitable policy responses.

2 Changes in data recording methods

Any changes in data collection and recording methods which lead to the inclusion of more events will clearly increase the number of disasters and the size of disaster losses. Such changes may reflect a shift in policy within an agency or be due to wider influences. For example, step increases in EM-DAT annual disaster totals exist immediately after 1964 (when OFDA was created) and after 1973 (when CRED was created). Such 'artificial' rises can be explained by institutional factors, but databases can also show 'false' trends that reflect wider influences.

The upper part of Figure 2.8 shows a long-term upward trend from 1790 to 1990 in annual volcanic eruptions, despite the fact that there is no physical reason for global volcanic activity to increase over this 200-year period. In fact, as

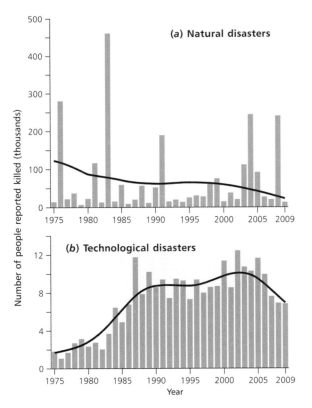

Figure 2.5 (a) Annual number of people reported killed by natural disasters 1975–2009, with trend; (b) annual number of people reported killed by technological disasters 1975–2009, with trend. After CRED EM-DAT. Copyright CRED 2009.

shown by the lower graph in Figure 2.8, the number of large eruptions – those most likely to be reported and archived – remained fairly constant. Therefore, the apparent upward trend is really a measure of volcanic hazard awareness and improved monitoring and reporting over the period. When the world was preoccupied with alternative matters, like war or economic depression, volcanoes became less newsworthy. However, several very large volcanic events stimulated media reporting for several years after the eruption. In a similar way, the high-profile Asian tsunami disaster in 2004 enhanced interest in a type of event that had attracted little previous attention.

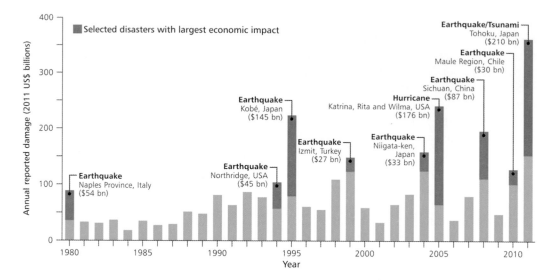

Figure 2.6 Annual total of reported economic losses in natural disasters 1980–2011. Losses attributable to individual high-impact disasters are identified. Figures in US$ billion at 2011 values. After CRED (2012) and UNISDR.

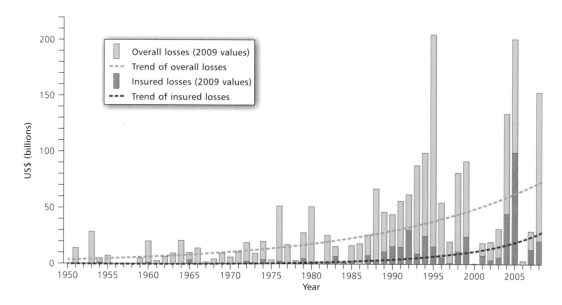

Figure 2.7 Annual total of overall losses and insured losses from Great Natural Catastrophes recorded 1950–2009, with trend. After Munich Re NATCATSERVICE (Geo Risks Research) at www.munichre.com (accessed 16 April 2011).

Political change towards democracy and more open societies also contributes. In the past, totalitarian regimes – such as those in the former USSR and China – routinely suppressed information about internal disasters. This was especially so for drought-related disasters because many famines were attributable to government incompetence in managing and transporting food supplies. Today the spread of digital technology and social networking via the media, such as the internet, Twitter and Facebook, severely limits any attempt to control the spread of disaster information. Cell phones are used by members of the public to record local disaster impacts and to disseminate the information worldwide. Digital data gathering and transmission from the most remote regions is now possible. Aid agencies collect still photographs, video and audio recordings for transmission to journalists who are unable to make field visits themselves. High-quality video can now be transmitted via a narrow-band satellite phone call to reach a global audience.

3 Changes in the physical environment

Few databases contain records suitable for long-term analysis, although proxy geological and historical data can be used to extend the time-scale for earthquakes, volcanic eruptions and floods. Over shorter periods, damaging geophysical phenomena vary and change in different ways and for different reasons. Tectonic activity is almost entirely natural and earthquakes, for example, are randomly variable without any complication from human activity. But some climate-related extremes, like hurricanes, show natural variations on annual or decadal time-scale variations. In addition, some climate-related hazards – like hurricanes or floods – may be subject to human influences of various kinds. As a result, identification of time trends in extreme geophysical events, and also any attribution of cause, is often difficult. Even if a genuine trend towards more

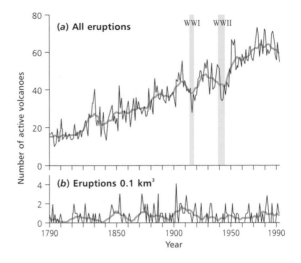

Figure 2.8 (a) Annual number of volcanic eruptions reported 1790–1990. This graph, for all reported eruptions, shows a clear upward trend, disturbed by two world wars. After L. Siebert, T. Simkin and P. Kimberly, *Volcanoes of the World*, © 2011 by the Smithsonian Institution. Published by the University of California Press. Reproduced with permission. (b) Annual number of volcanic eruptions of size 0.1 km³ or greater reported 1790–1990. This graph, for the largest eruptions only, shows no overall trend.

frequent or more severe events can be verified, it is not possible to establish a direct, one-to-one link with the incidence and scale of disasters. This is because disasters are the outcome of an interaction between hazardous physical processes and human communities. Therefore, any reported increase in disaster impacts can only be loosely attributed to changes in the physical environment because it is possible that at least part of any trend is explained by changes in human exposure and vulnerability. This reinforces the need to view disasters within the context of the coupled human-environment system (CHES).

About two-thirds of all events in disaster datasets are climate related. Therefore, any significant change in the frequency of these events is likely to influence the overall incidence of disasters. Figure 2.9 graphs the annual number of all climate-related disasters (wind-storms, floods, mass movements, extreme temperatures, droughts and wildfires) in the NATCAT record for 1950–2009. As previously stated, this dataset captures the largest events only. The sample is for a limited time-scale but is significant because these are the most damaging disasters. Overall, the graph shows an upward trend, particularly towards the close of the twentieth century, with a threefold increase in disaster occurrence over the period. This trend, broadly replicated in other datasets, has attracted much attention.

Many hydro-meteorological events are a function of near-surface physical processes that may have been subject to increasing human modification in recent decades. Quite apart from concerns about the growth in economic losses, greater frequency and magnitude of weather extremes is an expected outcome of global warming. Not surprisingly, some observers have concluded that the fingerprint of climate change – whether due to human actions or not – is on these data (see also Chapter 14, Section F). But caution is required. As already indicated, raw trends in disaster losses are unlikely to be driven solely by physical factors, not least because the rate of change in socio-economic conditions is usually faster than that for large-scale natural systems.

4 Changes in the socio-economic environment

As the world population grows, so more people are at risk and disaster mortality might be expected to rise; as the size of the world economy grows, and as price inflation continues, so the financial impact of disaster is likely to rise. Some human populations have become more vulnerable to hazards in recent years, despite positive steps taken to reduce disasters. To produce reliable conclusions, it is necessary to standardize deaths according to the number of people at risk (to compensate for population growth over time) and to standardize economic losses according to price (to compensate for changes in monetary value over time).

This process, called *normalization*, turns raw disaster data into a more consistent time-series by benchmarking socio-economic data to values in a particular year. In effect, the correction shows the losses as if all the disasters had occurred in that year. Economic losses are adjusted up (or down) to fit costs that applied in the selected base year in order to compensate for price inflation. Similar adjustments are applied to other human variables. For example, mortality data can be adjusted for changes in the size of population living within designated hazard zones, such as low-lying coasts; property damage can be adjusted to account for increases in the value of dwellings or increases in the number of dwellings over time. If losses are normalized for a period with more frequent hydro-meteorological extremes – and if little or no upward trend exists – the most likely explanation is that disaster reduction measures, such as improved flood protection works, have been successful.

For example, a data normalization exercise was conducted on twentieth-century hurricane-related losses in the USA (Pielke and Landsea, 1998). The raw information indicated a trend to rising damage and economic loss but, when adjusted for growth in the coastal population and the rise in exposed wealth, the losses in the 1970s and 1980s were found to be smaller than in some earlier decades (see also Chapter 9). More recent studies of disaster trends include those of earthquakes in the USA (Vranes and Pielke, 2009), wind-storms in Europe (Barredo, 2010) and meteorological hazards in Australia (Crompton and McAneney, 2008). Bouwer (2011) reviewed 22 studies of economic loss data extending over at least 30 years for hydro-meteorological disasters. The results were striking. In 14 cases the normalized data showed no trend – either upwards or downwards. Some upward trend existed in eight studies but, even in these cases, the rise did not persist throughout the data period.

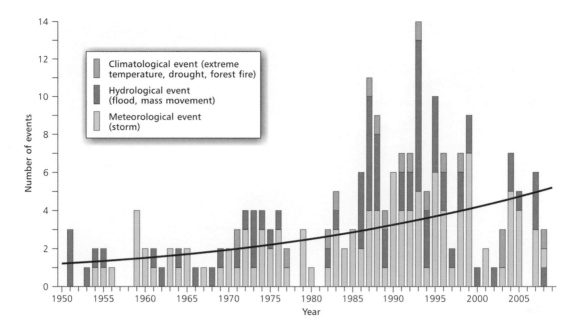

Figure 2.9 Annual number of Great Weather Catastrophes in meteorological, hydrological and climatological categories recorded 1950–2009, with trend. After Munich Re NATCATSERVICE (Geo Risks Research) at www.munichre.com (accessed 16 April 2011).

The present conclusion of most disaster experts is that, whilst the number of recorded weather-related disasters has probably increased in recent decades, there is no evidence of an associated trend in disaster losses when the data are corrected for societal changes. Most reported increases in loss are believed to result from economic, demographic, political or social influences that increase human vulnerability. In particular, there is – so far – little evidence of a definitive link between rising disaster losses and climate change. Of course, disaster data are historical, and present information may not remain valid, not least if climate change impacts become more prominent. Fresh data and new techniques, such as refinements to normalization methods proposed by Neumayer and Barthel (2011), could alter these conclusions in the future.

E EXPLAINING DISASTER: SPATIAL PATTERNS

Some years ago, a World Bank study concluded that more than half of the world's population was exposed to one or more natural hazards (Dilley *et al.*, 2005). The majority of these people live in countries with low or medium human development and economies that are still emerging. Of the 200 or so disaster-related deaths that occur on average each day, most are located in these nations. The cliché that in disasters the poor lose their lives while the rich lose their money is largely true. People in countries with low human development have little more than 10 per cent of the physical exposure to environmental hazards across the world but suffer over half the disaster-related deaths. Conversely, people in rich countries have about 15 per cent of the physical exposure to hazard but record less than 2 per cent

Plate 2.1 A woman cooks food outside a slum dwelling on stilts above a polluted waterway in Manila, Philippines. Poverty remains widespread, as compared with some other countries in the region, and is exacerbated by large inequalities in wealth. (Photo: Panos/Robin Hammond RHM02116PHI)

of the fatalities. Kim (2012) claimed that poor people (defined as those living on US$2 or less per day) have an exposure to natural disasters almost double that of others and that they experience about one-fifth more disasters than the non-poor.

Table 2.4 shows the differential impact of the five deadliest and the five costliest disasters recorded between 2000 and 2010. It is evident that deaths and damages do not correlate. Only one disaster – the 2008 earthquake in China – appears in both lists. This is because the most deadly disasters occur in the LDCs and the most costly disasters occur in the MDCs. The deadliest disasters produced an average of 150,000 fatalities, with losses of US$22 million; the five costliest

disasters produced an average of 17,000 fatalities, with losses of over US$60,000 million. This table not only reveals the strikingly different impacts of disaster in rich and poor countries but also the wide gaps in insurance cover that exist between wealthy people and those who live in poverty.

Clearly, disaster risks and disaster losses do not occupy the same geographical spaces. Poverty, resource depletion, marginalization, under-development and weak governance are all factors. As demonstrated by Bankoff *et al.* (2004), human vulnerability is a key factor (see also Chapter 3, Section F). Over 90 per cent of all recorded deaths are from wind-storms, earthquakes, floods and droughts, but there are strong regional relation-ships. The greatest loss of life from tropical

Table 2.4 Recorded deaths and economic losses in the five deadliest and the five costliest disasters in the period 2000–10

Year	Event type	Principal affected area	Fatalities	Overall economic losses US$ million (original values)	Per cent of losses insured
Five deadliest disasters					
2010	Earthquake	Haiti	222,570	8,000	2
2004	Earthquake, tsunami	Indian Ocean East Asia	220,000	10,000	10
2008	Tropical cyclone	Myanmar	140,000	4,000	0
2005	Earthquake	Pakistan	88,000	5,200	<1
2008	Earthquake	China	84,000	85,000	<1
Five costliest disasters					
2005	Hurricane	USA	1,300	125,000	50
2008	Earthquake	China	84,000	85,000	<1
2008	Hurricane	USA, Caribbean	170	38,000	49
2010	Earthquake, tsunami	Chile	520	30,000	27
2004	Earthquakes	Japan	50	28,000	3

Source: After Munich Re NATCATSERVICE (Geo Risks Research) at www.munichre.com (accessed 12 June 2011).

cyclones and floods – both absolutely and as a proportion of the population – occurs in the Asia-Pacific region. Fatalities from drought are concentrated in Africa. Individual countries are prone to specific hazards – Iran, Afghanistan and India to earthquakes; Bangladesh, Honduras and Nicaragua to tropical cyclones. China has generally high mortality, whilst the USA regularly records the highest damages. Thus, disasters strike differently, depending on whether vulnerable populations or material assets are most at risk.

EM-DAT records indicate that between 1991 and 2005 there were 47 natural disasters each responsible for causing at least US$1.5 billion of economic damage based on 2005 prices. The total amount of economic loss resulting from these events was US$1.18 trillion. Almost two-thirds of these losses (64 per cent) occurred in just three countries – USA 31 per cent, Japan 18 per cent and China 15 per cent. The remaining 36 per cent was distributed over 47 other countries. Clearly, large financial losses occur in the MDCs but these gross figures tell little about the relative impact on national economies or the implications for fur-

ther development. In fact, high-income countries suffer the lowest losses relative to wealth, and the greatest economic burden of disaster tends to fall on middle-income nations. Between 2001 and 2006 average disaster costs expressed in terms of national GDP were 0.3 per cent, 1 per cent and < 0.1 per cent respectively for low-, middle- and high-income countries (Cummins and Mahul, 2009). The high relative losses in the middle range can be attributed to the rapid growth of assets exposed to risk in some emerging economies. At more local levels, if disaster strikes in the MDCs, effective government relief schemes mean that it is not necessarily the very poor that suffer most. Rather, it is the people in the middle – not wealthy enough to cope on their own but with sufficient assets to render them ineligible for emergency aid.

These relationships apply to individual disasters. Figure 2.10 shows the top 10 disaster damages between 1991 and 2005 ranked in absolute terms (left side) and as a percentage of national GDP in the previous year (right side). It is notable that the two most expensive disasters – hurricane

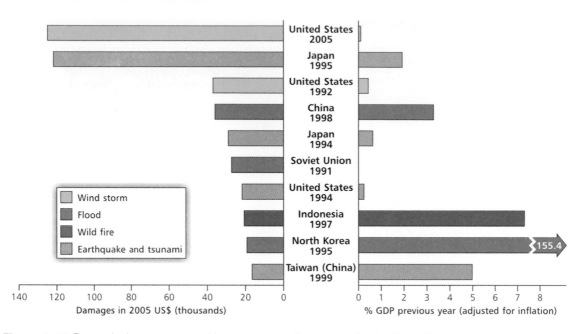

Figure 2.10 Economic damages reported by country and disaster type for the 10 costliest natural disasters recorded 1991–2005. Losses are expressed as total US$ billion at 2005 values and as a percentage of national GDP for the previous year, adjusted for inflation. Compiled by UNISDR from CRED EM-DAT.

'Katrina' and the Kobe earthquake – took only 0.1 and 1.92 per cent of GDP in the USA and Japan, respectively. In fact, the MDCs rarely suffer losses amounting to more than 5 per cent of GDP. In the LDCs, losses of up to 10 per cent are quite common. Hurricane strikes have caused economic output to fall significantly across the Central American and Caribbean regions (Strobl, 2012). In extreme cases, such as the 1995 floods in North Korea, disaster losses can exceed 150 per cent of GDP and have lasting effects. However, it is not impossible that a large earthquake in the Greater Tokyo area could create losses of up to US$3,000 billion, amounting to 25–75 per cent of Japan's GDP.

Global patterns in disaster impact do exist because, generally speaking, economic and social conditions improve with distance from the equator. Kummu and Varis (2010) found that human-induced pressure on natural resources is greatest between 5°N and 50°N, where 20 per cent of the world's population live under widely differing levels of socio-economic development and are exposed to various hazards. This latitudinal picture is reflected in a single global measure – the *Human Development Index* (HDI). This composite, nation-based statistic describes levels of human well-being according to societal variables like life expectancy, literacy, education and standard of living. In Figure 2.11, the HDI is used to group countries into four development categories – very high, high, medium and low. The two highest categories tend to occur relatively close to the poles in both hemispheres. There is considerable under-development in Africa. The largest block of medium development is in Asia, probably because vulnerability to disaster is often high in countries experiencing rapid change.

There has been a complementary search for a risk indicator that would enable individual countries to be ranked on a world disaster scale (Mosquera-Machado and Dilley, 2009). The *Disaster Risk Index* (DRI) rates countries on an eight-point scale (0–7) according to the risk of

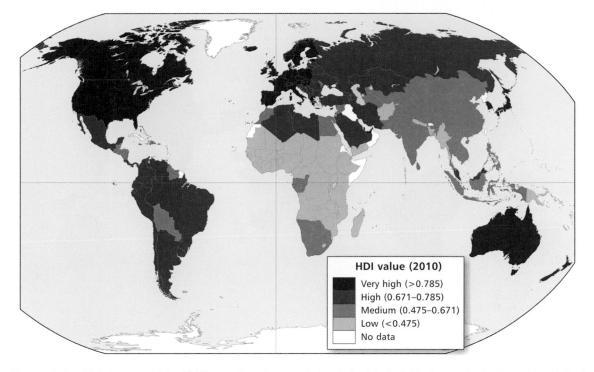

Figure 2.11 Global pattern of the UN Human Development Index derived for individual countries in 2010. After United Nations Development Programme at hdr.undp.org./en/statistics/data/mobility/map Reproduced with permission.

disaster-related death, based on data between 1980 and 2000. Mortality is used as a proxy measure of overall risk. At present, the index is limited to four hazard types – earthquake, tropical cyclone, flood and drought. These account for well over 90 per cent of all disaster-related deaths (Peduzzi, 2006). From Figure 2.12, it can be seen that the most extreme risks are mainly in small-island states and countries like Bangladesh and the Philippines, but a huge area of Asia, and some African nations, are also highly exposed.

Comparison of Figures 2.11 and 2.12 indicates that high human development (HHD) in North America, Europe and Australasia coincides with a low disaster risk, whilst the countries with low or medium human development, especially in the south-east Asian region, are more exposed. Broadly speaking, the following global pattern exists:

- More than 50 per cent of disasters occur in MHD countries.
- More than 50 per cent of disaster deaths occur in LHD countries.
- More than 75 per cent of disaster-affected people are in MHD countries.
- More than 50 per cent of the economic loss occurs in HHD countries.

The reasons are not always clear. For example, whilst nations with high HDI ratings tend to have the financial resources and democratic systems needed to reduce disaster impacts, many countries with medium-level ratings are at a greater risk than might be expected. This is due to rapidly changing socio-economic conditions. The rush for material growth creates more exposed wealth quickly but also produces greater social inequalities, as little attention is paid to health and safety issues.

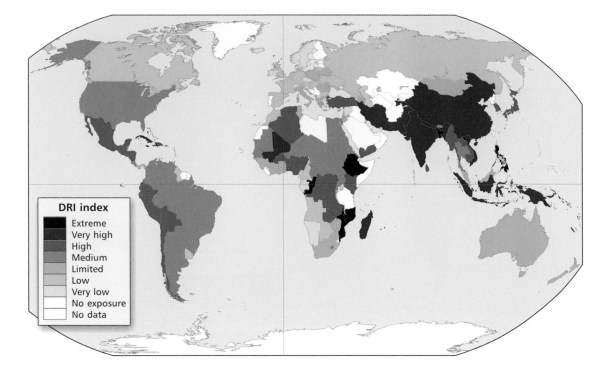

Figure 2.12 Global pattern of the UN Disaster Risk Index (DRI). Map kindly supplied by and used with permission of Dr P. Peduzzi.

In a survey of household wealth, Davies *et al.* (2009) found that Western Europe, North America and some Asia-Pacific nations together held almost 90 per cent of all global assets. The poorest countries were India, Pakistan, Indonesia and most of Central and West Africa, a group of the most disaster-prone countries in the world. Although there have been general increases in income over recent decades, wealth inequality has risen within a wide range of countries, including the OECD nations as well as the emerging economies of China, Russia and the European transition nations. For example, the gap within the UK has widened so that the wealthiest 10 per cent of society now earn 12 times as much as the poorest-paid. Future trends are difficult to predict, but the economic rise of China and India suggests that their people will capture an increasing share of the global wealth whilst the poorer countries in

Africa, Latin America and the Asia-Pacific region will continue to lack the resources necessary to invest in disaster reduction measures.

F MANAGING DISASTER

Disaster management is sometimes presented as an activity limited to the emergency phase of the event, a period seen by the media and others as chaotic and crisis ridden. In reality, effective disaster management depends on the implementation of a carefully planned sequence of actions over many years that embraces both pre-disaster protection and post-disaster recovery (Figure 2.13). These actions form a cycle; individual stages may overlap but they should operate as a closed loop in order to draw benefits from experience and post-disaster feedback.

1 Pre-disaster protection

This four-stage process aims to minimize disaster impact and involves:

- *risk assessment* – identification of the hazards to be faced, the accumulation of historical, statistical and other data and the preparation of loss estimates
- *mitigation* – steps taken in advance of disaster strikes aimed at decreasing or eliminating the loss. Various long-term measures, such as the construction of engineering works, insurance and land use planning are used
- *preparedness* – reflects the extent to which a community is alert to the disaster threat and includes hazard warning, provision of emergency shelters and the stockpiling of supplies
- *emergency plans* – to deal with short-term emergency procedures, including temporary evacuation and the availability of medical support.

2 Post-disaster recovery

This four-stage process helps the community back on its feet, hopefully in a better state to cope with future events by:

- *Relief.* This period covers the first 'golden' hours or days after the disaster. The rescue of survivors is followed by the distribution of basic supplies (food, water, clothing, shelter and medical care) to ensure no further loss of life.
- *Rehabilitation.* This happens in the first few weeks or months, when the priority is to start the community functioning again. An early and expensive task is the removal of disaster debris, such as building-rubble blocking roads or food spoiled due to power failure.
- *Reconstruction.* This is a much longer-term activity designed to recreate 'normality' after devastation. It includes improved disaster

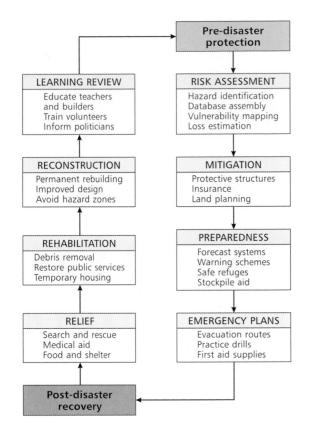

Figure 2.13 The reduction of risk through pre-disaster protection and post-disaster recovery activities. The time-scales needed for the actions shown range from hours (emergency evacuation) to decades (rebuilding damaged urban infrastructure).

preparation for future resilience, e.g. construction of hazard-resistant buildings and better land use planning.
- *Learning review.* This process allows the dissemination of greater disaster awareness amongst all stakeholders in the community and the training of emergency workers for the future.

Closure of the disaster-reduction loop through education of those at risk by managers at all levels is essential. At the community level, there is a need to understand the capabilities, and the limitations, of hazard mitigation. This can be done

Plate 2.2 People queue for food aid distributed through the UN World Food Programme in Port-au-Prince, Haiti, following destruction of much of the city by an $M_W = 7.0$ magnitude earthquake in January 2010. (Photo: Panos/William Daniels WDA00118HAT)

through the use of brochures, maps, videos, plus seminars, workshops and training exercises to improve disaster response. At a world level, international organizations and relief agencies require technical support in disaster management and need encouragement to pool resources and experience to cope with future disasters.

In practice, a fully integrated approach to disaster reduction is rarely achieved. It is difficult to quantify the combined risks from multiple hazards, especially those created by low-frequency/high-magnitude events. The risks may also be spread very unevenly between different communities and social groups. Estimating the costs of mitigation is also problematic, not least for the purpose of saving lives. The money spent can vary greatly in order to achieve the same level of risk. When funds are allocated, institutional weakness, lack of technical expertise and the poor enforcement of legislation weaken the effectiveness of disaster reduction strategies. These factors are a special problem in the poorest countries heavily dependent on external aid after disaster, where a rapid return to 'normality' is impossible. Such disadvantages can lead to low aspirations about the level of risk reduction that can be achieved. Even in more advanced nations, the dominant culture is still often based on emergency responses rather than a more proactive strategy that prevents – or at least reduces – disaster in the first place and also prepares the ground for a more sustainable future.

FURTHER READING

Kim, N. (2012) How much more exposed are the poor to natural disasters? Global and regional measurement. *Disasters* 36: 195–211. Quantifies the disaster burden carried by the poorest people in the world.

Kron, W., Steuer, M., Löw, P. and Wirtz, A. (2012) How to deal with a natural catastrophe database – analysis of flood losses. *Natural Hazards and Earth System Sciences* 12: 535–50. An excellent general overview of the pitfalls involved in the interpretation of disaster archives.

Miller A. and Goidel, R. (2009) News organisations and information gathering during a natural disaster: lessons from Hurricane Katrina. *Journal of Contingencies and Crisis Management* 17: 266–73. A specific example of media bias in disaster reporting.

Mosquera-Machado, S. and Dilley, M. (2009) A comparison of selected global disaster risk assessment results. *Natural Hazards* 48: 439–56. An attempt to compare two methods of quantitative disaster risk assessment.

Neumayer, E. and Barthel, F. (2011) Normalising economic loss from natural disasters: a global analysis. *Global Environmental Change* 21: 13–24. This paper clearly demonstrates the necessity for data normalization before explanations are offered for time trends of disaster.

WEB LINKS

Centre for Research on the Epidemiology of Disasters, Belgium www.cred.be

Munich Reinsurance Group, Munich www.munichre.com/topics/geo/

International Federation of Red Cross and Red Crescent Societies (IFRCRCS) www.ifrc.org

Disaster Risk Hotspots www.ldeo.columbia.edu/chrr/research/hotspots/

Complexity, sustainability and vulnerability

3

A INTRODUCTION

Synergy of purpose between researchers and practitioners across the field of environmental hazards is important for disaster reduction now and in the future. In recent years, much progress has been achieved by a greater focus on the global links between geophysical systems and socio-economic systems. In turn, these links have become absorbed within two umbrella concepts – *complexity science* and *sustainability science.* Complexity science can help repair divisions between earlier hazard paradigms by re-emphasizing disaster outcomes as interactions between natural, or quasi-natural, systems and human systems. It has also become clear that disasters are not simply short-term, localized events but have widespread roots and a capacity to disrupt the ecosystem services that support human life. Consequently, disaster reduction is best viewed within the remit of sustainability science, with an added appreciation of vulnerability and resilience to hazard.

B COMPLEXITY SCIENCE

Complexity theory evolved from an equation-based approach adopted in the natural sciences in the 1970s (Petley, 2009). It provides a general model applicable to physical and human systems and the ways in which they combine to create disasters. The basic idea is that any system – a river, the global atmosphere or a social grouping – acts as a set of components to produce a particular output. Previous attempts to understand such systems tended to simplify the model in order to simulate the key inputs, internal flows and outputs, as natural or social systems were viewed as machines that delivered an expected product. Thus, a river can be seen as a mechanism for delivering water from its catchment area to the sea. The interacting components within the drainage basin can then be manipulated to simulate the effects of changes in the materials entering or leaving the system, e.g. the effects of increased rainfall (a change in the input).

Complexity theory starts from the premise that systems are best understood by not grouping components together. This might suggest that many real-world natural and social systems become too complicated to study. Although these systems do consist of many individual elements, few have the ability to change the overall system performance in a substantial manner. In other words, the key output depends on the interactions between components rather than the components themselves. These interactions determine the state of the overall system and create *emergent behaviour* as the model output. Often the outputs can be forecast within acceptable limits of accuracy, even for highly complex systems like the global climate, rainforest biodiversity or economic performance. For example, a national economy functions due to countless decisions that are made by commercial organizations, government agencies and individuals, but the cumulative effect is normally a degree of stability with relatively minor changes in share values or other monetary measures.

The most efficient systems find an optimum balance between regulation and self-organization. Occasionally, systems do experience an unexpected shock, such as a stock-market crash, that can be seen as 'chaos'. There is evidence that the Earth's climate has changed abruptly in the past; societal systems have also experienced rapid transitions, like the collapse of communism across Eastern Europe and the Soviet Union around 1989 and the economic crises which started in 1929 and 2007–8. Rather than explaining these events as chaotic breakdowns, complexity theory assumes that a shift in total interactions within the system is responsible for the system's starting to operate – perhaps temporarily – in a new way. In some cases, the effect might be to accelerate a disturbance already underway and create a new order. But minor changes do not usually propagate throughout the system. It is more likely that change in one area will be damped out by other parts of the system through a process of self-organization.

C COMPLEXITY AND DISASTERS

The complexity paradigm is relevant because all disasters occur at the interface between natural, or quasi-natural, systems and human systems where the interactions are characterized by complexity. This creates the coupled human–environment system (CHES) that can be illustrated by the DNA model (Figure 3.1). The societal and physical systems are shown as two strands twisted together to form the well-known double helix. Linking the strands together are numerous interconnections

Human system
Physical system
Interconnectivity

Figure 3.1 The DNA model applied to complexity in disaster causation. The two strands are twisted together to form the double helix, indicating that the human system and physical systems are interlinked. Disasters arise from complex interactions between the two strands. After Petley (2009).

which serve to shape the structure in much the same way that the DNA structure forms the building blocks of life. Previous hazard paradigms concentrated on one strand or the other; the complexity paradigm gives them equal weight and stresses the links between them (Petley, 2009).

Disaster impacts result from the pattern of the social and physical strands and their inter-actions. During a hurricane, interactions take place within the social system (complexity within the affected population), within the storm (com-plexity within the atmosphere) and between the atmosphere and a human population (complexity in the coupled system). Comfort (1999), working with social systems complexity, argued that the interactions between individuals, emergency agen-cies and government bodies during a crisis can influence the overall effectiveness of the disaster response. Each decision taken may not, in itself, be exceptional, but particular coincidences of events can lead to different outcomes. Informa-tion flow and the application of local knowledge were identified as crucial elements of the disaster management structure and it was recommended that communication links should be physically robust and able to perform reliably under emer-gency stresses. More recently, Pelling (2003) sug-gested that such technical solutions should be part of a balanced approach, noting that effective disas-ter responses are most likely in communities with good social cohesion and a will to work towards a common goal.

The complexity paradigm provides a platform for disaster reduction that is interdisciplinary. It encourages natural and social scientists to work together and loosens some of the subject-bound paradigms. It is sometimes argued that the com-plexity approach is flawed because all the detailed disaster interactions are unknowable and disas-ters become unpreventable. However, the model suggests that when disasters occur it is because of a catastrophic chain of events. Intercepting events and breaking this chain could prevent, or reduce, the scale of evolving disaster.

D AN EXAMPLE OF COMPLEXITY: THE BAM EARTHQUAKE

On 26 December 2003 an earthquake struck southern Iran at 5:26 a.m. local time near the city of Bam (Figure 3.2). The magnitude of the earthquake ($M_W = 6.6$) was not particularly large – such earthquakes are recorded almost every week worldwide – but the disaster impact was huge. Most of the 140,000 residents were asleep at the time, with little chance of escape, and an estimated 26,000 people died (Bouchon *et al.*, 2006). There was almost total destruction of the ancient citadel of Bam, the world's largest adobe building complex, parts of which were 2,400 years old. About 70 per cent of all buildings collapsed completely, together with the three main hos-pitals and the fire station. It was estimated that 90 per cent of the building stock of the city was damaged by up to 60–100 per cent, whilst the re-maining buildings were damaged by 40–60 per cent (EERI, 2004).

What were the reasons for this exceptional loss?

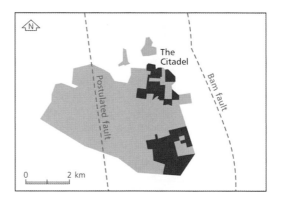

Figure 3.2 Location map of the city of Bam, Iran. The city was largely destroyed in the December 2003 earthquake. After Petley (2009).

1 Physically led explanations

The earthquake was fairly shallow, with a hypocentre about 7 km below the ground surface, and created a rupture along a 15 km stretch of the Bam Fault. The city was subjected to just 15 seconds of shaking, but a strong-motion seismometer located within the city recorded very high peak ground accelerations (Ahmadizadeh and Shakib, 2004). The earthquake damage zone was limited to an area of 16 km², another fact supporting the presence of intense localized shaking.

Not surprisingly, physical explanations dwelt on the severity of the ground shaking. Peyret *et al.* (2007) suggested that the rupture occurred not on the previously known Bam Fault but about 5 km to the west, at a location where no surface evidence of faulting existed. This would mean that the seismic waves occurred almost directly under the city. Conversely, Bouchon *et al.* (2006) argued that the rupture initiated on a fault to the south of the city and propagated seismic waves straight at the urban area. Whichever process prevailed, the net result was an unusually high intensity of ground shaking.

2 Socially led explanations

These concentrated on the poor quality of the building stock, including the widespread presence of adobe-constructed buildings in the Bam citadel

Plate 3.1 The ancient citadel of Bam, Iran, built of sun-dried mud bricks (adobe construction) and located on an elevated site, before the M_W = 6.6 magnitude earthquake in December 2003. About 70 per cent of all buildings collapsed. (Photo: Panos/Georg Gerster GGR00864IRN)

(EERI, 2004). However, Langenbach (2005) noted that, although the majority of these older buildings did collapse, only three people were trapped in the rubble within the citadel itself. Overall, most of the fatalities were recorded in buildings that were fewer than 30 years old. The traditionally built and unrepaired buildings fared much better than those either constructed or repaired in the late twentieth century.

Renovation of some traditional adobe buildings was found to be deficient, due to the use of cement containing too much sand (Kiyono and Kalantari, 2004). Adobe buildings have heavy roofs and, in the absence of any reinforcement of the walls, the roofs fall in if a single wall collapses. Many of these structures had been further weakened by termite activity, a situation reflecting more failures in the implementation and enforcement of the Iranian seismic building code. In turn, this neglect may have been due to the fact that the area had not suffered a large earthquake in historical times and was not considered to be at high seismic risk. In other words, a combination of code enforcement failures interacted with weaknesses in the building stock and earthquake mechanics to create a major disaster.

The emergency services were inadequately prepared for an effective response and suffered their own losses. Initial search-and-rescue operations were hindered by the destruction of key facilities, including the main fire station, which collapsed, crushing the fire engines and killing some fire-fighters. The three local hospitals, all the urban health centres and 95 per cent of the rural health centres were destroyed (Akbari *et al.*, 2004). One-fifth of health professionals in the area were killed; most of the remainder were incapable of providing support, due to injuries and post-traumatic stress. Many victims with severe injuries failed to receive urgently needed medical attention and died. During the month of January, night-time temperatures at Bam drop substantially below freezing. It is likely that a substantial number of the trapped victims who failed to

survive died of hypothermia during the first night (Moszynski, 2004).

The first rescue teams did not arrive until nightfall on the first day, 12 hours after the earthquake, and were unable to work properly until daylight the next morning. As often happens, most of the trapped people were rescued by other survivors working through the rubble with their bare hands. The subsequent arrival of international rescue teams was of limited practical value. A total of 34 international teams eventually arrived in Bam but, in total, these teams saved just 22 people. In contrast, it is estimated that local people recovered over 2,000 people from damaged buildings in the first few hours after the event. Further complications arose due to poor cooperation between the international aid agencies and the Iranian army. Under legislation passed in 2003, the Iranian Red Crescent Society was mandated to play the lead role in disaster response, but this led to tensions with the Iranian army, especially over the use of aircraft (IFRCRCS, 2004). Interviews with primary care nurses suggested that medical care was hindered by a lack of prior training and an inability of health workers to collaborate effectively, especially when overseas health teams were involved (Nasrabadi *et al.*, 2007).

A complexity-based perspective suggests that, for such a disaster to develop, a series of cumulative events has to occur. One way of representing this is by the Swiss Cheese model of disaster (Box 3.1 and Figure 3.3). In the case of Bam, some events in the chain – like the timing of the earthquake – could not have been modified by human action, but factors such as the termite infestation could have been addressed. If certain factors had operated in different ways – if the earthquake had occurred at another time of day, if the rupture had propagated from the north, if the buildings had offered greater resistance to seismic stress or if night-time temperatures had been less cold – the outcome would have been quite different.

Box 3.1 The Swiss Cheese model

The Swiss Cheese model of disaster was proposed by James Reason (1990) to explain technological and industrial accidents. Reason examined the strategies put in place by organizations to prevent accidents and concluded that they could be thought of as slices of Swiss cheese lined up one behind the other (Figure 3.3). Holes in the individual pieces of cheese were considered weak points in each line of defence and it was argued that an accident occurs only when the holes in all the slices align. If even one hole is out of line, then the defence works and a developing event is blocked.

This model of accident causation has been applied in the aviation industry, which is highly safety conscious (Petley, 2009). Many barriers are erected to prevent accidents, on the principle that no single component should cause the system to fail. Measures include conservative aircraft design, careful selection and training of pilots and well-established emergency procedures. This approach has shown that, even when an accident can be attributed to a single mistake, there is a surrounding framework that should prevent a disaster. The same reasoning can be applied to environmental disasters. For example, the impact of the hurricane 'Katrina' disaster would have been much lower if the hurricane had taken a different route away from New Orleans, if it had made landfall at low tide, if the levees had been properly built and maintained, or if New Orleans had been evacuated more promptly.

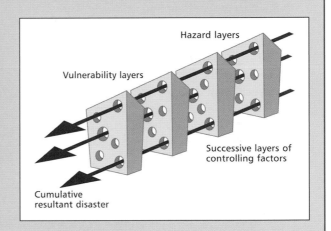

Figure 3.3 The Swiss Cheese model of disaster as proposed by Reason (1990). A disaster can occur only when several vulnerable circumstances arise simultaneously. The vulnerabilities are represented by holes in the cheese; disaster occurs when holes line up. Adapted from Petley (2009).

Complexity theory is even handed in the weight given to natural and human causation. It is a useful diagnostic tool for understanding the complicated nature of disaster but is perhaps less good in providing disaster-reducing solutions. In addition, complexity theory now has to be viewed alongside other major issues that influence disaster risk, such as the rise in global population, resource depletion and climate change.

E SUSTAINABILITY AND DISASTERS

The origins of sustainability science lie in the work of George Perkins Marsh, an American environmentalist who first drew attention to the damaging effects of human activities on some of the world's ecosystems (Marsh, 1864). Almost a century later, Gilbert White's work revealed that engineering schemes alone were not the answer to flood losses and that strategies based on conservation could play a part. More recently, sustainability science came to prominence aided by a seminal paper written – in part – by one of White's former students (Kates *et al.*, 2001). This paper outlined how efforts to satisfy the needs of an ever-rising global population, in an interconnected but unequal and increasingly human-dominated world, are undermining the Earth's life-support systems. The consequences of ecosystem loss can be found at differing scales of time and space but some evidence suggests that the capacity of the planet to absorb human-induced impacts may already have been exceeded with regard to climate change, biodiversity loss and the nitrogen cycle (Rockström *et al.*, 2009). All too often, these problems have been ignored by politicians and others because human progress is too closely identified with material prosperity based on economic growth. Comparatively little thought has been given to resources for the future.

Sustainable development aims to meet the legitimate needs of the current generation without compromising the ability of nature's life-support systems to meet the needs of later generations. Future progress is under threat whenever environmental hazards and disasters intervene to degrade natural resources, reduce productive capacity and limit development options. Disasters shock the CHES; they damage ecosystem services, disrupt lives and livelihoods and also delay the search for greater environmental and social justice in poor countries. A loss of agricultural production may threaten the security of food supplies; disruption of financial systems and money markets may take assets and energies away from development to support disaster recovery. However, such general effects are unlikely to be uniform. Employing provincial data for Vietnam, Noy and Vu (2010) found that although disasters with high fatality rates did result in lower output and material growth, the most costly disasters actually boosted the economy in the short term in those regions with access to reconstruction funds.

According to Turner (2010), a major research task is to clarify how the CHES responds to, and recovers from, the impact of natural disasters. Many observers believe that solutions will be found through greater collaboration between natural and social scientists. If the interacting processes are better understood, policies can be devised to reduce the uncertainties regarding natural system variability and expand the potential of ecosystem services to improve well-being for people in the future.

F VULNERABILITY AND RESILIENCE

1 Vulnerability

The concept of vulnerability emerged from research within the social sciences. Early applications were in the natural hazards field, but vulnerability theory has been extended from exposure to geophysical risk to embrace the human responses and adaptations to other threats. The concept now rests at the interface between science and policy making with respect to several themes, perhaps most prominently those related to global environmental issues, including climate change and perceived threats to sustainable livelihoods in the future (Adger, 2006). As shown by Bankoff *et al.* (2004) and Birkmann (2006), human vulnerability to environmental hazards arises from a mix of physical and socio-economic conditions. For example, poverty is a common factor, although not all poor people are vulnerable. In truth, powerful political, social and economic forces, including poverty, class, gender and ethnicity, interact to amplify disaster risk, especially for people in societies where disasters are not unusual or rare events but are part of a recurring pattern of loss from less extreme events.

Generally speaking, vulnerability has been viewed as a rather negative comment on societal ability to cope with risk. Given the widespread use of the term, it is perhaps surprising that little agreement exists either on its meaning or how it should be measured and applied in real life. Cutter *et al.* (1996) discovered no fewer than 18 separate definitions of vulnerability in the literature. Vulnerability was defined by UN/ISDR (2009) as:

the characteristics and circumstances of a community, system or asset that make it susceptible to the damaging effect of a hazard.

All vulnerable people are located in areas with some exposure to natural or technological hazards. Therefore, the term must capture both the physical nature of the risk as well as the special needs of disadvantaged people with limited access to resources. Vulnerability is a dynamic concept rather than a constant state; it is constantly in flux, due to changing interactions between geophysical and social processes through time and space. Above all, the concept has to anticipate a future state of risk and the societal outcome.

The highest levels of vulnerability tend to occur amongst the poorest people living in environmental settings, such as:

- Urban dwellers in the informal settlements and inner-city slums of the most rapidly expanding cities who inhabit unsafe structures on steep slopes or near dangerous industrial sites prone to hazards like earthquakes, landslides and fires.
- Rural dwellers, comprising almost three-quarters of the world's poor, located on marginal land, who suffer food insecurity from increasing natural resource degradation and climatic variability that brings floods, droughts and famines.

However, vulnerability is present in all nations and in all communities at all scales of time. Perhaps 50 million poor people live in informal settlements in Europe, whilst about 20 per cent of people in the USA suffer some personal disability and experience problems during an emergency. Vulnerability is associated with many factors, such as poverty, ill-health, resource depletion and marginalization. No single indicator is a sufficient guide. Due to their insecure place in society, vulnerable people are excluded from the normal information sources and decision-making processes that might be used to improve their well-being. Vulnerability is difficult to measure objectively in ways suitable for practical intervention because it is not solely an economic condition. Much present knowledge comes from

Plate 3.2 An old man made homeless by the $M_W = 7.6$ magnitude earthquake of October 2005 in Pakistan-administered Kashmir carries his meagre belongings through the streets of Muzaffarabad. At least 40,000 people were killed in this disaster. (Photo: Panos/Chris Stowers CST02029PAK)

the narrative case studies conducted into either reducing vulnerability or increasing resilience. Many of these are specific to one type of hazard and one location, often describing vulnerability in terms of losses after – rather than risks before – a disaster strike.

Research continues to find quantifiable measures of vulnerability that can be employed in a predictive way that helps to identify the most useful coping responses. Vulnerability indices are normally constructed from archived datasets of information, many of which will have been collected and assembled for other purposes. Some measures are objective single variables (such as hazard event frequency or the age of people at risk); others are less well-defined aggregate indices (such as the transparency of the local political system or the extent of local reliance on food aid).

United Nations-based researchers are developing a *Disaster Risk Index* (DRI), based on CRED data, in order to measure the relative levels of vulnerability between nations (UNDP, 2004). The index, based on over 30 socio-economic and environmental variables, aims to identify countries in the greatest need of disaster protection and development aid, based on the observed mortality disasters. In brief, physical exposure is expressed by the number of people located in each country combined with the recorded frequency of hazardous events in order to assess the risk of death during a representative time-span. Human vulnerability assessment is more complicated. Table 3.1 lists selected indicators, drawn from existing datasets, identified for possible inclusion in the DRI model. These include single factors, such as population density, level of unemployment and the number of hospital beds, together with aggregate indicators, including the *Human Development Index* (HDI), which employs factors like life expectancy, educational attainment, income and other measures.

As shown in Chapter 2, the DRI can be used to classify countries on an 8-point scale (0–7) according to the risk of disaster-related mortality.

The methodology is still under trial, but in a comparative study of observed and modelled risk, over 70 per cent of countries showed a difference of one class, or less, on this vulnerability scale (Peduzzi *et al.*, 2009). Some problems exist. For example, the DRI has proved over-sensitive to the effect of large events during short sampling periods. So far it has been limited to using disaster-related deaths as a proxy measure of vulnerability and cannot yet capture other impacts, such as loss of livelihood or homelessness. Some disaster types, like volcanic eruptions, drought and famine, have proved difficult to incorporate into the formula. Achieving a more composite, multi-hazard index is a challenge for the future.

For many nations, especially the LDCs, marked internal differences in wealth and well-being affect hazard vulnerability at the community and family level. Table 3.2 illustrates some differences that exist for hypothetical rural and urban households in poorer countries. Such inequalities reflect variations in the physical setting, social structure, economic status, health conditions, infrastructure assets and nature of local governance. Attempts have been made to capture this type of information. For example, Cutter *et al.* (2003) developed an *Index of Social Vulnerability* (SoVI) for use at the county level in the United States. Using factor analysis, over 40 socioeconomic and demographic variables were employed to produce 11 key factors that accounted for about 75 per cent of the variance in vulnerability. Selected variables are listed in Table 3.3. Other indices of vulnerability exist, including the Disaster Impact Index (DII) devised by Gardoni and Murphy (2010) and the Vulnerability and Capacities Index (VCI) from Mustafa *et al.* (2011).

The practical requirement is for robust but flexible measures that enable comparisons of vulnerability between households within a community and between communities within a region. Although scales of investigation vary, certain factors are known to raise vulnerability levels:

Table 3.1 Variables tested for use in the UNDP Disaster Risk Index

Category of vulnerability	Indicators
Economic	Gross Domestic Product (per inhabitant)
	Human Poverty Index (HPI)
	Total services (% of the exports of goods and services)
	Inflation (food prices annual %)
	Unemployment (% of total labour force)
Type of economic activity	Percentage of arable land
	Percentage of urban population
	Percentage of country in cropland
	Percentage of GDP dependent on agriculture
	Percentage of labour force in agricultural service
Quality of the environment	Percentage of land area in forest and woodland
	Percentage of irrigated land
	Degree of human induced soil degradation (GLASOD)
Demography	Population growth
	Urban growth
	Population density
	Age Dependency Ratio
Health and sanitation	Average calorie supply per capita
	Percentage of people with access to adequate sanitation
	Percentage of people with access to safe water
	Number of physicians (per 1,000 inhabitants)
	Number of hospital beds
	Life expectancy at birth (for both sexes)
	Mortality rate for those aged under 5 years
Political system	Transparency's CPI (index of corruption)
Early warning capacity	Number of radios (per 1,000 inhabitants)
Education	Illiteracy rate
	School enrolment
	Percentage in secondary education
	Labour force with primary, secondary or tertiary education
Level of human development	Human Development Index (HDI)

Source: After Peduzzi *et al.* (2009)

Table 3.2 Variations in vulnerability at the household and family level within LDCs

Household characteristics	Rural landowner	Rural labourers	Urban office worker, teacher	Urban squatter
Family size	7 members	5 members	5 members	6 members
Workers	4 men, 1 woman	1 man, 1 woman, 2 children	1 man, 1 woman	2 women
School-level education	3 men	0	5 (2 parents and 3 children)	0
Occupation	Farming, land renting, grain trading	Seasonal labouring, share-cropping	Office worker, teacher	Taking in washing
Income	Regular	No work, no pay	Regular, fixed income plus small pension	No work, no pay
Productive assets	Land, cattle, old tractor	Hand tools	Small savings	None
Credit source	Bank	Money lender	Bank	Money lender
Local contacts/ support network	Other farmers and traders, local officials	None	Local politicians	None
House construction	Brick walls and tile roof	Mud walls, thatch roof, mud floor	Brick walls and tile roof	Scrap metal, cardboard, plastic sheets
House ownership	Own house	Rented	Mortgaged	Illegal squat
Domestic facilities	Artesian well, electricity generator	Communal well, pit latrines, oil lamps	Electricity, piped water and sewage	Buy drinking water, communal baths and toilet, no drainage, illegal power connection
Location	Elevated flat site	Site near river, sometimes flooded	Paved street, regular garbage collection	Low-lying site or steep slope, no garbage collection
Access to facilities	Village school, clinic and shop	Village school and shop	Shops, school and health centre	Local doctor in emergency

- *Age.* The very young and the very old have limited mobility and need help to escape rapid-onset disasters.
- *Gender.* Women tend to have lower incomes and may be the sole carer for a large family in a single-parent household.
- *Disability.* People with special needs may not receive, or be able to respond to, warning messages or other information.
- *Poverty.* Poor people have a restricted access to all kinds of resources and are less able to recover from disaster.

- *Race.* Minority groups may have inadequate responses to disaster, due to language or cultural barriers.
- *Life expectancy.* Chronic disease and limited access to formal health services will hamper disaster recovery.
- *Occupation.* Self-employed or occasional workers may lose their job or tools and other resources in a disaster and lack capital for recovery.
- *Political system.* An absence of democratic processes and transparency in government will

exclude people from decision-making processes.

- *Education.* Low levels of attainment will hamper access to sources of, and limit understanding of, disaster-related information.
- *Food aid.* A high reliance on external resources of any kind will create problems if supply chains and infrastructure are lost in a disaster.

Vulnerability indices have been criticized for lacking scientific validity and usefulness in the policy sphere, other than when the physical context is clearly defined, such as on a river floodplain where the risk is specific (Hinkel, 2011). It is also true that other methodologies are available to aid agencies for assessing vulnerability in the field. Again, there is little agreement on

Table 3.3 Socio-economic aspects of vulnerability in disaster

Basic concept	Influence on disaster vulnerability
Socio-economic status – income and power	Poor people have a much more limited ability to absorb, and recover from, disaster losses
Gender	Women tend to have lower wages and also have family care responsibilities
Race and ethnicity	Minority groups can be hampered in disaster response by language and cultural barriers
Age	The very young and very old have limited mobility and may experience difficulty in moving quickly out of harm's way
Commercial and industrial development	The value and density of these buildings can indicate the economic status of a community
Employment loss	Those losing their jobs after disaster suffer a loss of income; large numbers of unemployed will delay the economic recovery process
Rural/urban mix	Rural residents tend to have lower incomes; extensive damage in high-density urban areas can reduce the efficiency of emergency evacuation procedures
Residential property	Expensive homes are costly and time consuming to replace; mobile homes and poorly built properties are easily destroyed
Infrastructure and lifelines	The loss of infrastructure (water and transport services) may be an impossible burden for small communities during the recovery period
Renters	People who rent often lack financial resources and access to financial aid information after disaster
Occupation	Self-employed people and migrant workers may lack the capital and physical resources to resume work quickly after disaster
Family structure	Families with large numbers of dependents (old and young) or single-parent households must combine these responsibilities with the recovery effort
Education	Lower educational achievements are likely to produce lower earnings and also limit personal access to and understanding of warnings and recovery information
Population growth	Rapid community growth can outstrip the provision of housing and essential social services (police and health care), especially for newcomers to the area
Medical services	Ready availability of, and access to, health care are important during the emergency and for post-disaster recovery
Social dependence	A high family or community dependence on social services implies some marginalization and a need for extra support
Special needs populations	People infirm, institutionalized, transient or homeless are at a high risk of being neglected in disaster and during the recovery period

Source: After S.L. Cutter *et al.* Social vulnerability to environmental hazards. *Social Science Quarterly* 84 (2003): 242–61. Copyright John Wiley and Sons 2003.

procedure, and conflicting results can be obtained – for example, between methods used for macroscale (regional) assessments as opposed to microscale (household) assessments. According to Darcy and Hofmann (2003), ideal judgements on vulnerability would be based on fairly objective 'outcome' indicators – such as mortality, morbidity or malnutrition – or even mental disorders (Salcioglu *et al.*, 2007). Unfortunately, this type of information is rarely to hand when decisions have to be taken quickly for distributing humanitarian aid.

2 Resilience

Unlike the concept of vulnerability, the concept of resilience began life in the biological sciences as a method for studying stability and change within ecosystems. Resilience is not the opposite of vulnerability but it does reflect a change in attitude from negatives to positives, namely: what can affected communities do for themselves and how can this capability be strengthened? Both concepts had early applications in the natural hazards field but have enjoyed wider currency. Like vulnerability, resilience is spread over a fragmented body of research and the term lacks agreed definition; Norris *et al.* (2008) found no less than 21 different definitions of resilience in the literature.

Resilience was defined by UN/ISDR (2009) as:

> *the ability of a system, community or society exposed to hazards to resist, absorb, accommodate to and recover from the effects of a hazard in a timely and efficient manner, including through the preservation and restoration of its essential basic structures and functions.*

An earlier interpretation identified two rather different forms – resilience and reliability (Timmerman, 1981):

- *Resilience* was seen as a measure of a system's capacity to absorb and recover from the impact

Table 3.4 Restoration of power supplies in Los Angeles following the Northridge earthquake in 1994

Time	Number of people without power
Initially	2,000,000
By dusk	1,100,000
After 24 hours	725,000
After 3 days	7,500
After 10 days	Almost all power restored

Source: After Institution of Civil Engineers (1995)

of a hazardous event. Traditional resilience was typical of the LDCs, where disaster is a more frequent part of life and group coping strategies are important. For example, nomadic herdsmen in semi-arid areas tend to accumulate cattle during years with good pasture as insurance against drought. However, resilience has been developed as a means of protecting public services in major cities. One example is the rapid recovery of the Los Angeles electricity supply following the Northridge earthquake in 1994 (Table 3.4).

- *Reliability* was suggestive of a reduced probability of system failure, especially for the protective devices put in place against disasters. This approach was associated more with the MDCs, where advanced technologies and building methods ensure a high degree of day-to-day reliability for most urban services. But extreme stress from an earthquake can disrupt urban networks on a massive scale and built environments have to be highly sustainable to withstand the stress of hazardous events in the future (Bosher *et al.*, 2007).

Klein *et al.* (2003) viewed resilience as a key tool in the successful adaptation to hazards for coastal mega-cities and the potential synergy between ecological and social resilience in coastal areas was also explored by Adger *et al.* (2005). For example, resilience along shorelines threatened by flooding could be improved by the preservation of coastal vegetation. This conforms to a 'living with hazards' strategy, similar to that for people exposed to river floods (see Chapter 11). Generally, resilience is more people centred and concentrates on the ability of a community – or any other societal system – not only to absorb short-term shocks but also to 'bounce back' after disaster. This view can be traced back to work on rural development in the 1970s and 1980s which led to the *sustainable livelihoods approach* summarized by Chambers and Conway (1992).

Resilience stresses the strengths of communities rather than their weaknesses. Studies of famine responses in the LDCs prove that local experience is crucial for household survival during drought (de Waal, 1989). Social cohesion and the networks of mutual support between families, friends and neighbours are important. In turn, such links often depend on gender, religion, caste or ethnicity. Spontaneous self-help groups regularly spring up in the immediate aftermath of disaster. For example, Mustafa (2003) described the importance of gender roles in Rawalpindi, Pakistan following a flood, when women were particularly active in relief work and urged more support from government. Disaster resilience is not confined to the LDCs. McGee and Russell (2003) showed how farmers and long-term residents in a rural community in Victoria, Australia, can withstand wildfire hazards, due to a culture of self-reliance based on experience and preparedness. Indeed, disaster management in Australia has shifted from a vulnerability-based approach towards working with local communities to build resilience and more sustainable communities in the future (Ellemor, 2005).

Resilience depends on local assets – or 'capitals' – to sustain livelihoods in the long term. These assets are most important in areas where people have little else to fall back on and include:

- natural capital – land, water, forests and minerals
- financial capital – savings, income, pensions, informal credit
- human capital – health, knowledge, practical skills
- social capital – family relations, networks, trust, mutual exchange of information and goods
- physical capital – infrastructure, roads, shelter, transport, sanitation.

Different combinations of assets can aid disaster from the emergency phase through to a longer period when adaptive policies can be implemented to make a community more resistant to later disaster shocks (Cutter *et al.*, 2008). This attitude reflects the growing application of resilience in climate-change studies where adaptation to a future state is a key theme.

Resilience theory recognizes that people's abilities to cope and to adapt are an under-tapped resource. This may lead to an examination of priorities in disaster reduction. For example, is it better to anticipate hazards and plan to reduce the losses, or to encourage measures that enhance the capability of people to 'bounce back' from the event? Both approaches are relevant, but perhaps too much attention has been given to protection or mitigation from above rather than the encouragement of local initiatives. Resilience is an inclusive approach because it accepts that disaster affects a whole community rather than a section of the population. It is part of a wider move away from vulnerability and emergency planning and towards fostering local capacities and resources so that, through greater political and social participation, a community can become more self-sufficient through improved disaster preparedness.

Three basic strands of resilience can be identified:

- the amount of stress that a community can absorb in a sustainable fashion
- the potential for self-organization and recovery within the community
- the ability to use local experience and skills to adapt and improve in the future.

Relevant factors include economic development, social capital, information and communication and community competence. Community competence, like the rest, is something that has to be built up over time. One way is through the improvement of resource rights and social justice. *Social justice* is the principle that all people should be heard on matters that affect their well-being, whatever their position in a community. If natural resources – such as land, water, fisheries – are in community ownership, or if private ownership rights are equitable and transparent, local people are more likely to feel they have an interest and to participate in the use of these resources.

Table 3.5 Community resilience indicators

Dimension	Candidate variables
Ecological	Wetlands acreage and loss
	Erosion rates
	Per cent impervious surface
	Biodiversity
	Number of coastal defence structures
Social	Demographics (age, race, class, gender, occupation)
	Social networks and social cohesion
	Community values/cohesion
	Faith-based organizations
Economic	Employment
	Value of property
	Wealth generation
	Municipal finance/revenues
Institutional	Participation in hazard-reduction programmes
	Hazard-mitigation plans
	Emergency services
	Zoning and building standards
	Emergency-response plans
	Inter-operable communications
	Continuity of operations plans
Infrastructure	Lifelines and critical infrastructure
	Transportation network
	Residential housing stock and age
	Commercial and manufacturing establishments
Community competence	Local understanding of risk
	Counselling services
	Absence of psychopathologies (alcohol, drug, spousal abuse)
	Health and wellness (low rates of mental illness, stress-related outcomes)
	Quality of life (high satisfaction)

Source: After *Global Environmental Change* 18(4), S.L. Cutter *et al*. A place-based model for understanding community resilience to natural disasters, 598–606, Copyright (2008), with permission from Elsevier.

Community involvement should help to prevent environmental degradation and encourage investment in sustainable practices that contribute to resilience. Disasters usually cause disproportionate harm to the disadvantaged. If there is less inequality and more social justice, with a smaller gap between privately owned resources and the poor, the growth of community bonds will enable people to pull together to recover from disaster.

Resilience indicators remain mostly theoretical, but some tests in real-life situations have been reported. UN/ISDR (2007) featured 16 case studies illustrating a range of practices used by local NGOs to build up community strength. Essentially, these were micro-scale initiatives that showcased the role of education in many aspects of disaster reduction – hazard awareness and preparedness, first-aid training programmes, small-scale insurance schemes, hazard warning systems, strengthening buildings, improved sanitation and better local decision making. Throughout, there was an emphasis on indigenous knowledge and the importance of schools, children and training to achieve a better future. These measures are undoubtedly helpful but the practice of resilience needs to progress beyond the micro-scale. Whilst it starts at the household and community levels, it needs to be scaled up to become part of regional, national and international policy making. Resilience cannot be pursued in a political vacuum. The dissemination of good practices, proven training methods and suitable technical skills needs to be integrated into the development agenda.

As with vulnerability, there is a continuing search for quantifiable indicators of resilience. Cutter *et al.* (2008) derived a framework for assessing the disaster resilience of a place and the *DROP model* can be used at the local or community level. The indicators were grouped under six headings (Table 3.5). Resilience is presented here as a dynamic process but certain indicators may be missing. A similar approach to model development, also based on publicly accessible data on economic and social capital, was adopted

by Norris *et al.* (2008). Here the aim was to reduce ill health, both physical and mental, after disaster. Sherrie *et al.* (2010) took this work forward to assess the capacities needed for resilience, but more work is required in order to inform the disaster reduction agenda.

G DRIVERS OF VULNERABILITY AND DISASTER

1 Economic factors: poverty and inequality

Human vulnerability is closely associated with absolute poverty and the economic gap between rich and poor. About 1.4 billion people live below the international poverty line of US$1.25 per day, including more than one-quarter of people in the developing countries. Almost half the world's population – more than 3 billion people – live on less than US$2.5 per day, whilst at least 80 per cent of the population live on less than US$10 per day. Figure 3.4 shows that, in the poorest countries, up to 80 per cent of the population have less than US$1 per day, as compared to less than 2 per cent of people in the richest countries. The broad latitudinal gradients found in Figure 2.11 (the Human Development Index map) are repeated with much poverty in Africa and the Indian subcontinent. About half the world's poor live in south Asia, with a further 25 per cent in Africa. In China, East Asia and the Pacific region, there are signs of a reduction in absolute poverty, but little change is evident in sub-Saharan Africa, where poverty blights 40 per cent of the total population.

Eight hundred million people in the world are both poor and malnourished, many being children under five years of age. Friis (2007) demonstrated that each year over 5 million children worldwide die because of an inadequate diet. In the poorest countries, many people do not have access to sufficient food to lead healthy, productive lives. Only one-quarter of people in Africa have access to safe drinking water; widespread ill

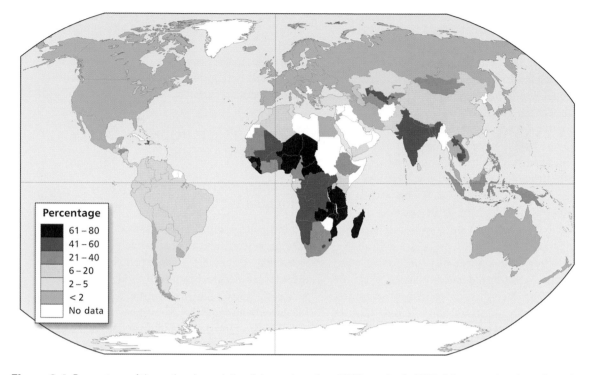

Figure 3.4 Percentage of the national population living on less than US$1 per day in 2007–8 for countries throughout the world. The highest regional concentrations are in Africa and Asia. Reproduced under Creative Commons Attribution-Share Alike 3.0 Unported license.

health from communicable diseases can prevent people from earning a living. Yet, in all countries where families survive by supplying labour and the oldest members depend on support from the young, the pressure for large families persists, especially in South Asia and sub-Saharan Africa.

Disaster impact is exacerbated by large gaps between rich and poor which create tension and lower resilience to disaster. Growth in per capita income will increase the competition for land and other natural resources and limit the ability of governments to invest sufficient money in education, social services and necessary infrastructure. Levels of wealth vary between, and within, nations. The poorest 40 per cent of the world's population accounts for only 5 per cent of global income; the richest 20 per cent of the world's population receives 75 per cent of the income. The combined wealth of the world's seven richest

people is probably more than the total GDP of the 41 most heavily indebted countries in the world with a total population of over 500 million.

The *Gini Index* measures household inequality from 0 (where everyone has the same income) to 1 (where one person has all the income). Most countries have a Gini Index between 0.25 and 0.6, with typical values for the MDCs like Australia 0.30, UK 0.34, Russia 0.42 and the USA 0.45. The lowest values (equal to or less than 0.25) are found in the Scandinavian countries. The highest values (more than 0.6) – which indicate the highest levels of internal inequality – are mainly in sub-Saharan Africa. The Gini Index has trended slightly upwards in recent years for some rich countries, and also because of rising prosperity in some newly emerging economies. For example, from the 1980s to the 2000s, the Gini Index for both the USA and China increased roughly from 0.3 to over 0.4. The future is difficult to predict. Some poor countries have economies with faster growth rates than richer nations, but the income and wealth gap may well increase with the spread of the knowledge economy.

It is easy to note that inequality is a key factor in disasters, but the reasons are less obvious. Relevant factors include the levels of poverty, the lack of effective insurance systems in poor countries and the difficulties of implementing building design codes. People lacking capital and other assets – such as land, tools and equipment – and with few able-bodied relatives capable of earning, are most vulnerable. Access to information, and the ability to mobilize support from outside the household, can be significant. Poor people may appear to have little to lose, but when hurricane 'Mitch' struck rural Honduras in 1998, the households in the lowest wealth quintile had their meagre assets reduced by 18 per cent, as compared with average losses of 3 per cent recorded for those in the upper quintile (Morris *et al.*, 2002). Most of the poorest people live in rural areas and have few earning skills or opportunities. National disparities in wealth continue to increase, thereby exacerbating vulnerability.

On the other hand, economic growth, especially in the wealthy countries, has increased the exposure of property to catastrophic damage. Along with the complexity and cost of the physical plant responsible for the world's industrial output, development has ensured that each hazard will threaten more assets unless steps are taken to reduce risks within cities and on industrial sites. Partly in response to the growing shortage of building land, some growth has occurred in areas subject to natural hazards, whilst man-made hazards involving the use of toxic chemicals and nuclear power have added to the loss potential. The availability of increased leisure time has led to the construction of many second homes built in potentially dangerous locations, such as mountain and coastal environments.

2 Social factors: population growth and demographics

The overall number of people exposed to hazard is increasing. The world population is expected to reach 9.3 billion in 2050. About 90 per cent of the growth will occur in developing countries – Africa +1.2 billion and Asia +1.8 billion. Here human vulnerability is already high through dense concentrations of population in unsafe physical settings. Continued population growth outstrips the ability of governments to invest in education and other social services and creates more competition for land resources, water and food.

Age and gender are important; the very young and the very old are most at risk. Better care and education for women would reduce birth rates and benefit many families. In the Bangladesh cyclone disaster of 1970, over half of all the deaths were suffered by children under 10 years of age, who comprised only one-third of the population (Sommer and Mosely, 1972). Work on earthquake disasters has shown that survivors over 60 years of age and females are most likely to have severe physical injuries. Females also suffer most from psychiatric stress disorders (Peek-Asa *et al.*, 2002; Chen *et al.*, 2001). However, children should not be seen solely as victims. There is evidence that child-led responses to disaster risk in developing countries can create positive behavioural changes amongst other people in their communities (Tanner, 2010).

The world has an increasingly elderly population. By 2050, one-third of the population in the MDCs will be over 60 years of age and disabled in some way. Similar demographic trends exist elsewhere. In China, where only about 10 per cent of the 2010 population were over 60, the proportion reaching that age by 2050 is expected to equal that in the MDCs. These residents have problems in emergency evacuation and survival in public shelters (McGuire *et al.*, 2007). Older people, especially widows in the LDCs, face difficulties in maintaining their livelihood after disaster. Ethnicity can be a factor when linguistic and religious divides threaten the security of minority groups. Elderly populations are also more vulnerable to infectious diseases, like an influenza pandemic. Greater mobility of people in the future is likely to increase risks, as migrants are often susceptible to infection from diseases, like malaria.

Socio-economic factors such as housing conditions can amplify risk. During July 1993 flash floods generated by monsoon rains struck a densely populated rice-growing area in southern Nepal and killed over 1,600 people (Pradhan *et al.*, 2007). A survey of over 40,000 residents showed that the fatalities were concentrated in certain groups. The crude fatality rate for all household residents was 9.9 per 1,000 persons but those most likely to die were children, females, those of low socio-economic status and those living in thatched houses (Figure 3.5). Over 70 per cent of houses were built of thatch and many were washed away.

Those living in thatched houses were over five times more likely to die than those in a cement/brick home.

As a result of work on disadvantaged people in the USA, Fothergill and Peek (2004) found that poor people were less likely to prepare for hazards or to buy insurance and less likely to respond to warnings. They are also more likely to die and suffer injury, have relatively high losses, more psychological trauma and have problems during the recovery and reconstruction phase. Elliott and Pais (2010) examined how socially disadvantaged people in the USA reacted when re-establishing their lives after hurricane 'Andrew' in 1992. In more developed urbanized areas (Miami) the long-term recovery effort tended to displace poorer residents from the hard-hit areas, probably because more wealthy people rebuild in high-value real estate areas and displace the poor. Conversely, in rural areas of south-west Louisiana, with fewer people and less property, the tendency was for socially vulnerable residents to concentrate in the affected area that was along the path of the storm.

3 Political factors: institutions and governance

The lack of strong central government is crucial in many countries because incompetence and corruption produce a weak organizational structure (everything from poor roads to untrained civil servants) and deficient welfare programmes (including inadequate housing and health provision and low nutritional status). Without a firm tax base, governments are unable to raise the revenue necessary for improvements in basic facilities such as water, sewage disposal and health care.

Some disaster losses are less tangible than others (Raschky, 2008). Economic development can provide the resources for disaster mitigation, such as early warnings and flood proofing, but in addition it normally provides better political institutions and governance. Democracies and

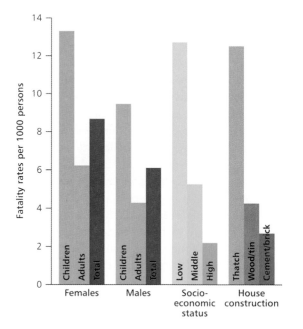

Figure 3.5 Socio-economic factors and fatality rates in flash floods during July 1993 in Nepal. Children were defined as 2–9 years of age, adults as 15 years or older. Socio-economic status was based on household ownership of land. Adapted from Pradhan *et al.* (2007).

well-ordered societies tend to fare better in disaster (Kahn 2005). For example, Toya and Skidmore (2007) found that losses were lower in those countries with higher levels of education, where the economic functions were more open and competitive, where the financial systems were more complete and where government was smaller. Generally speaking, the public demand for improved health and safety rises with per capita income. For poorer nations it may simply be too expensive to adopt engineering solutions such as flood control works and earthquake-resistant buildings.

In the global context, armed conflict, due to internal strife (tribal warfare, ethnic cleansing) or external warfare (border disputes with neighbouring countries), is important. Warfare is often a factor in creating large numbers of refugees and disrupting the distribution of food or other aid supplies. Since 1990 more than 70 million people have been displaced, either within their own countries or internationally. In some countries, the declared national government controls little more than the capital city, whilst the more remote centres and rural areas are left to fend for themselves.

Political upheavals can lead to greater vulnerability. Obvious examples include Eastern Europe and the former Soviet Union, where the collapse of communism removed the influence of the state with respect to health care, education and social provision. State paternalism has been replaced by an unregulated scramble towards free-market ideals in which the weakest members of society are ill equipped to compete. At the same time, the richest countries have reduced their commitments to internal welfare and to the international community. For example, in many western countries, health spending per person has declined since 1980 in real terms and the role of the welfare state has been deliberately reduced. For some years the volume of development aid declined, leaving aid agencies to fill the welfare role vacated by governments. This trend has been largely reversed with commitments to debt relief

and other measures. However, such pledges are subject to the vagaries of future political realities and there is little or no guarantee that increases in aid will continue in the longer term.

4 Environmental factors: biodiversity and agriculture

About 850 million people live in areas subject to severe environmental degradation. Loss of biodiversity reduces stability within natural systems and is no longer a local issue. Since 1900 over half the world's wetlands have been lost; since 1700 about 40 per cent of the forest cover has gone. Approximately 60 per cent of the Earth's ecosystem services have been degraded in the last 50 years alone. Annual economic costs due to deforestation and land degradation were estimated at US$2–4 trillion in 2008, or the equivalent of 3.3–7.5 per cent of global GDP (World Economic Forum, 2010). The per capita allocation of global freshwater resources fell from 17,000 m^3 in 1950 to 7,300 m^3 in 1995. On current trends, 90 per cent of available resources will be in use by 2030 and more than two-thirds of the world's population will experience serious water shortages. Existing water pollution damages the health of 1–1.5 billion people and is a factor in the deaths of millions of children aged under five years, largely due to outbreaks of infectious disease.

In some countries more than 80 per cent of the population is dependent on agriculture but poverty forces the adoption of unsustainable land use practices, such as deforestation, soil erosion and over-cultivation. Government attempts to increase food output frequently fail. In the tropics, capital-intensive plantation agriculture has displaced farmers from their land, whilst the construction of reservoirs for irrigation water reduces the seasonal inundation of low-lying land that is important for flood-retreat agriculture. Coastal areas have become more vulnerable to storm surge by the clearance of mangrove forests for fish farming, salt production and tourist development. Inland, the drainage of wetlands leads to

Plate 3.3 An area of deforestation in Para State, Brazil during 2005 on what is now the edge of the Amazon rainforest. The land has been cleared illegally for charcoal, which is sold to highly profitable pig iron-producing companies. Subsequent use of the land is likely to involve the raising of cattle or soybean cultivation. (Photo: Panos/Eduardo Martino EMA00024BRA)

a loss of common property resources such as fisheries and forests. As dietary habits change, traditional crops are likely to be replaced, with a consequent potential loss of biodiversity and genetic resources.

Many of the rural poor are dependent on traditional rain-fed agricultural production systems and are vulnerable to climate change. Unsustainable resource use is a major problem. For example, forests cover about 30 per cent of the land surface and contain over 50 per cent of all animal and plant species. Even the partial collapse of ecosystems can have major effects on agriculture. The unregulated competition for land and water resources – for example, between settled farmers and pastoralists – together with widespread illegal practices – such as logging timber

and open-cast mining – results in severe environmental degradation. The forested area of developing countries is expected to decrease by 20 per cent from 1980 to 2020 (OECD, 2003a). Soil fertility and biodiversity loss through increasing intensification of agriculture will make such areas less resilient when ecosystem disturbance through disaster takes place.

The decline of traditional agriculture and irregular slumps in market prices increase the threat of seasonal food shortages. In countries like Somalia, ravaged by over 10 years of near-continuous warfare and natural resource depletion, over 70 per cent of the population is undernourished. Food supplies are highly precarious (Hemrich, 2005). In many nations, most land holdings are too small to maintain livelihoods.

Over half the population may be malnourished and lack access to safe water or domestic sanitation. Such people suffer more from water-related diseases after floods, such as dysentery.

5 Geographical factors: urbanization and remoteness

In the first decade of the twenty-first century, more than half the world's population were estimated to be living in urban areas. Each year, some 20–30 million of the world's poorest people move from rural to urban areas, mostly driven by a perception of economic opportunity and a desire to escape tribal conflicts or disputes over land rights. Mega-cities create high hazard exposure (Mitchell, 1999) although Cross (2001) emphasized the risks for much smaller cities and rural communities that have fewer resources and are rarely a priority for central government support.

Over 1 billion people are urban slum dwellers living in poverty. According to Duijsens (2010), more than 90 per cent of these slums are in the global south. In 2010, the number of people resident in cities located in low- and middle-income countries was roughly equivalent to the global population in 1950 (IFRCRCS, 2010). Some of these cities are the largest in the world. One-third to one-half of these urban dwellers live in informal – often illegal – settlements, due to the high price of urban land and the inability of local administrations to keep pace with the rate of expansion. Rural migrants represent the poorest urban dwellers inhabiting unsafe sites in urban slums with inadequate water supplies and sanitation. Coupled with poor diets, the outcome is all too often endemic disease. Squatter settlements on the edge of large cities have particularly high levels of vulnerability. They tend to occupy steep slopes prone to landslides or flash floods and have population densities up to 150,000 people per km^2 housed in badly constructed buildings. These informal developments are characterized by high levels of poverty and child mortality.

Due to a shortage of building land, some urban growth is exposed to natural hazards. In the Indian subcontinent, for example, much of the rural–urban migration has been to cities with high earthquake and/or wind-storm risk. Global cities – like Mumbai, Rio de Janeiro and Shanghai – have high exposure to climate hazards (de Sherbinin et al., 2007). In the MDCs, coastal cities exposed to hurricanes have also grown rapidly, often with little consideration of the threats posed by storm surges and other perils. Others are located in seismically active areas, like the west coast of the USA and the north-east coast of Japan, where loosely compacted sediments or landfill sites will perform poorly under earthquake stress. A real concern in many MDCs is the potential for massively destructive fires to break out in the aftermath of large earthquakes. Such fires proved disastrous in the aftermath of the Kobe earthquake but many earthquake-prone cities have little or no preparedness for this threat.

It is estimated that, by 2025, there will be 26 mega-cities where residents number more than 10 million; some will house in excess of 20 million people. In many cases, cities provide a convenient and safe environment for living and offer many social and economic advantages. In wealthy nations, most people choose to live in cities and more than one-third of Thailand's GDP is generated by the 10 per cent of the population located in Bangkok. Figure 3.6 shows changes in the world's urban population from 1950 onwards. About 95 per cent of all the future population growth up to 2025 is predicted to occur in the urban areas of the low and middle income nations, notably in the Asian region. Most of the fastest growing urban centres in the world are in earthquake areas. By 2025, it is likely that one-third of the world's population will live in areas subject to seismic and volcanic activity (OECD, 2003a).

The most vulnerable people often live in relatively inaccessible regions away from the scrutiny of governments and where disaster reduction can be neglected. Such areas include the mountain villages of the Himalayas or the Andes and the

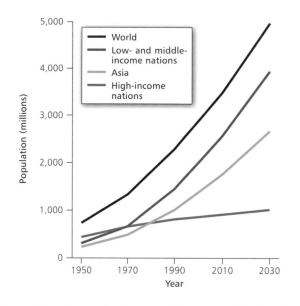

Figure 3.6 Observed and predicted future growth in the urban population globally and by region. The projections suggest that almost all future increases will take place in low- and middle-income nations and that Asia will experience massive urban growth. After IFRCRCS, *World Disasters Report* (2010). Reproduced with permission.

remote island communities of the Pacific Ocean inhabiting Small Island Developing States (SIDS). Typically, SIDS may be defined as territories less than 1,000 km² in extent with a population below 1 million. Some, like those in the Pacific Ocean, consist of hundreds of small, low-lying coral atolls; others, in the Caribbean, occupy steep volcanic sites. In general, SIDS are resource poor, with a combination of negative factors, such as:

- geographic isolation, with remoteness from markets and high transport costs
- small populations with limited skills; few economies of scale are available and out-migration is common
- a narrow economic base, often tied to natural resource exploitation (timber, minerals) or tropical export crops (bananas, sugar)
- limited land areas for freshwater storage or waste disposal

- fragile biodiversity subject to degradation through deforestation, soil erosion, exhaustion of fish stocks, crop epidemics and salt contamination
- dependence on foreign tourists or combinations of overseas investment and aid. Tourism can degrade the land by destruction of wetlands and mangrove forest.

Many SIDS are exposed to multiple environmental hazards. For example, Mayotte – a volcanic island in the Comoros archipelago – is prone to landslides, flash floods, storm surges and seismic movements. Its population has multiplied fourfold in less than 30 years, with many people living in slum dwellings on steep slopes or along riverbeds (Audru *et al.*, 2010). In the Pacific Islands region, between 1950 and 2004, more than 200 disasters were recorded. These took over 1,700 lives and created losses of US$6,526 million at 2004 prices (Bettencourt *et al.*, 2006). The 15 largest disasters inflicted about 80 per cent of the total damage, and economic and social impacts in selected nations are shown in Table 3.6. In disaster years, economic losses averaged almost 46 per cent of GDP in Samoa, 30 per cent in Vanuatu and 14 per cent in Tonga, whilst around 40 per cent of the population of Tonga and Samoa were affected. In fact, disaster damage averaged 2–7 per cent of GDP in all years because of the difficulty of making a sustainable recovery from individual events.

6 Global change factors

These take many forms. One is technical innovation, which can mitigate disaster through better forecasting and warning systems and safer construction techniques but can also create risk. High-rise buildings, large dams, building construction on man-made islands in coastal areas, the proliferation of nuclear reactors, more international travel are all examples of such trends. In the LDCs the introduction of low-level technology, such as the building of a new road through

Table 3.6 Estimated economic and social impact of disasters in selected Pacific Island nations 1950–2004

Country	Number of disasters reported	Total reported losses in US$ million (2004 prices)	Average population affected (%)		Average impact on GDP (%)	
			In disaster years	In all years	In disaster years	In all years
Fiji	38	1,174.6	10.8	5.1	7.7	2.7
Samoa	12	743.4	42.2	6.1	45.6	6.6
Vanuatu	37	384.4	15.5	4.5	30.0	4.4
Tonga	16	171.1	42.0	5.3	14.2	1.8
Guam	11	3,056.3	3.7	0.5	n/a	n/a

Source: After S. Bettencourt *et al.*, 2006 *Not If but When: Adapting to Natural Hazards in the Pacific Islands Region* (The World Bank, 2006). Copyright the World Bank 2006. Reproduced with permission.

steep, mountainous terrain, may increase landslides, and 'modern' concrete houses constructed to low standards may withstand earthquakes less well than traditional structures.

Increased global connections – through information flows and transport networks – permit 'echo-disruption' on the other side of the world. People have become much more mobile in recent years and expect to be transported around the world in the minimum elapsed time, irrespective of adverse environmental conditions, such as severe weather. Many people and business enterprises rely heavily on information technology. In December 2006 a fairly small sea-floor earthquake to the south of Taiwan caused damage to submarine cables providing internet access and telephone services. As a result, communication networks in Taiwan, Hong Kong, Japan, China, Singapore and South Korea were severely disrupted (Petley, 2009). Competition in commerce and industry has resulted in reduced levels of staffing and smaller operating margins. In turn, these apparent improvements allow less scope for corporate responses to environmental hazard.

Global interdependence affects most aspects of life. The functioning of the world economy works against the LDCs. Most of the Third World's export earnings come from primary commodities for which market prices have often been low. The LDCs have little opportunity to process and market their own produce and are dependent on manufactured goods from the industrialized nations. These goods are highly priced and may be tied to aid packages. The progressive impoverishment of the small-scale farmer, combined with a foreign debt burden that may be many times the national annual export earnings, takes resources away from long-term development, in a process that has been described as a transfusion of blood from the sick to the healthy. The cycle is reinforced when natural disaster destroys local products and undermines incentives for investment.

FURTHER READING

Byrne, D. (1998) Complexity theory and the social sciences. Routledge, London. A very useful general account.

Comfort, L.K. (1999) *Shared Risk; Complex Systems in Seismic Response*. Pergamon Press, Oxford. A pioneering treatment of social systems complexity and its implications for emergency response.

Cutter, S.L. *et al.* (2008) A place-based model for understanding community resilience to natural disasters. *Global Environmental Change* 18: 598–606. The use of available indicators to provide an objective picture of an elusive concept.

Kates, R.W. *et al.* (2001) Sustainability science. *Science* 292: 641–2. A model overview of this important theme.

Peduzzi, P., Dao, H., Herold, C. and Mouton, F. (2009) Assessing global exposure and vulnerability towards natural hazards: the Disaster Risk Index. *Natural Hazards and Earth System Sciences* 9: 1149–59. An example of the quantitative route to assessing disaster vulnerability.

Turner, B.L. (2010) Vulnerability and resilience: coalescing or paralleling approaches for sustainability science? *Global Environmental Change* 20: 570–6. An up-to-date overview of two key hazard concepts in their wider context.

WEB LINKS

A non-technical explanation of Chaos and Complexity theory http://complexity.orconhosting.net.nz/

Earthquake Engineering Research Institute report on the Bam Earthquake http://www.eeri.org/lfe/iran _bam.html

The United Nations Environment Programme Disaster Risk Index http://www.gridca.grid.unepch/undp/

The GINI Index www.data.worldbank.org/indicator/ SI.POV.GINI

The Human Development Index www.hdr.undp.org/ en/statistics/hdi

CHAPTER FOUR

Risk assessment and management

4

A THE NATURE OF RISK

The Chinese word for risk, 'wei ji', combines the two characters meaning 'danger' and 'opportunity'. Another interpretation is 'precarious moment'. Both translations show that risk is not a purely negative concept and that uncertainty involves a balance between profit and loss. A degree of risk is associated with almost every aspect of life. As observed by Adams (1995), if there is no uncertainty – if individuals know what an outcome will be – they are not dealing with risk in the first place.

In practice, people strive to assess and manage their own risks in order to reduce any adverse consequences. *Risk assessment* involves evaluating the significance of a particular threat, by either quantitative or qualitative means. *Decision theory* distinguishes between so-called risky prospects, where the probabilities of possible outcomes are thought to be known, and less-certain outcomes, where such information is unavailable. Quantitative assessments are generally based on estimates of the probability of an event together with the magnitude of its known adverse consequences, often expressed as:

RISK = Hazard Probability × Elements at Risk × Vulnerability

Quantitative risk assessment has not been attempted for all environmental hazards. Even when risks have been quantified, the level of uncertainty associated with the estimates of loss may be high. It is a process understood by a small minority of the general public. Therefore, it is important that risk assessments are communicated in a transparent and accessible manner to lay people and that care is taken to explain the extent of uncertainty attached to any estimate.

In terms of disaster reduction, the main practical goal is *risk management*, which aims to lower the known threats – from either natural or technological sources – whilst maximizing any related benefits. Potentially, almost every person, community and organization has something to contribute to risk management, but achieving the optimum balance between risk and safety involves controversial value judgements. There is often much difficulty in answering even basic questions, such as: What is an acceptable level of risk? Who benefits from risk assessment and management? Who pays for the process? What is meant by success or failure in risk reduction policy?

A sound approach to risk requires good science and good judgement. Neither risk assessment nor risk management can be divorced from choices. These choices are conditioned by individual beliefs and circumstances, including financial constraints, and also by wider attitudes in society. Since most people make decisions based on their personal assessment of a threat, *risk perception* has to be regarded as a valid element in risk management, alongside more scientific assessments. Distinctions are often drawn between *objective* and *perceived* risks. This is because the level of personal risk, perceived by the individual concerned, often differs from the results obtained by more objective assessments. Care must be taken to ensure that an objective risk analysis, perhaps based on financial models of costs and benefits, is not always assumed to be correct – or to lead to better outcomes – than assessments based on perception.

When dealing with the perceptions of individual people, risks are allocated into two categories:

- *Involuntary risks* happen to us without our prior knowledge or consent. As such, they are often seen as external to the individual. So-called 'acts of God', like being struck by lightning or a meteorite, are considered to be involuntary risks, as is exposure to some environmental contaminants. Occasionally, such risks are known to the individual, but then they are often seen as inevitable or uncontrollable, as in the case of earthquakes. Most of the hazards considered in this book represent involuntary risks to individuals or communities, due to their location in a hazard-prone environment.
- *Voluntary risks* are associated with activities that we elect to undertake as part of modern life, such as driving a car or smoking cigarettes. These risks, willingly accepted by individuals, are generally more common and controllable. Because they are undertaken on a personal

basis, they also have less catastrophe potential. The control of voluntary risk is exercised either through modifications of individual behaviour (stopping smoking or ceasing participation in a dangerous sport) or by government action (introduction of safety legislation, such as the requirement to wear a crash helmet when riding a motor cycle). Human-induced hazards, including risks from technology, are normally placed in this group.

In reality, the division between these two risk categories is less clear than it appears. For example, while cigarette smoking and mountain climbing are obvious cases of voluntary activities, the same cannot be so firmly stated for driving a car, which may be an essential form of transport for people in remote areas. The alternative to working in a dangerous chemical factory may be unemployment. In other words, a risk is more voluntary than another risk if its avoidance is associated with a greater personal sacrifice on the part of the risk-bearer. Some floodplain dwellers may elect to buy a home near a river because it is cheaper than an equivalent property in a safer part of town. Such a decision can be both voluntary and economically rational. These issues are further complicated by the poor levels of knowledge that most people have of the actual levels of voluntary risk. A limited understanding of risk means that, in many cases, individual decisions are not rational in terms of the statistical facts.

Most people react differently to voluntary risks, as compared to the risks imposed externally. In a pioneering study of public attitudes towards various technologies, Starr (1969) attempted a correlation between the risk of death to an individual, expressed as the probability of death per hour of exposure to a certain hazardous activity (P_f), and the assumed social benefit of that activity converted into a dollar equivalent. Figure 4.1 shows that there were major differences between voluntary and involuntary risks, with people willing to accept voluntary risks with a P_f value approximately 1,000 times greater than that of

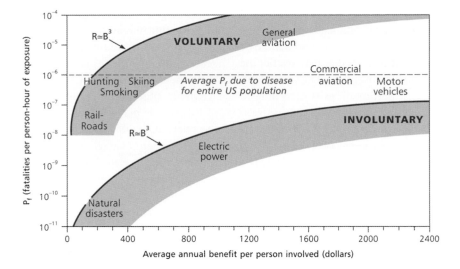

Figure 4.1 Risk (P_f) plotted relative to benefit and grouped for various voluntary and involuntary activities involving exposure to hazard. The diagram also shows the approximate third-power relationship between risks and benefits. The average risk of death from disease is indicated for comparison. After C. Starr, Social benefit versus technological risk. *Science* 165 (1969): 1232–8. Reprinted with permission from AAAS.

involuntary risks. Voluntary risks such as driving, flying and smoking were accepted even though they produced a risk of death of one in 100,000 or more per person per year, while the involuntary risks exposed people to a risk of about one in 10 million or less per person per year. Fell (1994) found that the acceptable level for risks perceived as involuntary varied between a frequency of 10^{-5} and 10^{-6} per year, as compared to between 10^{-3} and 10^{-4} per year for voluntary risks.

Starr also showed that the acceptability of risk from a given technology was approximately equal to the third power of the benefits. That is, the technologies with the greatest dangers also have the greatest benefits. Later workers, such as Slovic *et al.* (1991), suggested that such trade-offs between risks and benefits are not always made because perceived dangers influence attitudes more strongly than do perceived benefits. This situation is typified by so-called 'dread' hazards like nuclear power. It is also the case that perceived levels of risk can change quickly over time. For example, the perceived risk posed by tsunamis increased greatly after the 2004 Indian Ocean

earthquake, even though the actual risk worldwide continued to remain relatively low.

In summary, perceived risk is subjective and variable. The fact that individuals tolerate substantially more risk when the threat is associated with voluntary behaviour has been explained as an unrealistic belief in personal control. In other words, individuals rarely have the degree of control over events that they assume or would wish. Research, summarized by Sjöberg *et al.* (2004), challenged Starr's emphasis on voluntariness and opened up two other strands in risk perception – the *psychometric paradigm* (rooted within psychology and the decision sciences) and *cultural theory* (rooted within sociology and anthropology). The psychometric paradigm is based on studies of how people process information. It suggests that newly discovered risks, including 'dread' risks like nuclear power, greatly increase the seriousness with which the threats are perceived. The cultural theory was promoted by Douglas and Wildavsky (1982) on the premise that individual risk perception is conditioned by societal institutions, group cultural values and community ways of life. This approach has proved somewhat controversial, although cross-cultural studies have proved that people are influenced by their social context in ways that may restrict the scope for individual perception of hazards. Thus, variations in risk perception exist according to location, occupation and lifestyle, even between individuals of the same age and gender, as well as between nations (Rohrmann, 1994).

Others have taken an even wider view. The work of Giddens (1990) and Beck (1992) highlighted the fact that people in the MDCs today are faced with new and complex threats well beyond those posed by traditional 'natural hazards'. Many of these threats have been created by human activity, in particular the technological hazards linked to industrialization and the on-going modernization process. As a result, the concept of the *risk society* was born. Amongst other things, the risk society is characterized by much greater political and public concern about industrial practices, especially when major disasters like the Chernobyl and Fukushima nuclear accidents reinforce such views of technology.

When people believe they are well informed, whatever the source or reliability of the information, they are more likely to attempt risk assessment for themselves. Consequently, with the spread of information via the internet and other media, there has been a decline in the level of public trust given to figures in authority, such as industrialists, politicians and experts of many kinds. In turn, this distrust has been accompanied by a rise in support for concepts such as the *precautionary principle* (designed to minimize technological risks) and *sustainable development* (designed to secure future livelihoods). People currently living in the risk society attach greater importance to safety, and to a secure future, than did previous generations. As societies become richer and people enjoy longer, healthier lives, they perceive a definite value in an extended life and become more risk averse. One obvious expression of this has been the marked growth of an enthusiastic health and safety culture, and increased government regulation in everyday life, in most affluent countries.

Given that absolute safety is impossible to achieve, there is much sense in trying to determine the level of risk that is acceptable for any activity or situation. *Acceptable risk* is the degree of loss that is perceived by the community, or the relevant regulatory authorities, to be most relevant when managing risk. It is a much-misunderstood term. For example, it does not describe either the level of risk with which people are happy or even the lowest risk possible. Fischhoff *et al.* (1981) concluded that the word describes the 'least unacceptable' option because the associated risk is not really 'acceptable' in any absolute sense. As a result, the term *tolerable risk* is often used, i.e. the level of risk that is tolerated, rather than accepted. Tolerable risk is a dynamic concept because the actual level varies according to a wide range of factors. These include the severity of the risk itself, the nature of the potential impacts, the level of general understanding of the risk, the familiarity of the affected people with the risk, the benefits associated with the risk and the dangers and benefits associated with any alternative scenario.

It is important, when specifying the level of acceptable or tolerable risk, to be clear about the people *to whom* it is acceptable. Actual behaviour does not necessarily reflect the optimum choice. For example, in the case of a consumer buying a car, the act of purchase need not imply that the product is safe enough, just that the trade-off with other forms of transport is the best available. In this instance, the risk is tolerated rather than accepted. There are many factors that influence the consumer's choice of a car. Perhaps surprisingly, statistics on the safety of the vehicle are rarely prioritized, so, in most decisions, risk perception is just one element in the process.

To summarize, there is no fully objective approach to risk decisions and, since there is often uncertainty about the best way to manage hazards and risks, quantitative analysis is best viewed as a partial, rather than as a complete, function.

B RISK ASSESSMENT

Risk assessment involves three steps:

- The identification of hazards likely to result in disasters – what hazardous events may occur?
- The estimation of the likelihood of such events – what is the probability that it will happen?
- The evaluation of the social consequences of the hazard – what is the likely loss created by each event?

In reality, the process is more complex because there is an additional need to understand the magnitude of the event and how it may affect risk outcomes. For example, the probability of occurrence of an avalanche is related to the volume of snow; large avalanches occur less frequently and are, therefore, less probable. But, the frequency – volume relationship – may not be the full story because the threat posed could be influenced by the velocity of the avalanche flow, and the nature of the snow, as well as the volume.

Assuming that these problems can be overcome, the statistical analysis of risk is based on theories of probability whereby risk (R) is taken as a product of probability (p) and loss (L):

$$R = p \times L$$

If every event resulted in the same consequences, it would be necessary only to calculate the frequency of occurrence. But, as already indicated, environmental hazards have variable impacts. Therefore, an assessment of damaging consequences is required (Box 4.1). For many threats, especially technological hazards, the available data of past events are rarely adequate for a reliable statistical assessment of risk. In these cases *event* and *fault tree* techniques are used (Figure 4.2). These use a process of inductive logic, most often applied to industrial accidents, where a known chain of events must take place before a disaster can occur.

1 Magnitude–frequency relationships

Many natural hazards can be measured objectively on scientific scales of magnitude or intensity, e.g. earthquakes (M_W and Mercalli scales); tornadoes (Fujita scale); hurricanes (Saffir-Simpson scale). Unfortunately, such scales tend to measure just one physical factor that influences disaster impact. For hurricanes, the Saffir-Simpson scale relates to the maximum sustained

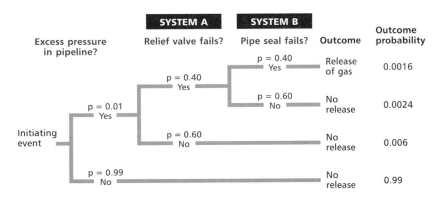

Figure 4.2 A probabilistic event tree for a hypothetical gas pipeline accident. The performance of safety systems A and B determines the outcome probability of the initiating event. Diagram courtesy of Dr J.R. Keaton, personal communication.

Box 4.1 Quantitative risk assessment

From experience, it is known that n different, mutually exclusive, events $E_1....E_n$ may occur. These events might be a series of damaging floods or urban landslides, but the effectiveness of the method depends heavily on the availability of a good database. Thus, the method is less satisfactory for rare natural events, such as large-magnitude earthquakes, or for some technological hazards, such as the release of radionuclides from nuclear facilities.

From historical data, it can be determined that event E_j will occur with probability p_j and cause a loss equivalent to L_j, where $_j$ represents any of the individual numbers $_1...._n$ and $L_1....L_n$ are measured in the same units, e.g. pounds sterling or lives lost. It is assumed that all the possible events can be identified in advance. Therefore, $p_1 + p_2....p_n = 1$.

After arranging the n events in order of increasing loss ($L_1<....<L_n$), the cumulative probability for an individual event can be calculated as $P_j = p_j + ...p_n$. This specifies the probability of the occurrence of an event for which the loss is as great as, or greater than, L_j, as shown in Table 4.1.

If we can categorize all possible events in terms of the property loss (expressed in pounds sterling), it may be possible to produce a risk analysis along the following lines:

Property loss(£)	Probability (*p*)	Cumulative probability (*P*) of exceedance
0	0.950	1.000
10,000	0.030	0.050
50,000	0.015	0.020
100,000	0.005	0.005

Table 4.1 Basic elements of quantitative risk analysis

Event	Prob-ability	Loss*	Cumulative probability
E_1	P_1	L_1	$P_1 = p_1 + ... + p_n = 1$
E_j	p_j	L_J	$P_j = p_j + ... + p_n$
E_n	p_n	L_n	$P_n = p_n$

Source: After Krewski *et al.* (1982)

Note: *Arranged in increasing order ($L_1 \leqslant ... \leqslant L_n$)

This theoretical example shows that there is a 95 per cent chance of no property loss and only a 2 per cent chance of a property loss of £50,000 or greater.

In some circumstances, it may be necessary or desirable to produce a summary measure of risk (R). This can be done by calculating the *total probable loss*:

$$R = p_1 L_1 ....+ p_n L_n$$

In this example, R would be £1,550. Alternatively, the *maximum loss* could be calculated. This is a rather extreme summary that ignores the probability of occurrence and takes the risk to be equal to the maximum loss, which, in this case, would be £100,000. Because of the skewed distribution, another way would be to take a given *per centile loss*, for example 98 per cent level of loss.

The same methodology can be applied when damaging events cause loss of life. For the above example, an appropriate tabulation might be:

Number of deaths	Probability	Cumulative probability
0	0.99	1.000
1	0.006	0.010
2	0.003	0.004
3	0.001	0.001

(After Krewski *et al.*, 1982)

wind speed only, whereas most damage is due to extreme wind gusts, storm surge or intense precipitation (Chapter 9). Even if scientific scales could incorporate all the damaging phenomena, the event alone is a poor guide to disaster impact because of moderating effects by local environmental and societal conditions. For example, an earthquake on a submarine fault might generate a tsunami; one in a mountain chain cannot. More significantly, disaster impact severity reflects the level of vulnerability in the communities where hazard strikes. Time of day can be important. At night more people will be indoors: perhaps protected from strong winds and rain, but vulnerable to earthquakes if buildings collapse. Vulnerability is not static, but varies over time as both human populations and the physical environment change (Meehl *et al.*, 2000).

As previously mentioned, the *magnitude* (size or intensity) of a hazardous process is usually inversely related to the *frequency* of its occurrence. For example, large earthquakes occur much less often than small ones and major disasters result from relatively rare, big events. The energy release from the 2004 Boxing Day earthquake, which killed about 250,000 people, was about 100 times that of the 2005 Kashmir earthquake, which resulted in 74,500 deaths. The five largest events during the twentieth century were responsible for over half of all the earthquake-related deaths. When the magnitude of an event is plotted against the logarithm of its frequency, it normally exhibits the relationship shown in Figure 4.3a. The *recurrence interval* (or return period) is the time that elapses, on average, between two events that equal, or exceed, a given magnitude. A plot of recurrence intervals versus associated magnitudes (Figure 4.3b) produces a group of points that approximates to a straight line on a semi-logarithmic graph.

The analysis of extreme events by probability methods relies on the assumption of *uniformitarianism* – a belief that past processes and events are a good guide for the future. It is most appropriate for hazards for which a long-term record

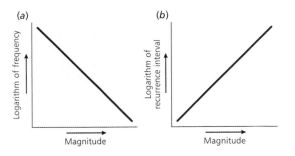

Figure 4.3 Generalized statistical relationships between the magnitude and (a) the frequency and (b) the return period for damaging natural events. A few very high-magnitude events are responsible for the majority of disaster losses.

exists, particularly hazards unaffected by human activity. For example, it is reasonable to assume that global tectonic processes, driven by large-scale geological forces, have remained fairly constant on human timescales. Probability analysis is less suitable for environmental processes known to have changed during historical times. Thus, there may be a shift in magnitude–frequency relationships for floods of a particular size if extensive deforestation has taken place over the drainage basin. Setting such limitations aside, probability-based approaches can be used to estimate the size of floods that might be expected once every year, once every 10 years, and once every 100 years and so on. But, whilst a 100-year flood has a *probability* of 1:100 of happening in any one year, and an estimated average return period of 100 years, in practice such a flood could occur next year, be exceeded several times in the next 100 years or not occur for 200 years.

Despite such limitations, probability-based estimates help engineers to design and build key defensive structures in hazard-prone areas. The list includes dams and levees for flood control, nuclear power plants protected against storm surges and hospitals reinforced against ground shaking in earthquake zones. Engineers plan for a selected *design event*, which is often the magnitude of the hazardous process that a structure is built to withstand during its expected lifetime. The actual return period for the design event varies according to the nature of the hazard and the vulnerability of the elements at risk. As an example, large dams on major rivers are often built to withstand the 1:10,000-year flood because the consequences of failure would be catastrophic for many downstream communities. On the other hand, in the UK, railway bridges are generally designed to withstand the 1:100-year flood event because the potential consequences of failure are much less catastrophic.

Magnitude–frequency relationships are used in other areas of hazard management. For example, a mortgage lender or an insurer might wish to know the magnitude–frequency relationships of flood risk for new houses built on a floodplain during the average mortgage span of 30 years. Figure 4.4 shows the risks of an event's being

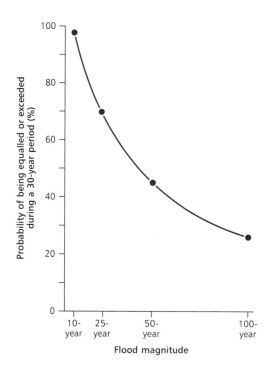

Figure 4.4 The probability of occurrence of floods of various magnitudes during a period of 30 years. This is the average duration of a standard property mortgage and the information on risk will be of interest to mortgage lenders and to property insurers.

equalled or exceeded during this period. It can be seen that an event as high as, or higher than, the 50-year flood has a 45 per cent probability of occurrence, but if the 100-year return period is chosen, the probability drops to 26 per cent. This is valuable information. If the probability of a claim's being made, and the likely cost of that claim, are known, then the insurance premium can be set appropriately. If the estimate of probable losses is too high, the premium will be high and may prove to be uncompetitive in the insurance market. If the estimate is too low, the insurance company stands to make a loss from unexpected claims.

2 The analysis of extreme events

Extreme event analysis is concerned with the statistical spread of maximum or minimum values, such as the strongest wind gust or the largest flood, at a given site. The process can be explained using an example of annual maximum wind gusts that might be employed to assess the potential for storm damage. In this case, data are available on the annual maximum wind gusts recorded at Tiree in western Scotland over the 59-year period from 1927 to 1985. The first step is to give a ranking (m) for these events, starting with m = 1 for the highest recorded wind gust, m = 2 for the next highest, and so on in descending order. The return period or recurrence interval Tr (in years) can be computed from

$$\text{Tr (years)} = (n + 1)/m$$

where m = event ranking and n = number of events in the period of record. The percentage probability for each event may then be obtained from

$$\text{P (per cent)} = 100/\text{Tr}$$

and the annual frequency (AF) is given by

$$1/\text{Tr (years)} = \text{AF}.$$

Figure 4.5 shows these wind gusts plotted using the return period calculation described above. The data fall on a straight line, illustrating once again the link between magnitude (gust speed) and frequency (probability). From this it is possible to estimate the return period corresponding to any desired gust speed and the speed that has a given return period. Caution is needed in extrapolating to gust speeds that are outside the range of the available data (about 100 knots in this case). This is because of uncertainty in the data beyond this point. Sometimes attempts are made to estimate a theoretical maximum value using physical principles, but these assumptions could be undermined by inadequate knowledge, changing climatic conditions or other factors.

When a dataset is too short to be representative, it may be necessary to extrapolate a design event, despite the risk of substantial error. It is for this reason that efforts are made in earthquake engineering and flood hydrology to extend the instrumental records, using historical documents and other proxy records to estimate the frequency and size of unmeasured extreme events. The situation is most difficult for very rare hazards, such as high-magnitude tsunamis, for which there is no statistically valid dataset. In these cases, the only viable approach is to examine the geological record for proxy evidence in order to create modelling scenarios for such events.

A major conclusion must be that the reliability of results from probability-based analysis depends heavily on the quality of the database. Ideally, each event in the record should be drawn from the same statistical population, should be independent and should follow a known distribution curve. For example, each of the maximum wind gusts in the Tiree dataset is independent – they are maximum annual gusts (so each value must be from a different storm event) but, because they are all caused by mid-latitude cyclones, they are drawn from the same statistical population. Other environmental phenomena are not necessarily independent. Earthquake occurrence is not random in time, as the magnitude of the event

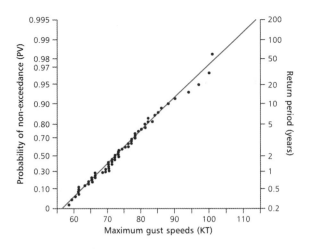

Figure 4.5 Annual maximum wind gusts (knots) at Tiree, western Scotland, from 1927 to 1985 plotted in terms of probability and return period. This example indicates a straight-line relationship, which offers limited potential for extrapolation for more extreme events.

depends in part upon the amount of strain energy that is stored up in the Earth's crust. When a large earthquake occurs, at least part of the strain energy is released. This reduces the immediate likelihood of another large event on the same section of fault until the strain energy has built up again. On the other hand, the stress may have been transferred onto other local faults, increasing the chance of an earthquake on a nearby fault.

Whilst it is sometimes assumed that the statistics are best described by a Normal distribution function, this is not always so. Daily rainfall data, for example, have a skewed, rather than a Normal, statistical distribution, with resulting complications for probability analysis. Other problems arise, as mentioned in the previous section, when past records are used for prediction purposes on the assumption that there will be no change in causal factors. This assumption, known as *stationarity*, ignores the possibility of wider environmental change. Changes to physical systems can occur naturally over very long time periods, but changes resulting from human activities are often more important. In terms of

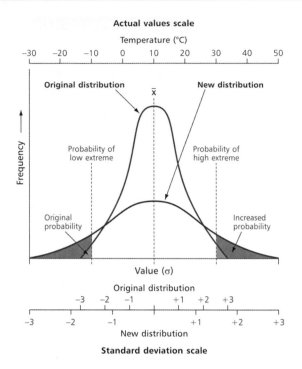

Figure 4.6 The effects of a change to increased variability on the occurrence of extreme events. Both upper and lower hazard-impact thresholds are breached more frequently as a result of the increased standard deviation, although the mean value remains constant. The example is provided on a temperature scale.

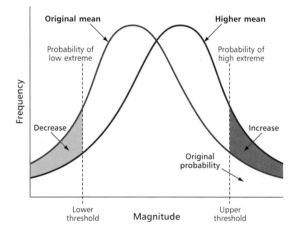

Figure 4.7 The effects of a change to an increased mean value on the distribution of extreme events. The shift results in an increased frequency of hazard impacts from 'high' magnitude events, with a corresponding decrease in the frequency of 'low' magnitude events.

near-surface geophysical processes, like floods, the relevant systems have almost certainly been affected to some degree by human activity over the last century or so. The prospect of climate change means that the existing statistical distributions are less likely to provide a reliable estimate of future events.

The consequences of such changes, when expressed in statistical terms, are complex. Changes in the frequency of hazardous events can be expressed most simply as shifts in the mean and standard deviation of the dataset. Figure 4.6 illustrates a climate-change situation in which the mean value remains constant but the variability, expressed by the standard deviation, increases. Thus, the frequency of both 'high' and 'low' extreme events increases relative to the thresholds which define the relevant social band of tolerance. This might simulate climate change that leads to both colder winters and warmer summers, as shown on the temperature scale. On the other hand, Figure 4.7 shows the consequences of an increase in the mean value but with no change in variability. This might simulate the effects of climate change in which a location undergoes a net warming without a major change to the weather patterns. In this case the frequency of 'high' extremes relative to the impact threshold rises, whilst the incidence of 'low' extremes falls. Needless to say, this effect would be reversed with a lower mean value.

In reality, environmental change might cause shifts in both the mean temperature and the variability. It might also alter the shape of the distribution. For this reason, accurate predictions of the effects of climate change on the occurrence of hazardous events are difficult to achieve with existing models. One complication is the non-linear relationships that exist between driving factors and the hazards themselves, such as between sea-surface temperatures and the formation of tropical cyclones (see Chapter 9). The probability function for most hazardous process is itself sensitive to changes in the mean value (Wigley, 1985). A shift in the mean value of only one standard deviation would cause an extreme event expected once in twenty years to become five times more frequent. Similarly, the return period for the one-in-a-hundred-year event would fall to only eleven years, an increase in probability of nine times. This is one reason why some researchers believe that the impacts of atmospheric hazards will increase significantly with only modest changes in climate.

Another challenge lies in understanding the changing sensitivity – or vulnerability – of societies to hazards. Some possibilities that give rise to increased risk of disaster are shown in Figure 4.8. Case (a) shows a constant band of social tolerance and constant variability of the hazardous element in question but a decline in the mean value of that element (perhaps a decrease in temperature). Case (b) represents a constant band of tolerance and constant mean but an increased variability (perhaps a trend to greater fluctuations in annual rainfall). Finally, in case (c), the variable does not change but the band of tolerance narrows and vulnerability increases (perhaps because population growth places more people at risk).

C RISK PERCEPTION AND COMMUNICATION

A distinction between *objective* (statistical) risk and *subjective* (perceived) risk has already been made. Objective risk assessment is the consequence of a scientific process. It follows a systematic procedure that seeks to exclude emotive elements due to personal preferences, in order to supply valid, reproducible results. Subjective risk assessment, on the other hand, does not result from a formalized process. It relies on personal views and experience and the resulting perceptions are not reproducible in a scientific sense. For example, an individual's views may change quickly, especially if they personally experience a disaster. But the two approaches are not polar opposites. Perceived risks may integrate a considerable body of scientific knowledge, whilst so-

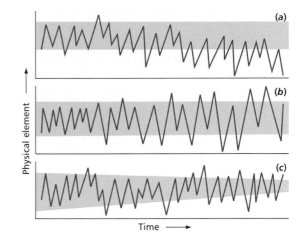

Figure 4.8 Changes in human sensitivity to hazard, due to variations in physical events and changes in societal tolerance, shown by the central bands on the graphs. In each case the magnitude and frequency of hazard increases through time. After de Vries (1985).

called 'objective' risk evaluations involve value judgements, such as the ways in which different impacts are compared.

The model of decision taking most widely employed in the hazards field assumes that perception acts as a personal filter through which the decision maker views the 'objective' environment. Faced with the complexities of natural and human systems, for which there is an imperfect knowledge base, the decision maker has to seek an optimum, rather than an ideal, outcome. Kates (1962) stressed that such choices are based on the individual 'prison of experience'. Consequently, hazard victims and hazard managers tend to respond to risk in different ways. It is generally considered that all individual perceptions of risk are equally valid and that, for any given threat, each individual has a right to choose their own response.

Differences between expert risk assessments and lay risk perceptions can lead to problems in the management of hazards (Table 4.2). A statistician might well rate voluntary and involuntary risks equally, whereas most non-experts

Table 4.2 Differences between risk assessment and risk perception

Phase of analysis processes	Risk assessment	Risk perception
Risk identification	Event monitoring	Individual intuition
	Statistical inference	Personal awareness
Risk estimation	Magnitude/frequency	Personal experience
	Economic costs	Intangible losses
Risk evaluation	Cost-benefit analysis	Personality factors
	Community policy	Individual action

show greater concern for involuntary risks. Additionally, an objective perspective would suggest that large infrequent events that take many lives can be seen as equivalent to frequent hazards that take only a single life at a time but which, over time, lead to a similar scale of loss. Conversely, in the perception of most lay people, the dramatic hazards that take many lives at a time are more significant. For example, in the UK, on average more people die each day in road accidents than die each year in rail crashes, but railway accidents attract more media coverage. This is partly because rail accidents are considered the result of involuntary risk and the events tend to produce striking images. Car accidents are perceived as the result of voluntary risk and the number of deaths per event tends to be low.

Resolving conflict between technical risk analysis and less formal risk perception is a headache for hazard managers. On the one hand, it is argued that the outcomes of objective analyses allow risks to be compared and balanced appropriately. In this way, rational economic decisions about expenditure on risk reduction can be made. Indeed, some analysts consider non-scientific perceptions of risk to be invalid simply because they arise from subjective influences. Others argue that risk is highly complex, going far beyond simple statistical estimates of mortality, morbidity or loss. Today, the incorporation of perceived risk into decision making is believed to capture a public view which is important for policy making within a democratic society.

The latter view accords with the view that lay perceptions have value because they tend to blend expert analysis with individual judgement based on personal experience, social context and other factors. The public is also aware that limitations apply to the information available to experts (Sjöberg, 2001). Some high-profile public-health fears in the UK include the risks to children of autism associated with the triple vaccine for measles, mumps and rubella (MMR), and the link between eating infected beef and contracting the illness CJD. In a 'risk society' the public is increasingly cautious about accepting scientific views about hazards. This is a recurrent issue in debates about global warming, when the – largely incorrect – assertion has been made that scientists over-emphasize the threats from anthropogenic global warming (AGW) in order to protect their sources of research funding.

During contentious decision making, serious breakdowns in trust between risk managers and the public can occur. In practice a balance must be achieved. Whilst community views should be taken into consideration, an emphasis on lay perceptions of hazard and risk well beyond the limits of objective analyses can lead to the waste of public resources in order to achieve very limited improvements in safety. Furthermore, perceptions of risk can sometimes be driven by unjustifiable prejudices, which are often amplified through the media and by politicians. In particular, where risk perception is used to drive hazard management, great care is needed to ensure that

Plate 4.1 Emergency response to a 2009 road traffic accident on Highway 2/E30 leading to Berlin near Hanover, Germany. Transport risks tend to be acceptable to the public, despite the relatively large cumulative death rate in road accidents. (Photo: Panos/Martin Roemers MRM01548GER)

the results do not disadvantage minorities in society.

Past conflicts, between technical assessors of hazard and risk and the public, demonstrate the need for good communication between the two interest groups. From a practical standpoint, improved communication should enable lay people to understand the results of objective analyses of risk and also help to inform scientists about the risks that cause most concern to the public. However, there are problems attached to communicating complex technical assessments of risk to the public (Slovic, 1986):

- People's initial perceptions of risk are often inaccurate.
- Risk information often frightens and frustrates the public.

- Strongly held, pre-conceived beliefs are hard to modify, even when the justification for those beliefs is incorrect.
- Naive or simplistic views are easily manipulated by presentation format. When it is stated that there is a 10 per cent chance of an event occurring, rather than a 90 per cent chance that it will not, opinions change.

The growth of online sources of information has made matters more complicated. Many people have access to a vast range of information and can undertake their own independent research using the internet. Whilst this is a valid and empowering trend, the quality of the information accessed is sometimes poor and data may be selected to reinforce initial misconceptions.

D RISK PERCEPTION IN PRACTICE

An individual's perception of risk is the result of complex interactions between the general attitudes taken by the community in which the person lives and any personal experience of dealing with the hazard in question (Garvin, 2001). The cultural environment provides the setting within which the risk is interpreted. For example, a person living in a community with strong religious beliefs may be more likely to view a hazard as an unmanageable 'act of God'. Past experience is important because people with personal knowledge of hazard events tend to have more accurate views regarding the probability of future occurrence. For example, people moving from rural areas to live in urban slums on the margins of large cities may be vulnerable to landslides because they are unaware of the threats posed by such slopes.

Personal experience can be a powerful incentive for hazard mitigation, as shown in an early study by Meltsner (1978). After the 1971 earthquake at San Fernando, California, 46 per cent of residents in San Fernando and nearby Sylmar took hazard-reducing measures to reduce future seismic hazards, but this dropped to 24 per cent for the rest of the San Fernando valley and fell to 11 per cent for the Los Angeles basin as a whole. Some may take the view that once any disaster, such as an earthquake, has occurred, the probability of a recurrence is reduced and there is less need to take further mitigating action. When direct hazard experience is lacking, as it is for most people, perceptions are moulded in other ways. The media – television specifically – is a powerful source of information. Given the extent of in-built bias in reporting disaster news, detailed in Chapter 2, and an increasing reliance on information from the internet, the hazard perceptions of lay people are likely to be shaped differently from more objective risk analysis outcomes. That having been said, media outlets provide an opportunity for scientists to influence community perceptions of risk, and there is awareness of the need for better public understanding of the science of risk.

Some lay people may perceive hazards differently from technical experts because of geographical location and aspects of personality. Early work on floods revealed that rural dwellers have hazard perceptions closer to statistically derived estimates than do urban dwellers, due to their greater levels of connection with, and reliance upon, the natural environment. The influence of personality is often classified according to the degree to which an individual believes that the impact of a hazardous event is dependent upon fate (it is externally controlled) or their own actions (it is internally controlled). Clearly, a range of views exists surrounding what is usually described as the 'locus of control'. Within this spectrum, three distinct types of perception can be identified. These are:

- *Determinism.* This pattern of behaviour, sometimes called the *gambler's fallacy*, exists when lay people find it difficult to accept the random nature of most hazardous events. This form of perception admits that hazards exist but seeks to place extreme events in some ordered pattern, perhaps associated with regular intervals of time or a repeating cycle. In the UK, for example, there is a common perception that a cold spell on the eastern seaboard of the USA precedes a similar cold spell in Britain, even though there is little evidence to support this view (Petley, 2009). For some earthquake sequences, this need not be an inaccurate perspective, but it does not fit the temporally random pattern of most threats.
- *Dissonance.* Although it takes many forms, dissonant perception represents denial or a minimization of risk. Typically, an event is viewed as a freak occurrence unlikely to be repeated. In extreme cases the existence of a past event may be denied completely.

Dissonance is a highly negative form of perception, often associated with people's having much material wealth at risk from a major disaster. In an early study, Jackson and Burton (1978) suggested that people living in areas subject to high levels of seismic hazard did not consider the hazard to be troublesome, partly because of the practical difficulty of coping with the consequences of a large earthquake and partly due to psychological problems in coming to terms with continuing vague threats. From this viewpoint, dissonance is an attempt to deal with on-going risks on a bearable day-to-day basis.

- *Probabilism.* Probabilistic perception is the most sophisticated view because it accepts that disasters will occur and that many events are random. It generally accords best with the views of officials charged with making decisions about risks. But in some cases the acceptance of risk is combined with a need to transfer the responsibility for dealing with the hazard to a higher authority, which may range from the government to God. Indeed, the probabilistic view has sometimes led to a fatalistic, 'acts of God' syndrome, whereby individuals feel no responsibility for hazard response and wish to avoid any actions or expenditure on risk reduction.

An important feature of public risk perception is *social amplification.* This happens when factors combine to create an exaggerated fear of a threat. It tends to occur when the threat is new to the individual, when people believe that the true magnitude of the risk is being hidden from them in some way, when there is a belief that the hazard cannot be controlled, when the individuals exposed to the hazard are considered to be highly vulnerable (e.g. if they are children) and when there is a feeling that experts do not understand the risks. On the other hand, risk perception may be reduced when individuals or groups are not able to relate directly to the hazard, when the level of media reporting about the hazard is limited or short term, when there are perceived benefits associated with the hazard, when there is a belief that the hazard is well understood and that the responsible individuals are trusted. In the UK, for example, there has been a perception that rail travel is more dangerous than the statistics suggest, partly because of high media interest when accidents do occur and partly due to a low level of public trust in the effectiveness of track maintenance.

Some of the factors that can increase or reduce public risk perception are listed in Table 4.3. Risks are taken more seriously if they are life threatening, immediate and direct. This means that an earthquake is rated more seriously than a drought. The type of potential victim can be significant, since risk perception is not restricted to purely personal concerns. Awareness is heightened if children are at risk or if the victims are a readily identifiable group of people. For example, threats to a school party would be greatly amplified. Level of knowledge can be important, particularly when related to the degree of trust in the sources of hazard information. This is a common feature in the perception of complex technological risks, especially if a lack of scientific understanding is combined with disbelief of opinions expressed by technical experts. Age is also a factor. Fischer *et al.* (1991) found that students gave relative emphasis to threats to the environment, whilst older people were more interested in health and safety issues. If technological hazards become more prominent, the public will increasingly view these risks as capable of some human control. Much weight is already given to the common hazards, like road safety, which can be significant for some countries. In New Zealand the death toll on the roads every six months exceeds the loss of life due to earthquakes throughout the recorded history of that nation.

Table 4.3 Twelve factors influencing public risk perception, with some examples of relative safety judgements

Factors tending to increase risk perception	Factors tending to decrease risk perception
Involuntary hazard (radioactive fallout)	Voluntary hazard (mountaineering)
Immediate impact (wildfire)	Delayed impact (drought)
Direct impact (earthquake)	Indirect impact (drought)
Dreaded hazard (cancer)	Common hazard (road accident)
Many fatalities per event (air crash)	Few fatalities per event (car crash)
Deaths grouped in space/time (avalanche)	Deaths random in space/time (drought)
Identifiable victims (chemical plant workers)	Statistical victims (cigarette smokers)
Processes not well understood (nuclear accident)	Processes well understood (snowstorm)
Uncontrollable hazard (tropical cyclone)	Controllable hazard (ice on highways)
Unfamiliar hazard (tsunami)	Familiar hazard (river flood)
Lack of belief in authority (private industrialist)	Belief in authority (university scientist)
Much media attention (nuclear plant)	Little media attention (chemical plant)

Source: Adapted from Whyte and Burton (1982).

E RISK MANAGEMENT

Risk management is a process whereby risk is evaluated to facilitate the introduction of hazard-reducing strategies. To some extent, there are differences between risk management for technological and natural hazards. For example, fault tree methods are more common for technological hazards, but, in both cases, an appreciation of the entire system involved is needed. In most countries, the prime responsibility for risk management lies with national governments, who create health and safety legislation, which, in turn, is implemented and enforced by regional authorities and specialized agencies. Whatever administrative structure exists, successful risk management depends on the use of effective and transparent methods that are cost-effective and acceptable to the stakeholders within the relevant community.

Management methods vary according to the type of risk and the needs of countries at contrasting stages of human and economic development. In an ideal world, there would be a clear set of agreed priorities for formal risk management, with the highest levels of risk addressed first. In order to develop such a priority list, a quantitative risk assessment of all relevant factors and consequences is required. This goal is almost impossible to achieve, due to a lack of data, the need to balance risks between high- and low-

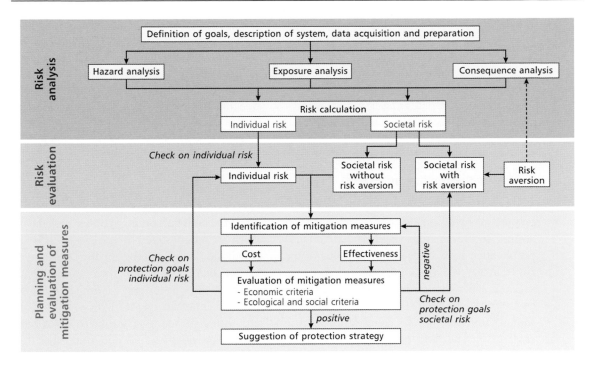

Figure 4.9 A sequential approach to natural hazard risk management as adopted in Switzerland. The process progresses through stages of analysis, evaluation and mitigation for individual and societal risks. After Bründl *et al.* (2009). Reproduced under Creative Commons Attribution 3.0 License.

frequency events, financial constraints and, as already discussed, the complexities attached to public perceptions of risk. In particular, top-down approaches may fail to capture the needs of some at-risk groups, especially in the LDCs, and there is now focus on more community-based participatory approaches to risk management and disaster mitigation.

1 The formal approach

Recent experience of large disaster losses, combined with reduced financial resources, has encouraged governments in Europe and elsewhere to adopt more rigorous and systematic methods of risk management, as exemplified by the procedures introduced for natural hazards reduction in Switzerland (Bründl *et al.*, 2009). As shown in Figure 4.9, this approach consists of a detailed,

three-stage procedure that includes risk analysis, risk evaluation and risk management.

- *Risk analysis* is the first step when a general hazard appraisal takes place using archived information from maps, terrain analysis, aerial photographs and satellite imagery. *Exposure analysis* then identifies and assesses the extent to which people or other local assets are at risk. *Consequence analysis* combines the hazard and exposure outcomes to provide estimates of expected damages or other losses from given events. *Risk calculation* is then conducted to determine the scale of expected losses for persons or socio-economic assets.

- *Risk evaluation* follows, so that the expected losses, expressed in terms of fatalities or economic damage, can be scaled against predetermined safety goals (e.g. with respect to fatalities) in order to determine what losses are acceptable and what are not. Intervention action is then prioritized accordingly.

• *Risk management* is the final phase of evaluation and planning, when a search is made for the most appropriate mitigation strategy based on a variety of economic, ecological and social criteria in use at the time.

2 The participatory approach

The notion that community risk assessment is best served by an inclusive approach has been around for some time (Pelling, 2007). Local people often have a good understanding of the risks that they face. This is especially so if they have previous experience of the hazard, like the rural communities exposed to volcanic threats near Mount Cameroon, West-Central Africa (Njome *et al.*, 2010). The participatory approach aims to respect the risk perceptions held by local stakeholders and to work jointly towards the application of indigenous skills and coping measures. In a study of inhabitants' perceptions of coastal threats in the flat coastal area of Benin, West Africa, Teka and Vogt (2010) found that risk perception differed according to factors such as age and ethnicity and concluded that risk management strategies should reflect these group-specific attitudes. Making use of a community GIS in Mexico, Krishnamurthy *et al.* (2011) demonstrate a way in which the local perceptions of hurricane risk can be factored into policy making for disaster reduction.

Case studies from developing countries show much potential for loss reduction by an understanding of social characteristics and vulnerabilities gained through active engagement with local people. In the immediate aftermath of disaster, local residents are usually the first responders, so better training, and reliance on family and community networks, will improve the search-and-rescue phase. The recovery process following disaster is crucial because this phase offers a window of opportunity to adopt initiatives designed to correct faults in development policies and weaknesses in infrastructure so as to make communities more resilient in the future. Once again, the communities themselves have skills, and sometimes the resources, to take on these tasks, even though these assets are often neglected by the outside world.

As previously indicated, the framework for risk management is usually set by government regulations operating at local, regional, national and international levels. For example, in the UK many everyday risks are managed through laws originating from both European and British parliaments which are administered by agencies such as the Health and Safety Executive and the Environment Agency. Enforcement may then be undertaken by those bodies, by the police (with respect to the management of risk on the highways) or by local authorities. In addition, the British Standards Institute provides a set of codes of practice that, although not legally binding, provide appropriate guidance to enable organizations to comply with the legislative requirements. Finally, some specialized industries have their own legislative frameworks and enforcement systems. For example, the aviation industry is covered by a specific set of laws, agreements and frameworks which, in the UK, are enforced by the Civil Aviation Authority.

The legal framework for risk management is supported by a range of other measures, such as the use of public information programmes that inform people about hazards, the purposes and nature of the regulatory framework and the actions that people can take to minimize their own risks. This advice may be backed up by economic instruments such as financial subsidies and tax credits for compliance, combined with fines for non-compliance. As an example, the authorities in an urban area with a high level of seismic risk might reduce risk by:

• enforcing a building code that requires all new structures to be able to withstand a specified earthquake risk. Ideally this building code will be enforced through legislation, with high

penalties (including demolition in extreme cases) imposed for non-compliance;
- providing tax incentives and subsidies to owners to encourage the retrofitting of existing buildings in order to meet the building code standards;
- educating the public about the building code and suitable measures for retrofitting buildings. Programmes that increase public awareness of the earthquake risk may be undertaken. Emphasis is often placed on teaching children how to react because this helps to protect some of the most vulnerable people in society and assists in the transfer of information to adults.

Despite its importance, risk management is only one of many goals in society. The resources required have to be balanced against other worthy demands. In many LDCs, for example, the management of hazard must be weighed against reducing poverty, improving health provision and low life expectancy and providing basic education. Generally speaking, the amount of risk-related government spending is small. In the UK direct public spending on health and safety regulations is only about 0.1 per cent of total central government expenditure (Royal Society, 1992). Even then, some of the investment in increased safety is likely to be traded off against other values, such as when spending on flood

---

Box 4.2 The ALARP principle

ALARP stands for 'as low as reasonably practicable' (Petley, 2009). The principle is applied to risk management on the assumption that society is faced with a hierarchy of risks from acceptable through tolerable to unacceptable (Figure 4.10). Risks in the unacceptable range (at the top of the diagram) are considered to be too great to bear and must be addressed, more or less regardless of cost. Risks in the tolerable range are then tackled, using the ALARP principle that states they should be reduced as far as is feasible within the wider economic and social framework. Finally, the lowest category, negligible (acceptable) risk, is not addressed through risk management because it would represent a misuse of resources.

The aim of risk management is to reduce all risk to the acceptable level. In reality, this is not achievable. Therefore a cost-benefit

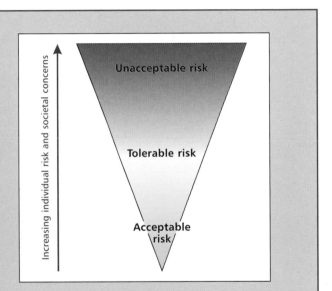

Figure 4.10 The ALARP approach to risk management. High-level (unacceptable) risks are at one end of the scale, lower-level (acceptable) risks are at the other end. The majority of tolerable risks should be managed and reduced as far as possible by practicable means. After Petley (2009).

calculation is used to enable the prioritization of resource use. In the UK, the ALARP approach is embedded in law, as a result of a legal ruling in the European Court of Justice in 2007.

defence works leads to greater property values and economic risk on floodplains, the so-called 'levee effect' (see Chapter 11). In effect, the aim of risk management is to reduce threats to an acceptable level that is compatible with other socio-economic demands (Helm, 1996) and is guided by some variant of the ALARP principle (Box 4.2 and Figure 4.10).

The principle of risk acceptance is especially relevant to the danger arising from hazardous industrial sites. Typically, the risks extend from individual operatives within a plant, up to wider societal consequences if the effects of explosions or dangerous pollutants travel off-site. Attempts have been made to scale these risks, and the consequent hazard severity, in terms of their level of acceptance. Table 4.4 shows a risk matrix for deaths and injuries for industrial accidents. According to Duijm (2009), practice in European countries has moved towards a consensus, so that, for individual or location-based risk, the probability of an accident should be less than 10^{-6} per year.

Many risk-management decisions are based on financial grounds. This means that there is a need to attribute an economic value to a human life, despite the fact that many people are uncom-

fortable with the notion. A number of approaches have been developed, of which the so-called *human capital* method is perhaps the best-established. This method works on the basis of an individual's lost future earning capacity in the event of accident or death. It is a relatively simple principle, which values the life of a child at the highest level, but it is flawed in that it places a zero value on those people who, for whatever reason, are unable to work. A better approach is *willingness to pay*, which seeks to determine how much people would be willing to pay in order to achieve a certain reduction in their chance of a premature death (Jones-Lee *et al.*, 1985). This is preferable because it measures *risk aversion*, i.e. the value people place on reducing the risk of death and injury, rather than on more abstract, long-term concepts. Willingness to pay can be assessed by questionnaires which ask the respondents to estimate either the levels of compensation required for assuming an increased risk, or the premium they would pay for a specified reduction in risk. Studies have found that the valuation of risk should include some allowance for the pain from, and aversion to, the potential form of death (e.g. high values for death by cancer) and that willingness to pay tends to decline after middle

Table 4.4 An example of a risk matrix for industrial accidents

Frequency classification	Frequency per year	Accident magnitude
Frequent Occurs several times during lifetime of installation	$1–10^{-2}$	Undesired event Minor material damage
Likely Will probably, but not necessarily, happen	$10^{-2}–10^{-4}$	Minor accident Minor occupational injuries on site
Not likely Could possibly happen	$10^{-4}–10^{-6}$	Serious accident Serious occupational injuries on site
Very unlikely Almost unthinkable	$10^{-6}–10^{-8}$	Major accident Fatalities on site, injuries to people off site
Extremely unlikely Frequency is under the limit of reasonability	$<10^{-8}$	Disaster Fatalities on and off site

Source: Adapted from Duijim (2009). Used with permission.

age, as the risk of mortality from natural causes increases.

F THE ROLE OF INFORMATION TECHNOLOGY

The contribution of information technology to the management of natural disasters has greatly increased over recent decades, as indicated by Cutter (2003), Tralli *et al.* (2005), Gillespie *et al.* (2007), Joyce *et al.* (2009) and Reddick (2011). These reviews reflect advances in geospatial technologies over time. For example, prominence has been given to the use of near real-time data by emergency services and the public during the emergency-response phase of disaster. In turn, this has led to more anticipatory forms of risk assessment. But information technology also has a place in hazard monitoring and the compilation of disaster databases for later analysis and preparedness planning. Many applications of information technology are hazard- and disaster-specific. For that reason, some parts of this topic – for example, developments in hazard forecasting and warning schemes – are reserved for Part Two of this book. The aim of this section is to provide a context in which to place more detailed treatments.

From the late 1970s onwards, increased computing capacity has offered fresh opportunities in project planning and real-time decision making in emergencies. By the early 1990s, relatively powerful and networked desktop computer systems were an integral part of disaster management operations, especially in the MDCs (Stephenson and Anderson, 1997). Drabek (1991) reported fairly wide use of PC-based decision-support systems in the USA, especially during the emergency phase, when damage assessment, route designation for evacuation and the availability of shelters are critical issues. At this time there was a concern for improved telecommunications and simulation modelling, plus the need for some integration of the emerging technologies. Although networked computer systems remain

Plate 4.2 A scientist from the United States Geological Survey Volcano Observatory on Hawaii Island uses a portable GPS receiver to track changes in the surface topography during an island-wide survey of volcanic activity. (Photo: Loren Antolik, USGS)

prone to power failure during disasters, the increasing reliability of portable radio-based transmission systems permits communication even when the ground-based infrastructure is destroyed. Notebook computers and PDAs can be carried into remote or devastated areas where GPS technology permits instantaneous location fixing and vehicle tracking. Satellite-based telemetry can then be deployed for imaging and field survey work.

1 Satellite remote sensing

Many forms of remote sensing have led to disaster reduction, especially in the LDCs (Wadge, 1994 and Table 4.5). In general, Earth observation satellites have supported pre-disaster preparedness through monitoring and mapping activities, while communication satellites have contributed to disaster warning and the mobilization of emergency aid (Jayaraman *et al.*, 1997). As already indicated, the specific application depends on the task and the hazard. For example, automated techniques are now in place for the detection of volcanic activity and wildfire by the remote sensing of excess heat. During the routine monitoring and land zoning of a volcanic cone, the imagery is unlikely to be time dependent, but during emergency operations, information is urgently required and must be available in all weather conditions.

The type of sensor used depends on the spatial or spectral resolution required. Radar data is needed when cloud obscures disaster areas, and a mixture of optical and infra-red bands is best for

Table 4.5 The use of remotely sensed imagery in hazard management

Hazard	Remote sensor	Nature of data	Uses
Storms	Geostationary satellites (5) (e.g. Meteosat)	Global, 5 km resolution, every half hour: gives cloud, water vapour	
	Polar-orbiting satellites (2) (e.g. NOAA)	Global, 1 km resolution, every 6 hours: gives cloud, temperatures	Storm tracking, weather forecasting
	Ground-based VLF (e.g. SFERIC service)	Global, time and position of lightning	Storm tracking
Floods	Landsat (SPOT, NOAA)	Near infrared discrimination of land/water	Flood extent mapping
	Satellite radar (e.g. ERS-1)	Water content from backscatter for soil/snow	Runoff/snowmelt models
	Ground-based radar	Intensity of rainfall	Weather forecasting/runoff models
Earthquakes	Satellite/airborne radar (e.g. ERS-1)	Interferometric mapping of surface deformation	Prediction of earthquakes?
	Differential GPS	Point monitoring of surface deformation	Prediction of earthquakes?
	Landsat/SPOT/Fuyo-1	Detection of topographic evidence for earlier faults and offsets	Estimate of earthquake recurrence
Volcanic eruptions	NOAA/TOMS	Eruption plume height, motion and SO_2	Aircraft warning, eruption monitoring
	Landsat (Thematic Mapper)	Size and temperature of emitted radiation	Eruption precursor/ monitoring
	Satellite/airborne radar	Deformation of volcano's surface	Eruption precursor/ monitoring
Drought/ pests	Meteosat and NOAA	Cloud temperatures and vegetation indices	African storm warnings, drought monitoring and pest migration prediction
Fires	NOAA (Thermal infrared)	Night-time thermal emissive anomalies give temperature and size of fires	Wildfire monitoring
Landslides	SPOT	Topography from stereopairs	
	Landsat	Spectral character of landslide ground	Landslide inventory and susceptibility mapping

Source: After Wadge (1994)

wildfire detection. Integrated techniques are increasingly used – for example Singhroy (1995) described the use of SAR (synthetic aperture radar) in assessing landslide and coastal erosion hazards. SAR technology is particularly useful for mapping the extent of surface floods. Other applications include lahar monitoring on the flanks of volcanoes (Kerle and Oppenheimer, 2002). Early detection of regional drought, before it grows into a disaster, is also possible. Because of the high cost of development and launching, Earth observation satellites have not been deployed as widely as communication platforms, although this situation may change with developments in small-satellite technology (da Silva Curiel *et al.*, 2002).

For many years, hurricanes have been tracked by geostationary satellites (e.g. Meteosat) that provide global cover between 50°N and 50°S at half-hourly intervals. Such repetition enables close monitoring of a storm as it moves towards landfall. As a result, no major tropical cyclone is likely to form without detection, but forecasting the future track of such storms remains problematic. A similar situation arises with smaller-scale phenomena like tornadoes. These storms are tracked by geostationary satellites, in combination with Doppler radar to monitor the rotation and the speed of forward movement of tornadoes. More recently, attention has been given to linking hurricane forecast tracks with up-to-date demographic information for the expected landfall area in order to target formal warning messages more precisely and forestall the evacuation of coastal populations not at risk.

Satellites provide a cost-effective, global coverage of volcanic activity through the detection of thermal anomalies and plume tracking. Similarly, large-scale drought monitoring is possible through changes in surface albedo and the application of a vegetation index (VI) that measures vegetation stress (Teng, 1990). This information can be used for a variety of purposes, ranging from a change in cropping patterns and irrigation practices early in a growing season to the late-season estimation of crop yields and their possible effects on food supplies (Unganai and Kogan, 1998). The mapping of flood-affected areas is also highly successful, due to differences in the spectral signatures for different types of inundation – standing water, submerged crops, areas of flood-water retreat, etc. In addition, the topographic information necessary for hazard zone mapping can be provided by instruments such as the SPOT and ERS satellites, which have stereo-imaging for this purpose. Over the last decade the availability of very high-resolution instruments, such as *Ikonos* and *Quickbird*, has allowed the identification of individual structures and even earthquake-induced cracks in the ground (Petley *et al.* 2006). This is now permitting the assessment of damage to be undertaken remotely.

There are limitations. Remotely sensed imagery needs filtering and correction, a process that is expensive and time consuming. The very high-resolution satellites typically image each area every few days only. In addition, the instruments are optical in character, meaning that they cannot penetrate cloud cover. Radar instruments can do this but the data resolution is often too poor to allow useful analysis for short-term damage assessment. High-resolution data are also very expensive to purchase. An attempt was made to address some of these issues by the establishment of the International Charter on Space and Major Disasters in 1999. Almost all of the satellite data providers are signatories to this charter, which allows member organizations (mostly government bodies, international agencies and the major NGOs) to acquire satellite data for disaster areas free of charge. In the aftermath of the 2005 Kashmir earthquake, the charter was used to allow the acquisition of data to assist with the relief operations in Pakistan and India. However, the effectiveness of such systems remains limited by difficulties in analysing the remotely sensed data in a timely fashion and the problems of communicating the analyses to end-users on the ground in an area with poor communication networks.

2 Geographical Information Systems

GIS technology provides a major resource for disaster mitigation and emergency managers. Local government offices and other agencies routinely hold archives of contours, rivers, geology, soils, highways, census data, phone listings and the areas subject to flooding, or other hazards, for their area. This archived information can then be integrated with dynamic layers of information on evolving floods and droughts which have been extracted from remotely sensed satellite data. GIS is used on PCs at an affordable cost to aid all aspects of disaster management, including land zoning decisions, warning of residents and the routing of emergency vehicles. GIS-based systems work best for those hazards that can be mapped at a suitable scale. For example, Emmi and Horton (1993) presented a GIS-based method for estimating the earthquake risk for both property and casualties which can be applied to disaster planning and land zoning in large communities, whilst Mejía-Navarro and Garcia (1996) demonstrated a GIS suitable for assessing a range of geological hazards, backed up with a decision-support system for planning purposes. However, as noted by Cutter (2003), most emergency responders (such as police, medical teams) are not always trained to use GIS and the practical support system needs to be transparent and user friendly.

Most success has been achieved with the monitoring and forecasting of meteorological and flood hazards. In turn, this has led to improved warning and evacuation systems. Dymon (1999) described how GIS models were used to calculate the height of the potential storm surge before hurricane 'Fran' reached the North Carolina coast in 1996. Emergency managers in the USA now use GIS information to identify the areas to be evacuated when a hurricane is forecast, whilst after the storm, detailed data on residential locations can help to verify insurance claims. Potential vulnerability to disaster, expressed by the location of the poorest groups, the elderly and women-headed households, can also be captured in a GIS

in order to promote better emergency responses in the future (Morrow, 1999). GIS can also be applied in the recovery period following disaster, when the need may be to devise alternative evacuation routes, plan the relocation of facilities and make other land use decisions. GIS and GPS technology is also starting to make a contribution to the alleviation of major humanitarian emergencies in the developing world (Kaiser *et al.*, 2003). Early applications were in the control of disease outbreaks, but later advances in Africa involve large-scale vulnerability assessment, mortality surveys, the rapid identification of basic disaster needs (such as water, food and fuel) and the mapping of population movements.

3 Social communication systems

Recent developments have taken place with respect to various social media – the internet, cell phones, Twitter and Facebook. The internet is widely used for the dissemination of official information about disasters and general advice on emergency procedures. For example, the US Federal Emergency Management Administration (FEMA) posts resources online, including certification courses in disaster-related fields. Less formal channels permit the rapid sharing of information, including images, and may well find more organized applications during emergencies in due course. For example, Twitter allows users to construct personal identities for dialogue and information exchange with other members of the public and also with organizations such as the emergency services and the police. This process may help to stimulate and direct emergency action and also increase public awareness.

Cell phones can be routinely programmed to receive automated alerts from an emergency operations centre. Cell networks may not always operate during a disaster, so this facility will be more useful prior to the event, perhaps to provide guidance on evacuation, or after the event to provide information in the recovery phase. Fire departments and other emergency personnel can now communicate over wireless networks to exchange text-based information about a developing situation. New techniques are constantly being trialled for capturing moving images in the field (Mills *et al.*, 2010). The literature on the disaster applications of social media sources is currently limited (see Smith, 2010; Yates and Paquette, 2010; Bedford and Faust, 2011) but these are innovations well worth watching into the future.

FURTHER READING

Beck, U. (1992) *Risk Society: Towards a New Modernity.* Sage Publications, New Delhi. This is the starting point for much hazard-related concern in western-style countries.

Fischoff, B., Lichtenstein, S., Slovíc, P., Derby S.L. and Keeney, R.L. (1981) *Acceptable Risk*, Cambridge University Press, Cambridge. Although dated, this remains a sound introduction to risk analysis.

Joyce, K.E., Belliss, S.E., Samsonov, S.V., McNeil, S.J. and Glassey, P.J. (2009) A review of the status of remote sensing and image processing techniques for mapping natural hazards and disasters. *Progress in Physical Geography* 33: 183–207. An authoritative and comprehensive overview.

Keeney, R.L. (1995) Understanding life-threatening risks. *Risk Analysis* 15: 627–37. A clear statement on disaster-type risks.

Kerle, N. and Oppenheimer, C. (2002) Satellite remote sensing as a tool in lahar disaster management. *Disasters* 26: 140–160. A useful demonstration of a specific remote sensing application.

Sjöberg, L. (2001) Limits of knowledge and the limited importance of trust. *Risk Analysis* 21: 189–98. An interesting development from Starr's early work.

WEB LINKS

Asian Disaster Preparedness Centre www.adpc.net/
International Charter on Disasters and Space www.disasterscharter.org/main_e.html
Pacific Disaster Centre www.pdc.org
Prevention Web – the Global Platform for Disaster Risk Reduction www.preventionweb.net/global platform/
Provention Consortium www.proventionconsortium. org/
UN International Strategy for Disaster Reduction www.unisdr.org/
World Bank www.worldbank.org

CHAPTER FIVE

Reducing the impacts of disaster

A SCOPING THE TASK

It is impossible to suppress environmental hazards at source. This is because humans have little control over the massive destructive processes of nature (Figure 5.1). In a single day, the Earth's atmosphere receives enough solar energy to power 10,000 hurricanes, 100 million thunder-storms or 100 billion tornadoes. Expressed relative to solar energy receipt (i.e. taking the daily global solar energy receipt as 1 unit), a very strong earthquake releases 10^{-2} units; an average cyclone 10^{-3} units. During its life cycle, and taking all atmospheric processes into account, a single large hurricane could release energy equivalent to about 200 times the current electrical generating capacity on the planet. The 1960 Chilean earthquake, the largest yet recorded, with a magnitude $M_W = 9.5$, radiated energy of about 1.1×10^{26} ergs. This is roughly equivalent to 2,600 megatons of TNT explosives or the energy that would be released by about 130,000 atomic bombs.

It is also impossible to avoid all potentially dangerous locations on Earth. This is because of the wide geographical spread of hazardous processes, poor understanding of the risks, land use

1	Earth's daily solar energy receipt
10^{-1}	
10^{-2}	Maximum earthquake potential
10^{-3}	Average tropical cyclone (10-day life) Chile earthquake 1960
10^{-4}	
10^{-5}	Krakatoa volcanic eruption 1883 Earth's average annual release of seismic energy
10^{-6}	New Madrid earthquake 1811 Mount St. Helens volcanic eruption 1980
10^{-7}	
10^{-8}	Electrical energy of average thunderstorm Nagasaki atomic bomb 1945 Average earthquake
10^{-9}	Average forest fire in USA
10^{-10}	
10^{-11}	Kinetic energy of average tornado
10^{-12}	
10^{-13}	Average lightning stroke

Figure 5.1 Energy release (in ergs) on a logarithmic scale for selected hazardous geophysical events compared with the Earth's daily receipt of solar energy.

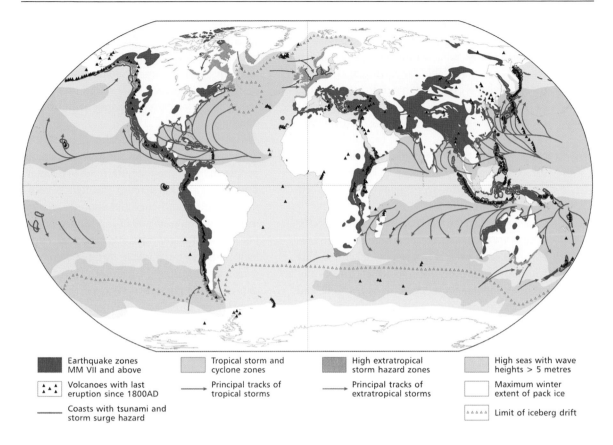

Figure 5.2 Simplified world map of selected natural hazards. After Munich Re at http://www.MunichRe.com (accessed 5 February 2012).

pressures and many other factors. Figure 5.2 shows the world pattern of some of the most damaging natural hazards, including earthquakes, volcanoes, tsunamis, storm surges, tropical cyclones and extratropical storms. Floods and droughts are omitted from this map, which indicates the global scale of the problem.

Even in adversity, people remain attached to place. If a disaster is viewed as a one-off, freak event by local politicians and other decision makers, it is likely that economic constraints and inertia will combine and the stricken community will rebuild on the same – or a nearby – site. Emergency relocation has rarely proved permanent, even for residents of small islands overwhelmed by disaster. Within two years of evacuation to Britain following

a volcanic eruption in 1961, a majority of the population of Tristan da Cunha, in the South Atlantic, had returned home. If there is a clear on-going threat, as in low-lying areas exposed to repetitive flooding, a policy of 'managed retreat' may be adopted and people will be moved to higher ground. But for large communities, such an explicit admission that hazard protection is impractical is uncommon.

Given these issues, *loss acceptance* is a common outcome. Some people are simply unaware that they occupy a hazardous location and do nothing to minimize their risk. Others attach an unduly low priority to natural hazards, as compared to day-to-day problems like inflation or unemployment, and also do nothing. Another factor is limited scientific knowledge. For example, due to a relative lack of understanding of the physics of the Earth's crust, reliable forecasting and warning schemes are unavailable for earthquakes. Sometimes the view is expressed that individuals should be free to assume whatever environmental risks they wish, as long as they accept the consequences of their decision. But lack of information and capital, rather than a properly calculated choice, forces so many people to locate in hazardous areas that few democratic governments can ignore their plight following disaster.

Therefore, in the vast majority of cases, specific actions are taken to protect people against hazards and reduce disaster impacts. These actions start with an identification of the known threats and an assessment of the degree of risk to life and property. Many tools are potentially available for the task, but the strategic choices fall into three main groups:

• *Protection.* These responses seek to reduce hazard impacts by adjusting damaging events to people. They apply scientific and civil engineering measures to exert some control over hazardous processes, often by creating special structures or by strengthening existing infrastructure to resist the physical stresses of the hazard. The scale of intervention varies from *macro-protection* (large-scale defensive works designed to protect whole communities) to *micro-protection* (strengthening of individual buildings).

• *Mitigation.* These responses aim to modify the loss burden, especially for the most vulnerable people, in the immediate aftermath of disaster. *Emergency aid* is delivered through government agencies and charitable bodies, using a mix of humanitarian and economic principles. For some disaster victims, financial compensation will be available through *insurance schemes* administered by either private companies or governments. All these measures attempt to spread the financial burden of disaster beyond those suffering loss. To that extent, they are *loss-sharing* – rather than *loss-reducing* – devices, but the potential exists for encouraging more loss-reducing responses in the future.

• *Adaptation.* These responses seek to reduce disaster vulnerability by adjusting people to damaging events. The key aim is to promote

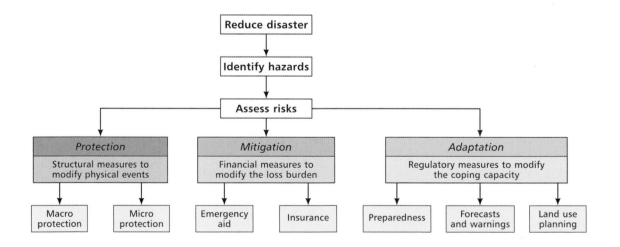

Figure 5.3 The three categories of disaster reduction strategy, with examples of the key measures adopted for protection, mitigation and adaptation.

changes in human behaviour with respect to hazards. Associated protective methods, or even partial relocation, may be necessary, but these measures are rooted in applied social science. The emphasis is on local coping capacity and improving resilience against future disaster strikes through a mix of *community preparedness programmes, forecasting and warning schemes* and improved *land use planning*.

These options are summarized in Figure 5.3. It is important to realize that they are not mutually exclusive choices. Each has individual strengths and weaknesses; no single measure can guarantee the elimination of risk. The physical nature of the setting may dictate an order of priority, but most measures work best in combination. For example, disaster aid is more efficient in association with insurance and hazard-resistant building construction; early warning systems and land use zoning go well together. Even when synergy is obtained through an array of measures and disaster impacts are reduced, a risk from rare events will still remain for many locations.

B PROTECTION: HAZARD RESISTANCE

Physical protection from the elements is a basic human need, but few man-made structures have been designed and built to withstand the most extreme forces of nature. Vernacular architecture using traditional building methods and local materials, such as mud bricks, wood and thatch, typifies many parts of the world where trial and error has resulted in some limited hazard resistance. For example, indigenous houses in Bali, lightly built with plant-matting walls and palm-frond roofs to ensure ventilation, have successfully withstood earthquakes. Wooden, loft-style stilt houses, called *palafitos* in Chile, are common along the banks of tropical rivers in South-East Asia, West Africa and South America and provide accommodation for fishermen above the

seasonal flood flows. Increasingly, the spread of manufactured products like concrete and modern construction methods has – with some exceptions – led to stronger buildings so that most disaster-related property damage is associated with older buildings.

Modern hazard resistance exists when structures are erected according to approved engineering methods and are maintained to satisfactory standards. This applies to large-scale public facilities as well as individual properties. For the vast existing stock of older buildings, the only option is retrospective improvement if present-day standards are to be met. All these measures are expensive. They rely on skills from civil engineering science and architecture, but they also have to comply with building codes and other regulations that depend on political initiatives, together with community acceptance. Where the greatest risks occur, entire communities may require protection.

1 Macro-protection

Purpose-built structures are widely used to protect people and property against surface or near-surface flows of potentially damaging semi-fluid materials such as rock falls, lava flows, lahars, mudslides and avalanches, as well as flood-waters from rivers or coastal sources (storm surge and tsunami). These defensive structures tend to act in two ways: either by holding back excess material (storing flood-water in reservoirs) or by diverting flows away from vulnerable sites (avalanche control walls). They exist at point locations (dams) or take a linear form (embankments and artificial channels). Protection against storm-force winds is undertaken on an individual property scale (see pages 103–105).

Most structures are built for flood control. Long embankment (levee) systems run alongside some of the world's great rivers to protect low-lying land from inundation. For example, embankments exist for over 1,500 km along the Red River in North Vietnam and for about

Plate 5.1 Part of the 17th Street Canal levee in New Orleans in August 2008, recently reconstructed after damage by flood-waters from hurricane 'Katrina' three years earlier. More than half of the city is below sea level and relies heavily on flood walls for protection. (Photo: Jacinta Quesada, FEMA 37676)

4,500 km through the Mississippi valley. Huge dams store flood-waters upstream. By the end of the twentieth century, there were over 45,000 large dams in the world, mostly in the USA, China and India. Many were constructed to supply irrigation water, but most can be regulated to provide some flood control. The Three Gorges dam on the river Yangtze in China is 175m high and almost 2km in length. It can hold 22 km³ of water and is expected to reduce major floods downstream from a frequency of 1:10 years to 1:100 years. To resist coastal flooding from the North Sea, the 1,400 km long coastline of the Netherlands has been transformed by a mix of soft defensive techniques (stabilized dunes, beach nourishment) and hard defensive techniques (sea walls, concrete embankments and tidal barriers). As indicated in

Figure 5.4, these devices protect almost one-third of the country that lies below sea level (Govarets and Lauwerts, 2009). An estimated 40 per cent of Japan's coastline is protected by concrete sea walls and breakwaters against storm surge and tsunami. Smaller structures have been erected against other similar types of hazard, and Box 5.1 shows how deflecting dams in Iceland have been effective in protecting property against snow avalanches.

Macro-scale engineering for flood protection has come under increasing scrutiny on grounds of financial, social and environmental acceptability.

The structural era 1930s–1950s

These early decades were almost completely dominated by 'hard' structures (reservoirs, levees, sea walls). Schemes were assessed on

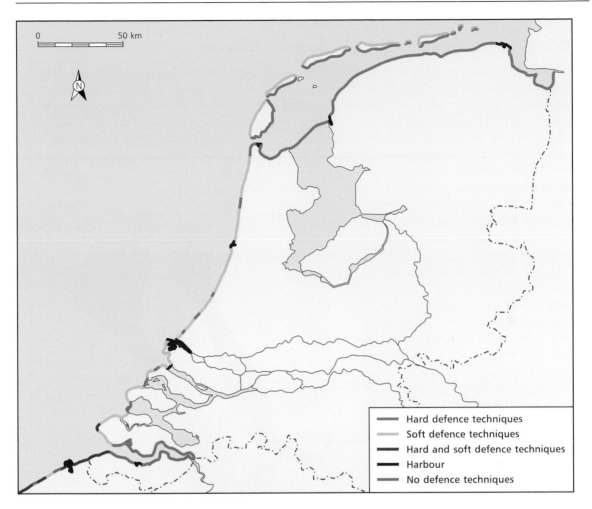

Figure 5.4 Flood defence along the mostly sandy beach coastline of Belgium and the Netherlands. Hard and soft techniques are often used in combination along the same stretch of coast. Only a small percentage of the coast remains unprotected. After A. Govaerts and B. Lauwerts, *Assessment of the Impact of Coastal Defence Structures*, OSPAR Commission, 2009, publication no. 435. Reproduced with permission.

civil-engineering criteria and cost-benefit grounds but little thought was given to community acceptance or environmental side-effects.

The floodplain management era 1960s–1980s

This period was characterized by a more mixed response, with the introduction of non-structural measures (flood warning, land use planning, insurance) designed to reduce human vulnerability to floods. Questions continued about the financial and ecological sustainability of large engineering projects.

The self-reliant mitigation era 1990s–?

More recently, local communities have been encouraged to live safely with floods in more sustainable ways. 'Softer' defensive works have been used to limit ecological damage and visual

Box 5.1 Avalanche-deflecting dams in Iceland

Avalanches threaten many communities in Iceland. On 26 October 1995 an avalanche containing about 430,000 m³ of snow struck the village of Flateyri in north-western Iceland and killed 20 people in an area previously thought to be safe (Jóhannesson, 2001). The avalanche was created by strong northerly winds blowing large quantities of snow from the plateau into the starting zones of the two avalanche paths of Skollahvilt and Innra-Bæjargil above Flateyri. Following this event two large deflecting dams, connected by a short catching dam, were built to divert future flows away from the settlement and into the sea (Figure 5.5). Each earth dam is about 600 m long and 15–20 m high and designed to intercept avalanche flows at angles of 20–25°. The purpose of the central catching dam, which is about 10 m high, is to retain snow and other debris that might spill over from two deflectors in a large event. The total holding capacity of the structure is around 700,000 m³.

The dams were completed in 1998. Since then they have successfully deflected two separate avalanches (February 1999 and February 2000) each with snow volumes over 100,000 m³, impact velocities of 30 m s⁻¹ and estimated return periods of 10–30 years. Estimated outlines of the avalanche run-out paths in the absence of the dams show that the Skollahvilt flow would have caused little loss, largely because houses destroyed in 1995 in this part of the village have not been rebuilt. But the 2000 avalanche from Innra-Bæjargil would have destroyed several houses. Although these two events are smaller than the design capacity of the deflecting dams, they provide a good example of the use of structures against moderately sized hazards.

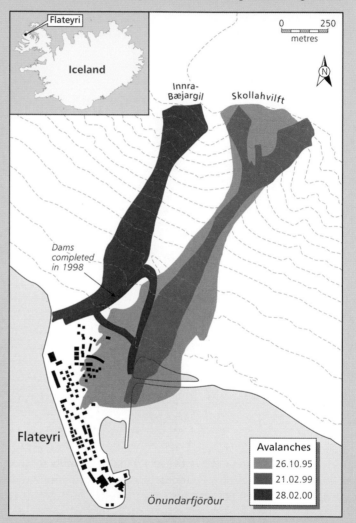

Figure 5.5 The effectiveness of deflecting dams in steering snow avalanches in 1999 and 2000 away from the township of Flateyri, north-west Iceland. The extent of the damaging 1995 avalanche that led to the construction of the dams is also shown. After Jóhannesson (2001). Reprinted from the *Annals of Glaciology* with permission of the International Glaciological Society.

intrusion. Stricter land use controls have been introduced and limited retreat from certain floodplains and shorelines has taken place.

2 Micro-protection

In most countries, the design and construction of key public infrastructure like dams, bridges and pipelines for water and gas supply is governed by legally enforceable regulations. The same is true for large industrial sites and government facilities. Such centralized regulation is rare for residential buildings, with the exception of house building codes for earthquakes and wind-storms. After a severe disaster even properly engineered structures may fail. There are many reasons for this. The formal standards apply to a specified event magnitude predicted to occur during the lifetime of the structure. Figure 5.6 shows the hypothetical example of a building designed to cope with a wind stress that occurs on average once in 100 years (1 per cent probability). Wind speeds just outside the design limits have little effect, but progressive stress beyond the performance envelope causes structural failure. Many buildings remain in use longer than their expected lifetime of, say, 50 years and therefore are exposed to higher risk. It follows that detailed structural inventories of the building stock should be maintained and routinely updated, but such information is rarely available because of a lack of qualified surveyors and the costs involved.

The construction methods used to reduce risk, whether for newly built or retrofitted structures, tend to be both hazard- and property-specific. More examples are provided in the relevant sections of Part Two but common features include:

- *Earthquakes.* Soft-storey collapse occurs when the ground floor is too weak to support the upper levels. This can be prevented by introducing beam columns or other types of support. Single or two-storey wood-frame domestic buildings need bolting to their concrete foundations so they cannot be shaken free or

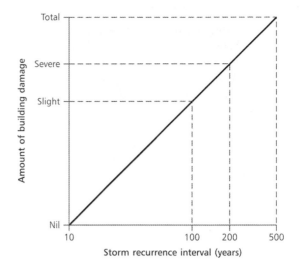

Figure 5.6 A theoretical illustration of the resistance of an engineered building to wind stress from storms of different return periods. It is important that the associated building codes are enforced if such performance standards are to be achieved.

displaced. Where soft soil exists beneath the structure, deep pilings can be driven down to underlying bedrock. Brick chimneys can be reinforced and braced onto structural elements to prevent collapse. Some *unreinforced masonry buildings* (URMs) are difficult to retrofit, although walls can be strengthened and tied to adequate footings while closets and heavy furniture can be strapped to the walls to protect contents.

- *Floods.* To protect against floods, walls can be made watertight, and flood-resistant doors and windows can be fitted. Another priority is to raise items at risk above the expected maximum water level. Some smaller wood-frame houses can be lifted above ground level to protect contents and basic fittings such as electrical supplies and heating boilers. Special attention is given to protect basements. Check valves ensure that the drains do not back up into the property and small walls may be erected around low-lying properties.
- *Hurricanes and tornadoes.* Here the priority is to maintain the structural integrity of the

property and prevent major damage when strong gusts create differential air pressures between the inside and the outside of a building. Measures include storm-resistant shutters to protect glass windows and reinforcement of doors. Care is needed to make roofing materials secure. This is achieved by bracing roof gables so that the roof is not torn off.

Most failures occur because of poor construction practices. A lack of financial resources and technical expertise, especially in the LDCs, often prevents safe design and building work. Political corruption is a recurrent problem and includes the abuse of building permits, illegal construction methods and non-observance of land use controls. In the 2001 Bhuj earthquake in India

over 450,000 traditional rural houses of rubble masonry construction were destroyed. Although anti-seismic codes of practice existed for modern buildings, non-enforcement of regulations and inadequate inspection methods also led to the failure of 179 high-rise reinforced concrete buildings in Amedabad, about 230 km away from the epicenter of the quake (Provention Consortium, 2007).

For many years, maximum protection was provided for public buildings and facilities expected to remain operative during emergencies (hospitals, police stations, pipelines). Schools, offices and factories, have also been strengthened in the belief that they will provide shelter for local residents. Hundreds of schools and hospitals have been retrofitted against earthquakes, but little

Plate 5.2 Office buildings in Concepcion, Chile designed and purpose built to withstand earthquake and tsunami stresses. These structures are elevated 10 metres above street level in order to allow a tsunami wave to pass below without causing damage. (Photo: Walter D. Mooney, USGS)

attention was paid to private homes. The defects of this policy were exposed when tropical cyclone 'Tracy' struck Darwin, northern Australia in 1974. This storm destroyed 5,000 houses or about 60 per cent of all homes in the city. Faced with the loss of electricity and other basic services, three-quarters of the population was evacuated to cities further south (Stark and Walker, 1979). This disaster demonstrated that residential housing is of equal importance to public buildings in most communities. When cyclone warnings are issued, public buildings close down and people seek shelter in their own homes.

There are examples of good practice and of better construction methods. In 1977, following a cyclone in the coastal area of Andhra Pradesh, India, 1,500 new houses were built using concrete block walls and reinforced slab roofs (Provention Consortium, 2007). Of these properties, 1,474 survived an even stronger cyclone in 1990. Similarly, 450 housing units were constructed in the Philippines after the passage of typhoon 'Sisang' in 1987, using concrete footings and steel post-straps bolted onto the core frame. These houses also withstood later typhoons with little damage.

3 Retrofitting

To provide a safer environment, hazard-resistance has to extend to the existing building stock. Retrofitting is the process of reinforcing all types of structures and their contents against loss. The process is mainly used against earthquakes, storms and floods. Most existing structures can be strengthened, but home-owners are reluctant to invest in protective measures, due to under-estimation of the risks and a preoccupation with short-term returns. These attitudes have consequences for the equitable distribution of disaster aid across nations and the provision of property insurance.

There is a general belief that retrofit measures are poor value for money and unaffordable. According to Hochrainer-Stigler et al., (2011), fewer than 10 per cent of earthquake and flood-prone houses in the USA have been subject to cost-effective improvements for disaster reduction, despite the financial incentive of reduced insurance premiums. Many owners are unwilling to pay because they think that any losses will be compensated by insurers or government aid. Some local authorities resist the introduction of building codes on the grounds that the costs of compliance and inspection will hamper inward investment and economic development. But, as commercial insurance against natural disasters becomes harder to obtain and as tax-payers increasingly rebel against paying for property owners who take no responsibility for hazard-reduction, better design, code enforcement and retrofitting will become more important.

All too often there is a lack of information about the most appropriate retrofit measures available. In Istanbul, Turkey, where earthquake risk is high, over half of all residents questioned had no knowledge of seismic strengthening (Eraybar et al., 2010). About one-third lacked plans for a safer home other than by purchasing a new, earthquake-resistant house. Without more information and some financial stimulus from government, property owners are unlikely to take action. Some local authorities do have schemes for the identification and strengthening (or demolition) of existing hazardous buildings. Remedial work is most necessary on low-value public housing; special regulations usually apply if unsafe buildings of historic significance are to be preserved.

Retrofitting has a role to play in the LDCs where very little disaster relief and development aid is spent on reducing physical exposure to loss. Case studies by Hochrainer-Stigler et al. (2011) of the costs and benefits of retrofitting residential buildings in several developing countries in Asia and the Caribbean region found that high returns on investment could be obtained by:

- strengthening doors and windows of middle-income homes to protect against hurricanes in St Lucia
- elevating high-income homes in Jakarta

- strengthening apartment buildings against earthquake risk in Istanbul
- building, and even replacing, brick homes on a firm plinth in Uttar Pradesh.

Costs are reduced if hazard resistance is factored in at the outset. Simple building changes could improve the cyclone resistance of *kutcha* (temporary non-masonry) houses in Bangladesh at an additional cost of only 5 per cent, whilst the introduction of comprehensive anti-seismic construction principles at the planning stage, including optimum layout and better design of structural connections, would increase building costs by less than 15 per cent.

To summarize, hazard-resistant measures for both existing and new homes do reduce disaster losses. It has been estimated that a retrofit policy in Los Angeles, California would produce a five-fold reduction in potential casualties from earthquakes. Valery (1995) estimated that losses in the 1994 Northridge earthquake in California would have been halved if all the damaged buildings had been built to the appropriate code. But few owners have taken action, partly because the area has not – to date – experienced large losses of life, and also because the number of high-rise buildings in multi-occupancy means that all property owners need to agree on any measures proposed. Kunreuther (2008) indicated that, if all residential properties in Florida were protected, the damage from a 1:100-year hurricane would be reduced by over 60 per cent. It seems clear that action should be taken to release the potential benefits of retrofitting.

C MITIGATION: DISASTER AID

Disaster aid stems from humanitarian concern about the losses in the stricken area and the need for victim support. The main purpose is to prevent further mortality and provide health care, subsistence and longer-term security for the survivors. For a severely affected, densely populated urban area, full recovery from disaster may take 10 years or more and involve overlapping stages that are often more chaotic than the chronological sequence presented below:

- *Emergency period*: 24 hours–3 weeks
  - Search-and-rescue operations (golden hours)
  - Emergency shelter
  - Community care (medicine, food, water)
  - Evacuation of survivors from unsafe areas

- *Relief period*: 2 weeks–6 months
  - Debris removal and refuse disposal
  - Clean-up of any toxic materials
  - Restoration of water and energy supplies
  - Demolition of unsafe buildings
  - Return of evacuees to homes declared safe
  - Distribution of emergency funds
  - Assessment of overall damage

- *Recovery period*: 5 weeks–10 years or more
  - Return to socio-economic functions
  - Permanent repairs to infrastructure
  - Preservation of heritage sites
  - Evaluation and update of local plans
  - Long-term planning for sustainability
  - Implementation of hazard-resistant measures

In order to prevent additional loss of life, clinical support and medical supplies must be delivered to disaster victims within the 'golden hours' immediately after the event. This period can be extremely short. Almost 90 per cent of earthquake victims brought out alive from collapsed buildings are rescued in the first 24 hours. International donations of medical supplies may arrive too late. A classic example was the Guatemala City earthquake of 1976, where the peak delivery of medical supplies came two weeks after the disaster, when most casualties had been treated and hospital attendance had fallen back to normal levels (Figure 5.7).

After initial search and rescue, the priority is for shelter, medical care and welfare. Some disasters, such as floods, create epidemics of diarrhoeal,

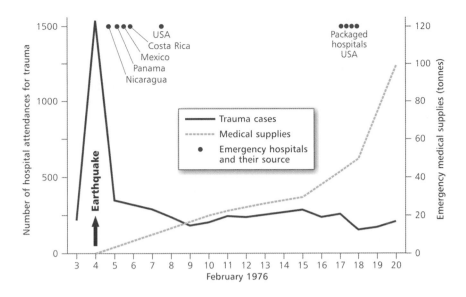

Figure 5.7 The daily number of disaster victims attending hospitals in Guatemala City in relation to the arrival of medical supplies and emergency hospitals after the 1976 earthquake. After J. Seaman *et al.*, Epidemiology of natural disasters. In M.A. Klingberg (ed.) *Contributions to Epidemiology and Biostatistics*, vol. 5 (1984), 145–56. Copyright 1984 S. Karger, Basel. Reproduced with permission.

Plate 5.3 Rescue workers recover a body from rubble created by the $M_W = 8.0$ magnitude earthquake at Hangwan, Sichuan Province, China in 2008. About 5 million people were made homeless in this disaster. (Photo: Panos/Qilai Shen QSH01973CHN)

respiratory and infectious diseases, whilst earthquakes are associated with bone fractures and psychological trauma. The use of local medical teams is preferred because they can be mobilized quickly and are culturally integrated with the survivors. This can be achieved only by preparedness. The rebuilding of lives and livelihoods includes everything from psychological counselling to practical help using family and other social networks. It is important to raise community morale, especially for those transferred to refugee camps, and to ensure that survivors are empowered with roles in decision making for the future. Special attention should be given to the most vulnerable groups, such as women, children and the elderly.

The relief phase begins a dynamic process of recovery with no clear end point. Emergency aid donations enable some people to return home, provide limited water and energy supplies, a restoration of links to the outside world and the restart of urban business functions. This phase typically merges into a much longer period of rehabilitation. The basic aim is to replace lost housing stock and infrastructure so that former levels of economic activity can be restored, but, as shown by Chang *et al.* (2011), shortages of manpower, building materials and other factors can impede recovery. Progress can occur more quickly if a pre-disaster response plan exists and adequate external finance is available. However, time is required to involve residents and other stakeholders in the many decisions to be taken, especially if the opportunity is grasped to plan for greater equity and resilience in the future.

Over a protracted recovery period, emergency aid and development assistance blend together and it becomes difficult to measure the real effectiveness of disaster appeals and responses. In addition, disasters in the LDCs often form part of more complex failings that require long-term solutions in health care, education and social welfare. Many aid donors recognize such needs. Global bodies like the World Bank are keen to make strategic investments that foster democratic institutions, building local capacity and sustainable rural development projects. Charitable bodies also increasingly stress the need for on-going disaster prevention rather than short-term aid. Alternatively, some governments wish to harmonize emergency aid with on-going trade and investment decisions.

Most humanitarian aid is raised by emergency appeals following disaster. According to Stoddard (2003), aid charities typically receive only one-quarter of their income from government funds and are highly dependent on public donations. As illustrated in Figure 5.8, various forms of support travel through a complex network of sources and intermediaries before they reach the recipients. *Governments* at several levels provide funding through the UN, the International Red Cross and Red Crescent Movement (IFRCRCS), NGOs, public–private partnerships and the government of the affected country. The *United Nations* manages some funding, provides direct support to aid recipients and also channels money to other delivery agencies (including NGOs) who are responsible for project management. *NGOs* provide direct support to aid recipients either through their international arms or via local offices. They also act as advocates for increased humanitarian action and are involved in policy making. The role of NGOs such as the Red Cross and Red Crescent Societies, Oxfam and Médecins Sans Frontières is vital. Individual charities have grouped together to optimize disaster appeals. For example, the Disasters Emergency Committee (DEC), formed in 1963 as an umbrella organization, coordinates the efforts of 14 UK aid agencies. *Private donors* may provide funds to all these bodies or pledge support directly to affected communities and individuals. *Military assistance* is a specific form of government aid, required when logistical transport or a peacekeeping role is required in the post-disaster period.

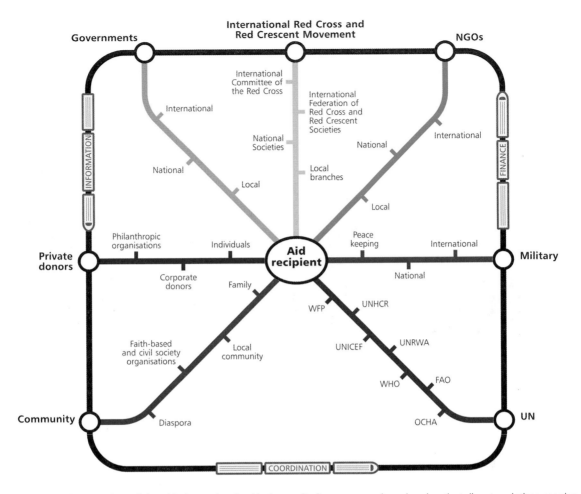

Figure 5.8 An overview of the aid players involved in humanitarian emergencies, showing that disaster victims receive assistance from many sources, often relayed through intermediaries, and that aid recipients have little opportunity to provide donors with direct information about their needs. The diagram does not intend to suggest linear connections or funding relationships. After Walmsley (2011) at http://www.globalhumanitarianassistance.org (accessed 27 March 2011). Reproduced with permission from Global Humanitarian Assistance.

1 Internal government aid

In the MDCs, disaster mitigation is normally achieved by spreading the financial load throughout the tax-paying population. Belgium and the Netherlands have designated national disaster funds with prior arrangements for aid distribution. Not all the financial assistance is given in direct grants; a substantial proportion is likely to be allocated as interest-free repayable loans. Most schemes incorporate a formula whereby the national disaster fund contributes at some agreed ratio to local spending once the disaster impact has exceeded a minimum threshold figure of loss.

In the USA, the president can issue formal disaster declarations (PDD) following a request from the relevant state governor. Such requests should be accompanied by damage assessments, but this procedure can be short-circuited in the interests of political expediency, especially when

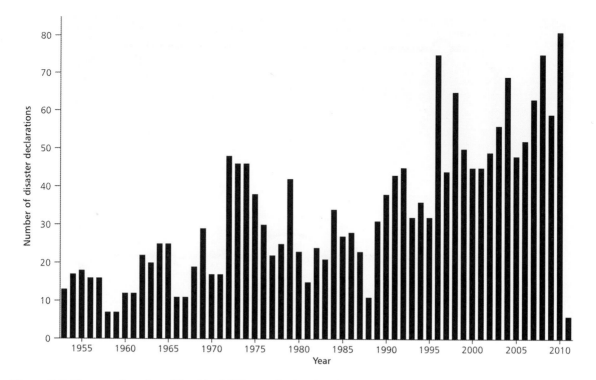

Figure 5.9 The annual number of Presidential Disaster Declarations in the USA 1953–2011. There has been a progressive rise in Declarations over the past half century. Data from FEMA at www.fema.gov/news/disaster-totals-annual.fema (accessed 2 March 2011).

media pressure exists (Sylves, 1996). A PDD routinely releases enough federal aid to cover up to 75 per cent of the costs of repairing or replacing damaged public and non-profit facilities, although this proportion has reached 100 per cent in some cases. As shown in Figure 5.9, the number of PDDs has risen steadily. The procedure has been criticized as unduly political in nature and for failing to meet real needs. For example, Schmidt-lein *et al.* (2008) demonstrated a weak statistical relationship between the annual number of PDDs and major hazard events across the country and claimed significant regional inequalities in the distribution of aid. It was suggested that more emphasis should be paid to the relative impact of events and that only severe events should qualify for support. To a degree, the increased PDDs reflect a sharp rise in economic losses in the 1990s, due to major disasters such as hurricane

'Andrew', 1992, Midwest floods, 1993 and the Northridge earthquake, 1994. Between 1980 and 2005, there were 67 weather-related disasters alone, each costing over US$1 billion. Hurricane 'Katrina' in 2005 is currently the most expensive natural disaster in US history.

This upward trend has fuelled a debate on the extent to which federal funds should provide disaster assistance. According to Barnett (1999), the system is:

- *expensive* because of the rapid rise in payouts in recent decades
- *inefficient* because it allows local governments to avoid a fair share of the costs, e.g. through a failure to enforce building codes or to insure public property
- *inconsistent* because equivalent losses are not always treated in the same way, e.g. localized

damage may not attain disaster-area status, thereby depriving victims of assistance
- *inequitable* because it allows a misallocation of national resources, e.g. when wealthy disaster victims are compensated by the general tax-payer.

These concerns, echoed in other developed countries, have led to tighter controls on central funds, together with a search for alternative strategies. Many governments now resist compensating privately insured losses. Further reforms include refusing disaster payouts to households over higher-income thresholds and tying aid to the enforcement of local building codes. All these efforts are designed to limit disaster costs and reserve aid for the most deserving people within a stricken community.

2 International aid

The LDCs rely heavily on external support after disaster. Before the creation of the International League of Red Cross and Red Crescent Societies (now the International Federation) in 1922, the transfer of aid was largely *bilateral* (directly government to government or indirectly through NGOs). As charitable bodies became more interested in overseas work, more agencies were set up – the UN Children's Fund (UNICEF) in 1946 and the FAO World Food Programme (WFP) in 1963. In 1972, the UN established the Disaster Relief Organization (UNDRO), based in Geneva. This initiative was reinvented several times because of under-funding, internal rivalry with other UN agencies and criticisms from some member countries. In 1992 a new Department of Humanitarian Affairs (DHA) replaced UNDRO; in 1997 the DHA was itself replaced by the Office for the Coordination of Humanitarian Affairs (OCHA). Along with other international bodies, such as the EU and the World Bank, these agencies receive *multilateral* donations which are not ear-marked.

OCHA has general coordinating and policy-development responsibilities, including a specific mission to mobilize humanitarian response to disasters. It works in partnership with national and international bodies in order to alleviate human suffering in major emergencies; to act as an advocate for the rights of people in need; to promote preparedness and prevention; and to facilitate sustainable solutions. Voluntary contributions from member states make up over 90 per cent of OCHA's budget, with the remainder coming from central UN funds. Key operating principles are that:

- Responsibility for people affected by emergency lies – first and foremost – with their respective states.
- States in need of aid are expected to facilitate the work of responding organizations.
- Humanitarian assistance must be linked to the principles of humanity, neutrality, independence and impartiality.

Other bodies support OCHA. For example, the USA maintains an Office of Foreign Disaster Assistance (OFDA), and in 1992 the EU took the first steps towards coordinating its member states through the European Community Humanitarian Office (ECHO). The EU is one of the world's biggest donors of humanitarian aid to disasters outside Europe. Like OCHA, it does not intervene directly on the ground but allocates funds through about 200 partners, including NGOs, UN agencies (mainly OCHA) and international organizations like the International Federation of the Red Cross and Red Crescent Societies.

Much of the financial support from the MDCs to developing countries comes from the Official Development Assistance (ODA) budget of countries within the Organisation for Economic Co-operation and Development (OECD). The UN-recommended figure for annual government spending on all overseas aid is 0.7 per cent of donor GDP, but the sums delivered are only about half this target. Disaster relief accounts for roughly 10 per cent of the total ODA budget. The amount varies from year to year but, as shown

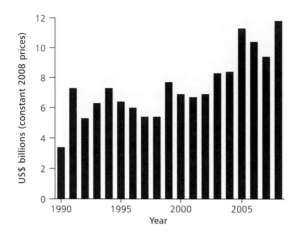

Figure 5.10 Annual total of humanitarian aid 1990–2008 in US$ billion from Development Initiatives based on OECD Development Assistance Committee (DAC) data. After Walmsley (2010) at http://www.globalhumanitarianassistance.org (accessed 27 March 2011). Reproduced with permission from Global Humanitarian Assistance.

in Figure 5.10, it has grown in real terms since 1990. Fewer than 10 countries donate over 90 per cent. The largest contributions are made by the United States, EU institutions, the UK and other European nations. The top donors per citizen are almost entirely in Europe. Over the years, a growing proportion of this aid has been donated bilaterally (Macrae *et al.*, 2002). The rise in donations, and in bilateralism, is attributable to some high-profile armed conflicts, plus a desire of donor governments to ear-mark and monitor their contributions more closely.

Much humanitarian assistance goes to war zones, but individual natural disasters can attract donations of US$1 million or more. The response to individual disasters varies greatly. The Indian Ocean tsunami of December 2004 created an unprecedented response, especially from voluntary bodies and private donors, with a total sum estimated at US$13.5 billion (Telford and Cosgrave, 2007). This event attracted great media interest, partly because the affected areas were familiar to donors through holiday experiences in that part of Asia. With perhaps 2 million people

adversely affected by the event, emergency relief was not limited by finance and a lot of the money was reserved for reconstruction projects extending over several years. Comparing disasters, and evaluating the efficiency of disaster aid, is a difficult task, especially when substantial funding occurs outside the appeal system. There is no central monitoring arrangement that covers all donations and tracks money through the system. Greater transparency is required on the delivery of pledges, but the most significant omission is the lack of feedback data on what has actually been received, and when, by the recipients on the ground.

Figure 5.11 shows the donor response during the first 17 weeks for four disasters that occurred between 2004 and 2010. The large increase for the Asia tsunami on the 75th day was due to the coincidental reporting of US$1.3 billion donations, mainly from private sources (Kellet, 2010). It is impossible to know how well these funds met local requirements or, more generally, how much money per person is necessary for equitable disaster relief. Even when funds are measured against the number of people 'affected' by disaster, that term tells little about the degree of assistance required. The differences can be striking. The US$6.2 billion of aid donated for the 2004 Asian tsunami amounted to US$2,670 per person 'affected'. This compares with US$110 and US$878 for each person affected by the Pakistan floods and the Haiti earthquake, respectively.

Disaster aid is well intentioned but no one really knows whether it is sufficient or reflects real needs. According to Olsen *et al.* (2003), the scale of donations depends on three factors – the intensity of media coverage, the degree of political interest and the strength of the international relief agencies in the country concerned. Sudden-onset disasters, like earthquakes and tropical cyclones, attract more funds than slow-onset disasters, like droughts and famine, irrespective of the number of survivors needing assistance. Journalists neglect the 'hidden' crises that arise from poverty and disease and concentrate on events with large body

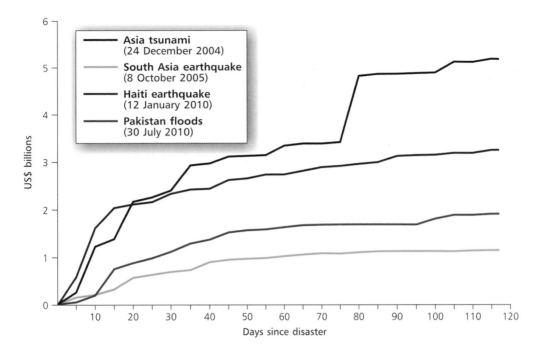

Figure 5.11 Cumulative donor response to appeals for aid in the periods following four major disasters. Generally the rate of funding declines with time. Data from OCHA FTS. Values in US$ billion. Global Humanitarian Assistance at http://www.globalhumanitarianassistance.org (accessed 27 March 2011). Reproduced with permission from Global Humanitarian Assistance.

counts and good photographic opportunities (Ross, 2004; IFRCRCS, 2006). Disaster aid is highly political and dependent on the priorities of the aid agencies. Drury *et al.* (2005) showed that the most important long-term influence on the allocation of US aid for overseas disasters was foreign policy. In European countries, disaster aid is raised most readily for former colonies.

Food aid began as a means of off-loading surplus agricultural production in North America and Europe. Many food donations, unacceptable for local religious or dietary reasons, have been sent to Third World countries along, with death-dated and clinically unusable drugs (Autier *et al.*, 1990). Over-generous donations of food aid can lower market prices and disrupt the local economy in some LDCs; in the longer term, food aid may deflect the receiving government from developing the local agricultural economy. Logistical difficulties, such as poor roads and lack of suitable transport, hinder the distribution of food and medical supplies to remote areas and delays occur through government bureaucracy and corruption. A recurrent dilemma for aid workers is whether to distribute more supplies to fewer victims or fewer supplies to more victims.

How can disaster relief become more efficient? According to Maxwell (2007), better local information and analysis would enable food aid to be delivered more efficiently to those with the greatest need. Quick and careful identification of at-risk groups is essential. This could be achieved through reliable early warning systems, especially for disasters involving food shortage. More training for staff and volunteers, plus the retention of experienced aid operatives, would also help, especially for responses in the 'golden hours'. Above all, international donors and aid agencies need to move from stereotypes of 'victims' and 'failed states' and be prepared to give more ownership of disaster relief to regional agencies and local communities. This is partly an issue of respect, but it can also be highly practical when international bodies lack expertise surrounding local needs.

There are alternatives to the traditional, resource-driven distribution of aid in the form of commodities like food, blankets and shelter materials. So long as safeguards are in place to restrict corruption, there is growing support for *cash-based aid* that – wherever possible – allows people to buy the goods they need for themselves. According to Mattinen and Ogden (2006), cash-based interventions provide more dignity and flexibility for disaster victims and also liberate aid from donor-driven priorities that may distort distant rural economies. This policy shift is reflected in the growing use of direct cash transfers by the World Bank to disaster-affected households as a complement to other relief and reconstruction efforts (Heltberg, 2007). *Cash-for-work schemes* that employ otherwise idle manpower on reconstruction projects – as in post-tsunami Indonesia – also represent a way forward (Doocy *et al.*, 2006).

D MITIGATION: INSURANCE

Disaster insurance exists when an asset is perceived to be at risk and the owner pays a fee (the premium), usually on an annual basis, to buy a contract (the insurance policy) that transfers the risk to a financial partner (the insurer). The insurer – a private company or the government – guarantees to meet specified costs in the event of loss or damage to the asset. By this means, the policy-holder is able to spread the cost of a potentially unaffordable disaster over several years. A commercial insurer assumes that either no loss will occur during the term of the policy or that, over time, the sum paid will total less than the premiums received. Commercial insurers pay claims out of investments and profits; government insurers pay claims out of tax revenues.

1 Commercial disaster insurance

This is important in the industrialized nations and about 80 per cent of all premiums for private property insurance are paid in America and Europe. Private companies cover (underwrite) property such as buildings against flood, storm or other specified environmental perils. Policy underwriters try to ensure that the property they insure is varied and spread over diverse geographical areas so that only a fraction of the total liability is at risk from a single hazardous event. By this means, the cost of payouts to claimants is distributed across all policy-holders. If premiums are set appropriately, they will cover costs. The insurance company makes profits largely by investing the money received from premiums.

Environmental hazards create special problems. Insurance claims after earthquakes or tropical cyclones cluster within short time scales and relatively small areas. The typical pattern of large claims following years with few losses also makes premium setting difficult. For example, in 1994 the insurance industry in California collected about US$500 million in earthquake premiums, but it paid out over US$15 billion over a period of more than four years for damage caused by the Northridge disaster (Figure 5.12). Unless a company accumulates a large catastrophe fund, it may not survive such demands and will go out of business (Born and Viscusi, 2006). Another problem is *adverse selection*, which occurs when the policy-holder base is narrow and dominated by bad risks. For example, only floodplain or coastal dwellers are likely to take out flood insurance, and this leads to a geographical clustering of risk. After hurricane 'Andrew' in 1992, nine insurance companies became insolvent and others attempted to quit the market in Florida.

The insurance industry can increase profitability in various ways:

- *Raising the premiums.* This is the most obvious method but is unpopular with the public. It may have other benefits if premiums become weighted to reflect the greater claims likely from the occupancy of high-risk areas.

- *Re-rating the premiums.* The detailed setting of premiums in line with local levels of risk is possible by harnessing Geographical Information Systems (GIS) to postcode districts. Insurers can then place individual policy-holders in different bands of hazard exposure and charge premiums appropriate to the risk.
- *Restricting the cover.* The size of claims can be restricted by applying a policy deductible (excess) or by capping policies at a maximum payout. In Japan the risk of huge losses from earthquakes in urban areas has led to limits on any one claim. Above an agreed threshold, the government shares costs with the policy-holder. As a last resort, a company can refuse to sell insurance cover in high-risk areas, although this is unpopular with the public and with governments.
- *Widening the policy-holder base.* This is done by spreading liability through a basket of cover rather than limiting cover to a single hazard. In the UK, the industry offers policies that include storm, flood and frost damage, as well as fire and theft. By this means, the uptake of household insurance – which is a requirement of all mortgage lenders – is high. Any losses arising from floods, for example, are subsidized by all policy-holders, including those with no flood risk.
- *Reinsurance.* Companies join together, or with government, to pass on part of the risk. For example, the primary company might agree to pay the first US$5 million of claims; for losses in excess of this sum the company would be reimbursed – perhaps up to 90–95 per cent – by its partners. The reinsurance market is international, and very high risks are spread through the world markets, although the rising cost of claims, and fears about factors such as climate change, make it difficult to obtain all the reinsurance that is required.
- *Reducing the vulnerability.* Insurers can offer lower premiums to policy-holders who reduce

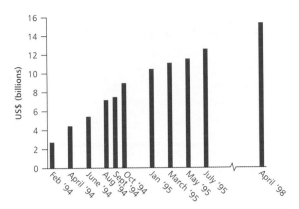

Figure 5.12 The accumulation of insured losses (US$ billion) after the Northridge earthquake on 17 January 1994. Six months after the event less than half the final total was known. Upward adjustments continued until April 1998, over four years after the disaster. After Munich Re (2001).

their risks by retrofitting against hazards. Cover for new properties can depend on the use of approved construction techniques such as anchoring the structure to the foundations to prevent slippage or using wind-resistant roofing and walling materials. But these measures are not widely deployed without government support through legislation and enforcement.

Some advantages and disadvantages of commercial insurance are shown in Box 5.2. Disaster claims are rising. Before hurricane 'Alicia' in 1988, no single disaster had cost the insurance industry more than US$1 billion (Clark, 1997). Times have changed. According to Munich Re (2006), the costly year of 2005 brought six major disasters that contributed overall losses of US$170 billion (out of a global total of US$212 billion) and insured losses of US$82 billion (out of a global total of US$94 billion). As in most years, windstorms caused most insured losses. In 2005, hurricane 'Katrina' – the sixth-strongest hurricane recorded since records began in 1851 – alone created total economic losses estimated at US$125 billion, with US$45 billion covered in the private

Box 5.2 Advantages and disadvantages of commercial insurance

Advantages

It guarantees the disaster victim compensation after loss. This is more reliable than disaster relief and appeals to those opposed to government regulation because it depends on individual choice and the private market.

It provides an equitable distribution of costs and benefits, provided that property owners pay a premium that fully reflects the risk and insurance payments fully compensate the insured loss.

Insurance can be used to reduce vulnerability. Provided that residents in hazardous areas pay the full-cost premium, there should be a financial disincentive to locate in such areas. The difficulty is that most residential development is by speculative builders and, until insurance premiums become high enough to make new hazard-prone properties impossible to sell, it is unlikely that developers will be deterred.

Existing home-owners can be encouraged to reduce their vulnerability, and enjoy lower insurance premiums, by strengthening their property and lowering the risk of loss.

Disadvantages

Private insurance may be unobtainable in very high-risk areas. In the USA the insurance industry has been reluctant to offer flood cover without government support. Even when available, landslide insurance normally covers the cost of structural repairs to property only and not that of permanent slope stabilization, because of the potential high costs.

There is frequently a low voluntary uptake of hazard insurance. Only 10 per cent of the buildings damaged in the 1993 Midwest flood were covered by flood insurance. Fewer than 20 per cent of the US$500 billion losses sustained in the USA between 1975 and 1994 were insured. Such under-insurance may benefit the industry when a major disaster strikes, and Japanese insurers survived the Kobe earthquake largely because only 3 per cent of affected home-owners had earthquake cover.

Even when insurance policies are taken out, a significant proportion of policy-holders will be under-insured for the full value of property at risk and are, therefore, unlikely to be fully reimbursed in the event of a claim.

Unless premiums are scaled directly to the risk, hazard-zone occupants do not bear the cost of their location. UK insurance companies have traditionally charged a flat-rate premium of buildings cover for all houses. This amounts to a subsidy from the low-risk to the high-risk property owners. Even if some link is attempted between premium and risk, the most hazardous locations will likely benefit from cross-subsidization through the company charging higher premiums than necessary in less hazardous areas.

Although insurance can be employed to reduce losses, the existence of moral hazard can often limit effectiveness. Moral hazard arises when insured persons reduce their level of care and thus change the risk probabilities on which the premiums were based. For example, some people may not move furniture away from rising flood-water if they know they will be compensated for any loss.

Table 5.1 The world's 10 costliest natural disasters (values in US$ millions)

Rank	Year	Event	Region	Economic loss	Insured loss
1	2005	Hurricane 'Katrina'	USA	125,000	45,000
2	1995	Kobe earthquake	Japan	100,000	3,000
3	1994	Northridge Earthquake	USA	44,000	15,300
4	1992	Hurricane 'Andrew'	USA	30,000	17,000
5	1998	Floods	China	30,000	1,000
6	2005	Hurricane 'Wilma'	USA	18,000	10,500
7	2005	Hurricane 'Rita'	USA	16,000	11,000
8	1993	Floods	USA	16,000	1,000
9	1999	Winter storm 'Lothar'	Europe	11,500	5,900
10	1991	Typhoon 'Mireille'	Japan	10,000	5,400

Source: Adapted from Munich Re (2005)

Note: Values are original losses, not adjusted for price inflation. Insured losses are more accurate than total economic losses.

market, and became the most expensive natural disaster to date (Table 5.1). However, preliminary loss estimates for the Great Tōhoku earthquake and tsunami in Japan were placed at US$250–300 billion, with insured property costs of US$20–30 billion.

The ratio of overall losses to insured losses is largely due to regional differences in economic development and insurance penetration. For example, relatively low insured losses following the Kobe earthquake reflect the limited take-up of private insurance in Japan. The extent of insurance cover is much lower in the LDCs, where disasters can create losses well over 10 per cent of gross domestic product (GDP), as compared with perhaps 2–3 per cent in western-style nations. Attempts are being made, via subsidized pilot schemes, to extend insurance cover within the LDCs (Linnerooth-Bayer *et al.*, 2005). On the other hand, threats from climate change and globalization are reducing the capacity of the industry to absorb large losses and some observers see partnerships between commercial insurers and national governments as the best way forward (Mills, 2005).

The USA illustrates many key problems. Insured losses of US$100 billion arising from a hurricane or an earthquake could stress the available reinsurance capital and bankrupt some companies (Malmquist and Michaels, 2000). High risks in the US insurance market result from urbanization, the coastward shift of population plus the failure of local governments to adopt and enforce stringent building codes. The use of federal disaster funds to compensate people with uninsured property has also contributed to the losses. According to Kunreuther (2008), the population of Florida was estimated to rise almost sevenfold between 1950 and 2010, from 2.8 million to 19.3 million. There is a high degree of insurance cover in the state, with most homes covered for wind-storm damage and about one-third of properties having flood insurance. The insured liability is almost US$1.9 trillion and about 80 per cent of the insured assets are at risk near the coasts.

Hurricane 'Katrina' exposed underlying problems facing the USA and other countries (Kunreuther and Pauly, 2006). The failure of property owners to protect or insure their homes suggests that they lack the resources to restore their assets when disaster strikes. Consequently, they fall back on poorly targeted public assistance provided by tax-payers. This aid is unlikely to extend equitably to low-income residents who cannot afford such measures. All recipients have little incentive to take responsibility for their own affairs. More research is required into the decisions taken by householders regarding risk reduction, but it appears that change will come about only through private-public partnerships that couple effective building codes and land use regulations with insurance cover. Incentives, such as long-term retrofit loans and subsidies for low-income families, could then be offered to make insurance more financially attractive and reduce the disaster aid cost to the national government.

2 Government disaster insurance

The creation of a national disaster fund can address flaws associated with commercial insurance. If made compulsory, nationwide cover not only widens the policy-holder base as far as possible but can also be used to raise public awareness of hazards and the need to strengthen buildings to approved standards. In theory, this would enable premiums to be related more closely to the risk. For example, government could legislate that properties not built to such standards were ineligible for state insurance. The National Flood Insurance Act (1986) was an early attempt by the US government to adopt this approach and shift some federal costs to state governments and the private sector.

Some countries dictate insurance cover for natural disasters through partnerships between government and the insurance industry. Spain has had a scheme for natural and technological disasters since 1954. In France, property and motor insurance has included mandatory cover for natural disasters since 1982, financed by a surcharge on private premiums and state reinsurance. New Zealand introduced government cover for earthquakes through the Earthquake and War Damage Act (1944). The scheme, operated by the Earthquake Commission (EQC), was extended to cover storms, floods, volcanic eruptions and landslips. Claims were financed by a surcharge on all fire insurance policies of 5 cents per NZ$100 of insured value (Falck, 1991). The

EQC could rate premiums according to risk, and refuse claims on poorly maintained properties, but political pressure ensured that almost every claim was met.

The present shift is towards persuading individuals to accept more responsibility for their disaster losses. In Turkey, government payout for earthquake loss has been replaced by a mandatory insurance scheme, although this becomes operative only when a property is sold and the new owner becomes liable. One of the most radical changes in public–private risk sharing occurred in New Zealand when the state scheme was reformed in 1993 (Hay, 1996). From 1996 onwards, the EQC withdrew disaster cover for non-residential property and, although insurance for residential property remained automatic for owners with fire insurance, the extent of cover was limited. The EQC retains a fund of some NZ$2.5 billion as a first call on disaster claims, and has reinsurance arrangements, but the New Zealand government remains responsible for any shortfall in disaster payments.

E ADAPTATION: PREPAREDNESS

Preparedness can enable a prompt and effective response to disaster. In theory, it involves the planning – and testing – of hazard reduction actions on all time-scales, ranging from seconds (earthquake or tsunami warnings) to decades (better land planning or measures to combat climate change). Greater risk awareness helps hazard zone occupants to identify threats and take appropriate actions, although there will always be a gap between what people are advised to do, what they say they will do and what they actually do in a stressful emergency situation.

Preparedness arrangements differ widely. In the MDCs the task may be devolved to existing bodies, like the armed forces or the police, or to dedicated agencies. For example, the Federal Emergency Management Agency (FEMA) has responsibility for the protection of the civilian population in the USA. In Australia, Emergency

Management Australia (EMA) develops and coordinates policy but public safety is managed at the regional level through the State Emergency Service units (Abrahams, 2001). These agencies depend heavily on volunteer bodies, like the local Bushfire Brigades, plus the police, fire and ambulance services. In developing countries, NGOs have an important role to play, due to their knowledge of local community needs (Luna, 2001).

Various stakeholder groups are involved (Figure 5.13). Loss-reducing measures include the activation of temporary evacuation plans and the distribution of stockpiled medical aid, food supplies and emergency shelters. It is important to pre-designate an operational control centre because many services – roads, water supplies or telephones – are unlikely to be fully available. Most importantly, training in self-help techniques – first aid, search and rescue and fire-fighting – should be given to communities at risk. Most disaster victims are rescued in the first 'golden hours' by other survivors, rather than by aid workers. Following the 1999 earthquakes in Turkey, about 50,000 people were rescued from damaged buildings; local people saved 98 per cent of them (IFRCRCS, 2002).

Proactive planning for disaster reflects a wider trend towards local resilience. Key elements include early-warning systems and emergency evacuation from hazard zones. Such measures have to be well organized but they also need to be 'people centred' if they are to empower the local population to respond effectively. For example, Chakraborty *et al.* (2005) suggested that successful hurricane evacuation along urbanized coasts in the MDCs should blend geophysical risk with social vulnerability assessment to ensure that extra help – such as public transport – is provided for disadvantaged groups. A tsunami warning plan for Galle in Sri Lanka identified five vulnerable population groups (women, children, people with disabilities, fishermen and workers in densely populated areas) when drawing up priority routes for evacuation from this coastal city (ISDR, 2006).

These groups are included in the first-priority escape routes shown in Figure 5.14.

Preparedness planning evolves over time, as in California, where there is a long history of raising earthquake hazard awareness. Some 40 years ago residents were poorly prepared to face a damaging event, but, following newspaper publicity in the San Francisco Bay Area, residents responded positively (Table 5.2). Once practical advice was available, a clear majority of those surveyed stored emergency equipment, together with food and water supplies. The proportion of those taking other steps, such as strapping water heaters to walls or purchasing earthquake insurance, was

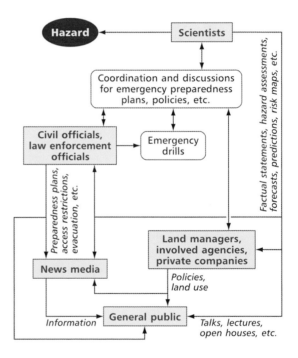

Figure 5.13 A typical set of stakeholder groups involved in hazard reduction planning. Each group plays a part in assembling and disseminating information aimed at preparedness in advance of a hazard strike. Adapted from Peterson (1996). With kind permission from Springer Science+Business Media: D.W. Peterson, Mitigation measures and preparedness plans for volcanic emergencies. In R. Scarpa and R.I. Tilling (eds) *Monitoring and Mitigation of Volcano Hazards* (1996). Copyright Springer 1996. Reproduced with permission.

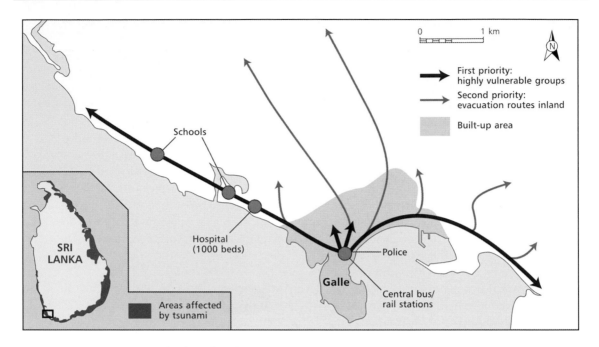

Figure 5.14 Evacuation map for Galle City, Sri Lanka, showing the first and second priority routes recommended for police enforcement during a tsunami emergency. This project was conducted within the UN Flash Appeal Indian Ocean Earthquake–Tsunami programme. After ISDR (2006). Reproduced with permission of UN HIC DSU Galle, Technical Committee – Early Warning DMC United Nations University for Environment and Human Security UNISDR and the Platform for the Promotion of Early Warning.

also encouraging, given that many residents were in apartments that precluded certain responses (Mileti and Darlington, 1995). The California Seismic Safety Commission (2009) published the actions that residents can take against hazards, listed in order of increasing cost and – therefore – decreasing likelihood of adoption:

- *Learning* – use available official sources to identify the risks and consider options.
- *Planning* – discuss cooperative emergency responses with neighbours, have a family plan of action, copy important documents.
- *Training* – practise first aid techniques, know how to shut off household water supplies and other utilities.
- *Organizing* – collect necessary equipment and supplies – first aid kit, flashlight and batteries, tinned food, bottled water, sandbags.

- *Securing contents* – raise key items above flood level, strap down heavy appliances, store hazardous materials safely.
- *Securing structure* – make property hazard resistant by raising floor level, make walls, windows and roof secure.

Community preparedness is a costly exercise. It ties up facilities and people that are apparently doing nothing, other than waiting for an event that nobody wants and many believe will not happen. In earthquake-prone urban areas, it is not unreasonable to plan for emergency sheltering of up to 25 per cent of the population. This requires usable buildings and the massive stockpiling of food, medical supplies and sanitation equipment. It is easy for the authorities to underestimate the physical devastation of an earthquake and the challenges for survivors. Figure 5.15

Table 5.2 The proportion of residents in the San Francisco Bay Area of California taking selected loss-reducing actions before and after newspaper publicity about increased earthquake risk

Preparedness action	Pre-publicity (%)	Post-publicity (%)	Increase
Stored emergency equipment	50	81	31
Stockpiled food and water	44	75	31
Strapped water heater	37	52	15
Rearranged breakable items	28	46	18
Bought earthquake insurance	27	40	13
Learned first aid	24	32	8
Installed flexible piping	24	30	6
Developed earthquake plan	18	28	10
Bolted house to foundation	19	24	5

Source: Adapted from Mileti and Darlington (1995)

Note: The postal sample in this survey consisted of 1,309 households and 806 usable questionnaires were obtained.

shows, for the area of Greater Memphis, Tennesse the estimated number of displaced households in relation to the lack of school premises available for shelter purposes following a projected $M_W = 6.5$ earthquake on the New Madrid fault (O'Rourke *et al.*, 2008). Experience suggests that advice needs to be distributed well in advance, both widely and often, to the public from a trusted government agency. Workshops, pamphlets, brochures, videos and other materials are important tools. In addition, public bodies and private sector companies can build hazard awareness into existing health and safety schemes, but any introduction at household level is difficult to monitor.

For developing countries, the lead agency for disaster preparedness is the UN Office for the Coordination of Humanitarian Affairs (OCHA). OCHA retains a stand-by team of voluntary emergency managers from over 60 countries for deployment anywhere in the world within 24–48 hours. It relies on specialist rescue and relief groups who supply equipment, transport and other services when disaster strikes. For example, the charity Oxfam has emergency stores with cooking equipment and material for constructing temporary shelters. Success depends on OCHA acting as a rapid link between aid donors and recipients so that an existing register of expertise can be matched as quickly as possible to the form of assistance required. Regional organizations exist. The Asian Disaster Preparedness Center, established in 1986, is an intergovernmental, non-profit body based in Bangkok. It works in cooperation with UNESCO to promote preparedness, with an emphasis on education and disaster awareness in schools and colleges.

One challenge in the developing nations is to ensure that schemes are compatible with prevailing social and cultural conditions. In the past, disaster planning has tended to follow military lines, with a stress on communications, logistics and security. These are important requirements, but the 'command and control' model, represented by a top-down approach, was not always appropriate for the LDCs. External aid has been perceived as an act of foreign policy on behalf of distant 'colonial' powers. In addition, military forces are unlikely to be completely sensitive when operating refugee camps or dealing with women and children. This approach has been overtaken by more inclusive and participatory responses. Military assistance may be vital – for example for air-lifting relief supplies into inaccessible areas – but this support is short lived,

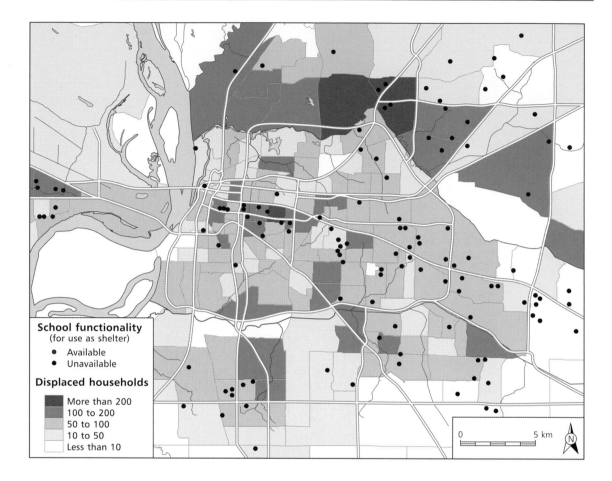

Figure 5.15 Map showing the expected location of displaced households and available disaster shelters in the greater Memphis, Tennessee area after a projected $M_w = 6.5$ earthquake on the New Madrid fault. After O'Rourke *et al.* (2008).

mainly because it has been diverted from on-going duties elsewhere.

The fostering of local preparedness is critical (UNESCO, 2007). For example, 22 states in India are classed as multi-hazardous because the 8,000 km long coastline is exposed to tropical cyclones, storms and floods. In the coastal areas of Tamil Nadhu, community-based preparedness has been pioneered with preliminary risk assessments by focus groups composed of village leaders, NGOs, active youth organizations, etc. Special care was taken to ensure that marginalized, under-privileged people were included. This venture is a considerable achievement, given the hierarchical caste structures and different socioeconomic backgrounds found in the villages. The practical outcome has been a Village Disaster Risk Management Training model (VDRMT). As shown in Figure 5.16, this framework encourages a comprehensive participatory disaster response owned by local people. It should embed sustainability for the future as local leaders are trained to incorporate disaster preparedness into villagelevel planning which, in turn, should be complemented by local government plans for development.

In Bangladesh, a Cyclone Preparedness Programme (CPP) began in 1973 in order to benefit

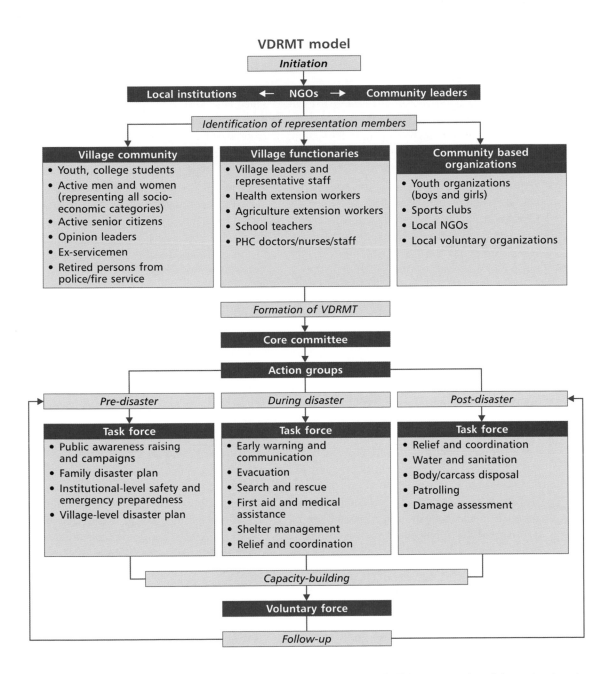

Figure 5.16 The Village Disaster Risk Management Training model (VDRMT). This conceptual model was developed as a capacity-building and disaster-reducing tool for rural communities in India, based on a participatory approach. After Anonymous (2007a).

the 11 million residents of the coastal region. The aim was to ensure that people are adequately alerted about an approaching storm and can move to safety in cyclone shelters or other buildings. The Bangladesh Meteorological Department transmits warnings to six zonal offices and 31 sub-district offices over high frequency (HF) radio. This message is passed on to some 274 village level unions by very high frequency (VHF) radio. Team leaders, typically responsible for one or two villages accommodating 2,000 to 3,000 people, then spread out on bicycles to disseminate cyclone warnings door-to-door using megaphones, hand sirens and public address systems. In total, the scheme covers about 3,500 villages through the use of 34,140 trained volunteers, including 5,000–6,000 females. Disaster-awareness training is given by the Red Cross and Red Crescent movement as part of the CPP programme. The system can evacuate over 300,000 people in 48 hours and has led to significant improvements. Whilst 300,000 people were killed by the 1970 cyclone, a similar event in 1991 claimed a much-reduced 140,000 victims and the reported fatalities in 2007 from cyclone 'Sidr' were 4,000–5,000. In Orissa, India, a programme of hazard preparedness and planning was introduced after a cyclone in 1999 (Thomalla and Schmuck, 2004). When tested in 2002, preparedness at the community level had improved, although interactions between government organizations and NGOs remained less than optimal.

F ADAPTATION: PREDICTIONS, FORECASTS AND WARNINGS

Disaster impacts can be reduced by forecasting the approach of many threats – from volcanic eruptions through to drought and malaria epidemics – and then warning communities at risk. Much prominence is given to sophisticated *forecasting and warning systems* (FWS) for hydro-meteorological hazards, due to advances in weather forecasting, communications and information technology. In the developing world, simpler methods may apply. For example, wind-up and solar-powered radios permit reliable access to hazard warnings for some of the world's poorest households. Following severe floods in Mozambique during 2000, when at least 700 people were killed, the local Red Cross Society integrated *Freeplay Lifeline* radios within a community-based early warning system for cyclones and floods (IFRCRCS, 2009). When floods returned in 2007 and 2008, death tolls were fewer than 30 in 2007 and 6 in 2008.

Most hazard warnings are based on forecasts, but for some threats (like earthquakes) preparedness has to be based on predictions. It is important to understand the difference between predictions, forecasts and warnings:

- *Predictions* are based on statistical theory and the historical record of past events. Because the results are expressed in terms of average probability, predictions tend to be long term, with no precise indication of when an event may occur. For earthquakes they may extend several years ahead. It is rarely possible to specify either the precise location or the magnitude of the event with much confidence.
- *Forecasts* depend on the detection and evaluation of a potentially hazardous event as it evolves. If the event can be monitored, it may be possible to specify the timing, location and likely magnitude of an impending hazard strike. Strictly speaking, forecasts are scientific statements and offer no advice as to how people should respond. They tend to be short term and the limited lead-time can limit the effectiveness of warnings.
- *Warnings* are messages advising people at risk about an approaching hazard and the steps that should be taken to reduce losses. All warnings are based on predictions or forecasts. Until recently, many agencies – including those responsible for national weather services – issued forecasts and warnings that contained only limited advice on safety and hazard limitation, although more advice is now attached to such messages.

FWS are very useful for hurricanes and floods, where emergency action, including evacuation, can avert disaster. Drought and tectonic hazards remain difficult to forecast, although some success is possible. For example, based on early warning indicators, the Philippine Institute of Volcanology and Seismology advised the government to evacuate residents within a 20-mile radius of Mount Pinatubo before the volcanic eruptions in June 1991. Although hundreds of people were killed and over 10,000 homes were destroyed, about 80,000 people were saved, together with an estimated US$1 billion in US and Filipino assets (OFDA, 1994).

Figure 5.17 shows a fully developed FWS with four key stages:

- *Threat recognition* covers the preliminary period when a decision is taken to establish a relevant monitoring programme leading to an FWS. To be effective, schemes need to be publicized among the community at risk and tested with mock disaster exercises. Ideally, feedback from this experience leads to design improvements in the system. Other revisions should occur as a result of hindsight reviews after disaster.
- *Hazard evaluation* includes several substeps, from observers first detecting an environmental change that could cause a threat, through to estimating the scale of the risk and the final decision to issue a warning. This requires a specialist agency, such as a national meteorological service, due to the need for continuous monitoring networks backed by scientific equipment and personnel. The priority at this stage is to improve the accuracy of the forecast and to increase the lead-time between issue of the warning and the onset of the hazardous event. In order to complete the process, and retain public confidence, stand-down messages should be issued when the emergency is over.
- *Warning dissemination* occurs when the message is transmitted from the forecasters to the hazard zone occupants. The message is likely to be formulated and conveyed by a third party through different communication methods, such as radio or television, and different personnel, such as the police or local residents. Once again, this stage contains several components, like the content of the message or the way in which it is conveyed, which are known to affect the eventual outcome.
- *Public response* is the phase where loss-reducing actions are taken. For example, over 2 million residents of the east coast of the USA evacuated inland following warnings of hurricane 'Floyd' in September 1999. From Figure 5.17, it can be seen that the response may be influenced directly through an input based on the public's knowledge of the evolving hazard. Various feedback mechanisms help to improve later editions of the warning. However, the response is largely determined by the nature of the warning message and the recipients' response.

To be effective, all FWS should be 'people-centred' (Basher, 2006). A realistic perception of the risk and prior knowledge of suitable actions will increase the quality of the response. An understanding of the socio-economic setting is as important as the accuracy of the scientific information because of a potential gap between the technical capacity of the forecast and the ability of a community to respond to the warning. For individuals, past experience of the same hazard raises the level of warning belief and there is evidence that women are more likely to accept the validity of a message than are men. Earlier adoption of anti-hazard measures, like retrofitting, will improve acceptance of the threat. Old and infirm people living alone will be least able to make an effective response to hazard warnings, either by protecting property or by evacuation, and special support should be made available. Often there is a reluctance to evacuate; perhaps because the message fails to specify this action, or

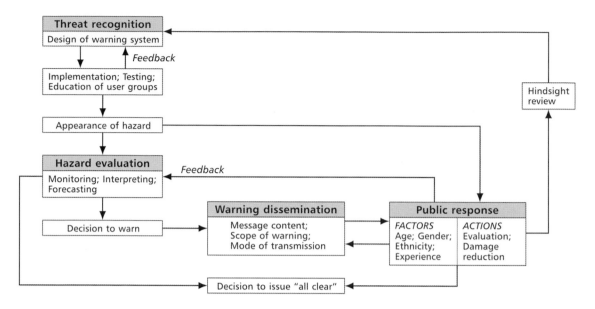

Figure 5.17 Generic model of a well-developed hazard forecasting and warning system with bypass and feedback loops. The key stages are threat recognition, hazard evaluation, warning dissemination and public response. The success of the scheme is highly dependent on the effectiveness of the community response.

because people believe they can cope or because they fear looting of an empty house. There is considerable natural attachment to the home environment. Family groups are more likely to evacuate than single-person households, often to the homes of relatives rather than to disaster shelters.

The decision to warn is crucial. In marginal situations, forecasters have to make quick decisions and can be caught between the dangers of issuing a false warning or issuing no warning at all. In the past, forecast agencies have assumed little responsibility for their products after they have been issued, an attitude sometimes reflecting a wish to avoid legal liability following either defective forecasts or the fear of offering inadequate advice about damage-reducing actions. Public confidence is most likely to be eroded in situations either where no warning is issued or when a false warning has been issued previously. Such mistakes are costly. The erroneous prediction of an eruption of the Soufrière volcano in

Guadeloupe in 1976 led to the evacuation of 72,000 people for several months. Today, false tsunami warnings for the Pacific Ocean occur because the exact nature of the triggering earthquake is unknown when the warning is issued.

The effectiveness of hazard response is influenced by the content of the forecast and the warning. Official warnings from a trusted source are judged the most reliable. Tiered warnings, incorporating a 'watch' phase before the 'warning' phase and a final 'all clear', tend to avoid gross errors, such as an unnecessary mass evacuation. But not all hazards (e.g. earthquakes) are suitable for tiered warnings. Pre-planning by the responsible authorities should ensure that all basic procedures are already in place, such as how the information is to be disseminated, the identification of the people at risk and the emergency bodies to be warned. There should also be alternative means to distribute messages in adverse environmental conditions which may include the loss of electrical power.

Real-time feedback within the system, including an accuracy check with the forecasters, and a response check on those being warned, is important. This is because the onward transmission of the message may be unnecessarily delayed, or even halted, by operators seeking confirmation on some aspect. This is most likely to happen with ambiguous messages. It is believed that the ideal warning messages should contain a moderate sense of urgency, estimate the time before impact and the scale of the event, and provide specific instructions for action by the recipients, including the need for spectators to stay clear of the hazard zone (Gruntfest, 1987). Continuing advice on the evolving situation, such as weather, roads and traffic-flow conditions, plus notification of the next warning update, is also helpful.

The mode of warning can be influential. For the general public, radio and television act as primary sources of information. The best warnings make the content personally relevant to those at risk. In this context, warnings delivered directly by other people living in the community are relevant. Although warnings via the mass media are likely to be believed, the initial message may simply alert people that something is wrong. Further information is then required. Some confirmation of the first warning received is almost always sought before action is taken, hence the advantage of tiered warnings. For example, confirmation may be sought by residents from members of the family or the police and responses often take place on the basis of group decisions.

G ADAPTATION: LAND USE PLANNING

The purpose of land use planning is to zone and control hazardous areas so that existing communities can be better protected and any new development can be steered away from dangerous sites. Success depends on an intervention in the market-driven process whereby hazard-prone land, formerly held in low-intensity uses such as forestry or agriculture, is converted into higher-intensity use such as housing. Land conversion is driven by competition and a desire for profits – all functions of population growth, urbanization and wealth creation. The result is a rise in land values and more urban infrastructure, so that greater losses are likely when disaster strikes. In the MDCs, more affluence and leisure time has led to second homes and recreational facilities in coastal and mountainous areas which can be hazardous. So far, regulatory zoning has been adopted mainly in the richer countries, but El-Masri and Tipple (2002) placed land use planning, along with improved shelter design and institutional reform, at the heart of sustainable disaster reduction for developing countries.

Land use can be regulated at all scales, from the regional-plan level through town zoning ordinances down to individual plot and sub-plot level. The policy is most useful in prohibiting new building in high-hazard areas, but conflicts with other community objectives and local vested interests must be anticipated. Typical opponents include landowners, real estate agents, developers and builders who may experience an erosion of assets or income. In addition, the designation of hazard zones within areas already developed for housing will be opposed by residents who anticipate a loss in market value of their property.

According to Burby and Dalton (1994), hazard-based land planning is most likely to succeed if the policy is backed by national government and has local community support. For example, whilst low-density zoning might be imposed in order to limit the potential property losses in one area, the builder concerned might be compensated by the issue of a permit for a high-density development in a safer area nearby. Land use planning can gain public support by guiding new development away from environmentally sensitive areas, such as wetlands, and by zoning some hazard-prone areas, such as river corridors, for outdoor recreation. Low levels of building density can be maintained by restricting development to large lots only or by dedicating unsafe areas to open-space uses, such as parks or grazing.

The main limitations on land use planning are:

- lack of knowledge about the hazard potential of smaller areas, such as individual building plots
- the presence of existing development
- the infrequency of many hazardous events and the difficulty of raising community awareness of risk
- high costs of hazard mapping, including detailed inventories of existing land use, structures and occupancy rates
- local resistance to land controls, on political and economic grounds.

Land use controls are most appropriate for growing communities with undeveloped land available. To this extent, they work best in areas that need them least. Where land development pressures are high, zoning will be more difficult, partly because hazard-prone land often appears desirable. Many landslide areas, coastal and floodplain sites have outstanding scenic views and can command high market prices, especially if there is no awareness of the risk. Under the ancient legal doctrine of *caveat emptor* ('let the buyer beware'), there is no obligation for the owner of such land to disclose any risks. In some countries, legislation requires the vendor to disclose information about known geological and hydrological hazards at an early stage in the purchase process so that the potential buyer can make an informed decision (Binder, 1998). But delay is common and the purchaser may disregard the information supplied. Disclosure is generally unpopular with local commercial interests and local authorities may not adopt regulations, in the belief that they will lose economic initiatives to more lenient communities nearby.

Effective land controls depend on the quality of the information available. An accurate delimitation of the hazard zone is crucial. All regulations must be seen as reasonable in terms of the development restrictions proposed and should be strong enough to withstand challenge in a court of law. Ideally, variations in risk should be identifiable down to the level of individual properties. For cyclones and earthquakes such precision is unattainable; the greatest accuracy is achieved with topography-dependent hazards like floods, landslides and avalanches. Therefore, accurate mapping is a key element in policy making, especially when the risks are high, but, even when care is taken, some uncertainty about risk zone boundaries will remain.

1 Macro-zonation

Macro-zonation (regional planning) shapes broad land use policy. For example, a probabilistic map of seismic shaking hazards in California could be used to identify areas suitable for retrofitting buildings with anti-seismic measures or introducing building codes for new development (Figure 5.18). Peak ground acceleration (PGA) values are relevant because they indicate the maximum forces that buildings should be able to withstand. Of course, the performance of individual buildings will vary for the same PGA value. Pre-1940 buildings will perform relatively badly and some earlier buildings are likely to be vulnerable to PGA values below 0.4g. On the other hand, post-1985 dwellings constructed to California earthquake standards have survived severe shaking of 0.6g with only limited chimney damage and some disturbance of house contents.

The sort of information required for detailed land use regulation can be illustrated for the island of Martinique in the West Indies, which suffered a major disaster when the Mount Pelée volcano erupted in 1902 (see also page 176). Planning policy here depends on estimating and mapping the maximum extent of hazards arising from a possible future eruption, and the likely effect on buildings, so that development can be limited in the most dangerous areas (Leone and Lesales, 2009). To achieve this, scientific views on future eruptions have to blend with political compromises in order to achieve economic progress and an acceptable level of risk. For example, the

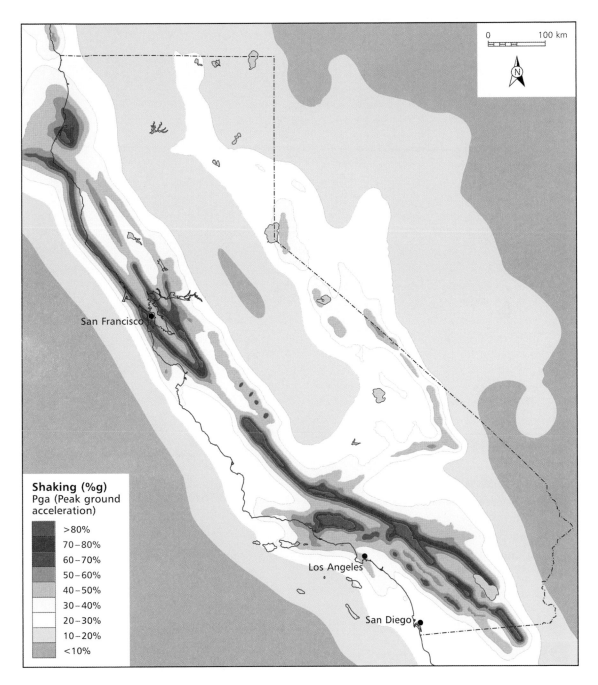

Shaking (%g)
Pga (Peak ground acceleration)

- \>80%
- 70–80%
- 60–70%
- 50–60%
- 40–50%
- 30–40%
- 20–30%
- 10–20%
- <10%

San Francisco

Los Angeles

San Diego

0   100 km

N

Figure 5.18 Map showing seismic shaking hazards from earthquakes in California relevant to the design of hazard-resistant buildings and the formulation of building codes. Shaking is measured by the peak ground acceleration expected in firm rock with a 10 per cent probability of being exceeded in 50 years. The unit 'g' is the acceleration of gravity. After California Geological Survey and USGS Probabilistic Seismic Hazards Assessment Model at http://www.conservation.ca.gov/cgs Reproduced with permission.

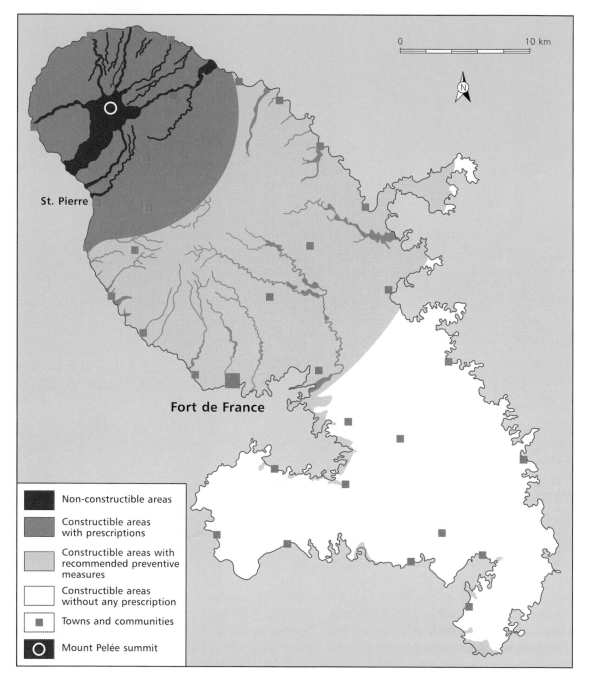

Figure 5.19 Proposed regulation map for volcanic risk reduction around Mount Pelée, Martinique. Such maps are intended to help planners and disaster agencies to reduce future losses by controlling land use and may also contribute to improving risk perception by the general public. After *Journal of Volcanology and Geothermal Research* 186 (3–4), F. Leone and T. Lesales, The interest of cartography for a better perception and management of volcanic risk: from scientific to social representations, 186–94. Copyright (2009), with permission from Elsevier.

residents of Martinique rate lava flows as a greater threat than the eruptive history of Mount Pelée would suggest. As a result, further development in the northern part of the island is a controversial issue. Four planning zones have been suggested (Figure 5.19). These are: no permanent construction should be allowed; buildings can be located in exceptional circumstances; hazard-proof structures can be built; and areas where no restrictions apply. Effective community engagement is required to carry forward plans like this.

Much of the world's population lives on or near coasts and many countries have adopted a policy of *Integrated Coastal Zone Management* (ICZM). The aim is to resolve issues arising from global climate change, including natural hazards, in a multi-disciplinary and sustainable fashion (Stojanovic and Ballinger, 2009). The settings range from small-island developing states (SIDS) like Samoa, where nearly 80 per cent of the population and most of the infrastructure is located along the coast (Daly *et al.*, 2010), to nations such as the USA, where the state of Florida is subject to severe coastal storms and flooding.

Under Florida's comprehensive state plan, local governments are required to define *coastal high-hazard areas* (CHHA) where specific land planning responsibilities exist for directing population away from the area, reducing hazard exposure for infrastructure and limiting public expenditures that subsidize development (Florida Department of Community Affairs, 2005). One such area is the low-lying Lee County CHHA. Figure 5.20a indicates the high degree of flood risk that is reflected in the insurance ratings produced under the National Flood Insurance Program. There is a spread of A zones (subject to still-water flooding from 1:100-year storm events) and, nearer the coast, V and VE zones where breaking waves of three feet or more in height can be expected. The CHHA, legally defined as the designated evacuation zone for a Category 1 hurricane, can then be identified (Figure 5.20b). In fact, the island of Sanibel, plus strips of land along the coast and the Caloosahatchee River, is

scheduled for evacuation to escape a tropical storm, a lower-level hazard than a Category 1 hurricane, whilst the whole area would be progressively evacuated in the event of a Category 4 hurricane.

2 Micro-zonation

Zoning bye-laws and local ordinances implement regional planning at the scale of communities and building lots. These local regulations can be used to control development through the provision of detailed reports on aspects such as soils, geological conditions, grading specifications, drainage requirements and landscape plans as well as specific hazard threats. Relatively large-scale maps (at least 1:10,000) are usually required for zoning in high-risk urban areas. Other regulations may apply when applications are made for development at the building-plot level. For example, subdivision regulations ensure that the conditions under which land may be subdivided are in conformity with the general plan.

Micro-zonation is most successful for relatively frequent surface flows of material guided by topography which can be located with sufficient precision to permit land regulation down to individual plot level. The Swiss have developed a generic approach that has been adopted in several other countries (Zimmermann *et al.*, 2005). As shown in Figure 5.21, it is based on the magnitude–frequency principle and combines the intensity and probability of the event to draw up three chief colour-coded zones with differing potentials for land development. For example, in relation to a river flood, the intensity would be measured by the expected maximum flood depth or the depth × velocity and the risk levels for an urban area would be typically interpreted as:

- *Red zone (high risk) – prohibiting.* New houses prohibited, existing buildings are allowed to be used, lives at risk even in buildings.
- *Blue zone (medium risk) – restraining.* New houses are allowed but must be proofed against

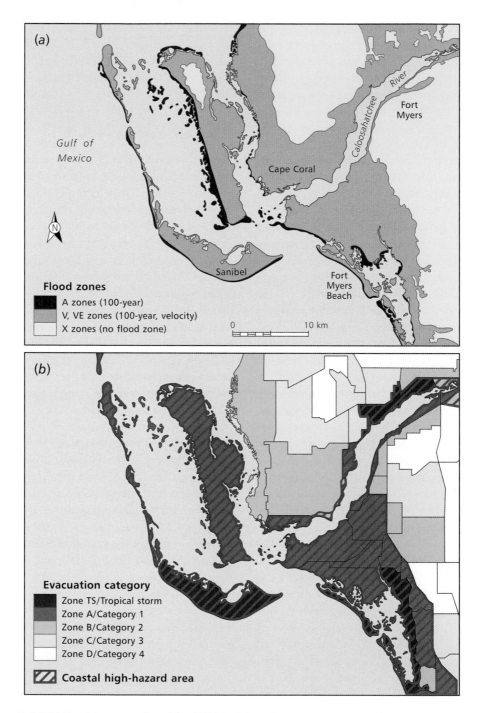

Figure 5.20 (a) Flood Insurance Rate Map (FIRM) of A and V rated zones in Lee County, Florida; (b) coastal high-hazard area of Lee County, Florida, based on Category 1 evacuation zone. The entire coastal area is zoned for different evacuation priorities depending on the forecast severity of the storm. After Florida Department of Community Affairs (2005).

disaster stresses, more detailed regulation may be required by the local authority.

- *Yellow zone (low risk) – warning.* Key public buildings (hospitals, schools) must be strengthened again hazard impact, residents to be warned and possibly evacuated.
- *Other (residual risk only).* No specific building or land use controls, public organizations (schools and hospitals) should have response measures and emergency plans ready in the event of disaster.

This type of zoning is suitable for debris flows in steep terrain that threaten development on the limited flat land available. In 1998 the Principality of Andorra in the Pyrenees adopted an Urban Land Use and Planning Law prohibiting new building development in zones exposed to natural

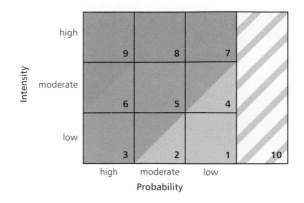

Figure 5.21 A matrix of the Swiss hazard zoning system. The system employs four main hazard zones scaled into 10 risk classes. Red – high risk; blue – moderate risk; yellow – low risk; hatched white/yellow indicates high intensity but very low probability. After Zimmermann *et al.* (2005). Reproduced with permission.

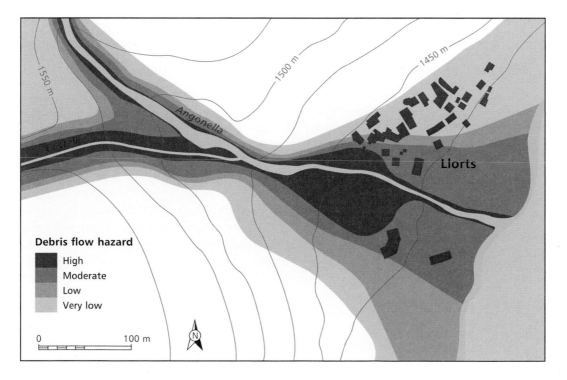

Figure 5.22 Debris flow hazard map of the alluvial fan at Llorts in the Pyrenean Principality of Andorra. Most of the village lies in a safe area but some existing development is exposed to moderate risk. After *Geomorphology* 78 (3–4), M. Hürlimann *et al.* Detailed debris flow hazard assessment in Andorra: a multidisciplinary approach, 359–72. Copyright (2006), with permission from Elsevier.

hazards (Hürlimann *et al.*, 2006). Geo-hazard maps have been produced of debris flows at the scale of 1:2000. The maps are based on a matrix analysis of flow intensity and estimated annual probability for the following recurrence intervals: high hazard <40 years, medium hazard 40 to 500 years, low hazard >500 years and very low hazard lacking flow evidence. The village of Llorts in the Pyrenees is built on part of a debris fan at the outlet of a 4 km² basin drained by three torrents (Figure 5.22). The highest part of the catchment reaches 2,600m above sea level; the fan apex is 250m long with a steep slope averaging 12°. Debris flows enter the apex of the fan and a high-hazard zone was identified in this presently undeveloped area. Although most of the village lies within the safe area, some buildings are exposed to moderate and low-level hazard.

Earthquake micro-zonation is less precise, but important because of the high disaster potential.

In this case the identification of active fault lines is crucial and building controls are then usually imposed over a set-back corridor running alongside the fault, as illustrated in Box 5.3 for an area in California.

For high-risk areas, the public acquisition of hazardous land is perhaps the most direct option for local governments. Once acquired, the land can be managed to protect public safety and meet other community objectives, such as open space or recreational facilities. But land acquisition is expensive and local authorities rarely have the resources for outright purchase. Another option is for a commercial agency to purchase land and then control development in the public interest, such as leasing it for low-intensity use. If public lands are available close to a hazard zone, and if the occupants are willing to relocate, it may be possible for privately owned parcels of hazardous land to be exchanged for safer areas. Any

Box 5.3 Earthquake hazard zoning in California

Seismic micro-zonation has been undertaken in the United States for many years – especially in California, where the 1972 Alquist-Priolo Earthquake Fault Zoning Act was passed to reduce the effects of surface faulting on residential property. This state law was enacted as a direct result of property damage arising in the 1971 San Fernando earthquake. It is concerned only with surface fault rupture, the situation when deep-seated ground movement breaks through to the land surface. The State Geologist is required to establish and map regulatory zones (Earthquake Fault Zones) around the surface traces of known active faults, i.e. a fault that has ruptured in the last 11,000 years. Local agencies must then control most types of proposed development within these zones, including all land divisions and most structures built for human occupancy. No new structure for human occupancy can be placed over a fault trace and any such must be set-back at least 50 feet. For residences erected before the designation of the regulatory zone, real estate agents must disclose to potential buyers that the property is within a fault zone. Because of low occupancy rates, the Act does not cover public facilities, like water pipelines, and generally does not apply to industrial sites, although local agencies can be more restrictive than state law requires.

Figure 5.23 shows a portion of the Alquist-Priolo Earthquake Fault Zone Map covering part of the creep-active Concord Fault located in downtown Concord in the eastern San Francisco Bay area. 'C' indicates the fault creep. The fault has a slip rate of about 3.5 mm yr¹. All zone boundaries are defined by straight lines drawn by joining up 'turning points' at locations easily identified on the ground, such as road junctions and drainage ditches. Most zones have an average width of about one-quarter of a mile.

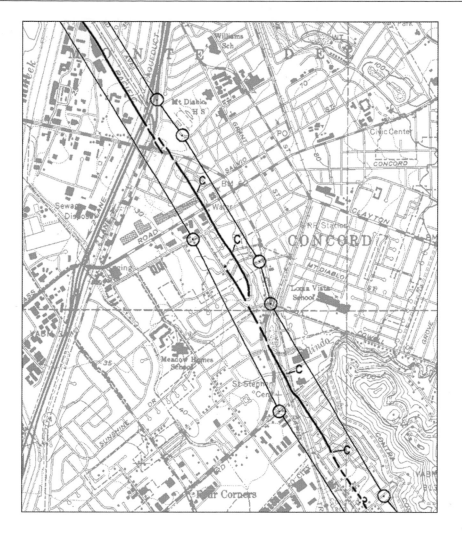

Figure 5.23 Portion of an earthquake fault zone map in California showing part of the Concord fault, and the surrounding zone of land regulation, in downtown Concord in the eastern San Francisco Bay region. This creep fault (indicated by 'C') is characterized by a slip rate of about 3.5 mm yr[1]. Reproduced with permission, California Geological Survey from Official Map of Alquist-Priolo Earthquake Fault Zones, Walnut Creek Quadrangle (1993).

movement of structures or occupants or the demolition of unsafe buildings is usually difficult, expensive and controversial. Relocation away from the area may destroy any potential the land might have to promote growth and generate local tax revenues. The purchase and demolition of buildings with historical or architectural importance will generate opposition from pressure groups.

Since effective hazard-reduction strategy depends on the understanding and cooperation of the community as a whole, public information is essential. Simple methods – like the posting of warning notices – highlight the threat. Publicity may be disseminated through conferences, workshops, press releases and the publication of hazard zone maps. Financial measures can also be applied. These work indirectly by altering the relative advantage of building at a dangerous location. For example, local government may decide to restrict investment in public facilities, such as roads, water mains and sewers, to hazard-free areas zoned for development. Any national scheme that provides grants, loans, tax credits, insurance or other type of financial assistance to encourage development can be adapted for local use. For example, tax credits can be offered on hazard-prone land that remains undeveloped or development funds can be approved for low-density development only. In the USA, land conversion is deterred by withholding federal grants and benefits from flood-prone communities not participating in the National Flood Insurance Program.

FURTHER READING

Basher, R. (2006) Global early warning systems for natural hazards: systematic and people-centred. *Philosophical Transactions of the Royal Society (A)*: 364: 2167–82. This is a very balanced and comprehensive interpretation.

Chakraborty, J., Tobin, G.A. and Montz, B.E. (2005) Population evacuation: assessing spatial variability in geographical risk and social vulnerability to natural hazards. *Natural Hazards Review* 6: 23–33. An account of some key practical issues surrounding successful emergency evacuation.

Chang, Y., Wilkinson, S., Brunsden, D., Seville, E. and Potangaroa, R. (2011) An integrated approach: managing resources for post-disaster reconstruction. *Disasters* 35: 739–65. Deals with an important but neglected topic.

Key, D. (ed.) (1995) *Structures to Withstand Disaster*, Institution of Civil Engineers, London. A clearly stated introduction with good examples.

Olsen, G.R., Carstensen, N. and Høyen, K. (2003) Humanitarian crises: what determines the level of emergency assistance? Media coverage, donor interests and the aid business. *Disasters* 27: 109–26. This remains a complex and often unanswered question.

Provention Consortium (2007) *Construction Design, Building Standards and Site Selection.* Guidance Note 12, Provention Consortium Secretariat, Geneva. This contains illustrations of effective practice in various parts of the world.

WEB LINKS

UN Office for the Coordination of Humanitarian Affairs www.unocha.org/

Emergency Management Agency Australia www.ema.gov.au/

European Commission Department of Humanitarian Aid www.ec.europa.eu/echo/index

Federal Emergency Management Agency USA www.fema.gov

International Committee of the Red Cross www.icrc.org/

Oxfam International www.oxfam.org/en/

United Nations Refugee Agency www.unhcr.org/

Part Two
THE EXPERIENCE AND REDUCTION OF HAZARD

Naturae enim non imperatur, nisi parendo.
(Nature, to be commanded, must be obeyed.)

Francis Bacon, 1561–1626

CHAPTER SIX

Tectonic hazards

Earthquakes and tsunamis

A EARTHQUAKE HAZARDS

Approximately 2 million earthquake-related deaths were recorded in the twentieth century. Most resulted from a few high-magnitude events with at least 50,000 fatalities each. 'High-magnitude' is a relative term, but, as a guide, the 10 largest earthquakes since 1900 were all above $M_W = 8.5$ (Table 6.1). Four were in the early years of the twenty-first century. Some relationships exist between earthquake magnitude and disaster impact but factors other than size are needed to explain the pattern of deaths and other losses. The $M_W = 9.1$ Sumatra earthquake of 2004 is amongst the 10 largest and the 10 deadliest earthquakes because of the many lives lost in the associated tsunami. High mortality has occurred with much lower-magnitude events. For example, the 1976 Tangshan earthquake in China had an official death toll of 255,000, although some estimates were of 655,000 or 750,000 deaths, whilst the 1920 Ningxia (Gansu) earthquake, also in China, is thought to have killed about 200,000.

The greatest losses of life and infrastructure exist when the intense seismic energy released along an earthquake fault coincides with high levels of hazard exposure and vulnerability. Most mortality is due to severe ground shaking and the collapse of buildings. For example, in the 1999 Chi-Chi earthquake in Taiwan, it was estimated that over 100,000 buildings collapsed, causing 86 per cent of the deaths (Liao *et al.*, 2005). Mortality was high at Tangshan in 1976 because the earthquake occurred at a shallow depth beneath a city of 1 million people at a time when most were asleep in structurally weak houses. Over 90 per cent of the residential buildings were destroyed. In urban areas, fire due to the rupture of gas and water pipes is an important secondary peril. Over 80 per cent of the property damage in the San Francisco earthquake of 1906, when about 3,000 people died, was due to fire. The 1923 Kanto earthquake, the worst natural disaster recorded in Japan until 2011, killed nearly 160,000 people in Tokyo and Yokohama. This earthquake occurred at a time when over a million charcoal braziers were alight in wooden houses to cook the midday meal. The resulting fires destroyed an estimated 380,000 dwellings. In the 2010 Haiti earthquake, extreme levels of poverty and low socio-economic development were important factors in the large loss of life and related impacts (Box 6.1).

Table 6.1 The 10 largest earthquakes in the world since 1900 and the 10 deadliest earthquakes on record

Ten largest earthquakes since 1900			Ten deadliest earthquakes on record			
General location	Year	Magnitude (M_W)	General location	Year	Number of fatalities	Estimated magnitude (M_W)
Chile	1960	9.5	Shensi, China	1556	830,000	8.0
Alaska	1964	9.2	Tangshan, China	1976	255,000	7.5
Sumatra	2004	9.1	Aleppo, Syria	1138	230,000	
Japan	2011	9.0	Sumatra	2004	227,898	9.1
Kamchatka	1952	9.0	Haiti	2010	222,570	7.0
Chile	2010	8.8	Damghan, Iran	856	200,000	
Ecuador	1906	8.8	Ningxia, China	1920	200,000	7.8
Alaska	1965	8.7	Ardabil, Iran	893	150,000	
Indonesia	2005	8.6	Kanto, Japan	1923	142,800	7.9
Assam-Tibet	1950	8.6	Ashkhabad, USSR	1948	110,000	7.3

Source: After US Geological Survey at http://neic.usgs.gov/neis/eqlists (accessed 8 April 2011)

Note: Some earthquakes have an epicentre off shore. The number of fatalities is often a rounded estimate, especially for pre-1900 events; even after 1900 some figures are unreliable. The figure for the 1976 Tangshan earthquake is the 'official' value authorized by the Chinese government, which may be an underestimate.

Earthquakes pose a threat to all societies. In urbanized Japan, the 1995 Kobe earthquake killed more than 5,300 people and made 300,000 homeless, whilst the Great Tōhoku earthquake of 2011 claimed more than 28,000 lives. Even for rural areas in LDCs, the economic costs can be high. In the 1993 earthquake at Maharashtra, India, the loss of agricultural assets exceeded 50 per cent and created severe difficulties for survivors seeking to regain their livelihoods

Table 6.2 The proportion of agricultural assets destroyed by the 1993 Mahrashtra earthquake

Livestock	%	Implements	%
Cattle	18.3	Buffalo carts	36.6
Buffalo	23.5	Tractors	48.9
Goats/Sheep	47.5	Ploughs	50.2
Donkeys	43.5	Pump Sets	47.8
Bullocks	12.9	Cattle Sheds	67.2
Poultry	65.3	Sprayers	62.3

Source: Adapted from Parasuraman (1995)

Note: The survey covered 69 affected villages with a population of 170,954 persons.

(Table 6.2). Large-magnitude earthquakes have great disaster potential because they shake the ground more severely, for a longer duration and over more extensive areas than do smaller events. But event impacts are influenced by local conditions. Geological factors, like steep slopes that cause landslides and alluvial soils that liquefy, enhance the effects of ground shaking. Most of the deaths in the 1920 Ningxia (China) earthquake arose from slope failures when loess deposits collapsed and buried entire towns. The time of day is often significant in determining human mortality. A 1992 earthquake at Erzinçan, Turkey claimed only 547 lives because it struck in the early evening, when many residents were at worship in mosques that were comparatively earthquake resistant. In contrast, the 2005 Kashmir earthquake killed over 19,000 children alone, largely because it occurred during school hours and many poorly constructed schools collapsed in the shaking.

Box 6.1 The 2010 Haiti earthquake: a long-term disaster

On 12 January 2010, a $M_W = 7.0$ earthquake struck the island republic of Haiti in the Caribbean Sea. The epicentre was 25 km west of the coastal capital of Port-au-Prince, home to over 2 million people. Due to the shallow 13 km depth of the earthquake, the resulting ground shaking was unusually severe and about 3.5 million people were subjected to Modified Mercalli scale intensities of VII to X – a range that can cause moderate to heavy damage to well-constructed properties. A total of 188,383 buildings collapsed. Of these, 105,000 were completely destroyed, including 8 hospitals while a further 22 hospitals were seriously damaged. The direct death toll was 222,570. About 2.3 million people, almost one-quarter of the national population, immediately fled their homes; 1.8 million were subsequently declared homeless. At the peak of the emergency, there were 1.5 million refugees in over 1,300 spontaneous settlements or camps. Damaged communication and transport systems affected the emergency response, together with confusion over lead responsibilities, failures to prioritize relief flights and delays in the distribution of aid supplies. Later in the recovery period, unpredictable movements of the survivors, disputes over land rights and fluctuating government policies further restricted progress. Later still, outbreaks of cholera spread to all 10 departments of the country (see Chapter 10).

This disaster would have taxed the resources of any country. However, the scale of devastation and the patchy recovery show that the earthquake exposed long-term socio-economic problems that accumulated over time and greatly amplified the disaster impact. Haiti is the poorest country in the western hemisphere. Most Haitians live on less than US$2 per day, with 80 per cent below the poverty line. More than two-thirds of the labour force lacks formal employment and is engaged in agriculture. The country is ranked about 150 out of more than 180 countries in the world on the Human Development Index scale and has been classed as economically vulnerable by the UN. Since the country gained independence from France in 1804, it has been under dictatorships for much of its history and democratic rule was restored as recently as 2006. It is also exposed to natural hazards other than earthquakes, including tropical cyclones, flooding and landslides.

Much of the damage was sustained in the capital, where about 85 per cent of the population was living in urban slums, mostly tightly packed, inadequately built concrete buildings. Half of these people had no access to toilet facilities and only one-third had piped water available. Haiti has no properly approved construction code and, before the earthquake, it was estimated that about 60 per cent of the houses were unsafe for occupation in normal circumstances. Following the extensive building collapse, there was 19 million m^3 of rubble in the streets of Port-au-Prince. The UN made an urgent Flash appeal for US$1.5 billion of disaster aid and by December 2010 just over US$1 billion had been received. It has been estimated that it will take US$11.5 billion and many years to fund and complete the reconstruction. The long-term challenge is to reinstate the capital and other areas in a more sustainable fashion. Previous attempts to improve the urban environment have faltered, due to a lack of public participation, low staff capacity and weaknesses in accountability, and it is hoped that the next few years will witness a better way forward.

B EARTHQUAKE BEHAVIOUR

Earthquakes are caused by sudden movements in rock, relatively near to the Earth's surface, along a zone of pre-existing geological weakness, called a *fault*. These movements are preceded by the slow build-up of tectonic strain that progressively deforms the crustal rocks and produces stored elastic energy. When the accumulated stress exceeds the resisting strength of the fault, it creates rock fractures. The sudden release of energy produces seismic waves that radiate outwards. It is the fracture of the brittle crust, followed by elastic rebound on either side of the fracture, which is the cause of ground shaking. The point of rupture (*hypocentre*) can occur anywhere between the Earth's surface and a depth of 700 km. The rupture propagates along the fault line, radiating seismic waves along the fault plane as well as from the hypocentre.

The size of an earthquake depends upon the amount of physical displacement on the fault. Generally speaking, larger fault movements – either vertically or laterally – and longer-length ruptures lead to bigger earthquakes. The 2004 Sumatra earthquake was very large because a major displacement of the fault (~15 m) took place over a long distance (1,600 km). It is also true that the most damaging earthquakes, accounting for about three-quarters of the global seismic energy release, are shallow-focus earthquakes (<40 km below the surface). For example, the 1971 San Fernando earthquake in California had only a moderate magnitude ($M_W = 6.6$) but, because it occurred only 13 km below the surface, near a highly urbanized area, the damage level was high.

The global distribution of earthquakes is far from random. About two-thirds of all large earthquakes are located in the so-called 'Ring of Fire' around the Pacific Ocean. This pattern is closely related to the geophysical activity associated with plate tectonics (Bolt, 1993). The Earth's crust is divided into more than 15 lithospheric

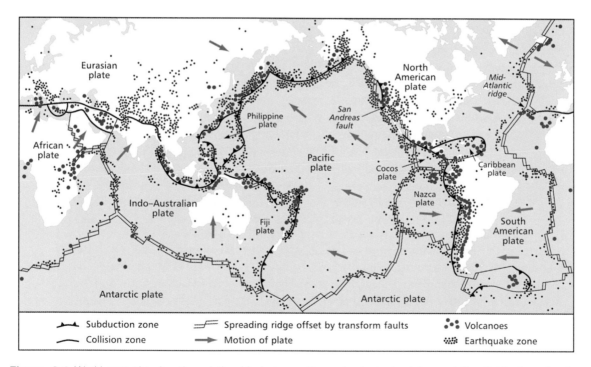

Figure 6.1 World map showing the relationship between the major tectonic plates and the distribution of active earthquakes and volcanoes. After G.W. Housner and quoted in Bolt (1999).

plates (Figure 6.1). The plates move across the globe at speeds up to 180 mm yr^{-1}, carried along by convection currents in the mantle. Most earthquakes occur when two plates collide, especially in the *subduction zones*, where one plate is forced beneath another. Sometimes earthquakes occur at weak points within plates. Although these intra-plate earthquakes account for less than 0.5 per cent of global seismicity, they are usually unexpected and pose a significant threat. During a few months in the winter of 1811–12, three large earthquakes decimated the town of New Madrid in Missouri, USA. The final quake is still considered to be the largest seismic event to have struck the contiguous states of the USA, even though the location was hundreds of kilometres away from a plate boundary.

1 Earthquake magnitude

Earthquakes are measured at the *epicentre*, the point on the Earth's surface directly above the hypocentre, and the magnitude is calculated on scales based on the work of Charles Richter. These scales describe the total energy released by the earthquake in the form of seismic waves radiating from the fault plane. The energy is calculated from seismographs that record the amplitude of ground motion during the earthquake. The original system, still known as the Richter scale, measures the *local magnitude* (M_L) of the earthquake. This scale is unsuitable for very large earthquakes. Today, seismologists employ a slightly different measurement based on *moment magnitude* (M_W). This method takes into consideration both the area of the fault that has broken and the amount of movement that has occurred. It is a more reliable measurement of the total energy released. The moment magnitude scale has been refined so that the resultant values are reasonably close to those of the original local magnitude scale, to ease comparisons.

It is important to understand that these scales are not linear. In Richter's original system, each point on the M_L scale indicated an order of magnitude increase in measured ground motion. Thus, an $M_L = 7.0$ earthquake produces about 10 times more ground shaking than an $M_L = 6.0$ event and around 1,000 times more ground shaking than an $M_L = 4.0$ event. Approximate energy–magnitude relationships show that, as the magnitude increases by one whole unit, the total energy released increases by about 32 times. The moment magnitude (M_W) scale measures this energy release. Thus, an $M_W = 6.0$ event releases about 32 times more energy than does an $M_W = 5.0$ event; an $M_W = 9.0$ event will release over 1 million times more energy than an $M_W = 5$ event. The scale has no theoretical upper limit but empirical evidence suggests that most shallow earthquakes need to attain a magnitude of at least $M_W = 4.0$ before damage is recorded on the surface (Bollinger *et al.*, 1993). Such events occur several times each day worldwide. Few cause significant loss (Table 6.3).

As already indicated, the amount of destruction caused by an earthquake depends on various factors including:

- *Duration of shaking* – in general, longer periods of shaking lead to more damage for the same magnitude of an event.
- *Distance from the fault* – as earthquake waves radiate outwards, their energy reduces with distance. Locations further from the fault tend to experience lower levels of shaking.
- *Local conditions* – many local conditions affect the nature of shaking. For example, soil and rock properties alter the characteristics of the earthquake waves, whilst topographic effects can be significant.
- *Population density* – if the population density is high, more people will be at risk from an earthquake.
- *Building quality* – much depends on the quality of local building. Poorly built, unreinforced structures with heavy roofs are most prone to collapse; people may survive better in lightweight buildings.

Table 6.3 Annual frequency of occurrence of earthquakes of different magnitudes based on observations since 1900

Descriptor	Magnitude (M_W)	Annual average	Hazard potential
Great	8 and higher	1	Total destruction, high loss of life
Major	7–7.9	18	Serious building damage, major loss of life
Strong	6–6.9	120	Large losses, especially in urban areas
Moderate	5–5.9	800	Significant losses in populated areas
Light	4–4.9	6,200	Usually felt, some structural damage
Minor	3–3.9	49,000	Typically felt but usually little damage
Very minor	Less than 3	9,000 per day	Not felt, but recorded

Source: After US Geological Survey at http://neic.usgs.gov (accessed 16 January 2003)

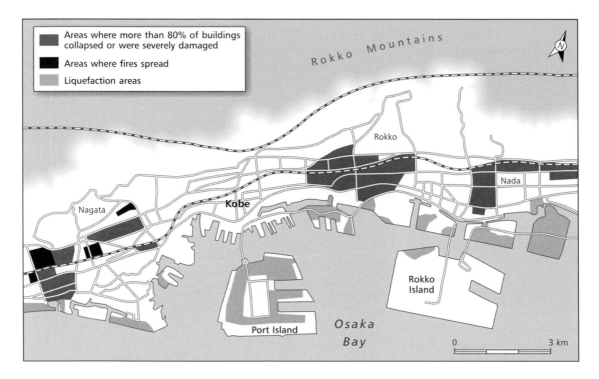

Figure 6.2 Map showing the pattern of damage following the 1995 earthquake in Kobe, Japan. Fires spread in the densely built-up areas of the city and liquefaction was widespread in the reclaimed industrial land along the shoreline. After Menoni (2001).

Box 6.2 The Modified Mercalli earthquake intensity scale

(Note: g is gravity = 9.8 m^{s2}).

Average peak velocity (cm s^{-1})	Intensity value and description of impacts		Average peak acceleration
	I.	Not felt except by a very few under especially favourable circumstances.	
	II.	Felt only by a few persons at rest, especially on upper floors of buildings. Delicately suspended objects may swing.	
	III.	Felt quite noticeably indoors, especially on upper floors of buildings, but many people do not recognize it as an earthquake. Standing automobiles may rock slightly. Vibration like a passing truck.	
1–2	IV.	During day felt by many, outdoors by few. At night some awakened. Dishes, windows, doors disturbed; walls make creaking sound. Sensation like heavy truck striking building. Standing vehicles rock noticeably.	0.015g–0.02g
2–5	V.	Felt by nearly everyone, many awakened. Some dishes, windows and so on broken; cracked plaster in a few places; unstable objects overturned. Disturbance of trees, poles and other tall objects sometimes noticed. Pendulum clocks may stop.	0.03g–0.04g
5–8	VI.	Felt by all, many frightened and run outdoors. Some heavy furniture moved; a few instances of fallen plaster and damaged chimneys. Damage slight.	0.06g–0.07g
8–12	VII.	Everybody runs outdoors. Damage negligible in buildings of good design and construction; slight to moderate in well-built ordinary structures; considerable in poorly built or badly designed structures; some chimneys broken. Noticed by persons driving cars.	0.10g–0.15g
20–30	VIII.	Damage slight in specially designed structures; considerable in ordinary substantial buildings with partial collapse; great in poorly built structures. Panel walls thrown out of frame structures. Fall of chimneys, factory stacks, columns, walls and monuments. Heavy furniture overturned. Sand and mud ejected in small quantities. Changes in well water. Persons driving cars disturbed.	0.25g–0.30g
45–55	IX.	Damage considerable in specially designed structures; well-designed frame structures thrown out of plumb; great in substantial buildings with partial collapse. Buildings shifted off foundations. Ground cracked conspicuously. Underground pipes broken.	0.50g–0.55g
>60	X.	Some well-built wooden structures destroyed; most masonry and frame structures destroyed with foundations; ground badly cracked. Rails bent. Landslides considerable from river banks and steep slopes. Shifted sand and mud. Water splashed, slopped over banks.	>0.60g
	XI.	Few, if any, (masonry) structures remain standing. Bridges destroyed. Broad fissures in ground. Underground pipelines completely out of service. Earth slumps and land slips in soft ground. Rails bend greatly.	
	XII.	Damage total. Waves seen on ground surface. Lines of sight and level distorted. Objects thrown into the air.	

Energy release alone is an imperfect guide to the socio-economic impact of earthquakes. For example, the 1995 Kobe, Japan earthquake was a moderate ($M_W = 6.8$) event. Huge losses were experienced because the shock waves reached a densely populated industrial port where buildings near the shore were founded on soft soils and landfill that induced severe shaking (Figure 6.2). In addition, there was much wooden housing built to withstand tropical cyclones rather than earthquakes. The heavy, clay-tile roofs, typically weighing two tonnes, collapsed and fires were readily started in the wooden structures. Over 90 per cent of the 6,400 fatalities occurred in these areas of suburban housing.

2 Earthquake intensity

Earthquake intensity is a measure of ground shaking that correlates better with disaster losses than with magnitude. Intensity is estimated on the *Modified Mercalli* (MM) scale, which allocates a numerical value to observations of the nature and extent of physical damage after the event (see Box 6.2). The scale ranges from MM = I (not felt at all) to MM = XII (widespread destruction). At first glance, the MM scale is less 'scientific' than other scales because it relies upon qualitative human descriptions rather than instrumental measurements. However, it captures important elements of earthquake impact. Another advantage is that, by using written accounts of past events, MM intensities can be assigned to earthquakes that occurred prior to the introduction of direct measurements. This allows the earthquake record to be extended back in history

C PRIMARY EARTHQUAKE HAZARDS

During an earthquake, the extent of ground shaking is measured by *strong motion seismometers*. These instruments are activated by strong ground tremors and record both horizontal and vertical ground accelerations caused by the

shaking (Box 6.3). Analysis of the data collected from these instruments shows that earthquakes produce four main types of seismic wave (Figure 6.3):

- *Primary waves* (P-waves) are vibrations caused by compression, similar to a shunt through a line of connected rail coaches. They spread out from the earthquake fault at a rate of about 8 km s^{-1} and are able to travel through both solid rock and liquids, including the oceans and the Earth's liquid core.
- *Secondary waves* (S-waves) move through the Earth's body at about half the speed of primary waves. These waves vibrate at right angles to the direction of travel, similar to a wave travelling along a flexed rope held between two people. S-waves, which cannot travel through liquids, are responsible for much of the damage caused by earthquakes, as it is difficult to design

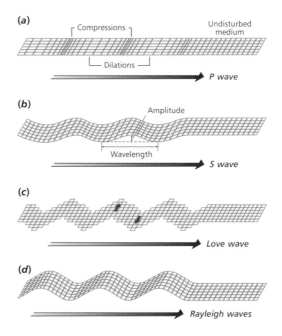

Figure 6.3 Schematic illustration of the four main types of earthquake waves: P-waves, S-waves, Love waves and Rayleigh waves. The S- and L-waves tend to have the largest amplitude and destructive force.

Box 6.3 Ground shaking in earthquakes

Information on ground motion is necessary to understand the behaviour of buildings in earthquakes. Ground acceleration is usually expressed as fractions of the acceleration due to gravity (9.8 m s^{-2}). Thus, 1.0g represents an acceleration of 9.8 m s^{-2}, whilst 0.1g = 0.98 m s^{-2}. If an unsecured object experienced an acceleration of 1.0g in the vertical plane it would, in effect, become weightless and could leave the ground. Values as large as 0.8g have been recorded in firm ground from earthquakes with magnitudes as small as M_W = 4.5, whilst the 1994 Northridge earthquake had localized peak ground motions of nearly 2.0g. Even very strong structures struggle to deal with such high vertical accelerations. The greatest damage is often generated by the Love waves, which cause horizontal shaking. Some unreinforced masonry buildings (URMs) may be unable to cope with horizontal accelerations as small as 0.1g.

Local site conditions influence ground motion. Significant wave amplifications occur in steep topography, especially on ridge crests. Ground motions in soil are enhanced in both amplitude and duration, compared to those recorded in rock. As a result, structural damage is usually most severe for buildings founded on unconsolidated material. In the Michoacan earthquake of 1985 the recorded peak ground accelerations in Mexico City varied by a factor of 5. Strong-motion records obtained on firm soil showed values of around 0.04g. This compared with observations from urban areas located on a dried lake bed where peak ground accelerations reached 0.2g. Similar effects were noted in the San Salvador earthquake of 1986. This had a very modest size (M_W = 5.4) but destroyed thousands of buildings as well as causing 1,500 deaths. The reason was rooted in layers of volcanic ash, up to 25 m thick, which underlie much of the city. As the three-second-long earthquake tremor passed upwards through the ash, the amplitude of ground movement was magnified up to five times.

The scale of destruction also depends on the frequency of the vibrations and the fundamental period of the structures at risk. The frequency of a wave is the number of vibrations (cycles) per second measured in units called Hertz (Hz). High-frequency waves tend to have high accelerations but relatively small amplitudes of displacement. Low-frequency waves have small accelerations but large velocities and displacements. During earthquakes, the ground may vibrate at all frequencies from 0.1 to 30 Hz. If the natural period of a building's vibration is close to that of the seismic waves, *resonance* can occur. This causes the building to sway. Low-rise buildings have short natural wave periods (0.05–0.1 seconds) and high-rise buildings have long natural periods (1–2 seconds). The P- and S-waves are mainly responsible for the high-frequency vibrations (>1 Hz) that are most effective in shaking low buildings. Rayleigh and Love waves are lower frequency and more effective in causing tall buildings to vibrate. The very lowest-frequency waves may have less than one cycle per hour and have wave-lengths of 1,000 km or more.

structures that can withstand this type of motion.

• *Rayleigh waves* are surface waves in which particles follow an elliptical path in the direction of propagation and partly in the vertical plane, much like water moving within an ocean wave.

• *Love waves* (L-waves) are similar to Rayleigh waves but with vibration solely in the horizontal plane.

The overall severity of an earthquake is dependent on the amplitude and frequency of these wave motions. The S- and L-waves are more destructive than the P-waves because they have a larger amplitude and force. In an earthquake, the ground surface may be displaced horizontally, vertically or obliquely, depending on the wave activity and the local geological conditions (Box 6.3).

D SECONDARY EARTHQUAKE HAZARDS

1 Soil liquefaction

An important secondary hazard associated with loose sediments is soil liquefaction. This is the process by which water-saturated material can temporarily lose strength and behave as a fluid when subjected to strong shaking. Poorly compacted sand and silt situated at depths less than 10 m below the surface is the principal medium. In the 2001 Bhuj earthquake, many reservoir dams were damaged by soil liquefaction in the water-saturated alluvial foundations. According to Tinsley *et al.* (1985), four types of ground failure commonly result:

- *Lateral spread* involves the horizontal displacement of surface blocks as a result of liquefaction in a sub-surface layer. Such spreads occur most commonly on slopes between 0.3 and 3°. They cause damage to pipelines, bridge piers and other structures with shallow foundations, especially those located near river channels or canal banks on alluvial floodplains.
- *Ground oscillation* occurs if liquefaction occurs but the slopes are too gentle to permit lateral displacement. Oscillation is similar to lateral spread but the disrupted blocks come to rest near their original position; lateral-spread blocks can move significant distances. Oscillation is often accompanied by the opening and closing of surface fissures. In the 1964 Alaskan earthquake, cracks up to 1 m wide and 10 m deep were observed.
- *Loss of bearing strength* occurs when a shallow layer of soil liquefies beneath a building. Large

deformations within the soil mass can cause structures to settle and tip. In the 1964 Niigata, Japan earthquake, four apartment buildings tilted as much as 60° in unconsolidated alluvial ground. Loss of bearing strength was a key reason for the high death toll in the 1985 Mexico City disaster, in which about 9,000 people died, even though the city was nearly 400 km from the fault rupture. Similar failures readily damage port facilities built on land reclaimed by dredged material like sand and silt.
- *Flow failure* is associated with the most catastrophic form of liquefaction because the slope fails at the surface as well as at depth. Flow failures can be very large and rapid, displacing material by tens of kilometres at velocities of tens – or even hundreds – of kilometres an hour. Such failures can happen on land or under water. The devastation of Seward and Valdez, Alaska in 1964 was largely caused by a submarine flow failure at the marine end of the delta on which these settlements were built. The harbour area was carried away and created water waves which then swept back onshore, causing further damage.

2 Landslides, rock and snow avalanches

Severe ground shaking causes natural slopes to weaken and fail. The resulting landslides, rock and snow avalanches are major contributors to earthquake disasters because many destructive earthquakes occur within mountainous areas. For example, more than half of all deaths recorded after large-magnitude ($M_W > 6.9$) earthquakes in Japan have been attributed to landslides (Kobayashi, 1981). Correlations between event magnitude and landslide distribution show that landslides are unlikely to be triggered by earthquakes of less than $M_W = 4.0$. However, the maximum area likely to be affected by earthquake-related landslides increases rapidly thereafter, to reach 500,000 km^2 at $M_W = 9.2$ (Keefer, 1984).

Central America is a region where landslides frequently occur after earthquakes (Bommer and Rodríguez, 2002).

There is considerable spatial variation in risk, due to differences in topography, rainfall, soils and land use. The greatest hazard exists when high-magnitude events (M_W =/> 6.0) create rock avalanches. These are large (at least 1 million m³) volumes of rock fragments that can travel for tens of kilometres from their source at velocities of hundreds of kilometres per hour. A notable mass movement occurred when an offshore earthquake (M_W = 7.7) triggered a massive rock and snow avalanche from the overhanging face of the Nevados Huascaran mountain, Peru in 1970 (Plafker and Ericksen, 1978). At an altitude of 6,654 m, Huascaran is the highest peak in the Peruvian Andes and its steep slopes have been the source of many catastrophic slides. In the 1970 disaster, a turbulent flow of mud and boulders, estimated at 50–100 × 10⁶ m³, passed down the Rio Shacsha and Santa valleys. It formed a wave 30 m high and travelled at an average speed of 70–100 m s⁻¹ in the upper 9 km of its course (Figure 6.4). The towns of Yungay and Ranrahirca, plus several villages, were buried under debris 10 m deep. In all, about 18,000 people were killed within four minutes after the original slope failure high on the mountain.

3 Tsunamis

The most distinctive earthquake-related hazard is the seismic 'sea wave' or tsunami. The word 'tsunami' comes from two Japanese words, *tsu* (port or harbour) and *nami* (wave or sea), an appropriate derivation, since these ocean waves inundate low-lying bays and coastal areas. If an earthquake rupture occurs under the ocean, or in a coastal zone, if the hypocentre is not deep within the Earth's crust and if the magnitude is large enough to create significant vertical displacement, then a tsunami may be generated. The result is a series of ocean waves that spread out from this point carrying very large volumes of water. When

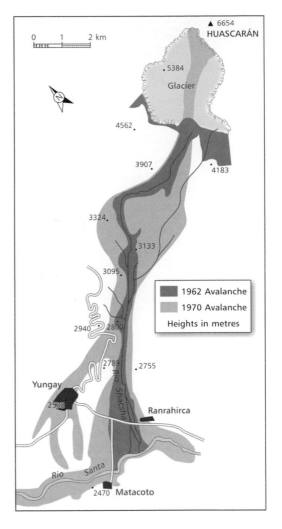

Figure 6.4 Map of the Mount Huascaran rock avalanche disasters in the Peruvian Andes in 1962 and 1970 showing the wide extent of the debris deposited in the 1970 event. After Whittow (1980).

the waves reach land, they are charged with debris and can flow inland for up to several kilometres over beaches and along estuaries, with considerable destructive force.

More than 2,000 tsunamis are known to have killed over 500,000 people worldwide. The 2004 Sumatra tsunami was the most deadly in recorded history. More than 227,898 people were killed and an estimated 1,126,900 more were made homeless

Plate 6.1 A tsunami wave generated by the $M_W = 9.0$ magnitude Great Tōhoku earthquake on 11 March 2011 crashes over a sea wall, carrying small boats, vehicles and other debris to devastate Miyako City, Iwate Prefecture, in north-east Japan. (Photo: Mainichi Shimbun/Reuters accessed on National Geographic Daily News website)

by the combined earthquake–tsunami event in several countries bordering the Indian Ocean. Large tsunamis causing fatalities are concentrated in the Pacific region, as demonstrated by 20 events recorded in the period 1995–2011 (Table 6.4).

A typical tsunami occurred in July 1998 on the north-west coast of Papua New Guinea. Following an earthquake ($M_W = 7.1$), a tsunami with maximum wave heights of 15 m overwhelmed a sand bar where several small villages were built some 1–3 m above sea level. All wooden buildings within 500 m of the shore were swept away, resulting in almost 2,200 fatalities (González, 1999). Many tsunamis are generated near Japan. Along the Sanriku coast of eastern Honshu, a tsunami wave 10 m high has been estimated to have a return period of about 10 years. A disaster in 1933 was caused by a submarine earthquake ($M_W = 8.5$) that produced a tsunami up to 24 m above mean sea level (Horikawa and Shuto, 1983). The death toll was 3,008, with 1,152 injured, together with 4,917 houses washed away and 2,346 otherwise destroyed. Smaller Japanese tsunamis have resulted in comparatively few deaths because of long-term hazard-mitigation schemes, but the 2011 Great Tōhoku disaster clearly demonstrated the continuing threat.

Tsunami behaviour

Over 60 per cent of all tsunamis originate around the Pacific Ocean and more than 80 per cent of damaging wave run-ups occur along the coasts around this basin. The first tsunami wave may not be the largest or most powerful. Most tsunamis arise from the following geophysical events:

- *Tectonic displacement of the sea-bed* is associated with large, shallow-focus earthquakes. These are most common when two tectonic plates collide and the oceanic plate slides beneath the continental plate to form a deep ocean trench. Large subduction zones exist along the island arcs and coastlines of the Pacific Ocean; in the Indian Ocean, the Indo-Australian plate is subducted beneath the Eurasian plate. When rapid vertical movement of the sea floor takes place, it displaces the water column above. In 2004 an accumulation of seismic stress near Sumatra caused the Earth's crust to deform downwards. When the fault ruptured, the crust rebounded upwards about 5 m. This raised a huge volume of water which flowed outwards to equilibrate sea level. The leading wave travelled quickly, arriving at the Indian and Sri Lankan coasts just 90 minutes after the earthquake, and reached the coast of Somalia after about seven hours.
- *Volcanic eruptions* can lead to high-mortality tsunamis with, or without, earthquake activity. It is likely that when volcanoes do trigger tsunamis, about one-quarter of the fatalities are due to the tsunami rather than the eruption. Approximately half of all these combined events are produced at the caldera of the volcano. Catastrophic consequences can result from the occasional collapse of volcanic islands (e.g. Krakatoa in 1883).
- *Major landslides* on coastal land or below sea level can create tsunamis with locally devastating powers. These events are often caused by earthquakes and result in the displacement of ocean water, due to large rock falls and debris slides into confined bays or lakes. For example, the submarine landslides following the Prince William Sound earthquake in 1964 generated a tsunami that killed some 80 people in Alaska.

Table 6.4 Worldwide recorded fatalities from tsunamis 1995–2011 (data from the National Geophysical Data Centre Tsunami event database)

Date	Magnitude	Location of deaths	Number
14 May 1995	6.9	Indonesia	11
9 Oct 1995	8	Mexico	1
1 Jan 1996	7.9	Indonesia	9
17 Feb 1996	8.2	Indonesia	110
21 Feb 1996	7.5	Peru	12
17 July 1998	7.1	Papua New Guinea	2,183
17 Aug 1999	7.6	Turkey	150
26 Nov 1999	7.5	Vanuatu	5
23 June 2001	8.4	Peru	26
26 Dec 2004	9	Indonesia	250,000
28 Mar 2005	8.7	Indonesia	10
14 Mar 2006	6.7	Indonesia	4
17 July 2006	7.7	Indonesia	664
1 Apr 2007	8.1	Solomon Islands	52
21 Apr 2007	6.2	Chile	3
29 Sep 2009	8.0	Samoa Islands	191
12 Jan 2010	7.0	Haiti	7
27 Feb 2010	8.8	Chile	124
25 Oct 2010	7.8	Sumatra	431
11 Mar 2011	9.0	Japan	13,232

Source: http://www.ngdc.noaa.gov/seg/hazard/tsu.shtml

Tsunamis are detected through the Deep-ocean Assessment and Reporting of Tsunamis (*DART*) programme that forms part of an almost worldwide early warning system. Surface buoys and sea floor monitors are strategically located to detect the water pressure changes that result from a tsunami and to transmit the information directly to observers onshore. Figure 6.5 shows the graph for DART station 21413, located 690 nautical miles south-east of Tokyo on 11 March 2011, which recorded the Great Tōhoku earthquake. It can be seen that, out at sea, with an ocean depth of about 5,800 m, the tsunami wave led to an immediate vertical displacement of the water surface amounting to about 1 m. The amplitude of this wave reached its maximum run-up height on reaching land.

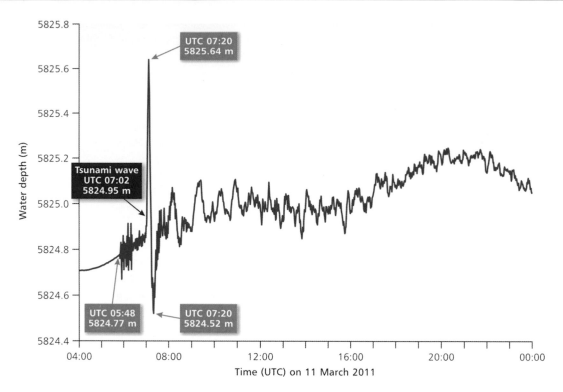

Figure 6.5 Changes in water level in the Pacific Ocean south-east of Tokyo, Japan, showing the large displacement that created the tsunami of 11 March 2011. Reproduced under Creative Commons Attribution-ShareAlike 3.0 Unported License.

At first, tsunamis behave as shallow-water waves, due to their exceptionally long (100–200 km) wave-lengths. This means that the initial velocity of forward travel is very high. The forward speed of a shallow wave depends upon the water depth according to the following function:

$$\text{Velocity} = \sqrt{gD}$$

where g = gravitational constant (9.81), D = depth of the ocean

So, if the depth of the Indian Ocean averages 3900 m:

$$\text{Velocity} = \sqrt{9.81 \times 3900}$$
$$= 196 \text{ m s}^{-1}$$
$$= 704 \text{ km per hour.}$$

The average recorded velocity of the 2004 tsunami wave from Sumatra was close to 640 km hr^{-1}, probably reflecting a slowing down of the wave as it approached coastal regions. Crossing the ocean, the wave was only about 60 cm in height and posed no hazard. But, as it neared the shore, the wave slowed down and started to increase in height, so that on Banda Aceh, where much of the damage took place, the wave reached a maximum height of over 30 m.

Figure 6.6 illustrates the evolution of a typical tsunami wave. In mid-ocean the great depth of the sea allows the leading edge to move forward rapidly. This movement keeps the wave amplitude low so that it may be detected simply as light swell in the open ocean. But, when entering shallower water, the speed reduces and the wave-length shortens. More of the wave energy is then put into a much smaller volume of water and the wave height increases. As a result, during its run-up phase over a low-lying coastal strip, the wave may reach 10 or 20 times its height in the open sea. If the tsunami passes into a confined gulf or bay, especially one that narrows inland, the amplitude will increase further.

Once an earthquake has been detected by the DART network of instruments, and its epicentre has been located, it becomes possible to predict with some accuracy the arrival time of the tsunami wave on a distant shoreline. Figure 6.7 is a computer-generated map of the passage of the 2004 Sumatra tsunami across the Indian Ocean from the epicentre. Such simulated travel times are normally accurate to within one hour of the actual arrival. In some cases the wave is preceded by a brief local retreat of the sea, leaving extensive areas of the seashore exposed. In the case of the Indian Ocean tsunami, some people recognized this feature as an indication of danger and managed to escape. But hundreds of others were attracted to the sea shore by this unusual sight and were exposed to the full force of the wave when it made landfall.

E PROTECTION

1 Environmental control

There is little immediate prospect of preventing earthquakes at source. Disaster reduction must, therefore, focus on lowering human vulnerability and limiting secondary hazards.

2 Hazard-resistant design

According to Key (1995), about 60 per cent of all earthquake-related deaths are caused by the failure of *unreinforced masonry structures* (URMs) in rural areas. The most vulnerable buildings are constructed from *adobe* material (sun-baked clay bricks). Adobe construction is common in arid and semi-arid regions because it is cheap, easily worked and widely available. In Peru, an

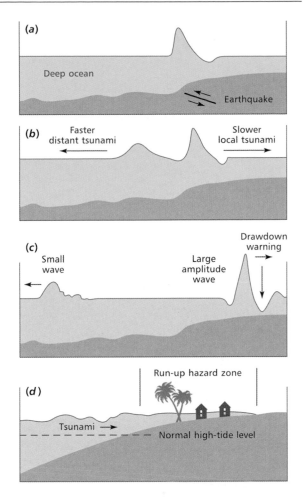

Figure 6.6 Typical evolution of a tsunami wave: (a) earthquake initiation, (b) wave split, (c) near-shore wave amplification, (d) coastal run-up zone. The vertical scale is greatly exaggerated on this diagram. Adapted from US Geological Survey (Western Coastal and Marine Geology) at www.walrus.wr.usgs.gov/tsunami (accessed 7 June 2003).

estimated two-thirds of all rural dwellers live in adobe houses; and in the 2003 Bam earthquake many such buildings collapsed, killing over 26,000 people (see Chapter 3). Houses built of rubble masonry are also prone to collapse. In the Maharashtra, India earthquake of 1993 the *pucca* houses – with thick granite walls and heavy timber roofs – led to more deaths than either the thatched huts or buildings poorly reinforced with concrete

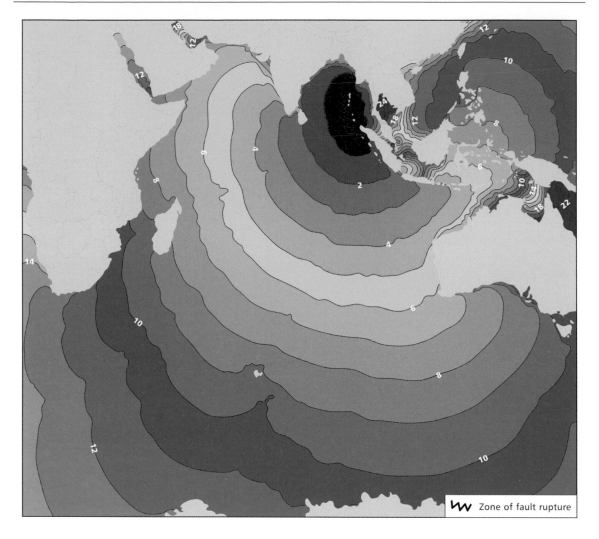

Figure 6.7 Progress of a tsunami wave across the Pacific Ocean after the M_W = 9.1 Sumatra earthquake at the Sunda trench on 26 December 2004. Isolines indicate one-hour time intervals. More than 4 million Indonesians live in tsunami-prone areas and need an effective monitoring system capable of delivering very short-term warnings. After NOAA at www.ngdc.noaa.gov/hazard/icons/2004 US government material not subject to copyright protection in the United States.

frames. Some traditional societies have deliberately employed 'weak' construction as a defence against earthquakes. Throughout tropical Asia, the indigenous house is lightly built with plant-matting walls and palm-frond roof (Leimena, 1980). During the M_W = 8.7 Nias earthquake in Indonesia in 2005, 1,300 people died in collapsed buildings, but the native wood-framed longhouses survived mostly undamaged. Other countries favour houses of wood-frame construction. Such buildings account for about 80 per cent of all houses in the USA. There is a clear seismic benefit because these structures tend to flex, rather than fail, when subjected to ground shaking, although they carry a high fire risk – an important secondary hazard in earthquakes.

Earthquake risk is also found in some large cities. Over time, URM buildings have been

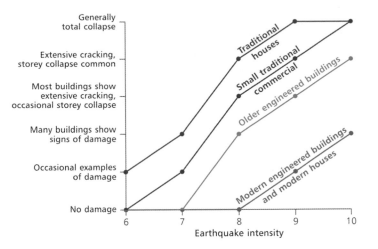

Figure 6.8 Relationships between earthquake intensity (Mercalli scale) and the damage for different types of building construction, based on the 1995 Kobe earthquake. Sharp difference in performance occured between traditionally constructed and modern engineered buildings. After Alexander Howden Group Ltd and Institution of Civil Engineers (1995).

supplemented by high-rise, reinforced concrete structures – mainly apartment blocks – built to accommodate the growing population. This means that most urban areas present a complex array of risk. Figure 6.8, based on the effects of the Kobe earthquake, shows general relationships between earthquake intensity and building damage for different types of structures. It also emphasizes how the risk of collapse rises with an ageing stock of buildings. Lack of building strength is a serious matter in the urban areas of the LDCs. In Kathmandu, Nepal, devastated by an earthquake in 1934, 70 per cent of the buildings are poorly designed and constructed. Even reinforced concrete structures here would perform badly in an earthquake (Petley, 2009).

The only way to achieve safe buildings in seismically active urban areas is through detailed risk assessments and the adoption of best construction practices. The first requirement is for a geotechnical engineer to assess the suitability of the location. Other things being equal, buildings on solid rock are less likely to suffer damage than those built on clays or softer foundations. This initial assessment should include geological investigations to avoid building near faults or

where the bearing strength of the foundation materials is inadequate (Box 6.4). Seismic *building codes*, now adopted in over 100 countries, stipulate the minimum construction standards required to minimize the risk of collapse. To be effective, a seismic code needs full legal status and procedures for up-dating the criteria over time. Above all, the code should ensure that the construction methods for the planned building are adequate to withstand the selected design earthquake and should also require regular site inspections throughout the construction period. Unfortunately, building codes are frequently neglected and bypassed, due to a lack of resources, imperfect technical knowledge and local corruption. In the 1999 Marmara earthquakes in north-west Turkey, 20,000 people were killed and 50,000 were injured, despite the fact that building codes have existed there since the 1940s. Most deaths were blamed on non-compliance with the codes, largely due to financial corruption.

The majority of building stock in most areas will pre-date the building codes. Retrofitting old buildings to meet new standards costs money and can be equally difficult to undertake. Much importance is attached to the earthquake

Box 6.4 Earthquake safety and buildings

The key to earthquake resistance is the appropriate choice of building design and construction methods. In this context, strong, flexible and ductile materials are preferred to those that are weak, stiff and brittle. For example, steel framing is a ductile material that absorbs a lot of energy when it deforms. Indeed, the spread of well-designed, steel-reinforced concrete buildings has been the primary factor in increasing earthquake safety for many decades. Glass, on the other hand, is a very brittle material that shatters easily. In practice, both types of material have to be incorporated into structures. Some otherwise well-designed structures collapse because of the failure of a single element which lacks sufficient strength or ductility. For example, buildings with flexible frames will often fail if the frames are in-filled with stiff, masonry brickwork.

The shape of a building will influence its seismic resistance. A stiff, single-storey structure (Figure 6.9a) will have a quick response to lateral forces, while tall, slender, multi-storey buildings (Figure 6.9b) respond slowly, dissipating the energy as the waves move upward to give amplified shaking at the top. If the buildings are too close together, pounding induced by resonance may occur between adjacent structures and add to the destruction. The stepped profile of the vertical mass of the building in Figure 6.9c offers stability against lateral forces. Most buildings are not symmetrical and form more complex masses (Figures 6.9d and 6.9e). These asymmetrical structures will experience twisting, as well as the to-and-fro motion. Unless the elements are well joined together, such differential movements may pull them apart. High-rise structures will be vulnerable if they do not have uniform strength and stiffness throughout their height. The presence of a *soft storey*, which is a discontinuity introduced into the design for architectural or functional requirements, may be the weak element that brings down the whole structure. Figure 6.9f shows a soft ground-floor storey, perhaps introduced to ease pedestrian traffic or car parking.

The weakest links in most buildings are the connections between the various structural elements, such as walls and roofs. Connections are important in the case of pre-cast concrete buildings, where failure often results from the tearing-out of steel reinforcing bars or the breaking of connecting welds. In the 1994 Northridge earthquake a number of multi-storey car parks failed when vertical concrete columns were cracked by lateral ground shaking to the point where they became unable to support the horizontal concrete beams holding up the different floors. Exterior panels and parapets also need anchoring firmly to the main structure in order to resist collapse. Architectural style can contribute to disaster if features like chimneys, parapets, balconies and decorative stonework are inadequately secured.

Difficult construction sites (Figures 6.9g and 6.9h) include localities near to geological faults and soft soils that amplify ground shaking. As far as possible these should be avoided or built up at low densities so that, for example, buildings cannot collide as a result of downward movement on slopes. Some slopes may have to be reformed by cut and fill to limit the threat from earthquake-related landslides (Figure 6.9i). Methods of building reinforcement include the cross bracing of weak components, placing the whole structure in a steel frame and the installation of special, deep foundations on soft soils (Figures 6.9j–l). Adequate footings are important. High-rise buildings on soft soils should have foundations supported on piles driven well into the ground. Wood-framed houses should be internally braced with plywood walls tied to anchor bolts linked into foundations 1–2 m deep. Some new buildings can be mounted on isolated shock-absorbing pads made from rubber and steel which prevent most of the horizontal seismic energy being transmitted to the structural components. The technique is expensive but provides maximum protection for the loose contents of buildings, thus making it attractive for hospitals, laboratories and other public facilities. In addition, base-isolated buildings need less structural bracing to withstand lateral forces, so that the reduction in construction materials offsets the extra cost of the isolation system.

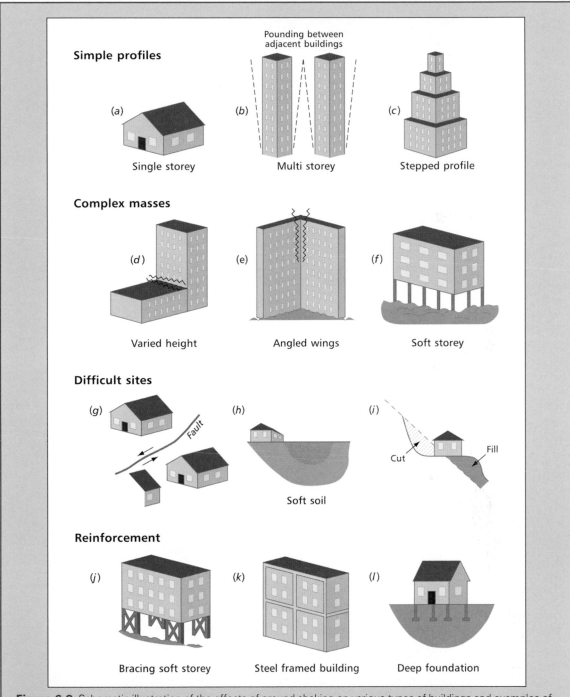

Figure 6.9 Schematic illustration of the effects of ground shaking on various types of buildings and examples of constructed measures designed to resist earthquake hazards: (a–c) simple building profiles, (d–f) complex building masses, (g–i) coping with difficult sites, (j–l) methods of building reinforcement.

resistance of public structures such as hospitals, dams, nuclear power stations and factories with explosive or toxic substances. Urban lifelines for transport, electric power, water supply and sewerage also need priority. Some commercial organizations take separate precautions, especially when insurance cover is unavailable unless there is compliance with standards in the building code. For example, the IBM manufacturing plant at San José, California was subjected to an early retrofit programme (Haskell and Christiansen, 1985). As a result, it was able to get all but one of its Santa Clara buildings back into full operation the day after the Loma Prieta earthquake of 1989.

In the USA, the *Uniform Building Code* is updated annually using a map of six seismic zones based on ground motion activity and recorded damage in previous earthquakes. The greatest threat is in California. Approximately three-quarters of the nation's seismic risk to general building stock lies in this state, where more than half of the population lives within 30 km of active faults. California has many URMs, built before the 1933 building code, now judged to be unsafe. The Unreinforced Masonry Building law passed by the state legislature in 1986 required all cities and counties in areas of high seismic hazard, including the metropolitan areas of Los Angeles and San Francisco, to identify all such buildings by 1 January 1990. At the time, the expected expenditure was estimated at US\$4 billion. The inventory has information on building use and daily occupancy loads. Superficially, there has been a high level of compliance with this law; over 98 per cent of the 25,900 URM buildings are in loss-reduction programmes (California Seismic Safety Commission, 2006). However, only 70 per cent of property owners have retrofitted in accordance with the appropriate building code. New measures

under consideration include incentives from local governments to promote more participation.

Even when properly applied, building codes cannot prevent all seismic damage. The codes may be based on an incomplete knowledge of local geology and calculations regarding the performance of buildings under stress may be flawed, especially where regulations are based on experience imported from elsewhere. Administrators and decision makers are not always convinced of the threats and may fail to see any financial advantage in the need for investment in earthquake security. At worst, building codes can lead to new building in hazardous areas if they create a false sense of security.

Civil engineering techniques are also used to limit the impact of secondary earthquake hazards. For example, slopes alongside transport networks in seismic zones are designed to withstand the maximum forces expected during the lifetime of the facility. In New Zealand, the design standard for road bridges is an earthquake with a 450-year return period; similar criteria apply to cut slopes and embankments. Special protection is necessary against tsunamis. Following the 1933 Sanriku tsunami off the north-east coast of Japan, government grants were provided for the relocation of some fishing villages to higher ground (Fukuchi and Mitsuhashi, 1983). This policy was ineffective, due to the limited availability of nearby land for resettlement and the natural desire of fishermen to remain close to the shore. Efforts then shifted to the construction of onshore tsunami walls, although these failed to prevent further losses in 1960. After this event, the government passed a special law to subsidize construction costs up to 80 per cent for sea walls designed to cope with wave heights equivalent to the 1960 event. Since then, engineers have continued to protect against

---

Plate 6.2 *(Facing page)* Examples of structural collapse in different types of buildings. (a) A 100-year old adobe building in Talca, Chile, after the Mw = 8.8 earthquake in February 2010. Adobe construction using clay, sand and straw is no longer authorized in Chile. (Photo: Walter D. Mooney, USGS) (b) A multi-storey residence in Port-au-Prince, Haiti in January 2010. The weight of concrete floors and roof proved too great for the strength of the supporting columns. (Photo: Walter D. Mooney, USGS)

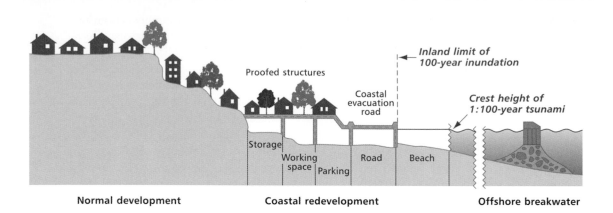

Figure 6.10 Depiction of tsunami protection works typical of the Sanriku coast, Japan, showing an offshore breakwater and some raised onshore development, including an emergency evacuation route.

even larger tsunamis and the highest walls stand up to 16 m above tidal datum level.

Japan employs tsunami breakwaters in addition to sea walls. Breakwaters do not take up valuable land and provide shelter for shipping, but they are expensive and can interfere with tidal circulations and damage the local fishing industry. More recently, some elaborate onshore tsunami walls designed to protect coastal property and transport links have been used, although these structures are visually intrusive. Figure 6.10 shows a tsunami defence scheme, including an offshore breakwater and an onshore tsunami wall combined with coastal redevelopment. These projects are expensive and give coastal communities the appearance of fortified towns, with limited views of the ocean for residents. The town of Taro has both inner and outer sea walls; the inner one is 10 m high and stretches 2.5 km across the bay. These defences were over-topped by the 2011 tsunami, which destroyed much of the town. About 80 km to the south, what is probably the world's deepest breakwater was completed in 2009, after a 30-year construction period and at a cost of US$1.5 billion, to protect the city of Kamaishi. These defences also failed in 2011. The whole issue of coastal engineering is currently under fresh scrutiny as Japan searches for solutions to the tsunami hazard.

F MITIGATION

1 Disaster aid

Earthquake disasters readily attract emergency donations because of the sudden loss of life, heavily publicized by television imagery. However, aid issues extend beyond the immediate crisis. In the aftermath of the 2004 Sumatra earthquake, more money was raised than could be used for emergency relief work and led to debate about the equitable distribution of funds. Six months after the event, a survey for Oxfam concluded that much of the aid had gone to businesses and landowners, thereby increasing the gap between rich and poor people in the area. There was evidence that the poorest people remained in the refugee camps for longer than others, that they experienced more difficulty with rebuilding their lives, and there was concern that their needs would be side-lined during the reconstruction phase (Petley, 2009).

The 'golden hours' after an earthquake are crucial for the recovery of survivors trapped in fallen buildings. About 17 specialist search-and-rescue teams are kept on permanent stand-by around the world with the capability to reach a disaster zone within hours of a request from the relevant country. These teams carry enough food

Table 6.5 The number of people who survived after being rescued from collapsed buildings, by day of rescue, following the Kobe earthquake on 17 January 1995

Date	17 January	18 January	19 January	20 January	21 January
Total rescued	604	452	408	238	121
Total who lived	486	129	89	14	7
Percentage rescued who survived	80.5	28.5	21.8	5.9	5.8

Source: After Comfort (1996)

and water supplies to remain self-sufficient for up to two weeks. This activity is highlighted by the donor authority and in the media, but there is little reliable data on the practical value of such external support. The UK's International Search and Rescue Group (ISAR) went in quick succession to Christchurch (New Zealand) and Honshu (Japan) following earthquakes in February and March 2011. It was reported that 62 people with nine tonnes of lifting gear went to New Zealand; 59 people, 4 medical staff and 2 sniffer dogs went to Japan. In neither case were any survivors rescued. This is not to say that there were no benefits. The teams helped to clear debris and bolstered morale as part of the humanitarian response, but prior training of local capacity would fit better with the emergency timescales involved.

Noji *et al.* (1993) observed that after the 1988 Armenian earthquake 67 per cent of all persons rescued were recovered in the first six hours. Fewer than 3 per cent of recoveries were achieved by specialist Soviet rescue teams flown in from outside the region and fewer than 1 per cent by overseas teams. The city of Kobe was similarly ill-prepared for the 1995 disaster. Search and rescue here was hampered by several factors, including a legal ruling that kept rescue dogs sent from overseas in quarantine until the fourth day after the earthquake (Comfort, 1996). Table 6.5 illustrates the decline in the numbers of people recovered from buildings over the first five days of this disaster and the progressive decline in the survival rate. Sometimes local self-help is vital.

After the Michoacan (Mexico City) earthquake of 1985, the official rescue service was so limited that residents in the most badly damaged area set up their own arrangements (Comfort, 1996). Landslides following earthquakes in more remote mountainous areas also cause problems for search-and-rescue operations and the delivery of aid (Box 6.5).

Once the emergency response phase is over, new decisions are required. The basic aim is to provide temporary support for as short a time as possible to avoid dependency. This means working closely with local food vendors and water providers, sourcing goods and other services in a way that minimizes competition with indigenous activities. In other words, a major question facing aid agencies is how best to manage the shift from short-term humanitarian support to longer-term intervention. In the first six months following the 2010 Haiti earthquake, the UK-based Disasters Emergency Committee spent about half of its appeal funds more or less equally between supplying water and sanitation and providing shelter. Other early needs included food and health care.

In the longer term, specific problems emerged regarding housing in Port-au-Prince, (Clermont *et al.*, 2011). During recovery in urban settings, the safe return of tented residents to permanent accommodation within their own neighbourhoods is a priority. If this is done on a well-planned, participatory basis, it is preferable to the creation of new settlements – often little more than camps – in areas lacking public services or job opportunities. However, this policy may conflict

Box 6.5 Problems of aid delivery in the aftermath of a major earthquake

The $M_W = 7.6$ earthquake of 8 October 2005 had a devastating impact on a large area of Pakistani Kashmir. According to official government statistics, the earthquake killed over 73,000 people in Pakistan, of which 19,000 were school-age children. A further 1,360 persons died in India. The earthquake left over 100,000 people with injuries. Over 780,000 buildings were damaged beyond repair, the vast majority (97 per cent) of which were houses. As a result, approximately 2.8 million people were left homeless (Petley, 2009).

A massive relief operation was initiated by the Government of Pakistan and by international agencies, such as the International Committee of the Red Cross and the World Food Programme. However, all assistance was hindered by two factors:

- A lack of preparedness meant that little planning had been undertaken for the relief operation.
- The Kashmir area is highly mountainous, with limited communication routes.

One of the main areas affected was the valley of the river Neelum. This area is accessed by a single road only, which crosses the fault in a zone with a steep river gorge. Landslides due to the earthquake blocked this road over a 20 km stretch and, in the after-shocks, landslides continued to occur on an hourly basis. Additionally, many of the most seriously damaged villages were located on the high slopes of the mountains, accessed only by small roads traversing very steep hillsides. In almost all cases, these roads were destroyed (Peiris et al., 2006). Aid delivery was extremely difficult. Even an assessment of the needs of the population was problematic, as the evaluation teams could not travel through the affected areas without helicopters, which were in short supply. Although the government mobilized 12 brigades of the engineer corps of the Pakistan Army, most of the major highways were not reopened for a month and reopening the Neelum Valley road took six weeks. Some minor roads remained closed three years later and are regularly damaged by landslides.

Logistical problems included restrictions on the supply of emergency medical care, food and shelters. The lack of shelter was a serious issue because winter in Kashmir is very cold. As a result, two unusual steps were taken in Kashmir. First, there was a major effort to move people into refugee camps close to the main roads in the valley floors. In most cases this meant moving people away from their home villages. This is undesirable because it increases disaster trauma and delays the rebuilding process, but, given the limitations of the transport system and the intense cold in the mountains, it was the only realistic solution. Second, there was much reliance on helicopters to deliver assistance. Helicopters were deployed from around the world – for example, the UK sent three RAF heavy-lift Chinook helicopters and the United States sent a further 12 to assist the large numbers used by the Pakistan Air Force. Additional assistance came from civilian agencies; the World Food Programme alone deployed 14 helicopters.

Fortunately, these humanitarian efforts, plus relatively benign winter conditions, combined to ensure the survival of most of the earthquake-affected population.

with the needs of the most vulnerable victims, who depend for longer on short-term shelters and require more health care and other support. Some NGOs favour an intermediate move into pre-fabricated transitional 'T' shelters because it often fits the timeframe of their budgets, but these shelters are almost always better suited to rural situations. These issues were prominent in Haiti, where it was difficult to erect 'T' shelters, due to disputes over land ownership and availability.

Long-term reconstruction rarely conforms to the original timetable. The 1988 Armenian earthquake made 514,000 people homeless and resulted in the evacuation of nearly 200,000 persons. Following the Soviet government's decision to accept international aid, over 67 nations offered cash and services amounting to over US$200 million. A reconstruction programme was announced to rebuild the cities within a two-year period on sites in safer areas, with building heights restricted to four storeys. But, during the first year, only two of the 400 buildings due for construction in Leninakan were completed and many people lived as evacuees many months after the disaster. Continuing aid and support is also required for less tangible purposes, such as the treatment of post-traumatic stress following earthquake disasters (Karanci and Rüstemli, 1995).

2 Insurance

Most earthquake risk is uninsured, largely because few private companies have the capacity to absorb the potential costs. Some countries have national insurance schemes that guarantee the house-holder a payout for earthquake-damaged property. Part of the central fund is invested to provide capital to cover claims and part is used to purchase reinsurance. Any shortfall is under-written by the government. Take-up is low. For example, the Taiwan Residential Earthquake Insurance Fund was established by the government in 2002 in response to the 1999 Chi-Chi earthquake, when fewer than 2 per cent of house-

holds had earthquake insurance. By 2007, the scheme covered only 25 per cent of households and it was estimated that the maximum take-up, when the scheme was fully mature, would still be less than 50 per cent (Petley, 2009). Given the low premiums that property owners appear willing to pay and the rising costs of earthquake-related damage, some observers recommend partner-ships between the public and private sectors. Such schemes could include a mandatory tax on occu-piers of very high-risk properties, but not all active faults are mapped and the insurance industry is rightly suspicious of the degree of compliance with local building codes in many areas.

The risks, and the potential costs, are high in California. The 1906 San Francisco earthquake caused property losses estimated at US$524 million at the time, a figure that would probably be in excess of US$100 billion today. The cost of domestic insurance premiums is rated according to the location of the property with respect to eight different seismic zones within the state, the type of construction and local soil conditions. Premium rates rise progressively from small wood-frame houses to masonry buildings con-structed on filled land. About 70 per cent of this 'catastrophic earthquake cover' is underwritten as an extension to the normal home-owner's policy by the California Earthquake Authority, a publicly managed organization. A basic home-owner's policy costs about US$500–US$2,000 per year. The average additional premium for earth-quake cover is around US$700 per year but could cost a further US$2,000–5,000 for an older property in San Francisco. All claims are subject to deductibles (the policy excess) usually amount-ing to 10–15 per cent. These costs are deemed too high by most Californians, especially those in the high-risk areas. Average take-up rate is estimated to be around 12 per cent.

On the whole, large businesses have a rather higher take-up rate for earthquake insurance than do householders but some commercial and industrial property may not be eligible for cover under government schemes. When the Northridge, California earthquake occurred in 1994 the insured losses totalled US$15.3 billion. At the time, only about 20 per cent of commercial firms carried insurance. A majority of these policy-holders failed to file claims after the event, sug-gesting that costs were absorbed within the organization. An important feature of business insurance is the cover for the interruption to work that occurs when either the property itself or the infrastructure that supports it – local roads or the electricity supply – is damaged. These indirect costs may be greater than the direct physical losses to the premises.

G ADAPTATION

1 Community preparedness

Community preparedness and disaster recovery planning is an important form of adaptation. Many schemes have been created in response to previous failures. Following the 1999 Marmara earthquakes in Turkey, the government estab-lished new emergency management centres in Istanbul and Ankara to prepare plans for miti-gating future disasters, but interest in prepared-ness planning is low in Turkey and elsewhere (Tekeli-Yeşil et al., 2010). During recent years, considerable resources have been invested in preparedness projects for tsunami-affected areas around the Indian Ocean. In the USA, earthquake preparedness is centred on concern about the potential for large-scale movements along the San Andreas fault in California, and in 1981 the Seismic Safety Commission established two preparedness projects for the Los Angeles and San Francisco areas.

Community preparedness is best developed at the local level within a framework established by state or national government. In some cases, it may be difficult to identify the areas at greatest risk. The unexpected devastation of the Kobe earthquake of 1995 was partly attributable to a belief that the Tokyo area was at greater risk. Consequently, the stockpiles of emergency foods

and medicines at Kobe were inadequate. The 2005 Kashmir earthquake in Pakistan was the first large earthquake to affect that country in living mem-ory. As a result, preparedness was inadequate. The emergency services were not trained for search-and-rescue operations and no contingency plan was in place to bring in assistance from outside the earthquake-affected area. Many people who survived the initial ground shaking were trapped under the rubble of their houses and injured survivors died because of a shortage of hospital beds and specialist treatment.

The creation of community preparedness is neither a short-term nor a simple process. As early as 1985, the California Legislature adopted a programme, 'California at Risk: Reducing Earth-quake Hazards 1987–1992', directed at involving local officials, city and county managers and others in an action plan for earthquake mitigation (Spangle, 1988). The basic checklist is shown in Table 6.6. Such a complex programme requires multiple pieces of legislation. In 1995 the Cali-fornia Seismic Safety Commission noted:

Although the Commission believes California's seismic safety practices for building and land use are among the best in the world, there remain weaknesses that result in unacceptable risks to life and the economy.

(CSSC, 1995)

To be successful, preparedness must engage the public. The degree of personal preparedness is best measured by safety improvements at the household level (Lindell and Perry, 2000). According to the California Seismic Safety Com-mission (2009), over 70 per cent of state residents have been affected by earthquakes, and over 80 per cent have received information about preparedness but the take-up of hazard-reducing measures is not high. Figure 6.11 shows the extent to which structural reinforcement and insurance has been adopted for two areas of high seismic risk, as compared with the remainder of the state. The take-up was slightly greater in the high-risk

Table 6.6 Earthquake safety checklist

Existing development

- inventory hazardous buildings
- strengthen critical facilities
- reinforce hazardous buildings
- reduce nonstructural hazards
- regulate hazardous materials

Emergency planning and response

- determine earthquake hazards and risks
- plan for earthquake response identify resources for response
- establish survivable communications system
- develop search-and-rescue capability
- plan for multijurisdictional response
- establish and train a response organization

Future development

- require soil and geologic information
- update and improve safety element
- implement Special Studies Zones Act
- restrict building in hazardous areas
- strengthen design review and inspection
- plan to restore services

Recovery

- establish procedures to assess damage
- plan to inspect and post unsafe buildings
- plan for debris removal
- establish programme for short-term recovery
- prepare plans for long-term recovery

Public information, education and research

- work with local media
- encourage school preparation
- encourage business preparation
- help prepare families and neighbourhoods
- help prepare elderly and disabled
- encourage volunteer efforts
- keep staff and programmes up to date

Source: After Spangle and Associates Inc. (1988)

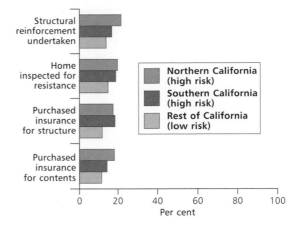

Figure 6.11 Structural measures and insurance adopted by residents of California to reduce losses from earthquakes. Data from CSSC (2009). Fewer than one in five residents have taken precautionary action. Reproduced with permission.

northern and southern California counties but fewer than 20 per cent of households had undertaken structural mitigation or purchased earthquake insurance. This pattern may reflect a need for more detailed advice about anti-seismic adjustments (Lindell and Whitney, 2000). Other aspects were more positive. For example, more than 60 per cent of all residents claimed that they knew how to stay safe in an earthquake and had made back-up copies of important documents. Over 65 per cent of Californians also claimed some training in first aid, but fewer than 5 per cent had learned this because of earthquakes. The report concluded that residents in the high-risk areas needed to do more to adapt to the risks that they faced.

After the Kobe earthquake in Japan, the road network failed because roads collapsed and were blocked by fallen debris. The arrival of medical assistance was delayed for several days. Better traffic management is clearly necessary after urban earthquakes, supported by heavy lifting equipment to clear the streets of rubble. Most disaster experts believe that urban earthquake survivors should be prepared to spend several days on their own and be given training in basic first aid, search-and-rescue and fire-fighting techniques. In these circumstances, preparedness at the family level is crucial (Russell *et al.*, 1995). In New Zealand, government advertising urges the population to have an earthquake-survival pack with enough supplies to last three days. Basic elements include canned or dried food, a portable stove, nine litres of bottled water per person, a first-aid kit, toiletries, torches and spare batteries, a radio, wind-proof and rain-proof clothing and sleeping bags.

Public participation in training schemes, such as earthquake drills, is a vital part of preparedness. If properly undertaken, emergency simulations can provide practical information on first aid and household evacuation as well as raise general hazard awareness. But such events are difficult to organize. One attempt to hold a coordinated drill throughout the San Francisco Bay area met with limited success (Simpson, 2002). In Japan, legislation first introduced in 1961 includes emergency drills in disaster prevention. A comprehensive exercise is organized annually in designated areas on *Disaster Prevention Day* (1 September), with clear emphasis on raising the awareness of children and other citizens. In 2001 the Tokai earthquake drill involved 1.5 million people from Tokyo and the surrounding area.

2 Forecasting and warning for earthquakes

A reliable forecast or prediction should specify that an earthquake of a given magnitude is likely in a certain area within a stated time-window. At the present time, such predictions are not possible and it is not even clear how such information could be applied for the maximum benefit.

Probabilistic methods

A record of the frequency of large earthquakes in any area can be used to estimate the future likelihood of similar events. In a country like New Zealand, where the pattern of earthquake activity does not correlate well with the surface geology, the historical record is useful (Smith and Berryman, 1986). Figure 6.12a employs data on shallow-focus earthquakes of M ≥ 6.5 from 1840 to 1975 to map the return periods for intensity MM VI and greater, the level at which significant damage begins. Figure 6.12b shows the intensities with a 5 per cent probability of occurrence within 50 years. This type of regional zoning has clear limitations because it is based on periods that are short on geological scales and fails to account for local ground conditions.

A major assumption implicit in statistical analysis is that earthquakes occur randomly through time. This may not happen, because fault lines can interact with each other. This point was illustrated following the tsunamagenic 2004 Sumatra earthquake, which led to increased stress concentrations on the adjacent fault lines. By February 2008, there had been seven subsequent earthquakes of $M_W \geq 7.0$ in the same area, including the March 2005 $M_W = 8.7$ Nias earthquake, which killed 1,300 people; the May 2006 $M_W = 6.2$ Java earthquake, which killed 5,800 people; and the July 2006 $M_W = 7.6$ Java earthquake, which killed 660 people. Further large events seem likely on unruptured sections of the fault, in particular near the island of Mentawi (Petley, 2009).

Seismic behaviour varies between individual fault segments. For example, the San Andreas Fault in California consists of both locked and creeping segments. Locked segments allow sufficient strain to build up to trigger major earthquakes, whilst creeping segments are characterized by more continuous sliding processes. Such creep appears to result from the presence of finely crushed rock and clay with a low frictional resistance which limits the build-up of stress. However, more competent rocks at greater depth may accumulate stress. Dolan *et al.* (1995) claimed that, in the Los Angeles region, far too few moderate earthquakes had occurred during the last 200 years to account for the observed accumulation of tectonic strain. It is possible that the historic record reflects a period of unusual

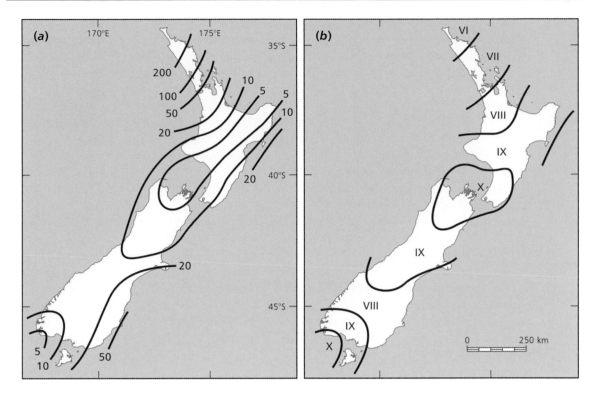

Figure 6.12 Earthquake prediction in New Zealand. (a) Return periods (years) for earthquakes of MM VI and greater; (b) intensities with a 5 per cent probability of occurrence within 50 years. After Smith and Berryman (1986).

quiescence but damaging strain may be accumulating. The US Geological Survey has calculated that the probability of at least one earthquake (M = 6.7 or more) striking between 2000 and 2030 in the San Francisco Bay area is 70 per cent (twice as likely as not).

Deterministic methods

Deterministic methods rely on the detection of earthquake precursors near the active fault. A number of different phenomena have been employed, including:

- *Seismicity patterns.* Some researchers have suggested that characteristic changes occur in background seismicity during the period before an earthquake, primarily due to shifts in the stress state of the fault as failure starts to develop.

- *Electromagnetic field variations.* Others have claimed that the development of a fault rupture might lead to variations in the Earth's magnetic field that can be detected.

- *Weather conditions and unusual clouds.* A few scientists maintain that distinctive cloud patterns can be observed along the line of earthquake faults prior to rupture. The reasons for this are unclear.

- *Radon emissions.* Post-earthquake analysis of borehole and soil-gas sensors indicate altered radon concentrations prior to an earthquake event. This is thought to result from the occurrence of cracking in the rock mass as the earthquake rupture begins and releases radon gas trapped within the rock.

- *Ground-water level.* There is some evidence that ground-water levels change prior to an earthquake, probably because of the cracking process outlined above.

- *Animal behaviour.* Anomalous animal behaviour has been widely observed and reported from some countries prior to large earthquakes.

The reliability of these techniques remains unproven because of inadequate scientific evidence. It is also difficult to distinguish normal variations in the above parameters from the conditions associated with an impending earthquake. For example, the water level in wells falls and rises in response to atmospheric pressure changes and rainfall amounts.

The US Geological Survey has conducted a long-term experiment to identify precursory phenomena. Near the town of Parkfield, California, a 25 km stretch of the San Andreas Fault is intensively studied because it slips with a fairly short recurrence interval of about 22 years to give moderate (M_W = 6.0) earthquakes and also because over 120,000 households are at risk from seismic activity in the area. The Parkfield prediction experiment involves monitoring the fault line through a network of sensitive seismographs, the use of tiltmeters to detect ground surface changes and geodetic lasers to measure any changes in distance across the fault. In September 2004 a M_W = 6.0 earthquake occurred on the fault and was recorded in detail by the instrument array. Subsequent analysis of the data revealed no precursory indications of this event (Park *et al.*, 2007).

To summarize, usable earthquake warning systems are not yet available. In Japan there are about 100 earthquake monitoring stations but no public warning has yet been issued. Neither the 1994 Northridge nor the 1995 Kobe earthquake was adequately anticipated. Indeed, both occurred on fault systems upon which the seismic potential was incompletely understood. A better scientific understanding of earthquake activity is vital because, although prediction and warning remains a long-term goal, it is not the only potential benefit

from this knowledge. For example, the 30-year prediction for the San Francisco Bay area could be exploited as a 'wake-up call' to reduce risks through preparedness programmes but there is little evidence that this might happen.

Some hope exists for very short-term warning systems. In Taiwan, the Central Weather Bureau, which is charged with the collection of seismic datasets, operates a nation-wide network of strong-motion instruments which deliver data to a central monitoring system in real-time. A map of the location of unusual earthquake activity can be produced within 22 seconds and the information is then relayed to vulnerable locations (Wu *et al.* 2004). At the current rate of data transfer, a city located 100 km from the earthquake epicentre would receive about 10 seconds' warning of the arrival of earthquake waves, based on a wave velocity of about 3 km s^{-1}. Whilst this is insufficient time to warn the population, automated shut-downs of computer servers, gas pipelines and nuclear power plants are possible. With better instrument arrays and data processing, warning times should increase and work is underway on an automatic shut-down system for the Taiwan high-speed railway network.

3 Forecasting and warning for tsunamis

Tsunami forecasting and warning systems are well established, although frequent false alarms and occasional threat denial have limited their effectiveness in the past. For example, a 1960 tsunami killed 61 people in Hilo, Hawaii, when residents who received a warning failed to evacuate from the coast. The first tsunami warning system was installed in 1948 for the Pacific Ocean, using a network of seismic stations to relay information to a warning centre near Honolulu, Hawaii. The international monitoring network is managed by the US National Oceanic and Atmospheric Administration. Following the tsunamigenic Alaska earthquake of 1964, the West Coast/ Alaska Tsunami Warning Centre was set up in 1967 to provide more localized warning for Alaska, British Columbia, Washington, Oregon and California. In 1996 the USA approved the National Tsunami Hazard Mitigation Program (NTHMP), a collaborative state–federal partner-ship designed to reduce tsunami hazards along US coastlines.

Such piecemeal development has been criti-cized by those who favour a comprehensive approach to tsunami warning in the Pacific basin

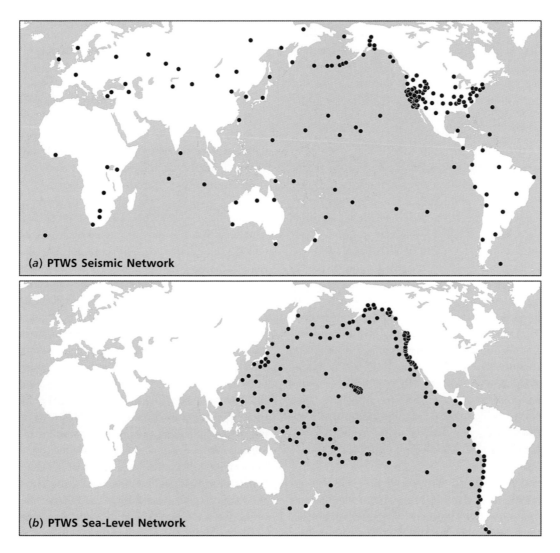

(a) PTWS Seismic Network

(b) PTWS Sea-Level Network

Figure 6.13 The Pacific Tsunami Warning System (PTWS) showing the network of monitoring stations for (a) seismic activity and (b) sea-level changes. The PTWS uses over 150 seismic stations and about 100 sea-level stations around the globe to evaluate the severity of a tsunami. Warning messages can be disseminated to authorities in more than 100 locations across the Pacific. After International Tsunami Information Center at http://www.tsunamiwave.info (accessed 5 January 2012). US government material not subject to copyright protection in the United States.

(Dohler, 1988). The Pacific Tsunami Warning and Mitigation System (PTWS) relies on about 150 high-quality seismic stations around the world (Figure 6.13a) to locate and assess earthquakes with tsunami potential. Data are also accessed from about 100 sea-level stations (Figure 6.13b) to verify the generation of a tsunami and to assess the scale of the threat. The PTWS disseminates information and warning messages to over 100 designated offices in countries bordering the Pacific Ocean. Two sub-centres provide regional alerts to the US west coast, Alaska, Canada and the north-west Pacific and South China Sea areas.

The system operates on two levels. The first level supplies warnings to all Pacific nations of large, destructive tsunamis likely to be ocean-wide. Following a high-magnitude earthquake (M = > 7.0), tide stations near the epicentre are alerted to observe unusual wave activity. If this is detected, information is issued in the form of watches and advisories and full tsunami warnings. The primary aim is to alert all at-risk coastal populations, within one hour, about the arrival time of the first wave, with an accuracy of ±10 minutes. More distant populations have longer to react and prepare for evacuation from the coast. The second level of cover is based on warning systems serving specific areas. Local tsunamis can pose a greater threat than ocean-wide events because they strike onshore very quickly. These systems rely on local data obtained in real time and typically aim to issue warnings within minutes for areas 100–750 km distant from the earthquake source.

The Japanese Meteorological Agency has maintained a warning service since 1952. This was updated in 1999. Previous warnings were based on the traditional method of tidal observations, calculation of earthquake location and magnitude, empirical estimation of tsunami wave heights, followed – if necessary – by the issue of a warning message for 18 coastal segments each several hundred kilometres long. The present method employs computer simulations of tsunamis gen-erated by various sizes and depths of earthquake. Once the location and magnitude of an earthquake are established, tsunami heights and arrival times are retrieved from a database containing about 100,000 simulations for 600 points around the Japanese coast. Wave heights and arrival times can then be forecast for 66 separate coastal segments. The new system was tested in the 2011 event and provides:

- the issue of an initial tsunami advisory or warning three minutes after an earthquake
- the issue of maximum wave heights and arrival times within about five minutes after the earthquake
- the subsequent issue of the times of high tides and continuing updates about the hazard situation.

The impact of the 2004 Indian Ocean tsunami stimulated interest in future protection for this region (Di and Jian, 2011). About 300,000 lives were lost, a figure that might have been reduced to about 15,000 if a system similar to the PTWS had been in place (Wang and Li, 2008). At a United Nations conference held in January 2005, the German government agreed, under the leadership of UNESCO, to supply such a system based in Jakarta, Indonesia. This country consists of 17,000 islands. About 60 per cent of the coastline is at risk from tsunamis because it lies at the meeting-point of three tectonic plates. The warning system consists of seismographic stations and a series of deep-ocean sensors, each of which relays information to the early-warning centre in Jakarta before transmission to 11 regional hubs across Indonesia (Strunz et al., 2011). The pilot version of the scheme was launched in November 2008; full ownership passed to the government in Jakarta in March 2011.

At present, warning times are very limited for tsunamis generated in coastal waters and the number of false warnings that are issued remains a problem for ocean-wide threats. One possible solution to the latter problem might lie in the remote sensing of the cold water that is lifted to the sea surface from deep in the ocean by submarine earthquakes. According to Lin *et al.* (2011), the arrival of the cold water at the surface emits an infrared radiation signal that was detected by satellite in both the 2004 Indian Ocean and the 2011 Tōhoku events. If this thermal anomaly proves to be a reliable signature of a tsunami, its detection could replace the DART system with a warning scheme that should be less expensive to maintain and that involves little time delay and is less likely to produce false warnings.

4 Land use planning

Land planning can guide earthquake-prone communities to a safer future. Land zoning based on seismic risk not only informs engineers about the design and construction methods to be employed to produce safer buildings, bridges and roads in a particular locality but it also promotes other hazard-reducing measures such as building codes, retrofit priorities and building insurance rates.

In California, local authorities are required by law to include seismic safety in land planning, although vague guidelines and limited oversight do little to ensure a consistent approach. Following the Northridge earthquake, the amount of damage to single-family homes was found to be related to the quality of the plans (Nelson and French, 2002). There is always some resistance to planning restrictions, especially by local bodies who seek to pass disaster costs on to a higher level of administration. Even when hazard information is available, it may not be applied to best effect. Californian state law requires that estate agents (realtors) advise all potential purchasers if residential properties are located near mapped fault lines. In practice, such information may not be disclosed until sale negotiations are well advanced. In any case, earthquake hazards are unlikely to be a major factor in decision making if other attributes – like an attractive view, schools, shops, investment potential – are important to buyers,

especially if they intend to relocate in a few years' time

Before land can be zoned and developed safely, the seismic risks have to be accurately mapped. A seismic hazard map shows the pattern of potential ground shaking hazard in future earthquakes. Probabilistic maps of seismic shaking hazard, such as that for California (see Figure 5.18), are the basis for more detailed mapping that allows contours of expected ground motion, typically measured by *peak ground acceleration* (PGA), to be viewed at scales suitable for local planning decisions. Figure 6.14 shows, for an imaginary location, the range of shaking – from strong to weak – expected in uniformly firm ground at various distances away from a known fault. Such maps indicate the chances of a particular value being exceeded in a given period of time. Thus, if we assume that the criteria for this map are a 5 per cent chance of occurrence over a time interval of 50 years, Hazard City would have a 5 per cent chance of a peak ground acceleration (PGA) of 0.1–0.2 being exceeded in a period of 50 years.

Maps like this need refinement because the shaking depends on the detail of the surface deposits. The highest seismic hazard exists above

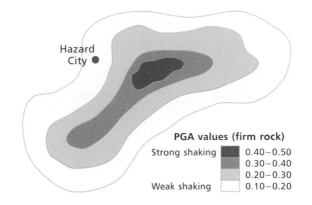

Figure 6.14 Hypothetical pattern of earthquake shaking in firm rock in the neighbourhood of a city. Assuming that the map is based on a 5% chance of exceedance over a 50-year period, Hazard City could expect peak ground accelerations of between 10 per cent and 20 per cent g during an earthquake. Such maps can be used to guide land use planning.

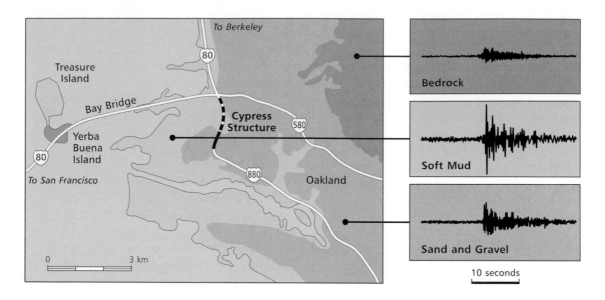

Figure 6.15 An illustration of variations in ground shaking due to surface geology. Deposits of mud and gravel shake to a much greater extent than bedrock. During the 1989 Loma Prieta earthquake, part of the elevated Cypress Structure freeway in the San Francisco Bay area collapsed, due to strong shaking in the soft muds on which it was constructed. After USGS Earthquake Hazards Programs at http://earthquake.usgs.gov/regional/nca (accessed 31 March 2011).

geologically young deposits, or man-made landfill, with the potential to liquefy into a sandy fluid during an earthquake. Such deposits exist in the low-lying sections of the San Francisco Bay area in the form of soft mud and sands and gravels. Figure 6.15 indicates the enhanced shaking in these deposits, relative to the bedrock, and the likely consequences for built structures. In the 1989 Loma Prieta earthquake, part of the Cypress Structure freeway approaching the Bay Bridge from Oakland collapsed and killed 42 people. Analysis of property losses revealed that enhanced ground shaking by amplified seismic waves in soft-soil deposits was directly responsible for about two-thirds of the total (Table 6.7; Holzer, 1994).

An attempt was made to map seismic vulnerability in Athens, Greece after a damaging earthquake in 1999 (Marinos *et al.*, 2001). Figure 6.16 shows proposed land zones, based on four class grades for both geology and building damage, relating to the municipality of Ano Liossia, less than 3 km from the fault rupture. It can be seen that severe damage (including building collapse) was mainly confined to the alluvial deposits in Zone 3, while moderate–severe damage occurred in the surrounding Zone 2. Little damage occurred above the rock formations of Zone 1, an indication that the proposed zoning is a guide to future

Table 6.7 Loma Prieta earthquake losses by hazard type

Earthquake hazard	Total damages (US$ millions)	Loss (% of total)
Ground shaking, normally attenuated	1,635	28.0
Ground shaking, enhanced	4,170	70.0
Liquefaction	97	1.5
Landslides	30	0.5
Ground rupture	4	0.0
Tsunami	0	0.0
Total	5,936	100.0

Source: After Holzer (1994), EOS 75 (26) Table 3, page 301, 1994. Copyright by the American Geophysical Union

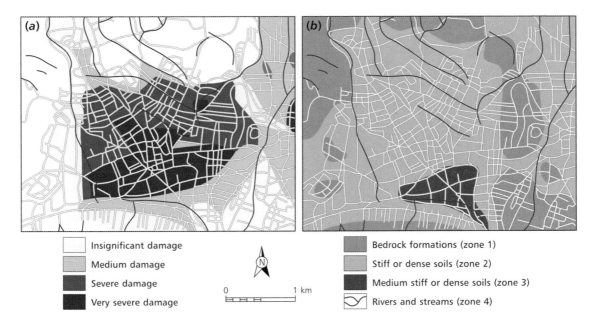

Figure 6.16 Earthquake hazard planning in the municipality of Ano Liossia, Athens, Greece. (a) Recorded building damage after the 1999 earthquake (M_W = 5.9); (b) four proposed seismic-risk zones in relation to surface geology. Zone 1 is lowest risk and the blank areas were excluded from the study. After Marinos *et al.* (2001). Reprinted from *Engineering Geology* 62, P. Marinos *et al.* Ground zoning against seismic hazard in Athens, Greece, 343–56. Copyright (2001), with permission from Elsevier.

seismic risk. In California, *set-back ordinances* are a major tool. Building set-backs are recommended where proposed development crosses known or inferred faults, and slope stability set-backs can be established where unrepaired active landslides, or old landslide deposits, have been identified. Set-backs can also be used to separate buildings from each other in order to reduce pounding effects where structures of different heights, resulting from different construction methods, are combined in close proximity.

The need for tsunami mitigation to be integrated into planning procedures along low-lying coastlines was emphasized by Preuss (1983) and by the Intergovernmental Oceanographic Commission (2008). In terms of tsunami protection, engineered sea walls remain relevant but, if the tsunami run-up height is over 4 m, they are of little benefit. These structures can be complemented with 'bioshields' or coastal tree planta-

tions. Although natural forest of mangrove and casuarinas is often severely damaged by tsunami waves, trees can provide a protective buffer for the land behind. Along the Sanriku coast of Japan, the planting of spruce forest is widely used to dissipate wave energy and filter out floating debris. Otherwise, residential areas, public facilities and transportation should be relocated to higher ground. Figure 6.17 is an illustration of the measures that could be incorporated into a comprehensive anti-tsunami scheme including physical structures and the provision of a coastal evacuation route. The presumption is always that new development will not be exposed to risk, but achieving a balance between safety and land function can be controversial, as in the case of the proposed Marine Education Centre for Wellington, New Zealand (Garside *et al.*, 2009).

Transportation planning is a difficult issue in view of the expected traffic volumes generated by

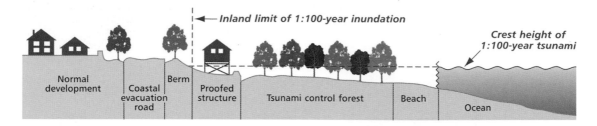

Figure 6.17 Typical example of coastal land planning for tsunami hazards. The beach and forest zones dissipate the energy of the onshore wave. Building development and the evacuation route are located above the predicted height of the 1:100 year event. After Preuss (1983).

mass evacuation after a tsunami warning. Many existing roads follow low-lying shores or valley sides which are likely to be inundated and can become congested with traffic. In some areas, like Hawaii, well-constructed high-rise coastal buildings can be used for vertical evacuation. Along the Sanriku coast, a few designated evacuation shelters and platforms have been built for this purpose. But they are expensive to construct and need to be several storeys high to remain

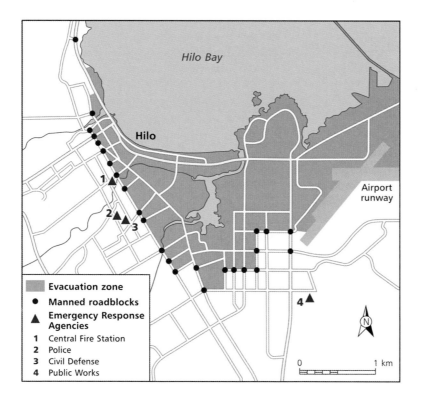

Figure 6.18 Part of the tsunami hazard map of the city of Hilo, Hawaii, showing the scheduled evacuation area and the location of the emergency-response agencies at the outer edge of this zone. Note the proximity of the airport to the evacuation zone. After Department of Emergency Management at www.oahuDEM.org (accessed 11 September 2011).

above the tsunami run-up. In turn, this increases the possibility of earthquake damage. A simpler and cheaper option is the creation of paths and steps leading up hillsides away from the shore, but the local topography may be unsuitable and some people with disabilities will be unable to use this type of escape route.

Despite these problems, an important element of tsunami land planning is the identification of evacuation zones. Figure 6.18 is part of the tsunami evacuation zone map for the city of Hilo, Hawaii, showing the evacuation area with the position of manned road blocks at its edge and the location of the emergency response agencies – civil defence, fire service, police and the public works department – just outside the zone. Wherever possible, people are encouraged to evacuate such areas on foot, so as to limit traffic congestion, and to follow signed routes where these are indicated.

FURTHER READING

Bilham, R. (2006) Dangerous tectonics, fragile buildings and tough decisions. *Science* 311: 1873–5. A readable and sound overview.

Clermont, C., Sanderson, D., Sharma, A. and Spraos, H. (2011) *Urban Disasters – Lessons from Haiti.* Disasters Emergency Committee, London. Very clearly shows the complex problems facing emergency-relief efforts in an impoverished part of the world.

Lindell, M.K. and Perry, R.W. (2000) Household adjustment to earthquake hazard: A review of research. *Environment and Behavior* 32: 461–501. Slightly dated but deals with ever-relevant issues.

Petley, D.N. (2005) Tsunami – how an earthquake can cause destruction thousands of kilometres away. *Geography Review* 18: 2–5. A good example of the globalization of hazard.

Strunz, G. *et al.* (2011) Tsunami risk assessment in Indonesia. *Natural Hazards and Earth System Sciences* 11: 67–82. An up-to-date account of on-going attempts at seismic-disaster reduction in a high-risk environment.

Tekeli-Yeşil, S., Dedeoğlu, N., Braun-Fahrlaender, C. and Tanner, M. (2010) Factors motivating individuals to take precautionary action for an expected earthquake in Istanbul. *Risk Analysis* 30: 1181–95. Ably illustrates the importance of public perceptions in risk assessment.

WEB LINKS

USGS earthquake program www.earthquake.usgs.gov/

Earthquake Engineering Resaerch Institute (EERI) www.eeri.org

NOAA Center for Tsunami Research www.nctr.pmel.noaa.gov/

International Seismological Centre www.isc.ac.uk/

The Tsunami Society www.sthjournal.org

Tsunami data and information www.ngdc.noaa.gov/

Tectonic hazards

Volcanoes

7

A VOLCANIC HAZARDS

There are some 500 active volcanoes in the world; about 50 erupt in an average year. Despite a dramatic appearance and high public profile, volcanoes create far fewer disasters than do earthquakes or severe storms. More than half the deaths recorded in the twentieth century occurred in just two events. The 1902 eruption of Mont Pelée, on the island of Martinique in the West Indies, killed 29,000 people in the port of Saint Pierre, leaving only two known survivors, whilst the 1985 eruption of Nevado del Ruiz in Colombia claimed a further 23,000 lives. It is normal practice to classify volcanoes as active, dormant or extinct. But the infrequency of major eruptive events is one of their most dangerous features and in 1951 Mount Lamington killed 5,000 people in Papua New Guinea, despite being considered extinct (Chester, 1993). To be prudent, all volcanoes that have erupted within the last 25,000 years should be regarded as potentially active. Most are located in geologically unstable areas and prone to multiple threats (Malheiro, 2006). On the other hand, volcanic terrain provides important natural resources, including geothermal energy, building materials and opportunities for tourism.

Although it is generally accepted that volcanoes killed fewer than 1,000 people per year during the twentieth century, Witham (2005) claimed that fatalities have been underestimated and drew attention to other important impacts, such as the large number of people evacuated in volcanic emergencies (Table 7.1). Historically, most volcano-related deaths have been due to indirect causes, like famine arising from crop failure caused by falling ash. Today, pyroclastic flows are the chief cause of death, whilst lahars (volcanic mudflows) are the chief cause of injuries and ash fall accounts for most people made homeless. Less explosive eruptions create lesser hazards. For example, only one person was killed in Hawaii over the past 100 years or so, despite the fact that, over the same period, some 5 per cent of the island was covered by fresh lava flows (Decker, 1986). However, there is concern about the increasing human exposure to volcanic hazards, especially in the growing cities of the LDCs (Chester *et al.*, 2001). The volcanic complex west of Naples, Italy, is now one of the most densely populated areas of active volcanism in the world, where some 200,000 people are at risk (Barberi and Carapezza, 1996). Many residents are vulnerable because they are poor, marginally employed and live in old, weakened buildings or unplanned housing (Chester *et al.*, 2002).

Volcanoes attract human settlement and it is the increase in exposed risk, rather than the frequency of eruptions, that explains the doubling of fatal eruptions from the nineteenth to the twentieth century (Simkin *et al.*, 2001). According to Small and Naumann (2001), 10 per cent of the world's population live within 100 km of a volcano active in historic times. The highest population densities at risk are in south-east Asia and Central America. In Europe, the Etna region contains about 20 per cent of Sicily's population, with rural population densities of 500–800 per km^2. Countries like Indonesia, located at the junction of three tectonic plates and with a population of over 150 million, face the greatest threat, and this nation has suffered two-thirds of all volcano-related deaths (Suryo and Clarke, 1985). In 1815 a massive eruption of the Tambora volcano directly killed 12,000 people and a further 80,000 persons later perished through disease and famine. The rare 'super' eruption at what is now Lake Toba in Indonesia some 75,000 years ago released an estimated 2,800 m^3 of debris (Rose and Chesner, 1987). A repeat event would have a catastrophic impact on the world today.

B THE NATURE OF VOLCANOES

The distribution of volcanoes is controlled by the global geometry of plate tectonics. Not surprisingly, seismic activity is often associated with volcanic eruptions, although most volcanic-related earthquakes are small (Zobin, 2001). Volcanoes are found in three tectonic settings:

- *Subduction volcanoes* are located in the zones of the Earth's crust where one tectonic plate is thrust and consumed beneath another. They comprise about 80 per cent of the world's active volcanoes and are the most explosive type, characterized by a composite cone and multiple hazards (Figure 7.1).

Table 7.1 Best estimates of the human impacts of volcanic hazards in the twentieth century (1900–99)

Human impacts	Number of events	Number of people
Killed	260	91,724
Injured	133	16.013
Homeless	81	291,457
Evacuated/affected	248	5,281,906
Any incident	491	5,595,500

Source: After Witham (2005)

Note: Each event may have had more than one consequence.

- *Rift volcanoes* occur where tectonic plates diverge and spread. They are generally less explosive and more effusive, especially when they occur on the deep ocean floor.
- *Hot spot volcanoes* exist in the middle of tectonic plates, where a crustal weakness allows molten material to penetrate from the Earth's interior. The Hawaiian Islands, in the middle of the Pacific plate, are an example.

All volcanoes are formed from molten material (*magma*) within the Earth's crust. Magma is a complex mixture of silicates which contains dissolved gases and, very often, crystallized minerals in suspension. As the magma moves towards the surface, the pressure decreases and the dissolved gases come out of solution to form bubbles. The bubbles expand and drive the magma further into the volcanic vent until it breaks through weaknesses to reach the surface. For a moderately large eruption, the total thermal energy released lies in the range 10^{15}–10^{18} joules, which compares with the 4×10^{12} joules liberated by a one kilotonne atomic explosion. There is no fully agreed scale for measuring the size of eruptions, but Newhall and Self (1982) drew up a semi-quantitative *volcanic explosivity index* (VEI) which combined the total volume of ejected products, the height of the eruption cloud, the duration of the main eruptive phase and several other items into a basic 0–8 scale of increasing hazard (Table 7.2). On average, an eruption with VEI = 5

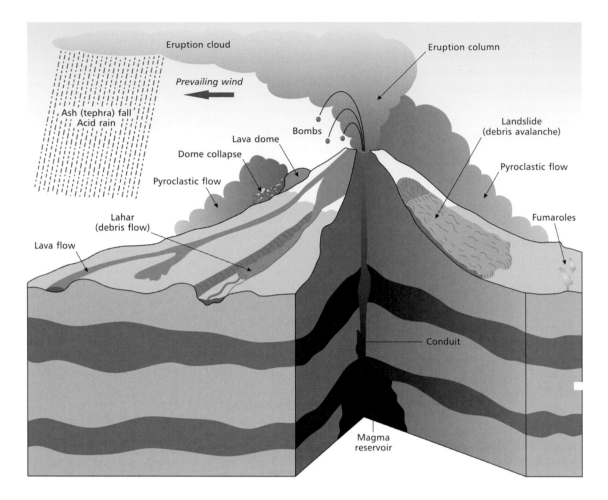

Figure 7.1 Section through a composite volcanic cone showing a range of possible hazards. Some hazards (pyroclastic and lava flows) occur during eruptions; other hazards (lahars) are more likely to occur after the event. After Major *et al.* (2001).

occurs every 10 years, and a VEI = 7 every 100 years.

The disaster potential of volcanic eruptions depends on the effervescence of the gases and the viscosity of the magma. High effervescence and low viscosity lead to the most explosive episodes. Thus, subduction zone volcanoes draw on magmas that are a mix of upper-mantle material and melted continental rocks rich in feldspar and silica. These *felsic* (acid) magmas produce thick, viscous lavas containing up to 70 per cent silicon dioxide (SiO_2) and lead to violent eruptions. On

the other hand, rift and hot-spot volcanoes draw on magmas high in magnesium and iron but low (< 50 per cent) in silica content. Such *mafic* (or basic) lavas are fluid, retain little gas and erupt less violently.

These characteristics allow a broad recognition of volcanic eruptions by type. Common examples include the *Plinian type*, which produces the most violent upward expulsions of gas and other materials associated with very viscous magmas (dacite and rhyolite). In the 1991 eruption of Mount Pinatubo, Philippines, a plume of tephra

Table 7.2 Selected criteria for the Volcanic Explosivity Index (VEI)

VEI number	Volume of ejecta (m³)	Column height (km)	Qualitative description	Tropospheric injection	Stratospheric injection
0	$<10^4$	<0.1	non-explosive	negligible	none
1	10^4–10^6	0.1–1.0	small	minor	none
2	10^6–10^7	1–5	moderate	moderate	none
3	10^7–10^8	3–15	mod-large	substantial	possible
4	10^8–10^9	10–25	large	substantial	definite
5	10^9–10^{10}	>25	very large	substantial	significant
6	10^{10}–10^{11}	>25	very large	substantial	significant
7	10^{11}–10^{12}	>25	very large	substantial	significant
8	$>10^{12}$	>25	very large	substantial	significant

Source: Adapted from Newhall and Self (1982)

Note: Column height: for VEIs 0–2 uses km above crater; for VEIs 3–8 uses km above sea level.

was ejected more than 30 km into the atmosphere. The *Peléan type* is also dangerous because the rising magma is trapped by a dome of solid lava and then forces a new opening in the volcano flank. This can produce powerful lateral blasts such as occurred during the eruption of Mount St Helens, USA in 1980, when much of the mountain top was destroyed and 57 people were killed. In contrast, the relatively low-hazard *Hawaiian type* of eruption jets highly fluid basaltic lava into the air from vents and fissures on the flank of the volcano to create *fire fountains*. The resulting surface flows of lava can travel miles from their source before cooling and hardening. This produces the low topographic profile associated with such volcanoes.

C PRIMARY VOLCANIC HAZARDS

These result from the products ejected by the volcanic eruption. A significant hazard feature is their long geographical reach away from the source (Figure 7.2).

1 Pyroclastic flows

These have been responsible for most volcanic-related deaths to date. They are sometimes called *nuées ardentes* ('glowing clouds') and result from the frothing of molten magma in the vent of the volcano. The gas bubbles then expand and burst explosively to eject a turbulent mixture of hot gases and pyroclastic material (volcanic fragments, crystals, ash, pumice and glass shards). Pyroclastic bursts surge downhill because, with a heavy load of lava fragments and dust, they are appreciably denser than the surrounding air. The clouds may be literally red hot (up to 1,000°C). They pose the highest risks when blasts are directed laterally (Peléan type) close to the ground, and are capable of travelling in surges in excess of 30 m s⁻¹ up to 30–40 km from the source. During the Mont Pelée disaster of 1902, the town of St Pierre – some 6 km from the centre of the explosion – experienced a surge temperature about 700°C in a blast travelling at around 33 m s⁻¹. People exposed to these surges are immediately killed by a combination of severe external and internal burns together with asphyxiation. The surge itself is preceded by air pressures with sufficient force to topple some buildings. Modern volcanic hazard planning, as at Vesuvius, Italy, anticipates emergency evacuation of people, and extensive building damage, within a radius of 10 km from the volcanic vent (Petrazzuoli and Zuccaro, 2004).

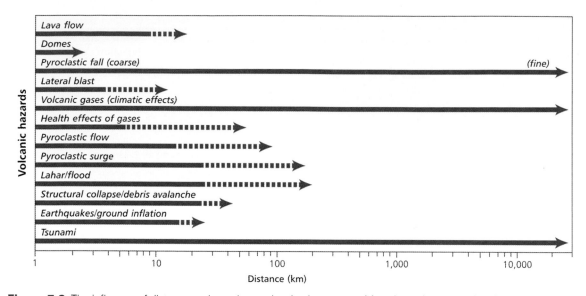

Figure 7.2 The influence of distance on hazardous volcanic phenomena. Many hazards are restricted to a 10 km radius of the volcano but the effects of fine ash, gases and tsunami waves can extend beyond 10,000 km. After Chester *et al.* (2001). Reprinted from *Environmental Hazards* 2, D.K. Chester *et al.* The increasing exposure of cities to the effects of volcanic eruptions, 89–103. Copyright (2001), with permission from Elsevier.

2 Air-fall tephra

Tephra comprises all the fragmented material ejected by the volcano that subsequently falls to the ground. Most eruptions produce less than 1 km³ volume of material but the largest explosions eject several times this amount. The particles range in size from so-called 'bombs' (>32 mm in diameter) down to fine ash and dust (<4 mm in diameter). The coarser, heavier particles fall out close to the volcano vent. A dry layer of ash 10 cm deep weighs about 65–100 kg m², whilst wet ash weighs twice as much. As a result, flat-roofed, un-reinforced buildings can collapse when ash accumulation approaches 1 metre. In some cases the tephra will be sufficiently hot to start fires on the ground. Depending on wind conditions, the finer dust may be deposited far away. Within six

hours of the modest eruption (VEI = 5) of Mount St Helens in 1980, ash clouds had drifted 400 km downwind.

Very large eruptions can influence the global climate. This happens when debris is emitted into the lower stratosphere, some 20–25 km above the Earth's surface, and forms a 'dust veil' over the planet. The maximum risk is from volcanoes in lower latitudes. After the 1883 eruption of Krakatoa, Indonesia, an aerosol cloud spread round the globe within two weeks. As shown in Table 7.3, the climatic effects can last from periods of a single day (by reducing the diurnal temperature cycle) to up to 100 years if a series of eruptions raises the mean optical depth of the atmosphere enough to cause decadal-scale cooling. Important changes in atmospheric chemistry, especially ozone depletion, can also occur. The main long-term effect is

Plate 7.1 *(Facing page)* Eruption of the Soufrière Hills volcano on the island of Montserrat in August 1997. Pyroclastic flows claimed several lives and forced the evacuation of more than two-thirds of the residents. (Photo: Panos/Andy Johnstone AJH00072MSR)

Table 7.3 The effects of large explosive volcanic eruptions on weather and climate

Effect	Mechanism	Begins	Duration
Reduction of diurnal cycle	Blockage of short-wave and emission of long-wave radiation	immediately	1–4 days
Reduced tropical precipitation	Blockage of short-wave radiation, reduced evaporation	1–3 months	3–6 months
Summer cooling of northern hemisphere, tropics and sub-tropics	Blockage of short-wave radiation	1–3 months	1–2 years
Stratospheric warming	Stratospheric absorption of short-wave and long-wave radiation	1–3 months	1–2 years
Winter warming of northern hemisphere continents	Stratospheric absorption of short-wave and long-wave radiation	6 months	1 or 2 winters
Global cooling	Blockage of short-wave radiation	immediately	1–3 years
Global cooling from multiple eruptions	Blockage of short-wave radiation	immediately	10–100 years
Ozone depletion, enhanced UV	Dilution, heterogenous chemistry and aerosols	1 day	1–2 years

Source: After Robock (2000). Reproduced by permission of American Geophysical Union.

net cooling of the Earth's surface, due to back-scattering of incoming short-wave radiation. For example, after the 1815 eruption of Tambora, Indonesia, with a VEI = 7, the year 1816 was called 'the year without a summer' throughout the northern hemisphere. The eruption of Mount Pinatubo in 1991 lowered surface air temperatures in parts of the northern hemisphere by up to 2°C in the summer of 1992, and during the winters of 1991–92 and 1992–93, raised temperatures by as much as 3°C, with implications for weather-sensitive activities like agriculture.

Although ashfalls account for fewer than 5 per cent of the direct deaths associated with volcanic eruptions, they create other problems. Heavy falls of *scoria* (cinder) blanket the landscape; even light falls of ash contaminate farmland, and create disruption and building damage in urban areas. The eruption of Mount Pinatubo in 1991 disrupted the livelihood of 500,000 farmers, as agricultural land up to 30 km distant was covered in ash. Wet ash conducts electricity and can cause the failure of electronic components,

especially with high-voltage circuits and transformers. Fine ash can clog air filters and damage vehicle engines; roads and airport runways become impassable if made slippery by ash deposits. Airborne dust reduces atmospheric visibility and poses a risk to air transport (Box 7.1).

3 Lava flows

Lava flows threaten human life when they emerge rapidly from fissure, rather than from central-vent, eruptions. As already explained, the fluidity of lava is determined by its chemical composition, especially the proportion of silicon dioxide (SiO_2). If silicon dioxide forms less than about half the total, the lavas are mafic and very fluid, as compared to the more viscous acid lava flows. This difference has led to the recognition of two types of lava flow:

• *Pahoehoe lava* flows are the most liquid, and cool leaving a relatively smooth, wrinkled surface.

Box 7.1 Airborne volcanic ash and aviation

Volcanic ash consists of minute particles of pulverized rock and natural glass. It is extremely abrasive, insoluble in water and conducts electricity when wet. The plume of ejected material can travel long distances in the atmosphere. These properties pose a distinct threat to jet aircraft flying through clouds of even well-dispersed ash, and the aviation industry is highly sensitive to the risks. For example, in May 2003, following the first historical eruption of Anatahan volcano in the western Pacific, several international flights were cancelled and Guam airport, 320 km south of the volcano, was disrupted (Guffanti *et al.*, 2005). Worldwide, more than 100 encounters of aircraft with volcanic ash have been reported, resulting in damage amounting to hundreds of millions of US dollars and posing severe risk to at least three large passenger jets.

The main threat arises when the small (< 2 mm diameter) fragments are ingested into a jet engine. This material can erode the turbine blades and can melt at the high operating temperature of the engines and adhere to critical parts, thus causing engine failure. Forward-facing aircraft surfaces, including the cockpit windows, are likely to be abraded and the ash may interfere with navigation and other electronic systems on board (Neal and Guffanti, 2010). Ash falling onto runways, particularly when wet, will reduce the safety of aircraft movements, especially the braking performance. Ash

Plate 7.2 The Eyjafjallajökull volcano in southern Iceland erupts and sends plumes of ash into the atmosphere just before sunset on 16 April 2010. This material was transferred southwards to create air traffic confusion over much of Western Europe. (Photo: AP Photo/Brynjar Gauti 1004151112787)

accumulations greater than ~1 mm require complete removal for an airport to resume normal operating capability.

In 1982, the International Civil Aviation Organization (ICAO) created the International Airways Volcanic Watch programme (IAVW). Nine Volcanic Ash Advisory Centres exist, with the aim of improving the detection of airborne ash worldwide and issuing warning messages to air traffic controllers, dispatchers and pilots. Some countries, like the USA, have made additional arrangements for their own air space. However, the system is not wholly reliable – pilot reports may be the first indication of an eruptive event – and the science remains under-developed, as illustrated by the 2010 Icelandic event.

In April 2010 an eruption of the Eyjafjallajökull volcano in Iceland produced an ash cloud that led to the cancellation of 95,000 flights over Western Europe, with a cost to the airlines of US$1.7 billion and lost productivity in the region estimated at over US$600 million per day (Chester and Duncan, 2010). This was a relatively small eruption, and the ash plume remained at comparatively low altitudes, but it was dispersed by northerly winds over several countries and disrupted airport operations as far away as Moscow and Athens. British air space was closed for five days by the Civil Aviation Authority (CAA), acting on advice from the Meteorological Office. This policy of closing air space was much criticized at the time as an over-reaction. However, subsequent tests showed that the particles of explosive ash that reached Europe were unusually sharp over their entire size range and that the edges remained sharp after two weeks of artificial abrasion in the laboratory (Gislason *et al.*, 2011). Scottish airports were partially closed once more by an ash cloud from another Icelandic volcano in May 2011.

According to Donovan and Oppenheimer (2011), such seemingly unusual incidents have to be placed in context and highlight the importance of multidisciplinary studies of environmental hazards. It is evident that aviation policy and operational responses need greater clarification. This applies not only to the detection and forecasting of ash cloud movements but also to the development of techniques to determine the true degree of aviation risk.

• *Aa lava* moves downhill in a blocky, slow-moving manner and leaves behind a rough, irregular surface.

On steep slopes, low-viscosity *pahoehoe* lava can stream downhill at speeds approaching 15 m s^{-1}. In the 1977 eruption of Nyiragongo volcano, Zaire, five fissures on the flanks of the volcano released a wave of lava that killed 72 people and destroyed over 400 houses. Around Mount Etna, Sicily, *aa*-type lava flows have done much damage to property and agriculture in the past. The city of Catania was partially destroyed in 1669. The greatest lava-related disaster in historic times occurred in 1783, when lava flowed out of the 24 km-long Lakagigar fissure in Iceland for more than five months (Thorarinsson, 1979). There was little direct mortality but more than 10,000 people, over one-fifth of Iceland's population at the time, died in the resulting famine.

4 Volcanic gases

Many gases are released from explosive eruptions and lava flows. The complex gaseous mixture commonly includes water vapour, hydrogen, carbon monoxide, carbon dioxide, hydrogen sulphide, sulphur dioxide, sulphur trioxide, chlorine and hydrogen chloride in variable proportions. Monitoring of the gas composition is difficult, due to the high temperatures near an active vent and because the juvenile gases interact with the atmosphere and each other. This constantly alters their composition and proportions. Carbon monoxide has caused deaths because of its toxic effects at very low concentrations but most fatalities have been associated with carbon dioxide releases. Carbon dioxide is dangerous because it is a colourless, odourless gas with a density about 1.5 times greater than air. When it accumulates in low-lying places disasters can occur; in 1979, over 140 people evacuating a village in Java, Indonesia walked into a dense pool of volcanically released carbon dioxide and were asphyxiated.

The release of carbon dioxide from previous volcanic activity can also create a highly unusual threat. In 1984 a cloud of gas, rich in carbon dioxide, burst out of the volcanic crater of Lake Monoun, Cameroon and killed 37 people by asphyxiation (Sigurdsson, 1988). Almost exactly two years later, in 1986, a similar disaster occurred at the Lake Nyos crater, also in Cameroon. This time 1,746 lives were lost, together with over 8,300 livestock, and 3,460 people were moved to temporary camps. The outburst of gas created a fountain that reached over 100 m above the lake surface before the dense cloud flowed down two valleys to cover an area over 60 km². These hazards are very rare. They are a function of high levels of carbon dioxide in the waters of these lakes, probably built up over a long period of time from CO_2-rich ground-water springs flowing into the submerged crater. Under normal circumstances the dissolved CO_2 would remain trapped below the water surface. In the case of Lake Monoun, the sudden gas release could have been due to disturbance of the water by a landslide originating on the crater's rim, but no evidence exists for such a trigger at Lake Nyos.

D SECONDARY VOLCANIC HAZARDS

1 Ground deformation

Ground deformation occurs as volcanoes grow from within by magma intrusion and new layers of lava and pyroclastic material accumulate on the surrounding slopes. The deformation may eventually lead to a catastrophic failure of the volcanic edifice and to mass movement hazards. For example, structural failure of the north flank of Mount St Helens in 1980 produced a massive debris avalanche that advanced more than 20 km down the North Fork of the Toutle River. In 2000 the almost total collapse of a new lava dome at the Soufrière Hills volcano, Montserrat generated pyroclastic flows and a number of lahars and debris avalanches in the surrounding valleys (Carn *et al.*, 2004). According to Siebert (1992), major structural failures of volcanoes have occurred worldwide, on average, four times per century over the last 500 years. Such instability is found on large polygenetic volcanoes, like Mauna Loa and Kilauea, Hawaii. Volcanoes like Etna are also prone to instability because of their complex construction of inter-bedded lavas and pyroclastic deposits lying on steep slopes, although relatively few deaths have resulted so far.

2 Lahars

After pyroclastic flows, lahars present the greatest risk to human life. They are often defined as volcanic mudflows composed of sand–silt sized sediments, although other volcanic material, such as pumice, can be transported too. Lahars consist of volcanic ash and rock up to at least 40 per cent by weight and create dense, viscous flows that can travel faster than clear-water streams. They occur widely on steep volcanic flanks in the wet tropics, and the term is of Indonesian (Javanese) origin. The degree of hazard varies greatly but, generally, the destructive potential tends to rise with flows containing the larger-size sediments, as shown at Popocatépetl volcano, Mexico by Capra *et al.* (2004). Most lahars result from heavy rainfall and excess water at the volcano surface. At Mount Pinatubo, on the island of Luzon in the Philippines, lahars can transport and deposit tens of millions of cubic metres of sediment in a day and threaten a local population estimated at 100,000 people. Hot mudflows can emerge from sources below ground. In May 2006 a 'mud volcano' in Indonesia caused several fatalities and the temporary displacement of about 25,000 people.

Lahars can be classified as:

- *primary* when they occur during a volcanic eruption and freshly fallen tephra is immediately mobilized by large quantities of water – sometimes resulting from the collapse of a crater lake – into hot flows.
- *secondary* when they are triggered by high-intensity rainfall between eruptions and old tephra deposits on the volcano slopes are re-activated into mudflows.

Some of the most destructive primary events arise from the rapid melting of snow and ice. This happens when hot lava fragments fall on snow and ice lying at the summits of the highest volcanoes. The water mixes with soft ash and volcanic boulders to produce a debris-rich fluid, sometimes at high temperatures, which then pours down the mountain side at speeds typically of 15 m s^{-1} that may reach >22 m s^{-1}. This a particular hazard in the northern Andes, where at least 20 active volcanoes straddle the equator, from central Colombia to southern Ecuador (Clapperton, 1986). The highest peaks are capped with permanent snow and ice and are structurally weak because of the action of hot gases over time. During an eruption in 1877, so much ice and snow was melted that enormous lahars, 160 km long, discharged simultaneously to the Pacific and Atlantic drainage basins. According to Tuffen (2010), ice thinning due to global warming may cause more explosive eruptions in the future because the removal of 100 m or more of ice thickness would likely reduce the load pressure on magma chambers. Any increase in magma–water interactions would create additional tephra hazards, including the possible collapse of the edifice of the volcano.

The second-deadliest volcanic disaster so far recorded resulted from lahars generated by the 1985 eruption of the Nevado del Ruiz volcano, Colombia, the most northerly active volcano in the Andes. This disaster illustrates the need to understand the history of a volcano, because large lahars had previously been recorded, notably in 1595 and 1845, when the surrounding population was relatively low (Wright and Pierson, 1992). More recent volcanic activity began in November 1984 but the main eruption did not occur until one year later. Large-scale glacier melting produced a huge lahar that rushed down the Lagunillas valley, sweeping up trees and buildings in its path (Sigurdsson and Carey, 1986). The town of Armero, 50 km downstream, was overwhelmed with a mudflow deposit 3–8 m deep that destroyed over 5,000 buildings and killed more than 23,000 people within minutes. A preliminary hazard-zone map had been completed one month before the 1985 eruption (Figure 7.3). This indicated fairly accurately the vulnerability of Armero to mudflows, as well as the likely extent of ash fall, but the limited emergency measures in place at the time were unable to provide an adequate response.

The accumulation of ash on volcanic flanks results in an increased threat of river flooding and sediment re-deposition, especially in countries subject to tropical cyclones or monsoon rains. For example, over 5,000 people were killed in a mudflow following the eruption of the Kelut volcano, Java in 1919. Tayag and Punongbayan (1994) reported that the 1991 eruption of Mount Pinatubo produced 1.53×10^6 m^3 of new lahar material. The most destructive lahars are those containing up to 90 per cent by volume of debris, often created by the breaching of dams holding back temporary lakes along channels and tributary valleys filled with sediment. These lake outbursts occur without warning, even when there is no rainfall. Figure 7.4 shows the lahar deposits that cover over 280 km^2 at Merapi volcano, Java. Most of these deposits lie in river channels and are 0.5–2.0 m thick, although some have depths greater than 10 m (Lavigne *et al.*, 2000). The sediments are quickly remobilized by tropical rainfall and eventually reach lowland rivers, where they reduce channel capacity and increase the risk of rivers migrating unpredictably across flood-plains.

3 Landslides

Landslides and debris avalanches are a common feature of volcano-related ground failure. They are particularly associated with eruptions of siliceous (dacitic) magma of relatively high viscosity with a large content of dissolved gas. This material can intrude into the volcano, as happened in May 1980 at Mount St Helens, USA. Swarms of small earthquakes ($M_W = 3.0$) and minor ash eruptions were followed by ground uplift on the north flank of the volcanic cone. Before the main eruption the bulge was nearly 2 km in diameter and large cracks appeared in the

cover of snow and ice (Foxworthy and Hill, 1982). On 18 May, when the surface bulge was 150 m high, an earthquake shook a huge slab of material from the over-steepened slopes and triggered a debris avalanche containing 2.7 km^3 of material.

4 Tsunamis

Tsunamis can occur after catastrophic eruptions. The structural failure of volcanic islands, creating potential debris flows in excess of one million cubic metres, can produce 'super' tsunami hazards. The most-quoted example is that of the island volcano of Krakatoa, between Java and Sumatra, in 1883 (VEI = 6). A series of enormous explosions, audible at a distance of almost 5,000 km, produced an ash cloud that penetrated to a height of 80 km into the atmosphere and was carried round the world several times by

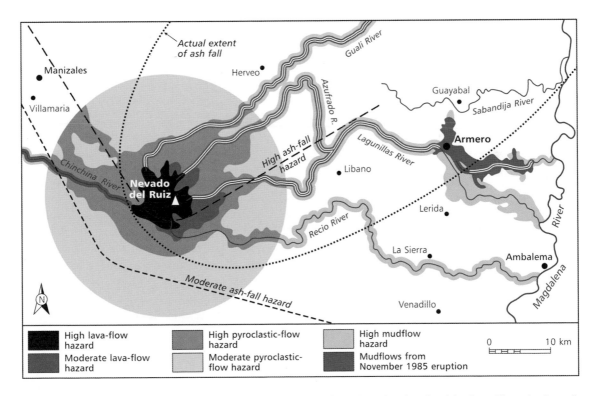

Figure 7.3 Hazard zone map for the Nevado del Ruiz volcano, Colombia, showing the risk of mudflows in the valley occupied by the town of Armero. The circle denotes a 20 km radius from the cone. After Wright and Pierson (1992).

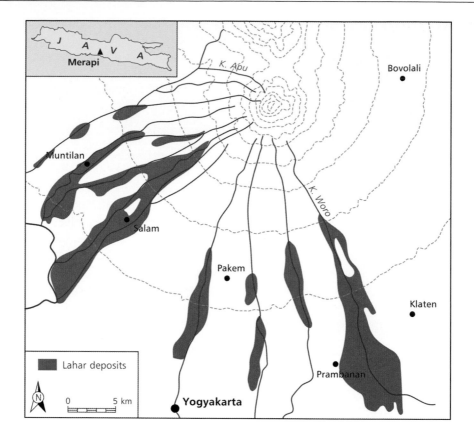

Figure 7.4 The distribution of lahar deposits on the slopes of Merapi volcano, Java. All the river courses shown have produced active lahars during historic times. After Lavigne *et al.* (2000). Reprinted from *Journal of Volcanology and Geothermal Research* 100, F. Lavigne *et al.* Lahars at Merapi volcano, central Java, 423–65. Copyright (2000), with permission from Elsevier.

upper-level winds. Such was the force of the eruption that the volcanic cone collapsed into the caldera. The resulting tsunamis swept through the narrow Sunda Straits, with onshore waves over 30 m high in places. It has been estimated that over 36,000 people were drowned.

It is believed that about 5 per cent of all tsunamis are due to volcanic activity. Perhaps 1 per cent of the total, including some of the largest events, are related to the collapse of volcanic ocean islands. According to Keating and McGuire (2000), there are 23 distinct processes capable of destabilizing volcanic islands. The high-

est risk of failure lies in the island arc volcanoes around the Pacific Ocean, due to their explosive nature and steep slopes. Mount Unzen, Japan created tsunamigenic landslides in 1792 that killed 14,500 people. Ward and Day (2001) postulated that a 500 km³ block slide could potentially collapse at La Palma in the Canary Islands, generating a tsunami that would devastate shores across the whole of the Atlantic basin, with waves 10–25 m high possible along the coasts of the Americas. This would be an extremely rare event.

E PROTECTION

1 Environmental control

There is no way of preventing major volcanic eruptions but attempts have been made to control the surface flow of lava:

- *Explosives* can be used in two situations. First, aerial bombing of fluid lava high on the volcano may cause the flow to spread and halt the advancing lava front by depriving it of supply. This method was first tried in 1935 on Hawaii, with only limited success, although modern technology may achieve better results Second, control of *aa* flows has been attempted by breaching the walls which form along the edges of the flow so that lava floods out and starves the advancing front of material. This method was used on the *aa* flow of Mauna Loa's 1942 eruption and in the 1983 eruption of Etna, when it proved possible to divert some 20–30 per cent of blocky flow (Abersten, 1984). These methods are not without risk and possible collateral damage. They would be used with great caution today.
- *Artificial barriers* can divert lava streams away from valuable property if the topographic conditions are suitable and local landowners agree. Barriers must be constructed from massive rocks, or other resistant material, with a broad base and gentle slopes. The method works best for thin and fluid lava flows that exert a limited amount of thrust. It is doubtful if diversion would work with the powerful blocky flows that attain heights of 30 m or more. In the Krafla area of northern Iceland, land has been bulldozed to create barriers to protect a village and a factory from flowing lava. During the 1955 eruption of Kilauea, Hawaii, a temporary barrier diverted the flow from two plantations but later flows took different paths and destroyed the property. Such uncertainty raises the possibility of legal action if lava is deliberately diverted onto

property that otherwise would have escaped. Permanent diversion barriers have been proposed to protect Hilo, Hawaii. Such walls would need to be 10 m high with channels capable of containing a flow about 1 km wide.
- *Water sprays* were first employed, in an experiment by the local fire chief, to control lava flows during the 1960 eruption of Kilauea, Hawaii. They were used on a larger scale during the 1973 eruption of Eldfell to protect the town of Vestmannaeyjar on the Icelandic island of Heimaey (Figure 7.5). Special pumps were shipped to the island so that large quantities of sea-water could be taken from the harbour. At the height of the operation, the pumping rate was almost 1 m^3 s^{-1}, effectively chilling about 60,000 m^3 of advancing lava per day. The exercise lasted for about 150 days. Soon after spraying began, the lava front congealed into a solid wall 20 m in height. Measurements of lava temperature confirmed that where water had not been applied the lava temperature was 500–700°C at a depth 5–8 m below the surface. In the sprayed areas an equivalent temperature was not attained until a depth of 12–16 m below the lava surface (UNDRO, 1985).

Physical protection against lahars depends on the construction of sediment traps and diversion barriers similar to those for other fluid flows. These structures are expensive and silt up with material over time. They can be located only where lahar paths are well defined, and do not work for major, destructive flows in deep valleys. Proposals for the diversion of lahars into wetland areas used for seasonal flood-water storage and fishing in the Philippines have proved controversial. The most ambitious attempts to control lahars at source have been undertaken on Kelut volcano on the island of Java (Box 7.2 and Figure 7.6).

As shown above (page 185), a different hazard occurs when high concentrations of dissolved carbon dioxide enter the bottom of deep, stratified lakes via underground springs. Sudden bursts of gas may produce deaths by asphyxiation in people

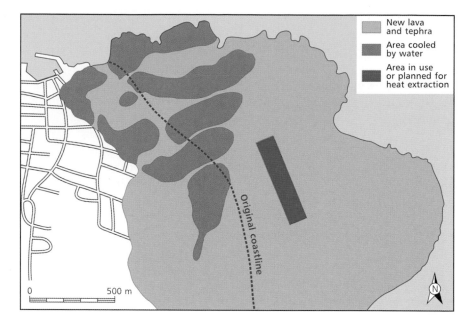

Figure 7.5 Simplified map of the eastern edge of the fishing port of Vestmannaeyjar, Heimay, Iceland after the eruption of Eldfell volcano in 1973, showing the new lava field beyond the earlier coastline and areas cooled by pumping sea water between March and June 1973. Planned heat-extraction zones are a reminder that gains, as well as losses, can result from disasters. After Williams and Moore (1983).

BOX 7.2 Crater-lake lahars in the wet tropics

Kelut volcano, eastern Java is one of the most deadly volcanoes in Indonesia. This is due to lahars produced by releases from the large crater lake, which in 1875 was estimated to contain 78×10^6 m³ of water. In the year 1586 about 10,000 people lives were lost, and in 1919 an explosive eruption threw some 38.5×10^6 m³ of water out of the crater lake and lahars travelled 38 km in less than one hour to claim 5,160 lives. To avoid a repetition of this disaster, Dutch engineers made an immediate start on a tunnel nearly 1 km long designed to reduce the volume of stored water from about 65×10^6 m³ to 3×10^6 m³. In 1923, with the existing crater already half full (22×10^6 m³) the plan was changed to seven parallel tunnels that would progressively lower the water level (Figure 7.6). This work was completed in 1926 and the lake volume was reduced to $<2 \times 10^6$ m³.

An eruption in 1951 created no large lahars but it did destroy the tunnel entrances and added to the water storage capacity by deepening the crater by some 10 m. Even with repair of the original lowest tunnel, the lake soon accumulated a volume of 40×10^6 m³ and became a serious threat once more. The Indonesian government started another low tunnel but stopped it short of the crater wall in the hope that seepage would help to drain the lake. This did not happen because of the low permeability of the volcanic cone. At the time of the 1966 eruption, the lake volume was about 23×10^6 m^3. Lahars killed hundreds of people and damaged much agricultural land. After this event a new tunnel, completed in 1967, was constructed 45 m below the level of the lowest existing tunnel and the lake volume was reduced to 2.5×10^6 m^3. Several sediment dams were also installed. When the 1990 eruption occurred, no primary lahars were recorded, although at least 33 post-eruption lahars were generated which travelled nearly 25 km from the crater. At the present time the lake is some 33 m deep and the 1.9×10^6 m^3 volume represents the lowest risk of primary lahar generation for many years (see The Free University of Brussels website, www.ulb.ac.be/sciences, accessed 22 July 2003).

A similar, but larger, problem emerged at Mount Pinatubo after the 1991 eruption created a 2.5 km-wide crater over 100 m deep and capable of holding over 200×10^6 m^3 of water. As a result, about 46,000 residents in the town of Botolan, 40 km north-west of the volcano, are at risk from a massive lahar, together with the people who have returned to the upper slopes since 1991. By August 2001 the threat of a breach in the crater wall, as a result of increasing water levels in the rainy season, persuaded the authorities to dig a 'notch' in the crater rim to drain water away from populated areas. The outfall started operating in September, accompanied by the short-term evacuation of Botolan as a safety measure. Despite this attempt at drainage, the lake level continued to rise. In July 2002, part of the western wall of the crater collapsed, with the slow release of about 160×10^6 m^3 of water and sediment. A more permanent solution will be required.

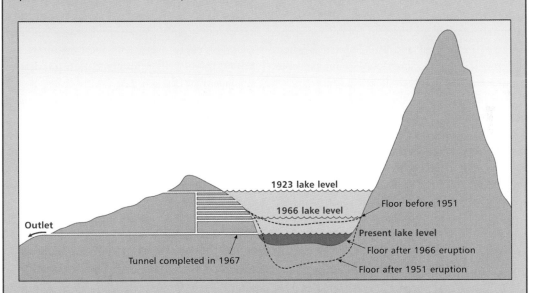

Figure 7.6 Diagrammatic section of the tunnel system constructed at Kelut volcano, Java to lower water levels in the crater lake and reduce the threat of lahars. After *Kelud volcano* at www.ulb.ac.be/sciences (accessed 21 July 2003).

nearby but the threat can be minimized by piping the CO_2-rich water to the lake surface so the gas can be released safely into the atmosphere. Controlled de-gassing of lakes Nyos and Monoun, in Cameroon, began in 2001 and 2003, respectively (Kling *et al.*, 2005). Success depends on the balance between artificial gas removal and natural recharge rates and, without the installation of more pipes and an increased abstraction rate, hazardous amounts of gas may remain within the lakes for some time.

2 Hazard-resistant design

If buildings remain intact during an explosive eruption, they can provide a temporary refuge. Even after a large eruption, some buildings beyond 2–3 km of the volcanic vent have resisted collapse from the pressure of pyroclastic flows (Petrazzuoli and Zuccaro, 2004). According to Spence *et al.* (2004), the buildings most likely to survive are those of recent masonry construction, or with reinforced concrete frames, so long as the door and window openings do not fail and allow the entry of hot gas and ash. Therefore, when warning times are too short for evacuation, people should be advised to seek shelter indoors.

Ash falls can cause the collapse of unstrengthened buildings, especially those with flat roofs. This is most likely if the ash is wet, because, whilst the bulk density of dry ash ranges from 0.5 to 0.7 t m^{-3}, that of wet ash may reach 1.0 t m^{-3}. After the 1991 eruption of Mount Pinatubo, ash falls accumulated to a depth of 8–10 cm in Angeles City, about 25 km from the volcano, resulting in the collapse of 5–10 per cent of the roofs. The only defence against ash fall failure is to make an inventory of building design and type, with a view to retrofitting existing structures and building new ones to higher standards, including the introduction of suitably pitched roofs (Pomonis *et al.*, 1993).

F MITIGATION

1 Disaster aid

Volcanic disasters impose their own character on humanitarian relief. This is mainly because an eruptive spell can continue over months or years, creating a need for support that blurs the distinction between emergency aid and longer-term development investment. Evacuation is a common response. For example, the 1982 Galunggung, Indonesia emergency was created by no fewer than 29 explosive phases occurring over a six-month period, and led to the evacuation of over 70,000 people. In January 2002 a stream of lava from the Nyiragongo volcano, Democratic Republic of Congo devastated about one-third of the city of Goma, killing at least 45 residents and forcing 300,000 others to flee across the border to Rwanda. Many evacuees soon returned but had little to eat for several days because the approach roads were blocked by lava. More than one month later, 30,000 people remained dependent on aid in temporary camps. Volcanic disasters on small islands tend to overwhelm local resources and pose extra difficulties for evacuation and disaster management (see Box 7.3).

Aid has not always been well coordinated. After the Cameroon volcanic gas disaster of 1986, more than 22,000 blankets (five for each displaced person) and over 5,000 gas masks (without some necessary components and cylinders) were supplied (Othman-Chande, 1987). Most food aid was unsolicited and unusable because it was unfamiliar in local diets and could not be stored adequately in the tropical climate. Since then, the performance of aid agencies has improved. For example, after the Goma disaster, clean drinking water was declared a priority in order to prevent outbreaks of cholera at two refugee camps housing about 13,000 people, and health-care staff provided medicines and other support for local clinics.

Box 7.3 Emergency response in Montserrat following the volcanic disaster starting in 1995

In July 1995, the Soufrière Hills volcano, located in the south of the small Caribbean island of Montserrat, began a prolonged phase of unpredicted eruptive activity that continued for over a decade. It was characterized by several phases of dome building and subsequent collapse, accompanied by multiple hazards, including extensive ash fall and lahar deposits. Only 19 lives were lost, but most of the island's infrastructure was destroyed, with economic losses of about £1 billion. By December 1997 almost 90 per cent of the population of over 10,000 people had been forced to relocate, more than two-thirds had been evacuated from the island and the island's GDP had declined by 44 per cent.

Montserrat is a self-governing Overseas Territory of the UK, and after the disaster became totally dependent on British support (Clay *et al.*, 1999). In the absence of an emergency plan, the UK government and the Government of Montserrat had to learn to work together in all aspects of disaster response during a period plagued by uncertainties about both the severity of the volcanic threat and responsibilities for the delivery of aid. An initial emergency plan was prepared in the first few days. Following various evacuations of people to temporary shelter in public buildings, such as schools and churches, relief rations were distributed to 4,000–5,000 refugees. But delays did occur.

Plate 7.3 The spire of a church protrudes from ash, mud and rock deposited by pyroclastic flows that destroyed Plymouth, the former capital of the island of Montserrat, during multiple eruptions of the Soufrière Hills volcano in the 1990s. (Photo: Panos/Steve Forrest SFR00093MSR)

The chronology of repeated evacuations from Plymouth, the island's capital, illustrates the uncertainty: first evacuation 21 August 1995, re-occupied 7 September; second evacuation 2 December 1995, reoccupied 2 January 1996; final evacuation 3 April 1996. Later that month, a state of public emergency was declared and residents were offered voluntary relocation off the island – either to a neighbouring Caribbean island or to the UK with full rights to state benefit and accommodation for a two-year period. By August 1997 about 1,600 refugees were still in temporary shelter and, even in late 1998, 322 people were still housed in these conditions.

In June 1997 pyroclastic flows resulted in fatalities and the volcanic risk was re-assessed, with more than half the island being placed in an exclusion zone. Around this time, more permanent arrangements began to appear, many aimed at securing the longer-term future of the island. These included the construction of an emergency jetty to aid evacuation; new directly built housing, aided by subsidized soft mortgage schemes; strengthening of the scientific capability of the Montserrat Volcano Observatory and the publication of a draft Sustainable Development Plan. Inevitably, there were further delays and, by November 1998, only 105 out of 255 planned houses had been built. In 1999 the UK provided an assisted return passage scheme for those who had left the island. Up to March 1998, the UK government had spent £59 million in emergency-related aid, with an estimated total expenditure of £160 million over a six-year period.

This emergency highlighted a need for pre-disaster planning and swifter investment decisions in volcanic crises, especially where governance is shared between different authorities. More regional cooperation, in this case between the Caribbean countries, would help with monitoring of volcanic activity and the raising of awareness and preparation for future events.

G ADAPTATION

1 Community preparedness

The cost of monitoring volcanoes and pre-disaster planning is small compared to the potential losses. Following steam-blast explosions in early April 1991, scientists began intensive on-site monitoring at Mount Pinatubo that led to the successful forecasting of the major eruption of 15 June 1991. More than 5,000 lives were saved. Property losses of at least US$250 million were prevented and the total costs of the operation were calculated at some US$56 million, showing that savings in property damage alone amounted to five times the overall investment (Newhall *et al.*, 1997).

The main elements of volcanic preparedness are shown in Figure 7.7. The length of time available for the alert phase differs widely. Some volcanic activity may start several months before an eruption; in other cases, only a few hours may be available. For effective evacuation, it is essential that the population at risk is advised well in advance about evacuation routes and the location of refuge points. These directions have to be flexible, depending on the scale of the eruption (which will influence the pattern of lava flow) and the wind direction at the time (which will influence the pattern of ash fall). Some local roads may be destroyed by earthquake-induced ground failures and steep sections of highways can become impassable with small deposits of fine ash that make asphalt very slippery.

Unfortunately, the infrequency of volcanic activity induces poor hazard awareness and low levels of community preparedness. Surveys of residents on Hawaii (Gregg *et al.*, 2004) and Santorini (Dominey-Howes and Minos-Minopoulos, 2004) revealed a relatively poor understanding of volcanic risk and, in the case of the

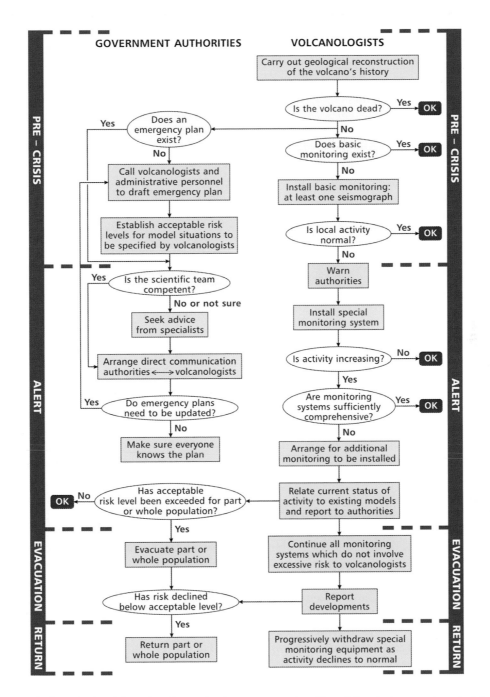

Figure 7.7 Organizational flow chart for a volcanic emergency plan. Close liaison between volcanologists and government authorities should ensure an effective disaster response. After UNDRO (1985).

latter island, no emergency plan existed. Hazard awareness is especially low for rare volcanic threats such as those posed by *jökulhlaups*. This is an Icelandic term for the sudden discharge of melt-water containing ice blocks and volcanic sediments following the eruption of sub-glacial volcanoes. Discharges up to 300,000 m^3 s^{-1} have been recorded for peak flows. People in southern Iceland are at risk from such a potential outburst because the highly active Katla volcano, lying beneath the Myrdalsjökull icecap, has erupted about twice per century since the year 874. Questionnaire surveys and a full-scale evacuation exercise in 2006 showed that many residents thought that the volcano is no longer active, were unaware of the *jökulhlaup* risk and were unlikely to react to warnings or adopt hazard mitigation measures (Bird *et al.*, 2009; Jóhannesdóttir and Gisladóttir, 2010).

All too often, emergency planning for volcanic hazards occurs after a disaster. Before the Nevado del Ruiz disaster in Colombia in 1985, there was no national policy for volcanic hazards (Voight, 1996). Similar policy failures occurred at the Galeras volcano, Colombia, where hazard mapping was held up by the unwillingness of the authorities to accept either the concept of disaster or the cost of mitigation (Cardona, 1997). The successful evacuation of densely populated areas requires adequate transportation. Evacuees need support services, including medical treatment (especially for dust-aggravated respiratory problems and burns), shelter, food and hygiene.

Because volcanic emergencies can last for months, the 'temporary' arrangements for refugees sometimes function from one crop season to another. For example, a total of 26,000 people were evacuated in 1999 from the slopes of Tungurahua (Ecuador) because of the possibility of an eruption; some remained in special accommodation for over one year (Tobin and Whiteford, 2002). Prolonged relocation is never popular. During the Tungurahua emergency, the evacuated population of Baños, a town heavily reliant on tourist revenue, organized a return to their homes while the town remained under an evacuation order so that they could regain their livelihoods (Lane *et al.*, 2003). Another feature of volcanic eruptions is that ash fall has the potential to disrupt communities several hundreds of kilometres away. Hazard mitigation specialists have an uphill task persuading these people that they are at risk.

Given adequate monitoring, warning of certain eruptive phases is possible. This makes adaptive responses such as public education, access controls and emergency evacuation procedures important (Perry and Godchaux, 2005). Evidence from the western USA suggests that, whilst residents do acknowledge the threat, there is little prioritization in the actions taken (Perry and Lindell, 1990). People at risk in the Philippines are offered training in the recognition of precursory signs of volcanic activity, such as crater glow, steam releases, sulphurous odour and drying vegetation (Reyes, 1992). In Ecuador about 3 million people live within the two main volcanic mountain ranges prone to lahars. The principal threat is in the Chillos and Latacunga valleys, where an ever-growing population of some 30,000 has settled on lahar deposits from the 1877 eruption (Mothes, 1992). Again, public education programmes, including field trips and evacuation exercises, help to raise awareness.

2 Forecasting and warning

Most volcanic eruptions are preceded by a variety of environmental changes that accompany the rise of magma towards the surface. An understanding of lava dome inflation led to the forecasting of a repetitive cycle of eruptions during 1996–97 at the Soufrière Hills volcano and most of the residents were safely evacuated (see Box 7.3). UNDRO (1985) classified some of the physical and chemical phenomena that have been observed before eruptions (Table 7.4). Unfortunately, such phenomena are not always present and the most highly explosive volcanic eruptions are particularly difficult to forecast.

Table 7.4 Precursory phenomena that may be observed before a volcanic eruption

Seismic activity
- increase in local earthquake activity
- audible rumblings

Ground deformation
- swelling or uplifting of the volcanic edifice
- changes in ground slope near the volcano

Hydrothermal phenomena
- increased discharge from hot springs
- increased discharge of steam from fumaroles
- rise in temperature of hot springs or fumarole steam emissions
- rise in temperature of crater lakes
- melting of snow or ice on the volcano
- withering of vegetation on the slopes of the volcano

Chemical changes
- changes in the chemical composition of gas discharges from surface vents (e.g. increase in SO_2 or H_2S content)

Source: After UNDRO (1985)

Real-time measurements of volcanic processes are made with GPS technology and satellite imagery (Kervyn, 2001). In some cases, local rainfall measurements may be useful (Barclay *et al.*, 2006). These monitoring programmes provide the best hope of developing reliable forecasting and warning systems, but only about 20 volcanoes worldwide have well-equipped local observatories (Scarpa and Gasparini, 1996). This situation has been alleviated by advances in remote sensing, notably the observations made by 10 Earth Observing System (EOS) satellites that monitor changes in volcanic activity involving features such as thermal anomalies, plume chemistry and lava composition (Ramsey and Flynn, 2004).

Specific techniques apply to the following processes:

- *Earthquake activity* is common near volcanoes and, for predictive purposes, it is important to gauge any increase in activity in relation

to background levels. This requires a study of records going back many years. During a period of high alert, these records can be supplemented by data from portable seismometers. Most volcano-related earthquakes are less than $M_W = 2$ or 3 and occur less than 10 km beneath a volcano. A prior seismic signature has been incorporated into a tentative earthquake swarm model for the prediction of volcanic eruptions (McNutt, 1996). As shown in Figure 7.8, the onset and subsequent peak of a 'swarm' of high-frequency earthquakes reflects the fracture of local rocks as magmatic pressure increases. This phase is followed by a relatively quiet period, when some of the pressure is relieved by cracks in the Earth's crust, before a final tremor results in an explosive eruption.

- *Ground deformation* is sometimes the forerunner of an explosive eruption, although the information is not easy to fit into a forecasting model. The method is also difficult to employ for explosive subduction volcanoes because they erupt so infrequently and there is a lack of comparative data. In rare cases, such as the 1980 event at Mount St Helens, the deformation was large and easily visible, but it is usually necessary to detect movements with survey equipment and *tiltmeters*. These instruments operate like a woodworker's level. They are very sensitive but can record changes in slopes only over short distances.

- *Electronic distance meters (EDM)* can measure the distance between benchmarks placed on a volcano in order to pinpoint when magma is rising and displacing the ground surface. This technique is usually less available and requires a series of visible targets on the volcano. There is a wide range of capability for these instruments, but short-range (<10 km) to medium-range (<50 km) EDMs are typically employed. Short-range EDMs can measure distances with an accuracy of about 5 mm but the trick is to place benchmarks in the right places and make frequent observations.

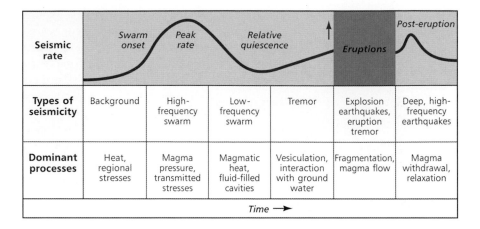

Figure 7.8 The stages of a generic volcanic-earthquake-swarm model. The precursor earthquake swarm, which can be used as a warning sign, reflects the fracturing of rocks in response to growing magmatic pressure. After McNutt (1996). With kind permission from Springer Science+Business Media: S.R. McNutt, Seismic monitoring and eruption forecasting of volcanoes: a review of the state of the art and case histories in R. Scarpa and R.I. Tilling (eds) *Monitoring and Mitigation of Volcano Hazards* (1996), 99–146. Copyright Springer 1996. Reproduced with permission.

- *Global Positioning Systems* (GPS) rely on about 24 satellites that orbit the Earth twice each day at altitudes around 20,000 km and constantly transmit information that allows the position of ground-based receivers to be determined. Normally 5–8 of these satellites are in view from any point on Earth, provided that there is an unobstructed sky view. Since a ground-level accuracy of a few centimetres is necessary to detect the build-up of pressure from rising magma, GPS receivers are commonly placed at several locations on the volcano to minimize any variations in signal transmission through the atmosphere to individual receivers.

- *Thermal changes* might be expected as magma rises to the surface and increases the surface temperature. Many volcanoes have erupted without any detectable shift in temperature, but this approach has proved useful where a crater lake exists. An early example occurred at the Taal volcano, Philippines when the temperature increased from a constant 33°C in June 1965 to reach 45°C by the end of July. The water level also rose during this period. In

September 1965, a violent eruption occurred. Such ground observations can be supplemented and confirmed by thermal imaging from satellites.

- *Geochemical changes* can be detected in the composition of the gases issuing from volcanic vents. In addition, the rate at which a volcano releases gases is related to variations in the amount of magma within the sub-surface systems. Direct field sampling of gases escaping from surface vents is the usual method employed, but remote sensing has been an important tool for the surveillance of volcanic activity for over 40 years (Goff *et al.*, 2001; Galle *et al.*, 2003). Juvenile gases show considerable variation over short periods and distances and it is difficult to judge how gas samples represent more general conditions in the volcano. Larger-scale visual observations of steam emissions or ash clouds depend on meteorological conditions. Sulphur dioxide injected high into the atmosphere can be measured by on-board satellite instruments and the behaviour of volcanic plumes can be monitored by weather satellites (Malingreau and Kasawanda, 1986; Francis, 1989).

- *Lahars* have been monitored for years via visual sightings by local people and, more recently, with the aid of video cameras located in remote valleys. An automated detection system now exists using a series of acoustic-flow (AFM) monitoring stations placed downstream from the volcano (see USGS http://volcanoes. usgs.gov/activity/methods/hydrologic/lahar detection.php). At each location, a seismometer detects ground vibrations from an approaching lahar. The amplitude of vibration is sampled every second and information – including an emergency message – can be transmitted to a base station further down the volcano's flank and closer to a population centre. Short-term warnings and emergency evacuation are then possible.

Although there is no fully reliable forecasting and warning scheme for volcanic eruptions, some success has been achieved. At Mount Pinatubo in 1991 about 1 million people, including 20,000 American military personnel and their dependants, were within 50 km of the volcano. After intensive on-site monitoring of early explosions, a 10 km-radius danger zone was declared and followed by an urgent warning of a major eruption. In total, over 200,000 people were evacuated from a 40 km-radius before the eruption, about three times more people than had been previously evacuated in a volcanic emergency. Fewer than 300 people died. For small volcanic islands, and for coastal communities, off-shore evacuation may be necessary, as in the case of Montserrat (see Box 7.3). The proposed evacuation plan for the area surrounding Mount Vesuvius envisages the evacuation of 700,000 people. This scenario is based on the 1631 eruption, when earthquakes were experienced for at least 15 days prior to the eruption. As shown in Figure 7.9, evacuees would potentially be transported to a wide range of destinations in the country. Even a partial evacuation of residents and the closure of tourist areas

would paralyse the region and incur enormous costs.

Since the 1991 eruption of Mount Pinatubo, lahars have destroyed the homes of more than 100,000 local people and several new towns have been built on higher ground (Newhall *et al.*, 1997). Some traditional Indonesian villages are provided with artificial mounds so that people can quickly climb to a safer level while a lahar passes. Over 1 million people live on the slopes of the Merapi volcano, central Java, and secondary lahars are triggered by rainfall of about 40 mm in two hours. Lahars are very rapid-onset, short-duration events, lasting between 30 min and 2 h 30 min, and have average velocities of 5–7 m s^{-1} at elevations of 1000 m (Lavigne *et al.*, 2000). Reliable forecasting of lahars is impossible because of the short lead-time and variations in rainfall intensity during the monsoon season, but monitoring is a necessary step to the better understanding of these hazards.

3 Land use planning

Land zoning of high-hazard volcanic areas, plus the selection of safe sites for emergency evacuation and new development, depends on probability assessments of dangerous activity. To determine the likelihood of an event, all previous eruptions require accurate dating on geological time-scales using techniques such as radiocarbon dating, tree-ring analysis, lichenometry and thermoluminescence. Volcanic-hazard maps can then be prepared to show the likely area at risk. Such mapping is hampered by limited knowledge of the size of future eruptions and the extent of pyroclastic surges or lahars. Environmental conditions at the time of eruption will be important; the amount of seasonal snow cover will affect lahar and avalanche hazards, whilst the speed and direction of the wind will determine the airborne spread of tephra.

Because of these problems, many zoning maps are restricted to just one or two volcanic hazards. For example, the island of Hawaii has nine hazard

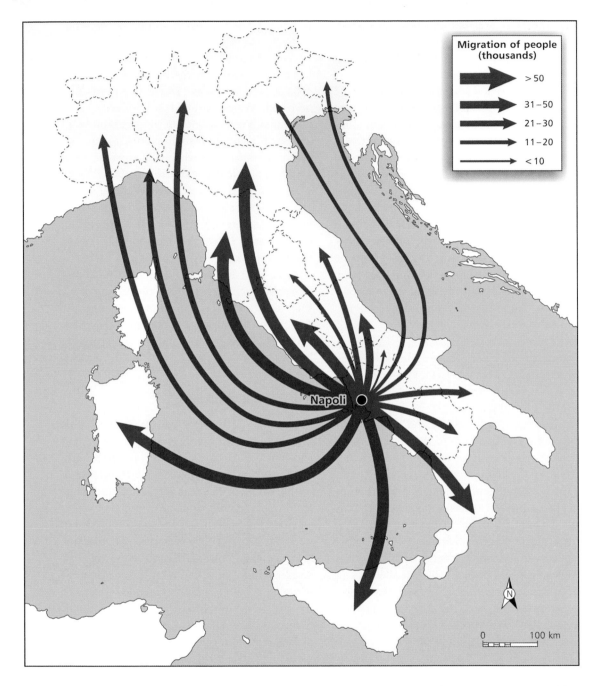

Figure 7.9 Proposed destinations for evacuees according to the preparedness plan for a major eruption at Mount Vesuvius, Italy. The logistical challenges of this exercise would be formidable. After Chester *et al.* (2002). Reprinted from *Journal of Volcanology and Geothermal Research* 115, D.K. Chester *et al.*, Volcanic hazard assessment in Western Europe, 411–35. Copyright (2002), with permission from Elsevier.

zones ranked on the probability of land coverage by lava flows, based on the location of volcanic vents, the topography of the volcanoes and the extent of past flows. Zones 1–3 are limited to the active volcanoes of Kilauea and Mauna Loa, while zone 9 consists of Kohala volcano, which last erupted over 60,000 years ago (Figure 7.10). This map ignores hazards other than lava flows. In comparison, the combined lava flow and ash fall risk assessment available for the island of Tenerife provides a more comprehensive picture (Araña *et al.*, 2000).

Most land planning maps undergo revision and updating. Figure 7.11 illustrates the third version of the hazard zones modelled for the Galeras volcano, Colombia (Artunduaga and Jiménez, 1997). The volcanic hazard extends

for some 12 km around the vent, based on the assumption that future eruptions will come from the active cone, that the geological record of the last 5,000 years is reliable and that the data collected during on-site monitoring (1989–95) are representative. Three zones are identified:

- *High-hazard zone* – this is restricted to areas of pyroclastic flows. Within 1 km of the active cone, there is a 78 per cent probability of encountering a ballistic fragment between 0.4 and 1 m in diameter.
- *Intermediate-hazard zone* – this area could experience pyroclastic flows in the largest eruptions and there is also a lahar threat.
- *Low-hazard zone* – this area is predicted to receive tephra falls.

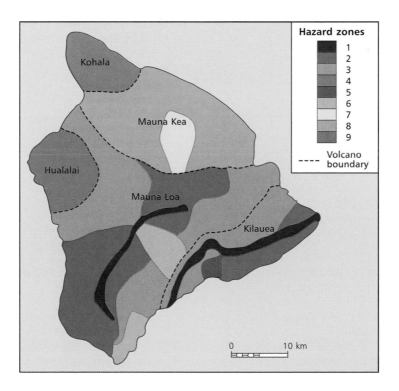

Figure 7.10 The island of Hawaii zoned according to the degree of risk from lava flows. Zone 1 is highest rated, Zone 9 is lowest rated and the change between zones is gradual rather than abrupt. All property destroyed in the last 30 years was in Zone 2 located within 12 km of the vent of Kilauea. After US Geological Survey at http://pubs.usgs.gov/gip/hazards/maps.html (accessed 26 February 2003).

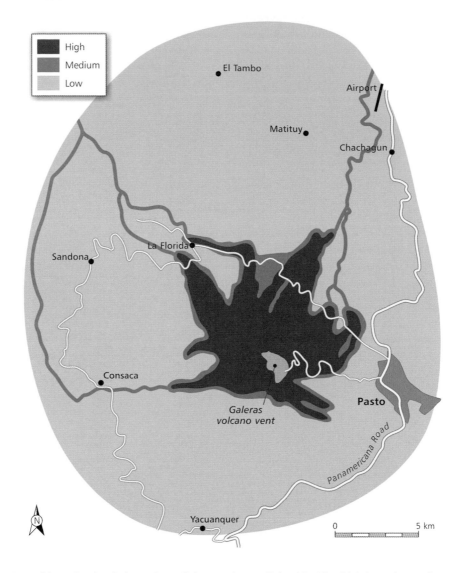

Figure 7.11 Map of volcanic hazards at Galeras volcano, Colombia. The high-hazard zone is exposed to pyroclastic-flow deposits; the low-hazard zone is exposed to ash-fall deposits. After Artunduaga and Jiménez (1997). Reprinted from *Journal of Volcanology and Geothermal Research* 77, A.D.H. Artunduaga and G.P.C. Jiménez, Third Version of the hazard map of Galeras volcano, Colombia, 89–100. Copyright (1997), with permission from Elsevier.

It is clear that land planning should attempt to restrict future development of the area between the volcano and the city of Pasto, with a population of around 250,000 people, and that preparedness planning should include the likely effects of ash fall at the airport some 21 km distant.

The value of comprehensive volcanic hazard mapping can be demonstrated with reference to the Mount St Helens, USA eruption of 1980 (Crandell *et al.*, 1979). Figure 7.12a shows that pyroclastic flows were expected to flow down the upper valleys for up to 15 km, with mudflows and

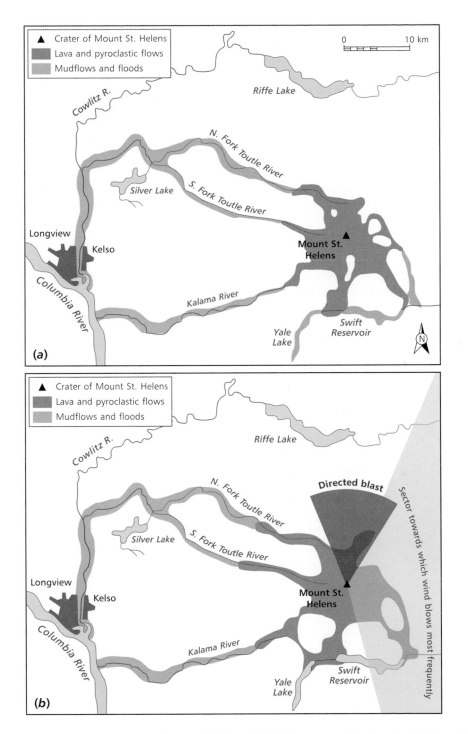

Figure 7.12 Volcanic hazards around Mount St Helens, USA. (a) as mapped before the 1980 eruption; (b) modifications after the 18 May eruptions, showing the area at risk from future directed (lateral) blasts. After Crandell *et al.* (1979) and Miller *et al.* (1981).

floods continuing downstream for many tens of kilometres. Indeed, it was predicted that lahars up to 110×10^6 m^3 in volume could reach the local reservoirs and create additional flooding if storage was not reduced in advance of the wave. Tephra deposits were predicted to occur over a 155° sector extending away from the volcano from north-northeast to south-southeast, based on the wind direction experienced for 80 per cent of the time. Some ash fall was assumed to reach a distance of 200 km, within the range of the town of Yakima. This scenario was generally accurate, apart from the magnitude of the landslides and the severity of the lateral blast (Figure 7.12b). Mudflows, laden with logs and forest debris, were channelled down the valleys and a flood surge entered the upper Swift reservoir. Since the water level had been lowered previously, the added volume did not over-top the dam and flooding along downstream parts of the Lewis River was avoided. On the other hand, noticeable ash falls occurred as far away as Nebraska and the Dakotas, while at Yakima the depth of tephra reached a disruptive 250 mm.

FURTHER READING

Araña, V. *et al.* (2000) Zonation of the main volcanic hazards (lava flows and ashfall) in Tenerife, Canary islands: a proposal for a surveillance network. *Journal of Volcanology and Geothermal Research* 103: 377–91. A typical example of precautionary land use planning.

Bird, D.K., Gisladdóttir, G. and Dominey-Howes, D. (2009) Resident perceptions of volcanic hazards and evacuation procedures. *Natural Hazards and Earth System Sciences* 9: 251–66. This paper demonstrates the practical difficulties facing communities that are unlikely to respond effectively in a volcanic emergency.

Chester, D. (1993) *Volcanoes and Society*. Edward Arnold, London. A comprehensive and reliable, hazard-based account.

Kling, G.W. *et al.* (2005) Degassing Lakes Nyos and Monoun: defusing certain disaster. *Proceedings of the National Academy of Sciences of the USA* 102: 14185–90. An interesting example of a very rare type of volcanic hazard.

Sparks, R.S.J., Biggs, J. and Neuberg, J.W. (2012) Monitoring volcanoes. *Science* 335: 1310–11. A concise update on the status of volcano monitoring and research in relation to the needs of hazard forecasting.

Witham, C.S. (2005) Volcanic disasters and incidents: a new database. *Journal of Volcanology and Geothermal Research* 148: 191–233. A detailed independent survey of the volcano hazard worldwide.

WEB LINKS

International Volcanic Health Hazard Network www.ivhhn.org

US Disaster Center Volcano Page www.disaster center.com/volcano

Volcanic Ash Advisory Centre: http://aawu.arh.noaa. gov/vaac.php

CHAPTER EIGHT

Mass movement hazards

A LANDSLIDE AND AVALANCHE HAZARDS

Mass movement is a common hazard in mountainous areas. It results from the displacement of surface materials down-slope under the force of gravity and exists in most environments where slopes are present. The movements vary greatly in size (ranging from a few cubic metres to over 100 cubic kilometres) and in speed (ranging from millimetres per year to hundreds of metres per second). Rapid mass movements generally cause the greatest loss of life, whilst slower down-slope movements create significant economic costs. It is convenient to classify mass movements according to the material forming most of their mass. *Landslides* consist mostly of rock and/or soil; *snow avalanches* are formed predominantly from snow and/or ice. Mass movements are usually triggered by natural processes, notably earthquakes, intense and/or prolonged rainfall and snowmelt, but some damaging landslides occur in materials deposited by humans. These include mining waste, landfill or garbage and, overall, people play an increasing role in the causation of this hazard.

In the past, the losses associated with mass movements have probably been underestimated. This is partly because many events occur in rural mountainous environments remote from population centres. Also, a substantial proportion of the cumulative loss results from numerous small mass movements, rather than large incidents. The process is often attributed to the trigger event, such as an earthquake or a rain-storm, rather than to the mass movement itself (Jones, 1992). In recent years, mass movements have attracted more attention. Figure 8.1 shows the steep rise in the annual number of publications on landslides since 1990, a feature aided by the introduction, in 2004, of an international journal dedicated to landslide study. Typically, mass-movement investigations are conducted either on an intensive site-specific basis for risk assessment on individual slopes or as a regional-scale evaluation to identify trends and hazardous zones relevant to wider land use planning.

Estimates of loss vary widely. According to the EM-DAT database, about 9,000 people died as a result of mass movements in the 1990–99, decade but the Durham Fatal Landslide Database shows that between 2002 and 2007 alone there were approximately 44,000 fatalities (Petley, 2009). Most of the deaths were in geographically distinct regions: the Pacific Rim, Central America and the

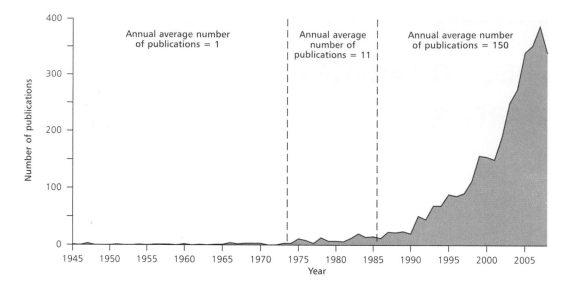

Figure 8.1 The annual number of landslide publications in the international literature 1945–2008. This steep increase is probably similar to the rise in research output for all environmental hazards over the same period. After Gokceoglu and Sezer (2009). With kind permission from Springer Science+Business Media: C. Gokceoglu and E. Sezer, A statistical assessment on international landslide literature (1945–2008). *Landslides* 6 (2009): 349–51. Copyright Springer 2009. Reproduced with permission.

Caribbean, mainland China, south-east Asia and along the southern edge of the Himalayan Arc. These areas have hilly or mountainous terrain, active tectonic processes (including uplift and earthquakes), intense rainfall events (associated with tropical cyclones, El Niño/La Niña events or monsoon weather patterns) and comparatively large populations of poor people who live in unsafe locations. Little information is available about the exact causes of landslide deaths. However, following a tropical storm over the volcanic islands of Chuuk State, Micronesia, when over 250 landslides and debris flows killed 43 people, 90 per cent of the recorded deaths were attributed to suffocation (Sanchez *et al.*, 2009). This is similar to the pattern of mortality associated with snow avalanches.

Mortality from mass movements is comparatively low in most MDCs. For example, Italy has the highest fatality rate from slope failures in Europe. Deaths average 60 per year and around 80 per cent occur in fast-moving events (Guzzetti,

2000). However, the economic losses are high. In the USA, Canada and India the estimated costs of landslides exceed US$1 billion per year (Schuster and Highland, 2001) whilst the direct damage caused by landslides between 1945 and 1990 exceeded $15 billion in Italy. Indirect losses are poorly quantified but include damage to transport links, electricity transmission systems and gas and water pipelines, flooding due to landslide dams across rivers, impaired agricultural and industrial production, loss of trade and a reduction in property values. It is generally accepted that mass movement risks and losses have increased over time. The reasons for this trend are unclear, mostly because landslide disasters are complex events with physical triggers, such as earthquake activity and high rainfall, combined with societal causes such as rapid economic development, population growth and urbanization.

In China, about 80 per cent of major landslides occur on the tectonically active flank of the Tibet Plateau. Since 1980, there has a general increase in

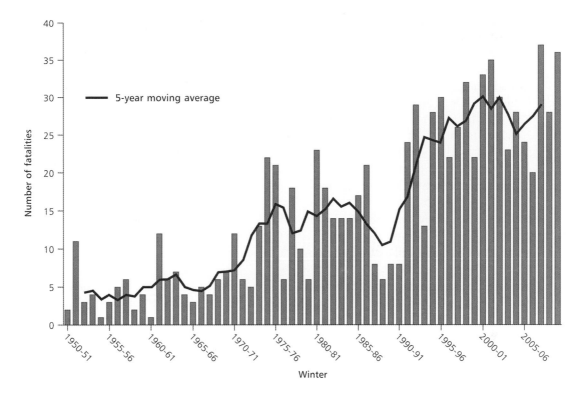

Figure 8.2 The annual number of avalanche fatalities in the USA over the 1950/51 to 2009/10 winter seasons. The rising deaths reflect growing participation in winter-sports activities. Over half the victims were mountain climbers or back-country skiers. After Colorado Avalanche Information Center at http://avalanche.state.co.us/acc/acc_images (accessed 27 March 2011). With permission of the Colorado Avalanche Information Center.

landslide activity, mainly attributed to more construction work and a shift in climatic conditions, causing about 1,000 fatalities per year (Runqiu, 2009). The expansion of many cities in LDCs has forced more people to live in unregulated *barrio* settlements located on steep slopes. Hong Kong suffered a major increase in landslide-related fatalities during the 1970s, primarily due to the growth of illegal communities of immigrants on unstable slopes, although government action to move these people onto safer terrain reduced the death toll. The 1999 landslides at Vargas on Venezuela's northern coast, resulting from exceptionally heavy rainfall triggered by La Niña conditions, killed up to 30,000 people and created economic damage amount-

ing to US$1.9 billion, 30 per cent of which was to infrastructure (IFRCRCS, 2002). Fatalities were concentrated in settlements that had developed over the previous 30 years on debris fans deposited by earlier landslides. Petley *et al.* (2007) found considerable variation in landslide fatalities from year to year in Nepal. A strong control was exercised by heavy rains linked to the strength of the summer monsoon, but an upward trend outside the monsoon cycle was tentatively attributed to land use disturbances arising from rural road construction.

Snow avalanches tend to occur in Arctic and temperate regions whenever snow is deposited on slopes steeper than about 20°. Each year, avalanches claim more than 150 lives worldwide.

The USA annually suffers up to 10,000 potentially damaging avalanches, although only about 1 per cent affect life or property. The worst avalanche disaster in the USA occurred in 1910 in the Cascade Range, Washington, when three snowbound trains were swept into a canyon with the loss of 118 lives. In February 1999 two avalanches struck the towns of Galtür and Valzur in Austria, leading to the loss of 38 lives. The risk posed by avalanches is more severe in Europe than in North America because the population density is higher in the Alps than in the Rockies. This highlights the ways in which the development of mountain areas for winter recreation over the last 50 years has increased avalanche risk. About 70 per cent of avalanche fatalities in the MDCs are associated with the voluntary activities of ski touring and mountain climbing. Many countries have recorded an increase in the number of fatalities since the early 1950s, as illustrated for the USA in Figure 8.2. Most victims are male skiers or backpackers around 30 years of age and fatalities cluster in the late winter months, when snowfall accumulations are at a maximum.

Apart from the growth of winter recreation, the increase of transportation routes through mountain areas has led to more avalanche-related mortality. The construction of roads and railways leads to the removal of mature timber that, if left intact, would help to stabilize the snow cover and protect people and infrastructure in the valley bottom. For example, the Trans-Canada Highway runs under nearly 100 avalanche tracks in a 145 km section near Rogers Pass, and at least one vehicle is under an avalanche path at any given time. Much less is known about avalanche hazards in mountainous areas of the LDCs, such as the Himalayas, but the threats are serious. In the Kaghan valley, Pakistan, avalanches pose a threat to local residents over-wintering on the valley floor, and 29 people were killed in single event in the winter of 1991–92 (de Scally and Gardner, 1994), whilst in December 2005, 24 people were killed by a single avalanche in the Northwest Frontier Province of Pakistan.

B LANDSLIDES

The term 'landslide' describes down-slope movements of soil and/or rock under the influence of gravity. Although many landslides occur through the process of rock or soil sliding on a distinct surface, this is not always so. In practice, there is a wide variety of types of movement, including falling, sliding and flowing. The type of movement depends upon the angle of the slope, the nature of the materials and the various stresses that act upon them.

Landslides are concentrated in five major types of terrain (Jones, 1995):

- *Upland areas subject to seismic shaking*
  Earthquakes in hilly or mountainous terrain often trigger large numbers of landslides. For example, in the 1999 Chi-Chi earthquake in Taiwan over 9,200 large landslides were triggered in 35 seconds (Hung, 2000). In the years after a large earthquake further landslides may occur, as slope materials have been destabilized by the shaking.

- *Mountainous environments with high relative relief*
  Mountain areas are subject to high levels of rock falls and landslides, due to the steep terrain, deformed rock masses and high rainfall totals. Rock avalanches are a particular hazard. They involve massive quantities of rock debris, with volumes over 100×10^6 m^3, able to travel very large distances. In New Zealand, for example, some rock avalanches with volumes in excess of 100 km^3 have travelled more than 15 km. On average, about one massive rock avalanche occurs worldwide each year, generally in high, tectonically active mountain chains such as the Himalayas, the Rockies or the Andes.

- *Areas of moderate relief suffering severe land degradation*
  Human activity and land degradation is the cause of many events. North Korea is prone to

Table 8.1 Classification of landslides

Type of movement	Bedrock	Type of material	
		Engineering soils	
		Mainly coarse	Mainly fine
Falls	Rock fall	Debris fall	Earth fall
Topples	Rock topple	Debris topple	Earth topple
Slides			
Rotational	Rock slump	Debris slump	Earth slump
Rotational (few units)	Rock block slide	Debris block slide	Earth block slide
Translational	Rock slide	Debris slide	Earth slid
Lateral spreads	Rock spread	Debris spread	Earth spread
Flows	Rock flow (deep creep)	Debris flow (soil creep)	Earth flow (soil creep)
Complex	Combination of two or more principal types of movement		

Source: After Varnes (1978)

landslides because the population has been forced to remove almost all of the forests on the hills, to burn as firewood. South Korea, on the other hand, has similar terrain, but a proactive policy of slope management and afforestation has resulted in a much lower incidence of landslides.

- *Areas with high rainfall*
  Intense or prolonged rainfall is the most common trigger of slope instability and areas subject to very high rainfall totals are inevitably susceptible to landslides. This process is especially active in humid tropical areas, where rock weathering can penetrate tens of metres below the ground surface. In Malaysia the weathered material can extend to a depth of 30 m or more and landslides are common in the intense rainfall during the passage of a tropical cyclone. Areas affected by monsoonal rainfall, such as the Indian subcontinent, are also vulnerable to landslides. In temperate regions landslides tend to result from the disturbance of more shallow surface layers, brought about by seasonal rainfall in the form of either high winter totals or convectional storms in summer.

- *Areas with thick deposits of fine-grained materials*
  Fine-grained deposits, such as loess and tephra, are weak, and vulnerable to the effects of saturation. As a result, they are prone to landslides. Notable areas at risk include the loess plateaux of Gansu in northern China. In the 1920 earthquake, flowslides in the loess triggered by ground shaking are estimated to have killed over 100,000 people, the largest landslide disaster in recorded history. Areas mantled in tephra on and around volcanic sites are also at high risk.

Landslides are classified according to the materials involved and the mechanism of movement (Table 8.1).

1 Rock falls

Rock falls involve the movement of material through the air and occur on very steep rock faces.

The blocks that fall often detach from the cliff face along an existing weakness, such as a joint, bedding or exfoliation surface. The fall starts as an initial slip along a joint or bedding plane which then transitions into falling, due to the steepness of the cliff. The scale of rock falls varies from individual blocks through to *rock avalanches* that are hundreds of millions of cubic metres in size. Whilst the largest rock falls clearly have the greatest hazard potential, an individual block the size of an egg-cup can be fatal if it hits a person on the head. In 1903 a major rock fall occurred in Alberta, Canada and partially destroyed the small town of Frank. This failure took place across the bedding planes in a steep anticline formed in the well-jointed limestone of Turtle Mountain. Ground-water seeping into the joints had partially dissolved the limestone and the joints

Plate 8.1 The Ferguson rockslide above Highway 140 in the Merced River canyon, California. This slide became active in 2006 and is closely monitored because it is near a main entrance to Yosemite National Park. There is a threat to all road traffic between mileposts 103 and 104 as well as recreational users of the river in the run-out path below. (Photo: Mark Reid, USGS)

were further weakened by a combination of frost action and mining operations. About 70 people were killed.

The triggering of rock falls is complex. Earthquakes are an important factor because seismic waves can literally shake blocks off a cliff. For example, in the 1999 Chi-Chi earthquake in Taiwan, the road system was severely damaged by rock-fall activity in the epicentral zone, which greatly hindered the government response. Rock falls are also triggered by the presence of water in joints and fissures, which can loosen material. During winter periods, the process of freeze–thaw, in which water repeatedly freezes and expands in cracks in a rock face, is particularly important. Rock fall activity often increases during heavy rain and in periods when the temperature passes though many freeze–thaw cycles in spring-time in high-altitude areas. In some high mountain areas, permanent ice serves to hold fractured blocks on the slope, although recent atmospheric warming has led to some thawing of the ice and more rock fall activity (Sass, 2005). In some cases, there is no obvious trigger. In May 1999 an unexpected rock fall occurred at Sacred Falls State Park in Oahu, Hawaii. A mass about 50 m^3 detached from the walls of a steep gorge and fell 160 m into the valley below, striking a group of hikers and killing eight people (Jibson and Baum, 1999).

2 Landslides

Many landslides involve the sliding of a mass of soil and or rock, usually along a slip surface which either forms as the slide develops or results from the activation of an existing weakness, such as a bedding plane, joint or fault. There are many ways in which the strength of the material comprising the slope is exceeded by down-slope stress, to create instability. Sometimes the sliding occurs because of the deformation of a weak layer within the rock or soil, in which case a shear zone is formed.

In a slope, the forces that cause movement are called *shear forces*, and are due to gravity trying to

pull the mass down the slope. There are two main forces that resist the movement of the landslide. These are:

- *Cohesion*. This is the resistance arising from the 'stickiness' of particles or from interlocking. So, for instance, sandstone gets some of its strength from the cement that glues the particles of sand together. As a result of cohesion, sandstone cliffs are often vertical, to substantial heights. Beach sand, however, has no bonds between the particles and rests at a much lower angle. In a sandcastle, the initial wetness of the sand generates a suction force between the particles that holds them together – this also is a form of cohesion.
- *Friction*. This arises from the resistance of particles to sliding across each other. In a pile of dry sand the gradient of the slope is sustained by the friction between the sand grains. The magnitude of the friction force depends upon the weight of the material above the surface, just as it is more difficult to move a chair with someone sitting in it than when it is empty.

Friction and cohesion together provide the resistive forces that maintain stability in a slope. Movement of the landslide occurs when the shear forces exceed the resistive forces. The crucial relationship between the resistive forces and the shear forces varies continually in most slopes for the following reasons:

- *Weathering*. Through time, weathering of the rock or soil mass can reduce its strength. In particular, weathering can attack the bonds between particles that provide cohesion and so weakens the rock or soil. As this occurs the slope becomes progressively less stable.
- *Water*. When the rock or soil forming the slope has water in its pore space, the overall weight of the slope increases, which slightly increases the shear forces. More importantly, this water provides a buoyancy force to the landslide mass

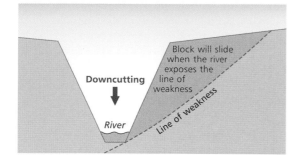

Figure 8.3 Down-cutting by rivers, construction activity or erosion by glaciers can cause landslides by exposing a weak layer of rock that permits sliding to occur. After Petley (2009).

that acts to reduce the friction, in much the same ways as a car skids once it starts to aquaplane. Thus, as the slope gets wetter it usually becomes less stable.

• *Increased slope angle.* In some situations the slope angle may increase, perhaps due to erosion of the toe of the slope by a river or through cutting of the slope by humans, perhaps for road construction or other developments. This serves to increase the shear forces.

• *Earthquake shaking.* During earthquakes, the magnitude of the forces varies as shaking of the slope occurs. This can both increase the shear forces and reduce the resistive forces, thereby triggering failure.

Quite often the surface on which movement occurs has a planar form. In this case, a *translational landslide* will result. This occurs because the landslide has activated an existing plane of weakness, such as a bedding plane. Commonly, down-cutting by a river causes an inclined bedding plane to be exposed in the river bank (Figure 8.3). The block on the slope is then free to move when the resisting forces become sufficiently weak. This is a significant hazard during road construction in mountain areas, when cutting of the slope to create a bench for the road often exposes inclined bedding surfaces or joints (Petley, 2009). Translational landslides are often rapid. Once sliding starts, the materials in the shear zone lose their cohesion as inter-particle bonds are broken, and their frictional strength is reduced as the shear surface becomes smooth or even polished. This allows the landslide to accelerate rapidly, sometimes reaching very high speeds.

A specific hazard is the *landslide dam,* caused by large quantities of rock and other debris

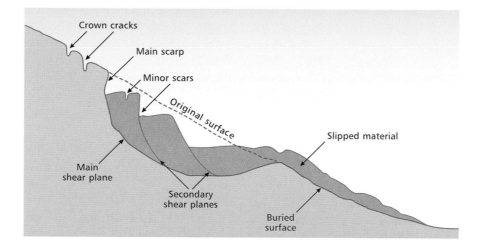

Figure 8.4 The characteristic profile of a rotational landslide. Surface changes, such as the prior opening of crown cracks, can be used as warning signs and lead to evacuation of people downslope.

blocking river flow in a narrow valley, or a gorge, to create a lake. A dual problem ensues. Valley-bottom land is flooded upstream and may damage agriculture and other infrastructure, whilst the unregulated release of the stored water in a flood wave may create even greater losses downstream. A typical example is the Tsatichhu dam, Bhutan when the Kurichuu River was blocked by $7–12 \times 10^6$ m^3 of material in September 2003 (Dunning *et al.*, 2006). The lake was released by the collapse of the dam almost a year later when a flood peak of 5,900 m^3 s^{-1} was recorded 35 km downstream. Another example is the La Josefina landslide disaster, Ecuador, in 1993 (Morris, 2003). Here the Paute River was blocked for 33 days by a massive landslide that killed some 200 people. Upstream, 300 ha of land flooded and 14,000 people were evacuated to safety downstream. Explosives were used to clear the channel and released a peak flood estimated at 10,000 m^3 s^{-1}. This destroyed several bridges and flooded the small town of Paute, where reconstruction subsequently took place on the previously occupied flood-prone sites.

In other instances, sliding occurs on a surface with a curved form and will produce a *rotational landslide*. This type of landslide is most commonly seen in comparatively homogenous materials, such as clays, and in horizontally bedded rock masses. The mobile block rotates as movement occurs, leading to a characteristic set of land forms (Figure 8.4). These landslides tend to be less rapid than translational slides, even though the same processes of loss of cohesion and friction apply. This is because, so long as the block stays intact, the geometry of the movement usually prevents rapid forward acceleration. In some cases, especially where the landslide is formed of weak materials – such as strongly weathered bedrock – the mobile block breaks up to form a flow. This commonly occurs during very intense rainfalls in New Zealand, where it has left a landscape covered in a pattern of scars and depositional features. Intact rotational landslides tend to cause substantial amounts of property damage

but few fatalities. About 400 houses in the town of Ventnor on the Isle of Wight, England are built on an active rotational landslide. Fortunately, the rate of movement is low and loss of life is unlikely, even though the estimated annual cost of damage caused by ground movement exceeds £2 million.

3 Debris flows

Debris flows are movements of fluidized soil and rock fragments acting as a viscous mass. They occur when loose materials become saturated and start to behave as a fluid rather than a solid. Flows most commonly occur in very heavy rain but, in most cases, the flow actually starts as a different type of landslide. For example, in tropical environments, the heavy precipitation associated with the passage of tropical cyclones, when rainfall totals often exceed 600 mm day^{-1} and intensities reach 100 mm hr^{-1} (Thomas, 1994), can trigger large numbers of small, shallow translational landslides in the soil that mantles the hill slopes. In some cases, the initial movement of the landslide allows the saturated soil mass to break up, changing the movement into a debris flow. Debris flows often accelerate rapidly down the slope, disrupting and entraining soil and other material. In this way, landslides of just a few cubic metres can turn into debris flows with a volume of tens of thousands of cubic metres and cause high levels of loss.

Debris flows follow existing stream channels. Consequently, the areas likely to be affected may be predictable. However, steep rock gullies provide little resistance to the flow and allow the mass to reach high rates of movement. Problems can arise when the flow reaches the foot of the slope, where the flow starts to spread out. When this occurs in inhabited areas, the amount of damage can be large because the density of a debris flow is greater than that of a flood (typically by a factor of 1.5 to 2.0) and the rate of movement is rapid. As a result, debris flows claim the majority of lives lost in landslides.

Many tropical cities, such as Rio de Janeiro and Hong Kong, are at risk from both landslides and debris flows. Jones (1973) documented the effects of exceptionally heavy rainfall, often linked to stationary cold fronts, around Rio de Janeiro, Brazil. In 1966, landslides produced over 300,000 m³ of debris in the streets of Rio and more than 1,000 people died when many slopes, over-steepened for building construction, failed. One year later, further storms hit Brazil and mud-flows caused a further 1,700 deaths and some disruption of the power supply for Rio. In February 1988 further debris flows took at least 200 lives and made 20,000 people homeless (Smith and de Sanchez, 1992). Most victims in this area live in unplanned squatter settlements on deforested hillsides (Smyth and Royle, 2000). Rural environments are not immune; most of the 30,000 fatalities in the 1999 Venezuela landslides resulted from debris flows down the river valleys.

C LANDSLIDES: CAUSES AND TRIGGERS

Scientists tend to differentiate between the *causes* of a landslide, which are factors that render a slope susceptible to instability, and the *trigger* of a landslide, which is the final event causing failure. Causes and triggers serve either to decrease the strength of the slope materials or to increase the shear forces.

Causes

Causes of landslides tend to be long term and include:

- *Weathering* of the slope materials may serve to reduce their strength through time until they are no longer strong enough to support the slope during periods of high pore-water pressure. Weathering often occurs as a front that moves down through the rock or soil from the surface, but it can also occur preferentially along joints and fractures in the slope or even deep in the slope, due to the circulation of hydrothermal fluids.

- *An increase in slope angle and removal of lateral support.* Landslides in undeveloped terrain are often caused by river erosion at the base of the slope. This increases the angle of the slope and decreases support to the upper layers. Of course, human activity can have the same effect. For example, Jones *et al.* (1989) described how cutting of a road into the base of a slope in Turkey left 25 metre-high faces in colluvium standing at an angle of 55° but supported only by a 3 metre-high masonry wall. Eventual collapse of this slope led to the 1988 Catak landslide disaster in which 66 people died.

- *Head loading* is a common cause of human-induced slope failure. This occurs when additional weight is placed on a slope, often through the dumping of waste material or the emplacement of fill for house or road construction. This increases the forces driving the landslide and may also increase the slope angle. Head loading can also occur naturally, for instance when a small slope failure flows onto material further downslope, thereby increasing its weight and rendering it more likely to fail.

- *Changes to the water table* can destabilize a slope. Sometimes climate or weather conditions increase the level of the water table, rendering a slope more vulnerable to intense rainfall events. Human activity can produce similar effects. For example, in the case of the Vaiont landslide (Box 8.1 and Figure 8.5) a rise in the water table as the lake filled led to the destabilization of the slope. Leaking pipes are a problem in urban areas when small movements along a slope crack drains, water supply pipes or main sewers.

- *Removal of vegetation* either by wildfires or by human activities like logging, overgrazing or construction is important. Trees are particularly good at limiting landslides, partly because of the retaining strength of the roots and partly

Box 8.1 The Vaiont landslide

The Vaiont disaster is the worst landslide disaster in recorded European history (Petley, 2009). It was triggered in October 1963 by the filling of a reservoir constructed for hydroelectric power generation. The landslide, which had a volume of approximately 270 million m^3, slipped into the reservoir at a velocity of about 30 m s^{-1} (approx. 110 km h^{-1}), displacing about 30 million m^3 of water (Figure 8.5). This water wave swept over the dam and crashed onto the village below, killing about 2,500 people.

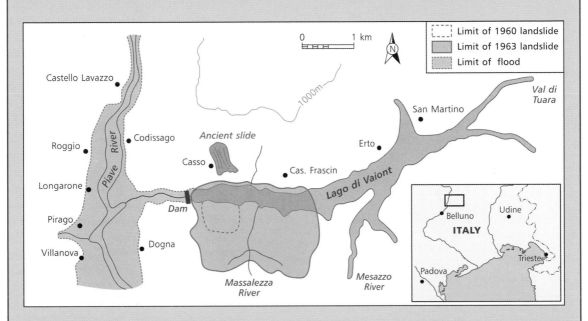

Figure 8.5 A map showing the area of land disturbed in the major Vaiont landslide of 1963 and the downstream area inundated by the subsequent flood wave. After Petley (2009).

The dam site managers were aware of the landslide and had been monitoring the movement of the slope since 1960. During 1962 and 1963, they deliberately induced movement of the landslide by raising and lowering the lake level with the intention of causing the mass to slide slowly into the lake. Although this would have led to a blockage of that section of the reservoir by the landslide mass it was believed that the volume of water in the unblocked section would be sufficient to allow the generation of electricity. A bypass tunnel was constructed on the opposite bank so that, when the reservoir was divided into two sections, the level of the lake could still be controlled. Unfortunately, the catastrophic nature of the landslide was not anticipated and these plans never came to fruition.

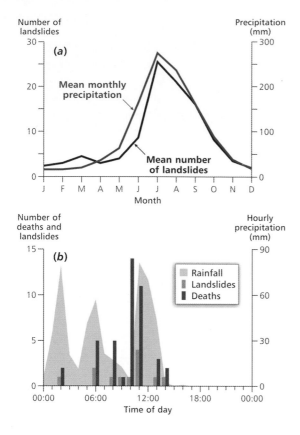

Figure 8.6 Landslide activity in relation to rainfall in the tropics. (a) Average monthly precipitation and the number of rainfall-induced fatal landslides in South Asia 2003–9. After Petley (2010). D.N. Petley, On the impact of climate change and population growth on the occurrence of fatal landslides in South, East and SE Asia. *Quarterly Journal of Engineering Geology and Hydrogeology* 43(4) (2010): 487–96. Reproduced with permission of the Geological Society. (b) Total hourly rainfall, number of fatalities and the timing of associated landslides, State of Chuuk, 2 July 2002. After C. Sanchez *et al.*, Risk factors for mortality during the 2002 landslides in Chuuk, Federated States of Micronesia. *Disasters* 33 (2009): 705–20. Copyright John Wiley and Sons 2009. Reproduced with permission.

because of the role that trees play in controlling water movement on slopes. Clear-felling often leads to an increase in landslide activity, although the effects may be delayed for a few years until the roots rot.

Triggers

Triggers of landslides are necessary for most events and include:

- *Increase in pore water pressure* in the slope materials to the point where shear stress exceeds shear resistance. In most cases, this occurs because of intense storms or prolonged periods of rain. Important relationships between rainfall and landslides over different time-scales are illustrated in Figure 8.6. Figure 8.6a shows the average monthly total of rainfall-induced fatal landslides during the years 2003–9 across South Asia, together with the monthly precipitation. A strong seasonal pattern emerges, reflecting the dominant role of summer monsoon rainfall in the creation of landslides in this region. Figure 8.6b illustrates hourly rainfall totals recorded during a tropical storm on 2 July 2002 that triggered fatal landslides in Chuuk, Federated States of Micronesia. Once again, the timing of the rainfall is closely associated with the triggering of landslides and the resulting fatalities.

- *Earthquake shaking.* During powerful earthquakes, such as the 2005 event in Kashmir, vertical ground accelerations can exceed 1 g. This means that landslide material instantaneously becomes weightless, reducing friction on the base to zero. According to an analysis by Keefer (1984), earthquakes of magnitude 4.0 and greater are able to trigger slope failure, and earthquakes with a magnitude of greater than about $M_L = 7.0$ are able to generate thousands of slope failures in hilly areas.

- *Human activity.* In some cases, human activity is the trigger. This is common in quarries, where excavation and blasting can destabilize a slope to the point of failure. Human-induced failures are common in mountain areas where road construction and slope cutting has taken place. This is a cause of mortality amongst road maintenance teams and road users in the Himalayas. For example, in December 2007 in Hubei Province, China a rock fall triggered by

Plate 8.2 The Las Colinas landslide was one of thousands of slope failures triggered by the M_w = 7.6 earthquake that hit El Salvador in January 2001. Ground shaking at this location was amplified by ridge topography at the edge of a steep escarpment. Weak deposits, mainly of volcanic origin, quickly lost strength and a rapid flow slide swept into Santa Tecla city, a residential suburb of San Salvador. Many houses were buried and about 580 lives were lost. (Photo: Ed Harp, USGS)

the construction of a road tunnel buried a bus; 33 passengers and 2 construction workers were killed.

For a small group of cases, it is impossible to determine the final landslide trigger. Investigations of the 1991 Mount Cook landslide in New Zealand, in which the country's highest mountain lost some 10 m from its peak, failed to reveal any trigger event (McSaveney, 2002). The failure may have occurred as a result of a time-dependent process or a triggering mechanism not yet understood. The 1979 Abbotsford landslide at Dunedin in New Zealand is a good example of how physical and human factors may combine to increase landslide risk and thus prevent attribution to a single cause. This urban landslide, which destroyed 69 houses across an area of 18 ha at a cost of NZ$10–13 million, resulted from at least three contributory factors (Hancox, 2008):

- Unfavourable geology – the site was on a 7° dip slope composed of Tertiary sediments with very weak clay layers.
- A rise in the water table – this was due to increased rainfall over the preceding 10 years and leakage from a Dunedin City Council water main at the top of the slope.
- Quarry activity – the excavation of some 300,000 m³ of sand from a quarry at the toe of the slope.

D SNOW AVALANCHES

Just like landslides, an avalanche risk exists when the shear stress exceeds the shear strength of the material, in this case a mass of snow located on a slope (Schaerer, 1981). The strength of a snow-pack is related to its density and temperature. When compared to other solids, snow layers have the ability to undergo large changes in density. Thus, a layer deposited with an original density of 100 kg m^{-3} can densify to 400 kg m^{-3} during a winter, largely due to compression by the weight of over-lying snow, pressure melting and the re-crystallization of the ice. This densification increases the strength of the snow. However, the shear strength decreases as the temperature warms towards 0°C. As the temperature rises further, such that liquid melt-water is present in the pack, the risk of movement within the snow blanket increases.

Most snow loading on slopes occurs slowly. This gives the snowpack an opportunity to adjust by internal deformation, because of its plastic nature, without any damaging failure. The most important triggers of pack failure tend to be heavy snowfall, rain, thaw or some artificial increase in dynamic loading, such as skiers traversing the surface (Box 8.2). For failure to occur in a hazard-ous snowpack, the slope must also be sufficiently steep to allow the snow to slide. Therefore, avalanche frequency is related to slope angle, with most events occurring on intermediate slope gradients of between 30° and 45°. Angles below 20° are generally too low for sliding to occur and most slopes above 60° rarely accumulate sufficient snow to pose a major threat.

Most avalanches start at fracture points in the snow blanket where there is high tensile stress, such as a break of ground slope, at an overhanging cornice or where the snow fails to bond to another surface, such as a rock outcrop. Avalanches are most likely to run during, or immediately follow-ing, a heavy snow-storm. This is because the weight of new snow places extra stress on the snowpack, especially if it does not bond well to the pre-existing surface. Snow is a good thermal insulator and diurnal temperature changes may have little influence on pack stability. However, a warm front bringing in milder weather can raise temperatures sufficiently to induce melting in the surface layers.

Three distinct sections of an avalanche track can usually be identified. These are the *starting zone*, where the snow initially breaks away, the *track* or path followed and the *run-out zone*, where the flowing snow decelerates and stops. Because avalanches tend to recur at the same sites, the risk can often be detected from the identification of previous avalanche paths in the landscape. Clues in the terrain include breaks of slope, eroded channels on the hillsides and damaged vegetation. In heavily forested mountains, avalanche paths can be identified by the age and species of trees and by sharp 'trim-lines' that separate the mature, undisturbed forest from the cleared slope. Once the hazard location is determined, a range of potential adjustments is available, some of which are shared with landslide hazard mitigation.

Snow avalanches can exert high external load-ings on built structures. Using reasonable estim-ates for speed and density, it can be shown that maximum direct impact pressures should be in the range of 5–50 t m^{-2}, although some pressures have exceeded 100 t m^{-2} (Perla and Martinelli, 1976). Table 8.2 provides a guide to avalanche impact pressures and the likely damage to man-made structures. The Galtür disaster in Austria, which occurred in February 1999, was the worst in the European Alps for 30 years. In this event, 31 people were killed and seven modern buildings were demolished in a winter sports village previ-ously thought to be located in a low-hazard zone. A series of storms earlier in the winter deposited nearly 4 m of snow in the starting zone, a previ-ously unrecorded depth. By the time the highest level of avalanche warnings was issued, the snow mass in the starting zone had grown to approx-imately 170,000 tonnes. During its track down the mountain, at an estimated speed in excess of 80 m s^{-1}, the avalanche picked up sufficient snow

Box 8.2 How snow avalanches start

Snow avalanches result from two different types of snowpack failure:

- *Loose-snow avalanches* occur in snow where inter-granular bonding is very weak thus producing behaviour rather like dry sand (Figure 8.7a). Failure begins near the snow surface when a small amount of snow, usually less than 1 m³, slips out of place and starts to move down the slope. The sliding snow spreads to produce an elongated, inverted V-shaped scar.

- *Slab avalanches* occur where a strongly cohesive layer of snow breaks away from a weaker underlying layer, to leave a sharp fracture line or *crown* (Figure 8.7b). Rain or high temperatures, followed by re-freezing, create ice-crusts which may provide a source of instability when buried by subsequent snowfalls. The fracture often takes place where the underlying topography produces some upward deformation of the snow surface, leading to high tensile stress, and the associated surface cracking of the slab layer. The initial slab which breaks away may be up to 10,000 m² in area and up to 10 m in thickness. Such large slabs are very dangerous because, when a slab breaks loose, it can bring down 100 times the initial volume of snow.

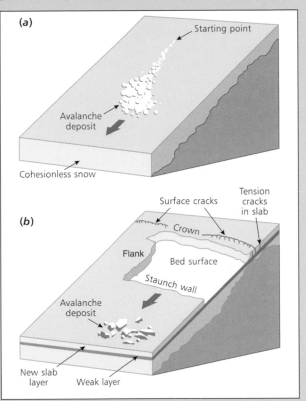

Figure 8.7 The two most common types of snow-slope failure: (a) loose-snow avalanche; (b) slab avalanche. Slab avalanches normally create a greater hazard because of the larger volume of snow released.

Avalanche movement depends on the type of snow and the terrain. Most avalanches start with a gliding motion but then rapidly accelerate on slopes greater than 30°. It is common to recognize three types of avalanche motion:

- *Powder avalanches* are the most hazardous and are formed of an aerosol of fine, diffused snow behaving like a body of dense gas. They flow in deep channels but are not influenced by obstacles in their path. The speed of a powder avalanche is approximately equal to the prevailing wind speed but, being of much greater density than air, the avalanche is more destructive than windstorms. At the leading edge its typical speed is 20–70 m s⁻¹ and victims often die by inhaling snow particles.

- *Dry flowing avalanches* are formed of dry snow travelling over steep or irregular terrain with particles ranging in size from powder grains to blocks of up to 0.2 m diameter. These avalanches follow well-defined surface channels, such as gullies, but are not greatly influenced by terrain irregularities. Typical speeds at the leading edge range from 15–60 m s^{-1} but can reach speeds up to l20 m s^{-1} whilst descending through free air.
- *Wet flowing avalanches* occur mainly in the spring season and are composed of wet snow formed of rounded particles (0.l to several metres in diameter) or a mass of sludge. Wet snow tends to flow in stream channels and is easily deflected by small terrain irregularities. Flowing wet snow has a high mean density (300–400 kg m^{-3} compared to 50-150 kg m^{-3} for dry flows) and can achieve considerable erosion of its track despite reaching speeds of only 5–30 m s^{-1}.

Table 8.2 Relationships between impact pressure and the potential damage from snow avalanches

Impact pressure (tonnes m²)	Potential damage
0.1	Break windows
0.5	Push in doors
3.0	Destroy wood-frame houses
10.0	Uproot mature trees
100.0	Move reinforced concrete structures

Source: After Perla and Martinelli (1976)

to double the original mass. By the time it reached the village the leading powder wave was over 100 m high, with sufficient energy to cross the valley floor and reach the village.

E PROTECTION

1 Landslides

The design and construction of measures to prevent slope failure is a routine task within geotechnical engineering. For example, within the 1,100 km² area of Hong Kong, over 57,000 slopes have been engineered to prevent failure. Similarly, the railway agency in the UK has to maintain over 16,000 km of earthworks designed to prevent slope failures (Petley, 2009). Methods of slope

protection are well developed and include the following:

- *Drainage.* As slope failures are generally linked to the presence of high water pressures in a slope, drainage is a key tool for improving stability. The aim is either to prevent water from entering a critical area of slope by installing gravel-filled trench drains around that area or to remove water from within a slope by installing horizontal drains. Simple methods can work well. Holcombe and Anderson (2010) reported the success of a community-based, low-cost scheme to improve slope stability in the Eastern Caribbean. Following the construction of a network of open drains, and a 1:100-year rainfall event, no landslides were recorded on previously unstable slopes occupied by densely populated urban communities. Problems can arise through a lack of maintenance. Drains can become blocked with fine particles or even by animals using them as burrows. In addition, small amounts of movement in a slope can cause drains to become cracked or broken and so leak water into a slope at critical locations.
- *Regrading.* In many cases, the landslide threat can be minimized by reducing the overall slope angle. This can be achieved by excavating the upper parts of the slope or by placing material at the toe, an approach used during road

construction in upland areas. In some cases, good results can be achieved by removing the natural slope soil or rock and replacing it with a lighter material.

- *Supporting structures.* Supporting structures such as piles, buttresses and retaining walls are widely used for slopes adjacent to buildings and transportation routes. For example, the UK rail network has over 7,000 slopes supported by retaining walls. Although effective, this is an expensive and visually intrusive approach and there is a move towards measures that sit within the soil or rock rather than on the surface. Examples include soil nails and rock bolts. In addition, structures can be designed to deflect small landslides around vulnerable facilities. For example, diversion walls are often constructed around electricity pylons in mountain areas to control localized debris flows.

- *Vegetation.* Vegetation on slopes performs several functions. Tree and plant roots help to bind soil particles together and provide stability. The vegetation canopy protects the soil surface from rain splash impact, whilst transpiration processes reduce the water content of the slope. In recent years, bioengineers have developed new ways of controlling shallow landslides in soil and for preventing soil erosion. It is important to ensure that the vegetation can thrive in the chosen location, and preference is normally given to local species. This approach is considered to be more environmentally aware than traditional engineering methods. The capital costs of bioengineering are usually less than those for conventional structures, although maintenance costs tend to be higher.

- *Other methods* include the chemical stabilization of slopes and the use of grouting to reduce soil permeability and increase its strength. On some construction sites, moving soil has been frozen temporarily while soil-retaining structures were completed, but this is a very expensive option. In many tropical countries, shallow and localized slope failures are covered in plastic sheets to reduce the impacts of rainfall until full stabilization can be achieved.

Whilst engineering remains predominant in landslide protection, it is increasingly pursued alongside planning legislation to limit new development on dangerous slopes. In the USA, urban planning is managed by the *Uniform Building Code*. This specifies a maximum slope angle for safe development of 2:1, which approximates to a 27° angle, as well as minimum standards for soil compaction and surface drainage. Similarly, in New Zealand, there is a legal requirement that proposed new building must have a *resource consent* for which slope stability is a major component. The success of these schemes depends on the availability of technically trained inspectors to enforce the regulations. This is often a problem, especially where local corruption also exists, but long-term success can be achieved. The city of Los Angeles introduced a grading ordinance as early as 1952. Before this date more than 10 per cent of all building lots were damaged by slope failure, but more recent losses at new construction sites have been estimated at fewer than 2 per cent.

2 Avalanches

Two main physical techniques are used to protect against the hazard posed by snowpacks.

Artificial release

Artificial release is accomplished through the use of small explosive charges to trigger controlled avalanches. This technique is used surprisingly often; in the USA about 10,000 avalanches are artificially released each year. The main advantages are:

- The snow release occurs at predetermined times when the downslope areas – perhaps containing recreational facilities and roads – can be closed.

- Measures to allow snow clearance can be put in place before the avalanche occurs, thereby minimizing inconvenience.
- The snowpack can be released safely in several small avalanches, rather than allowing the build-up of a major threat.

Explosive charges are most effective when placed in the initiation zone, or near the centre of a potential slab avalanche, when the relationship between stress and strength within the snowpack is delicately balanced. These requirements can be met only through close liaison with a snow stability monitoring and avalanche forecasting service. In some cases, dedicated teams are dropped by helicopter into the initiation zone in order to place the charges. Needless to say, this is a hazardous task with respect to both the handling of explosives and the possibility of the team's triggering an avalanche with unexpected consequences. Alternatively, it is possible to use military field guns to fire explosives onto the slope from a safe zone, in order to protect key facilities. For

example, the Rogers Pass in the Canadian Rockies funnels both the Canadian Pacific rail route and the Trans-Canada highway through the Selkirk Mountains of British Columbia. At this location, Parks Canada and the Canadian Armed Forces work together to trigger avalanches with field artillery.

Defence structures

Defence structures are a common adjustment to avalanches throughout the world. In Switzerland alone, the total amount spent on avalanche defence structures in the period 1950–2000 was approximately €1 billion (Fuchs and McAlpin, 2005). There are four main types of avalanche defence structure (Figure 8.8):

- *Retention structures* are designed to trap and retain snow on a slope and thus to prevent the initiation of an avalanche or to stop a small avalanche before it can develop fully. Above the starting zone, *snow fences* and *snow nets* are used to hold back snow. On ridges and gentle

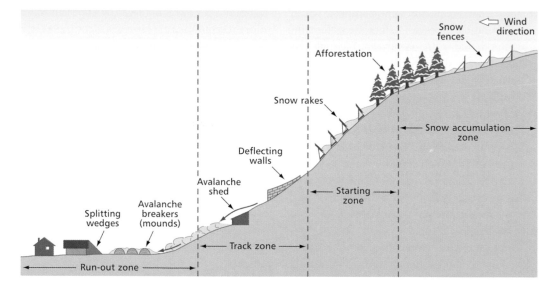

Figure 8.8 Idealized slope section showing avalanche hazard reduction measures. Snow retention is achieved by defensive structures in the snow-accumulation and starting zones, often complemented by a mature forest cover. Deflection of the avalanche away from people and infrastructure, and lowering the destructive potential of the flow, is achieved by structures in the track and run-out zones.

Plate 8.3 A combination of snow fences and mature forest cover protect valley development and surrounding infrastructure from avalanches in the Austrian Alps. (Photo: ALMIDI.NET/J.W. Alker (© SuperStock 1848–450211 by imagebroker.net)

slopes, large volumes of snow can be intercepted and retained in this way. In the starting zone *snow rakes* or *arresters* are used to provide external support for the snowpack, thus reducing internal stresses. They may also stop small avalanches before they gain momentum. The earliest structures were massive walls and terraces made of rocks and earth. Today they are made of combinations of wood, steel, aluminium and/or pre-stressed concrete. Such structures are effective but they do have a negative impact on the visual appearance of the landscape, an important consideration in areas dependent on tourism. Their use is also less effective in regions with large seasonal accumulations of snow, like the north-west USA, where snowfall may bury the structures and so limit their effectiveness.

- *Redistribution structures* are designed to prevent snow accumulation by drifting. In particular, they are used to prevent the build-up of cornices that often break off steep slopes and initiate an avalanche.
- *Deflectors and retarding devices.* The easiest way to control an advancing avalanche is to guide it along a gently curving path, so deflectors built of earth, rock or concrete are placed in the avalanche track and run-out zone. However, the scope for lateral diversion is limited, and changes of direction no greater than 15–20° from the original avalanche path have been proved to be most successful. In addition, wedges pointing upslope can be used to split an avalanche and divert the flow around isolated facilities like electricity transmission towers or isolated buildings. Towards the run-out

zone, on ground slopes of less than 20°, other structures – earth mounds and small dams – are useful to obstruct an avalanche as it loses energy. Braking mounds are widely used against dense, wet snow avalanches.

- *Direct-protection structures* such as avalanche sheds and galleries provide the most complete avalanche defence. They are designed to allow the flow to pass over key built facilities, and avalanche sheds typically act as protective roofs over roads or railways. They are expensive to construct and need careful design to ensure that they are correctly located and can bear the maximum snow loading on the roof.

Some techniques for managing avalanche hazards mimic the natural protection offered by mature forests. Therefore, where possible, it is desirable to plant forests on avalanche-prone slopes and avoid the need for unsightly, maintenance-intensive structures. As shown in Figure 8.8, forest cover is often used in combination with structures, with structures retaining snow on the upper slope and trees providing further safety above transport routes or settlements. Allowances have to be made for differences in forest structure and terrain roughness, but, while large avalanches released far above the forest are unlikely to be completely stopped by tree cover, small avalanches triggered close to the tree-line are expected to experience significant deceleration.

A major difficulty is that avalanche tracks offer poor prospects for successful tree planting and growth. Gully erosion by previous events means that avalanche-prone slopes are often characterized by thin soils with limited water retention. Furthermore, young trees may be destroyed by avalanches before they can provide stability to the snowpack. Therefore, expensive site preparation, coupled with soil fertilization, is frequently required and it may be necessary to stabilize the snow in the starting zone while the tree cover establishes itself. Sometimes it can take over 75 years before slower-growing species have reached

the point where they are strong enough to resist avalanche forces. In some cases, the natural forest may already have been removed to allow the introduction of economic activities, such as skiing, and attempts to re-establish tree cover may not be wholly welcome.

F MITIGATION

1 Disaster aid

Mass movement disasters rarely attract substantial disaster aid, due to the relatively small scale of the losses, although the occurrence of multiple mass movement events has become more common. Large-scale relief operations were required in the aftermath of the 1997 hurricane 'Mitch' disaster in Nicaragua and Honduras, the 1999 Vargas landslide disaster in Venezuela (Box 8.3), the 2005 Kashmir earthquake (which induced thousands of landslides), the 2006 Leyte landslide disaster in the Philippines and the 2007 landslide disaster in North Korea. In the case of the Leyte landslide, a large rock slope collapsed onto the town of St Bernard, burying a town of 1,400 people, including 268 pupils in a school. Search-and-rescue teams were dispatched from the Philippines, Taiwan, the UK and the USA, although few people, if any, were rescued by them. The low success rate achieved by international emergency responders re-emphasizes the need for the prior establishment of local capability (Petley, 2009).

2 Insurance

In many countries, including the UK, private insurance against mass movement hazards is unavailable because the industry fears the prospect of many high-cost claims. Unavailability of insurance can discourage development in hazardous areas but, because information about landslide hazards is not widely disseminated, many people remain unaware of the risk. Limited insurance is provided by some government schemes. For example, in the USA some cover is

Box 8.3 The Vargas landslides

On 14–16 December 1999, a huge rain-storm struck Vargas state in Venezuela. It deposited approximately 900 mm of rainfall over a three-day period and triggered many landslides in the hills. The landslides transitioned into a series of debris slides and mud flows which struck urban areas located on alluvial fans beside the coast. Whilst the precise death toll of these landslides will never be known, the best estimate is that about 30,000 people lost their lives and the economic losses reached US$1.8 billion (Wieczorek *et al.* 2001).

More than 8,000 homes were destroyed, displacing up to 75,000 people. Over 40 km of coastline was significantly altered. In the aftermath of the disaster, the national government attempted to evacuate about 130,000 people from the northern coastal strip (IFRCRCS, 2002). The government used its own resources, and an unexpected opportunity, to attempt a permanent re-location of people from the coast, where living conditions were poor and population densities exceeded 200 km^2, to less crowded parts of the country. By August 2000, some 5,000 families were re-settled in new – if sometimes unfinished – houses; 33,000 remained in temporary accommodation. Many evacuees opposed the resettlement programme and drifted back to their original location. In 2006 the population of the area had reached the pre-disaster level, leaving the area highly vulnerable to a repeat of the disaster.

provided through the National Flood Insurance Program, which requires areas subject to 'mudslide' hazards associated with river flooding to have insurance in order to qualify for federal aid. Unfortunately, technical difficulties in mapping 'mudslide' hazard areas have led to limited use of this provision. A more successful example exists in New Zealand, where the government-backed Earthquake Commission (EQC) provides limited coverage for houses and land, in cooperation with private insurers. The EQC pays out for landslide damage to residential property on a regular basis. Interestingly, in the period 2000–7, the EQC paid out more to cover damage from landslides than it did for earthquakes, due to a marked upward trend in landslide claims.

Generally speaking, legal liability forms a basis for financial compensation after landslide losses. American jurisprudence recognizes civil liability for death, bodily injury and a wide range of economic losses associated with landslides. In most MDCs, the legal defence of 'act of God'

carries decreasing credibility because of assumptions that landslide risks are reasonably well understood, that access to relevant information for particular sites is available and that a wide range of landslide mitigation measures can be applied. As a result, court judgments have tended to identify developers, and their consultants, as responsible for damage related to mass movements. In some areas, local planning agencies have shared the liability because it has been successfully argued that the issue of a permit for residential development implied the warranty of safe habitation. Litigation can be costly. For example, legal claims arising from the landslide-induced failure of the Ok Tedi tailings dam in Papua New Guinea in 1984 included a law suit for over US$1 billion for costs for direct damage and US$4 billion for compensation arising from pollution of the Fly River (Griffiths *et al.*, 2004). Both cases were settled out of court, showing that litigation is an inadequate substitute for proper hazard-reduction strategies.

G ADAPTATION

1 Community preparedness

Mass movements kill in two chief ways. Some victims die through fatal trauma (collisions with rocks, trees or ice slabs) but the main cause of death is suffocation. Humans die quickly if buried beneath snow, soil or rock because they are unable

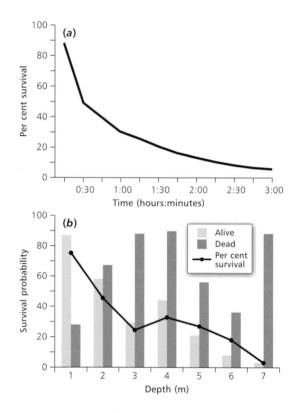

Figure 8.9 Survival after an avalanche. (a) The percentage chance of survival against time for US avalanche victims buried in the snow. After 15 minutes the survival rate is almost 90 per cent, but it falls to less than 30 per cent after 1 hour. After Colorado Avalanche Information Center at http://geosurvey. state.co.us/avalanche (accessed 4 March 2003). Reproduced with permission of Colorado Avalanche Information Center. (b) The probability of survival in US avalanches 1950–2006 in relation to depth of burial. After Colorado Avalanche Information Center at http://avalanche.state.co.us/acc/acc_images (accessed 31 May 2011). Reproduced with permission of Colorado Avalanche Information Center.

to breathe under a depth of such material greater than 30 cm. Even shallow burial can be fatal. In addition, victims are at risk from hypothermia in the case of avalanches and from drowning in the case of water-rich debris slides. Most of the survivors from mass movement disasters are rescued quickly and are often saved by some small-scale physical protection, perhaps the shelter offered by a building or a vehicle.

Almost 90 per cent of people who are trapped by an earthquake – and survive – are rescued within the first 24 hours. Information from the USA suggests that the survival rate for avalanche victims buried in the snow declines even more rapidly with time (Figure 8.9a). After only 15 minutes, almost 90 per cent of avalanche victims are still alive, but this falls to about 30 per cent after 60 minutes and to fewer than 5 per cent after 3 hours. The time taken in locating people buried deeply in the snow, and then removing the snow cover, also explains why survival rates for avalanche victims also decline with the depth of snow (Figure 8.9b). At depths in excess of 1.2 m, the survival rate drops to 20 per cent or less. *Avalanche beacons*, sometimes called transceivers, are standard equipment for some ski-slope patrollers because they represent the best means of finding a buried victim, as long as a device is also carried by the victim. Collapsible ski-pole probes are used to locate buried bodies.

The need for a locally based rapid-response search-and-rescue capability is crucial for all mass movement hazards. In Canada, local offices of Parks Canada regularly monitor the stability of relevant snowpacks and issue avalanche risk warnings in collaboration with the British Columbia Ministry of Highways. The provincial government coordinates most local search and rescue in the Canadian Rockies and the Royal Canadian Mounted Police have men and dogs trained for avalanche rescue work. Specialized weather forecasts are available from the Atmospheric Environment Service and avalanche awareness is promoted through bodies like the Canadian Avalanche Centre with technical courses, films and videos. Although these methods may increase the general knowledge of avalanche threats, Butler (1997) found that local residents were often unaware of the danger at a given time. A full avalanche search is a complex operation, but the increasing use of avalanche airbags and digital transceivers by winter sports enthusiasts will reduce the chances of burial and increase the chances of being found, respectively.

Such advanced arrangements are rare for landslides. In Hong Kong, the government's Geotechnical Engineering Office (GEO) provides a 24-hour, year-round service to provide advice on the actions to be taken when landslides occur. Activities are coordinated from an emergency control station that is manned whenever landslide warnings are issued. Officers from GEO are then on permanent stand-by to attend the sites of landslides in order to assist in rescue, recovery and management. This system has been effective, but there are few examples elsewhere.

2 Forecasting and warning

Considerable effort has gone into the development of warning systems for both avalanches and landslides through the following approaches.

Site-specific warnings based upon surface movement

Most rainfall-induced landslides are preceded by a period of slow movement, called *creep*. This phenomenon has been employed in some warning schemes and Hungr *et al.* (2005) described how site-specific changes in earth movement have been successfully used to predict the time of final failure. Slope movement is monitored with field instruments such as inclinometers, tiltmeters, theodolites and electronic distance recorders, supplemented by Global Positioning Systems (GPS) and radar satellite (InSAR) techniques (Casagli *et al.*, 2010 and Box 8.4). Advances in technology allow data to be sent in real time to a computer which can then compare the observed movement against predetermined

trigger factors – often based on the rate of movement or the acceleration – so that a warning can be issued.

In the past, traffic routes were protected from avalanches by avalanche sheds, but now avalanche warning systems are available for key transportation corridors. These systems detect the movement of an avalanche high on a slope before operating a set of barriers or traffic controls that close the road or railway. Surface movement of the snow is detected using trip-wires, radar, geophones or wire-mounted tiltmeters. Although these systems are expensive, they have been deployed in both North America and Europe. Evidence suggests that the site-specific nature of avalanche paths, plus the economic importance of the traffic route, are crucial factors in optimizing such schemes (Rheinberger *et al.*, 2009). Figure 8.11 depicts an avalanche management scheme operated along a 14 km corridor of Idaho State Highway 21 that crosses 56 avalanche tracks (Rice *et al.*, 2002). Automatic avalanche detectors, using tilt switches, are suspended from a cableway near to the road over the most active avalanche track. When these switches exceed a preset threshold, the system can initiate a call by radio telemetry to alert the highway authority of the avalanche and can advise road users of the blockage immediately, either by activating flashing warning signs or by automatically closing snow gates at each end of the corridor.

General warnings based on weather conditions

For landslides, historical precipitation and landslide records are combined to identify the rainfall level (intensity and duration) at which past landslides have been initiated over the area in question. Such regional alerts are mainly relevant to the recognition of precipitation thresholds that trigger shallow landslides (Brunetti *et al.*, 2010). For example, in 1986 the United States Geological Survey trialled an early landslide warning system for the San Francisco Bay region, based upon six-hour forecasts of rainfall duration and intensity

Box 8.4 The Tessina landslide warning system

The Tessina landslide is a 3 km long earth flow located in the Dolomite mountains of northern Italy (Figure 8.10). It has been active since 1960 but the rate of movement, and the volume of material involved, increased in 1992 (Petley, 2009). Given that the village of Funes was located on the margin of the landslide, there was a risk that a substantial movement would cause the landslide to over-run the settlement, creating casualties and large economic losses. In response, the Italian government research agency CNT-IRPI designed and implemented a landslide warning system (Angeli *et al.*, 1994).

The warning system consisted of two elements:

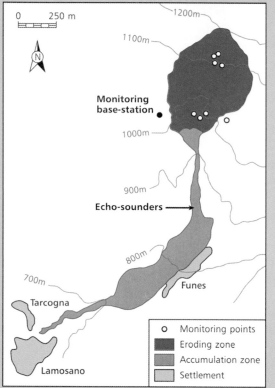

- In the source area of the landslide, 13 survey prisms were installed on the landslide. On the margin of the active movement, in stable ground, a robot theodolite was installed, powered by solar cells. Every 30 minutes this instrument measures the location of each of the prisms. The data are then stored on a computer, which also determines the level of movement. Located alongside the prisms are two wire extenometers that also measure the displacement of the landslide, again feeding data to the central computer unit.
- About 100 m above the village, two tiltmeters were installed approximately 2 m above the landslide. These consist of 2 m long steel bars containing a device to measure the angle at which the bar is hanging. If a flow were to come down the slope, the bars would tilt. The instruments are designed such that if the bar is tilted at more than 20° for over 20 seconds, the central computer would be alerted. A back-up echometer located on one of the wires provides an indication of any rapid height change on the surface of the flow that might indicate a slide event.

Figure 8.10 A map showing the area of the Tessina landslide threat in the Dolomites of Northern Italy. The warning system has protected the town of Funes for many years. After Petley (2009).

The central computer is programmed to compare all information against pre-set thresholds. If these thresholds are exceeded, an alarm is sounded in the local fire station. In addition, the fire-fighters have access to three video cameras situated at key locations that offer an additional opportunity to verify the indicated movements.

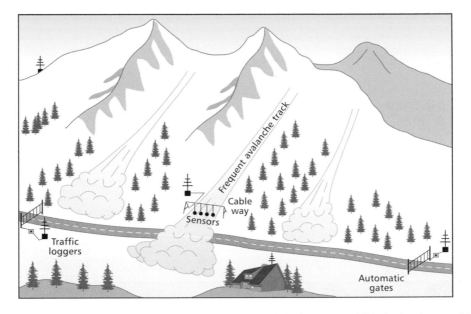

Figure 8.11 Avalanche hazard management on mountain roads in the western USA. Avalanches reaching the highway are detected by sensors suspended from a cableway above the track. The relevant stretch of road can then be closed for snow clearance by the highway agency alerted by telemetry. Adapted from Rice *et al.* (2002). Reprinted from *Cold Regions Science and Technology* 34, R. Rice Jr *et al.*, Avalanche hazard reduction for transportation corridors using real-time detection and alarms, 31–42. Copyright (2002), with permission from Elsevier.

(Keefer *et al.*, 1987), but the maintenance costs proved to be prohibitive.

One of the most effective systems is that operated in Hong Kong. This uses a network of 110 rain-gauges scattered across the Territory, together with analysis of Doppler radar data, to issue warnings of the occurrence of landslides. The warning threshold is based on the accumulated rainfall over the previous 21 hours plus the forecast rainfall for the next 3 hours. Using GIS technology, the forecast rainfall is used to calculate an estimated number of landslides across the Territory. This number is the basis for issuing a warning, which is released to the media, along with advice to the public to maximize safety. But the physical relationships are not entirely straightforward. Antecedent soil moisture conditions are an important factor and threshold values vary considerably with the size of the area. For exam-

ple, threshold rainfall amount and durations vary by three orders of magnitude over the USA and by over one order of magnitude across smaller areas such as a county (Baum and Godt, 2010).

Avalanche warning systems using forecasts and predictions have existed for many years. Forecasts are used in the day-to-day management of winter sports facilities, whilst predictions aid long-term land zoning. Avalanche forecasting involves the testing of the stability of snow tests, with an emphasis on the detection of weak layers. The results are then evaluated in conjunction with weather forecast information. Regional avalanche schemes are often computer aided. The method introduced in Switzerland in 1996 relies on model calculations of the snowpack, inputs from about 60 weather stations and a GIS-based mapping system of avalanche tracks to provide daily forecasts for areas of about 3,000 km^2 (Brabec *et al.*, 2001). In conditions of severe risk, it is normal practice to clear ski slopes and to restrict traffic on dangerous sections of highway or railway track.

3 Land use planning

The recurrence of mass movements at the same topographic site means that mapping is important for hazard mitigation, if only through the qualitative identification, and avoidance, of unsafe sites

(Parise, 2001). Most nations possess sufficiently accurate topographic and geological databases to create slope maps and digital elevation models that, together with field surveys, produce fairly reliable risk assessments. These methods are appropriate for small vulnerable nations. For example, Figure 8.12 illustrates the relatively large areal extent of landslides mapped on Tonoas island, Chuuk State, Micronesia following tropical storm Chata'an in 2002 (Harp *et al.*, 2009). The largest debris flows travelled several hundred

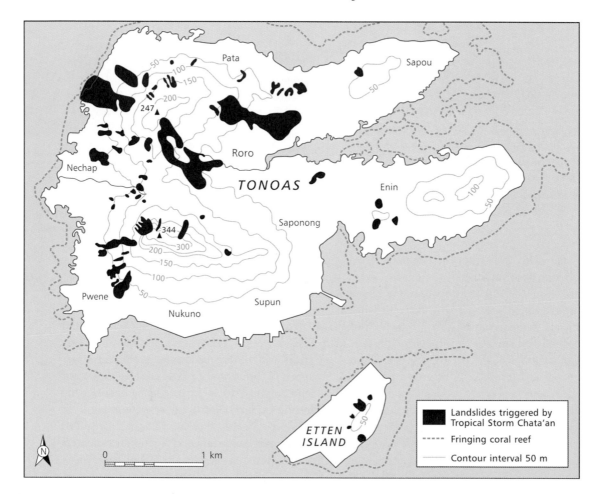

Figure 8.12 The widespread distribution of landslides on Tonoas Island, Federated States of Micronesia in 2002. These debris flows and mudslides were triggered by intense tropical rainfall during the passage of typhoon Chata'an and disrupted several small coastal communities. After Harp *et al.* (2009). Reprinted from *Engineering Geology* 104, E.L. Harp *et al.*, Mapping of hazard from rainfall-triggered landslides in developing countries: examples from Honduras and Micronesia, 295–311. Copyright (2009), with permission from Elsevier.

metres and crossed coastal flatlands to reach populated areas. Such landslide-inventory maps can be used to identify safer zones suitable either for the temporary evacuation of people during future storms or for the permanent relocation of residents.

Remote sensing has been used for many years to produce preliminary maps of both landslide and avalanche tracks (Sauchyn and Trench, 1978; Singhroy, 1995). More comprehensive post-disaster assessments have become available (Tsai *et al.*, 2010), although difficulties remain with the use of space-borne sensors for hazard mapping in steep mountain areas (Buchroithner, 1995). Reconnaissance information can be followed up with low-level air photography. Vertical aerial photographs at scales of 1:20,000 to 1:30,000 are often suitable, especially if taken when tree foliage and other vegetation cover is at a minimum. For example, many avalanche tracks function as landslide gullies during the spring and summer seasons. The recognition, and mapping, of less frequent hazards is not such a routine matter. One such hazard is the break-off of large ice masses from overhanging glaciers, which then fall onto the starting zones for snow avalanches, although surveys after the event can be used to compile hazard maps and safety plans (Margreth and Funk, 1999).

The pressure for building land on the edge of many cities means that the application of development restrictions based upon simple criteria, such as slope angle, is no longer adequate. More sophisticated land use planning approaches based upon the assessment of susceptibility and hazard are required and two basic approaches have been employed:

1 *Geological techniques* involve producing a very detailed risk map of landslides in a defined study area. Apart from mapping previous slides, information is collected on causal factors such as rock types, slope angle, the presence or absence of vegetation and the rainfall distribution. Attempts are then made, often using GIS, to correlate the location of the landslides with the possible causal factors. However, such approaches have had mixed success, sometimes because they overestimate the area likely to be affected by slope failure (van Asch *et al.*, 2007).

2 *Geotechnical techniques* use mathematical slope-stability equations to determine the likelihood of slope failure. In recent years, it has become more usual to attempt this using GIS (Petley *et al.*, 2005). These techniques require quantitative estimates of parameters such as the strength of the soil, the angle of the slope and the depth of the water table. Geological and topographic maps are then used to determine the spatial distribution of the key parameters. Unfortunately, general figures from the literature have to be assigned to factors such as the soil strength. This is a major weakness because the values of these parameters can vary considerably.

Despite their faults, such techniques are widely used for general guidance when planning new development on potentially dangerous slopes. When an area is identified as being medium or high hazard, a more detailed geotechnical investigation should be undertaken to assess the risk and any measures needed to render the site safe. Some results have been impressive. For example, in 1958 the Japanese government enacted the 'Sabo' legislation to mitigate landslides and debris flows triggered by typhoon rainfall. In 1938 nearly 130,000 Japanese homes were destroyed and more than 500 lives were lost in landslides. In 1976 – the worst year for landslides in that country for two decades – only 2,000 homes were lost and fewer than 125 people died. Similar evidence exists for Hong Kong, where hillside development was not properly regulated until the 1970s (Morton, 1998). A slope safety scheme was introduced in the 1990s and the rolling average annual fatality rate, which peaked at about 20 during the 1970s, fell sharply (Figure 8.13).

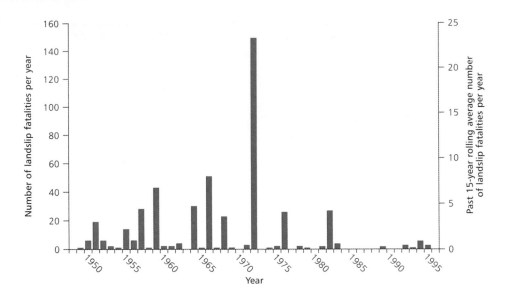

Figure 8.13 The reduction of landslide incidence by hazard management in Hong Kong, 1948–96. After Morton (1998).

However, systematic approaches to landslide hazard reduction remain the exception rather than the rule. A major problem arises when potential slope failures are identified in areas already used for housing. There is often a demand for mitigation to be undertaken at national or local government expense, although most administrations fund emergency works only. This is because the cost of permanent mitigation is considered to be the responsibility of householders, even though building insurance rarely covers landslides.

As with landslides, the most effective mitigation of avalanche hazards is through land use planning based on the identification of site-specific risk. In Switzerland, avalanche zoning laws were mandated by the government as early as 1951. Zoning begins with the collection of historical avalanche data, supplemented with terrain models and an understanding of avalanche dynamics, to determine detailed degrees of risk. Where sites are near existing settlements, avalanche frequency will be a matter of local knowledge. At more remote locations, other methods

are necessary, for example the use of satellite imagery and a digital elevation model (Gruber and Haefner, 1995). The long-term pattern of avalanche activity can be compiled from dendrochronological information. For example, Muntán *et al.* (2009) reconstructed avalanche paths in the Pyrenees over a 40-year period. The scarring of tree rings in tracts of forest damaged by previous events, but remaining standing in the avalanche track, can produce reliable frequency estimates over the past 200 years or so (Hupp *et al.*, 1987). Where trees have been destroyed, close inspection of the residual damaged vegetation, including height and species, can be a useful guide. Table 8.3 shows how this evidence can be used when initial mapping is undertaken at a scale of about 1:50,000.

In British Columbia snow avalanche atlases are used as operational guides for highway maintenance personnel. The maps are accompanied with a detailed description of the terrain and vegetation for each avalanche site, together with an assessment of the hazard impact. Where avalanches threaten settlements, larger-scale maps

Table 8.3 Vegetation characteristics in avalanche tracks as a rough indicator of avalanche frequency

Minimum frequency (years)	Vegetation clues
1–2	Bare patches, willows and shrubs; no trees higher than 1–2 m; broken timber
2–10	Few trees higher than 1–2 m; immature trees or pioneer species; broken timber
10–25	Mainly pioneer species; young trees of local climax species; increment core data
25–100	Mature trees of pioneer species; young trees of local climax species; increment core data
>100	Increment core data needed

Source: After Perla and Martinelli (1976)

Table 8.4 The Swiss avalanche zoning system

High-hazard (red) zone

- Any avalanche with a return interval <30 years
- Avalanches with impact pressures of 3 tm^{-2} or more and with a return interval up to 300 years
- No buildings or winter parking lots allowed. Special bunkers needed for equipment

Moderate-hazard (blue) zone

- Avalanches with impact pressures <3 tm^{-2} and with return intervals of 30–300 years
- Public buildings that encourage gatherings of people should not be erected
- Private houses may be erected if they are strengthened to withstand impact forces
- The area may be closed during periods of hazard

Low-hazard (yellow) zone

- Powder avalanches with impact pressures 0.3 tm^{-2} or less with return intervals >30 years
- Extremely rare flowing avalanches with return periods >300 years.

No-hazard (white) zone

- Very rarely may be affected by small air blast pressures up to 0.1 tm^{-2}
- No building restrictions

Source: After Perla and Martinelli (1976)

(from 1:25,000 to 1:5,000) are needed. Two important parameters are always difficult to determine in avalanche hazard studies. These are the length of the run-out zone, which determines whether or not a particular site will be reached by flowing snow, and the impact pressure at any given point, which determines the likely level of damage. Computer modelling of avalanche dynamics is increasingly sophisticated but the precision required for the model to be reliable remains high (Brabolini and Savi 2001). As the scale resolution of terrain data and the models improve, such methods grow in importance.

The avalanche hazard maps found in most countries normally adopt a three-zone, colour-coded system (Table 8.4). These schemes need regular updating. For example, following the 1999 avalanche disaster at Galtür, Austria, the exclu-sion zone for buildings, previously drawn up for a 1 in 150-year event, was extended, and revised regulations required all new buildings to be reinforced against specified avalanche pressures. In addition, snow rakes were installed for the first time in the starting zone and an avalanche dam was constructed across part of the run-out zone on the valley floor.

FURTHER READING

Hewitt, K. (1992) Mountain hazards. *Geojournal* 27: 47–60. A good general account of problems in high terrain.

Hancox, G.T. (2008) The 1979 Abbotsford landslide, Dunedin, New Zealand: a retrospective look at its nature and causes. *Landslides* 5: 177–88. An excellent, reflective case study of a typical urban landslide disaster.

Harp, E.L., Reid, M.E., McKenna, J.P. and Michael, J.A. (2009) Mapping of hazard from rainfall-triggered landslides in developing countries: examples from Honduras and Micronesia. *Engineering Geology* 104: 295–311. A very well-illustrated report on practical issues relevant to many other tropical nations.

Rice, R. Jr *et al.* (2002) Avalanche hazard reduction for transportation corridors using real-time detection and alarms. *Cold Regions Science and Technology* 34: 31–42. A high-tech approach to risk reduction in avalanche-prone areas.

Runqiu, H. (2009) Some catastrophic landslides since the twentieth century in the south-west of China. *Landslides* 6: 69–81. This paper places the hazardous nature of China in sharp focus.

Smyth, C.G. (2000) Urban landslide hazards: incidence and causative factors in Niterói, Rio de Janeiro State, Brazil. *Applied Geography* 20: 95–117. A recurrent and on-going problem in this area and many other parts of the world.

WEB LINKS

The International Consortium on Landslides www.iclhq.org/

The Durham University International Landslide Centre www.landslidecentre.org/

The United States Geological Survey landslides hazard program www.landslidesusgs.gov

Current avalanche information from the United States Forestry Service www.avalanche.org/

A consortium of avalanche hazard management organizations in Canada www.avalanche.ca/

Colorado Avalanche Information Center www.geosurvey.state.co.us/avalanche

The Swiss Federal Institute for Snow and Avalanche Research www.slf.ch/welcome-en.html

CHAPTER NINE

Severe storm hazards

9

A ATMOSPHERIC HAZARDS

Most environmental hazards are atmospheric in origin. Only a portion of the world's population lives near active geological faults or on unstable slopes, but all are exposed to weather-related extremes. Some individual weather elements, like temperature, constitute a direct hazard to human welfare when they create physiological cold stress or heat stress. But storm-related disasters mostly occur when extreme atmospheric conditions combine adversely, or interact with other environmental factors, to create risk.

- *Severe storm disasters* are the most common threat. Table 9.1 shows that, although all severe storms have some features in common, each type has its own mix of damaging conditions. Synergy is important. For example, a blizzard – defined by the US National Weather Service as snow falling or blowing in wind speeds over 16 m s^{-1} causing visibility less than 44 m for at least 3 h – creates a much greater hazard than that of snowfall or wind speed alone.
- *Weather-related disasters* occur when atmospheric hazards – especially hydro-meteorological events – are amplified by other

environmental conditions, like steep topography or human vulnerability. For example, excessive rainfall can produce landslides and floods; insufficient rainfall can produce droughts and famines. Approximately half of all environmental disasters, and over two-thirds of disaster-related deaths, are weather and climate related. The potential effects of climate change on storms and on weather-related disasters are considered in Chapter 14.

The significance of atmospheric hazards is not always appreciated. Pielke and Klein (2005) showed that the impacts of tropical cyclones, especially those related to flooding, were underestimated in the record of US disasters maintained by FEMA. Even so, in 2005 three hurricanes – 'Katrina', 'Rita' and 'Wilma' – struck the coastline along the Gulf of Mexico, killing over 1,500 people and costing insurers and the federal disaster-relief budget a record sum of over US$180 billion. Hurricane 'Katrina' became the most expensive insured disaster in the world, with losses of US$46.3 billion, as compared to the estimated US$35.5 billion cost of the 9/11 terrorist attacks in New York.

Severe storms account for almost 80 per cent

Table 9.1 Severe storms as compound hazards showing major characteristics and impacts

Tropical storms	Mid-latitude storms			
Tropical cyclones	Tornadoes	Hailstorms	Winter cyclones	Snowstorms
Wind	Wind	Hail	Wind	Snow
Rain	Pressure drop	Wind	Rain	Ice
Storm surge and waves	Updraughts	Lightning	Flooding	Glaze
Coastal erosion	Building damage	Building damage	Landslides	Wind
Flooding	Agricultural losses	Agricultural losses	Coastal erosion	Blizzards
Landslides			Building damage	Transport disruption
Saline intrusion			Agricultural losses	Building damage
Building damage				Agricultural losses
Agricultural losses				
Transport disruption				

Note: All these storms are responsible for causing deaths and injuries, especially tropical cyclones and tornadoes

of the global cost of property insurance claims following natural disasters. Until recently, some storm loss models used by the insurance industry were limited to wind speed assessments because rainfall was viewed as a minor factor (Munich Re, 2002a). Following several high rainfall-related losses, the industry began to distinguish between 'dry' storms and 'wet' storms and pay more attention to the effects of storm surge in coastal areas. According to Rauch (2006), severe storms accounted for 40 per cent of all recorded natural disasters between 1950 and 2005. They were responsible for 38 per cent of the total economic losses and 79 per cent of the insured property losses. Over the period, the cumulative cost of severe storms, based on 2005 values, was estimated at US$1,700 billion and US$340 billion respectively for total and insured losses.

B THE NATURE OF TROPICAL CYCLONES

An annual average of 90 tropical cyclones form across the globe, but variability is high from year to year and within individual ocean basins. Tropical cyclones are defined as non-frontal, low-pressure, synoptic-scale systems with strongly organized convection that form over warm oceans. The term 'tropical cyclone' is used in the Indian Ocean, Bay of Bengal and Australian waters, whilst the same storms are called 'hurricanes' in the Caribbean, Gulf of Mexico and the Atlantic Ocean. In the region of greatest frequency, which is the north-west Pacific near to the Philippines and Japan, they are known as 'typhoons'. According to Landsea (2000), in an average year, about 86 *tropical storms* (winds of at least 18 m s^{-1}), 47 *hurricane-force tropical cyclones* (winds of at least 33 m s^{-1}) and 20 *intense hurricane-force tropical cyclones* (winds of at least 50 m s^{-1}) are recorded worldwide.

The term 'tropical cyclone' is rather general. The minimum mean speed threshold for *hurricane* winds is set at 33 m s^{-1}. Such winds blow around very low-pressure centres with strong isobaric gradients. As the storm evolves from the initial closed circulation with only moderate depth and heavy showers to a fully developed hurricane, many authorities recognize the intermediate stages of a *tropical depression*, characterized by maximum mean wind speeds below 18 m s^{-1}, and a *tropical storm*, with maximum mean wind speeds from 18 m s^{-1} to 32 m s^{-1}. A hurricane's severity can be classified according to central pressure, wind speed or ocean surge on the Saffir/Simpson scale (Table 9.2). A Category 4 or 5 system releases more power in one day than the USA uses in a year.

There are three main hazards associated with tropical cyclones.

- *Strong winds* cause most of the structural damage. The atmospheric pressure at the storm centre often falls to 950 mb. The deepest low ever recorded was 870 mb, when typhoon 'Tip' hit the Pacific island of Guam in October 1979 with sustained surface wind speeds of 85 m s^{-1}. Wind gusts ranging from 90 km h^{-1} up to 280 km h^{-1} can occur round the centre of the most severe storms. The inertial force of the wind experienced when a structure is perpendicular to the moving mass of air is proportional to the wind speed, so the damage potential increases rapidly with storm severity. As shown in Figure 9.1, the destructive energy of a Category 5 hurricane, with wind speeds around 70 m s^{-1}, can be up to about 15 times greater than the damage potential of a tropical storm with wind speeds around 20 m s^{-1}.
- *Heavy rainfall* creates freshwater flooding and landslides, as illustrated by hurricane 'Mitch'. At any one location, the total rainfall during the passage of a tropical cyclone can exceed 250 mm, all of which may fall in a period as short as 12 hours. Higher falls are likely if there are mountains near the coast. The most intense rains have been recorded on La Reunion Island, with a 12-hour fall of 1,144 mm and a 24-hour fall of 1,825 mm in January 1966. Rainfall associated with a decaying cyclone can continue inland a long way from the coast, to cause flooding.
- *Storm surge* is a raised dome of sea-water perhaps 60–80 km across, which often causes the greatest losses in the LDCs through deaths by drowning and salt contamination of agricultural land. Swell waves move outward from the storm centre, perhaps three to four times faster than the storm itself, and can act

Table 9.2 The Saffir/Simpson hurricane scale

Scale	Central pressure (mb)	Wind-speed (m s^{-1})	Surge (m)	Damage
1	>980	33–42	1.2–1.6	Minimal
2	965–979	43–49	1.7–2.5	Moderate
3	945–964	50–58	2.6–3.8	Extensive
4	920–944	59–69	3.9–5.5	Extreme
5	<920	>69	>5.5	Catastrophic

as a warning of its approach to coastlines over 1,000 km distant. The wind-driven waves pile water up along shallow coasts to a maximum height that depends on the intensity of the tropical cyclone, its forward speed of movement, the angle of approach to the coast, the submarine contours of the coast and the phase of the tide. Typically the height is 2–5 m above normal tide level (Figure 9.2). Confined bays with low-lying coasts, such as the Gulf of Mexico and the Bay of Bengal, are at special risk. A further increase in sea level is due to low atmospheric pressure, at a rate of 260 mm for every 30 mb fall in air pressure. The likely

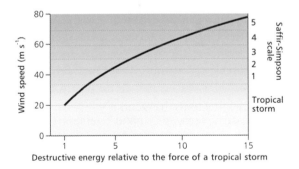

Figure 9.1 The destructive energy of hurricane wind speeds, as compared to a tropical storm. Damage potential is proportional to wind energy, so winds of 65 m s^{-1} – typical of a Category 4 hurricane – have about 10 times the destructive power of the airflow in a tropical storm. Adapted from from R.A. Pielke and R.A. Pielke, *Hurricanes: Their Nature and Impacts on Society* (1997). Copyright John Wiley and Sons 1997. Reproduced with permission.

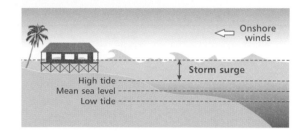

Figure 9.2 The nature of the storm surge hazard. Strong onshore winds drive water inland well above high tide levels, to threaten low-lying coastal property.

maximum surge was created by an 1899 cyclone in the Bathurst Bay area of Australia, with an estimated height of 13 m.

C HOW TROPICAL CYCLONES DEVELOP

Tropical cyclones depend for their existence on heat and moisture, so they form over warm oceans with sea-surface temperatures (SSTs) of at least 26°C. In fact, tropical cyclones originate mainly over the western parts of the main ocean basins, where no cold currents exist. Figure 9.3 illustrates the main regions of formation and the land areas most affected by the storm tracks. Tropical cyclones do not form in the eastern South Pacific Ocean, nor do they occur in the South Atlantic Ocean, because of low temperatures and unfavourable upper winds. That being said, the highly variable incidence of these storms can be linked to perturbations in the tropical ocean–atmosphere system, such as El Niño Southern Oscillation events (see Chapter 14). Most storm systems decay rapidly over land areas, although some remain dangerous for thousands of kilometres, with sufficient energy to cross mid-latitude oceans and threaten higher-latitude coasts, such as the northeast of the USA.

The meteorological evolution typically begins with a small, low-pressure disturbance, perhaps a vortex near the Inter-Tropical Convergence Zone. If surface pressure continues to fall, perhaps by 25–30 mb, this creates a circular area with a typical radius of only 30 km with strong inblowing winds. The disturbance can then develop into a self-sustaining hurricane if certain environmental conditions are satisfied:

- The rising air, convected over a wide area, must be warmer than the surrounding air masses up to 10–12 km above sea level. This warmth comes from latent heat taken up by evaporation from the ocean and liberated by condensation in bands of cloud spiralling around the low-pressure centre. There must also be high atmospheric humidity up to about 6 km. If the rising air has insufficient moisture for the release of latent heat, or if it is too cool in the first place, the chain reaction will not start. This is why cyclones form only over tropical oceans with surface temperatures of 26°C or more.

- Hurricanes need vorticity to give the low-pressure system an initial rotation. Therefore, they do not develop within 5° latitude of the equator, where the Coriolis force is almost zero and inflowing air will quickly fill a strong surface low. But, between 5–12° north and south of the equator, the air-flow converging on a low is deflected to produce a favourable spiral structure.

- The broad air current in which the cyclone is formed should have weak vertical wind shear because wind shear inhibits vortex development. Vertical shear of the horizontal wind of less than 8 m s^{-1} allows the main area of convection to remain over the centre of lowest pressure in the cyclone. Although this is not a difficult condition to satisfy in the tropics, it does explain why no cyclones develop in the strong, vertically sheared current of the Asian summer monsoon. Most cyclones occur after the monsoon season, in late summer and autumn, when sea-surface temperatures are at their highest level.

- In combination with the developing surface low, an area of relatively high pressure should

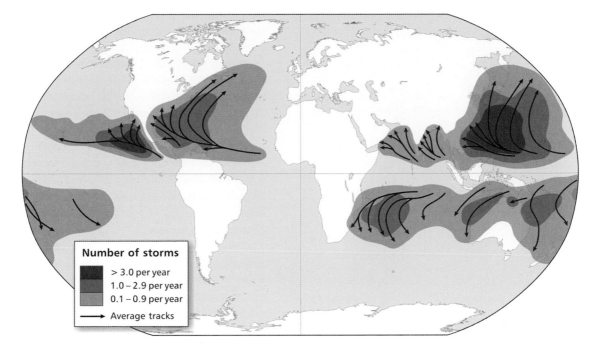

Figure 9.3 World map showing the location and average annual frequency of tropical cyclones. The importance of the western North Pacific region and the way in which storm tracks curve polewards to make landfall on populated coastal areas in the USA and elsewhere is evident.

exist above the growing storm. As this happens rarely, few tropical disturbances develop into cyclones. If high pressure exists aloft, a strong divergence or outflow of air is maintained in the upper troposphere. Crudely stated, the system acts like a suction pump, drawing away rising air and strengthening the sea-level convergence.

Mature tropical cyclones can be regarded as thermodynamic heat engines where the energy derived from evaporation at the ocean surface is lost partly by thermal radiation where the moist air rises and diverges and partly by surface friction as it moves over the sea. As the wind speed increases, and the storm intensifies, these energy losses grow relative to the energy gain and a theoretical upper limit to storm development is set. The wind velocity increases towards the eye and the lowest central pressures produce the highest-velocity winds. Because the lowest-recorded central pressures exist in the North-West Pacific, the upper storm limit is probably higher here than elsewhere. A ring-like wall of towering cumulus cloud rises to 10–12 km around the 'eye' of the storm.

Most of the rising air-flows outward near the top of the troposphere, as shown in Figure 9.4 (vertical section), and acts as the main 'exhaust area' for the storm. The release of rain and latent heat encourages even more air to rise and violent spiralling produces strong winds and heavy rain. A small proportion of the air sinks towards the centre, to be compressed and warmed in the 'eye' of the storm. The warm core helps to maintain the system because it exerts less surface pressure, thus maintaining the low-pressure heart of the storm. Recent evidence suggests that some hurricanes may be intensified by 'eyewall replacement', a process by which the original eyewall clouds are

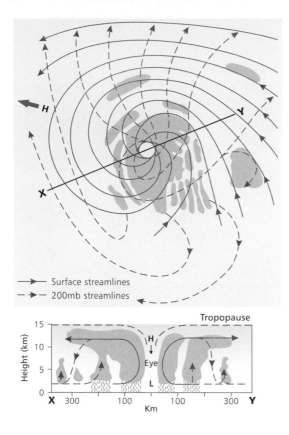

Figure 9.4 A model of the areal (above) and vertical (below) structure of a tropical cyclone. Spiral bands of cloud and rainfall are indicated in the vertical section X–Y across the system. The streamline symbols refer to the upper diagram. After R.G. Barry and R.J. Chorley, *Atmosphere, Weather and Climate* (Methuen 1987). Reproduced with permission.

replaced by the formation of a new eyewall further out from the centre of the storm (Houze *et al.*, 2007). Although all storm systems move westward at about 4–8 m s^{-1}, driven by the upper-air easterlies, they eventually recurve erratically towards the pole.

D TROPICAL CYCLONE HAZARDS

Tropical cyclones are responsible for most 'windstorm' deaths, mainly due to drowning in the storm surge. According to Peduzzi *et al.* (2011), the number of people exposed to tropical cyclone hazards in an average year has grown from about 73 million in 1970 to 123 million in 2010. Many are in the LDCs, where more than 100 million people live in coastal zones at elevations less than 10 m above mean sea level. Over 85 per cent of the mortality occurs in Bangladesh and India. Like other hazards, tropical cyclones bring benefits as well as losses. For example, although they account for one-quarter of the cost of natural disasters shouldered by the Australian economy, they are often instrumental in ending drought in that country and elsewhere.

The greatest risk exists in three landscape settings:

- *Densely populated deltas in the LDCs.* Bangladesh is the most vulnerable nation, with some 20 million people exposed to the cyclone hazard, mainly in rural communities along the fertile delta at the head of the Bay of Bengal. About 10 per cent of all tropical cyclones form in the Bay of Bengal and this area averages over five storms per year, with about three reaching hurricane intensity. The two deadliest storms of the twentieth century were recorded here (Table 9.3), partly because there is little rising ground offering safe refuge from the storm surge. In November 1970 up to 300,000 people died, and damage of US$75 million occurred, when wind speeds reaching 65 m s^{-1} created a storm surge 3–9 m in depth. In the absence of an effective warning, and with no evacuation plan, most of the survivors sought refuge by clinging to trees. On 29 April 1991, in the early pre-monsoon part of the cyclone season, the south-east coast of Bangladesh was again struck by a powerful tropical cyclone. At least 139,000 people were killed by the 6-metre-high storm surge and up to 10 million were made homeless as the poorest houses made of mud, bamboo and straw were washed away. The greatest devastation occurred on the islands of silt near the head of the Bay. On Sandwip

Island, where 300,000 people lived, 80 per cent of the houses were destroyed.

- *Isolated island groups.* The Japanese, Philippine and Caribbean island groups are all at risk from tropical cyclones, as well as some remote island communities in the Pacific Ocean. The Caribbean lies in the path of most Atlantic hurricanes. Quite apart from the resulting fatalities and the impacts on tourism, the agricultural sector of these islands is particularly vulnerable to damage, with the defoliation of banana and other tree crops by strong winds and the washing away of food crops in heavy rain. Future harvests can be affected by salt contamination of the soil from storm surge. Commercial crops like bananas, grown for vital foreign exchange, suffer considerable damage.
- *Highly urbanized coasts in the MDCs.* The greatest damage potential exists along the Gulf of Mexico and the Atlantic coastline of the USA. The deadliest natural disaster in US history occurred in September 1900, when more than 6,000 people were killed by a storm surge in Galveston, Texas. The regional death toll exceeded 12,000 (Hughes, 1979). At that time, Galveston's highest point was less than 3 m above sea level and nearly half of all dwellings in the city were destroyed. In late August 2005, hurricane 'Katrina' – the sixth-strongest Atlantic hurricane ever recorded and the third-strongest to make landfall in the USA – struck south-east Louisiana. Some 1,600 people died, despite the issue of evacuation orders covering 1.2 million residents of the Gulf Coast. Some 275,000 homes were damaged or destroyed. New Orleans was extensively flooded and the coastline was devastated up to 150 km inland, thus creating what was then the world's costliest natural disaster (Box 9.1).

The hazard potential of tropical cyclones is related to storm intensity and human exposure. For example, although intense tropical cyclones (winds of at least 50 m s^{-1}) account for only one-fifth of all hurricanes making landfall in

Table 9.3 The world's 10 deadliest tropical cyclones in the twentieth century

Year	Location	Number killed
1970	Bangladesh	300,000
1991	Bangladesh	139,000
1922	China	100,000
1935	India	60,000
1998	Central America	14,600
1937	Hong Kong (China)	11,000
1965	Pakistan	10,000
1900	United States	8,000
1964	Vietnam	7,000
1991	Philippines	6,000

Source: Adapted from CRED and NOAA data

the USA, these severe storms account for over 80 per cent of all hurricane-related damage. Hurricane 'Andrew', a Category 5 storm on the Saffir-Simpson scale (Table 9.2), had sustained winds of 74 m s^{-1} at landfall, which created most of the loss, plus a storm surge of 4.5 m. In high-risk Florida about 28,000 residential structures, including some 5,000 mobile homes, were destroyed. The storm killed 65 people and made 250,000 homeless. The entire east coast of the USA is vulnerable, as in 1972, when hurricane 'Agnes' moved north from the Florida panhandle, causing at least 118 deaths and more than US$3 billion in damage, mainly due to inland floods (Bradley, 1972).

Tropical cyclones hit hardest at poor people in poor countries. When hurricane 'Fifi' struck Honduras in 1974 it produced landslides on the steep hills where most of the peasants had relocated after being forced off more fertile valley land. Several thousand lives were lost. Hurricane 'Mitch', another Category 5 storm, devastated much of Central America in October 1998. It was the fourth-strongest hurricane ever recorded in the Atlantic basin, with sustained wind speeds of 80 m s^{-1} and intense rainfall that created many

Box 9.1 Hurricane 'Katrina': lessons for levees and for lives

New Orleans is built on the Mississippi delta, between Lake Pontchartrain to the north and the main distributary of the river to the south. Less than half of the city is above sea level; most of the area lies on sinking alluvial and peat soils between 0.3 and 3.0 m below sea level (Waltham, 2005). Rainwater drainage is routinely pumped from low-lying areas into Lake Pontchartrain, but the city relies on the surrounding wetlands and barrier islands, plus a complex system of artificial floodwalls and levees, for protection against hurricane storm surge. Over many years, these defences have been weakened. Levee construction and dredging have limited sediment supply for delta renewal, leading to the loss of some 75 km^2 of wetlands each year, and the barrier islands along the Louisiana coast are eroding at rates of up to 20 m per annum. As a result, the entire delta is subsiding, New Orleans is sinking further below sea level and the natural coastal buffer has been largely destroyed. Human vulnerability compounded this picture. Before 'Katrina' there was a lot of unemployment in the city, due to the on-going closure of port functions in the city and a contraction of the local oil industry. About 25 per cent of all families were living in poverty and clear ethnic inequalities existed, making an efficient emergency response unlikely. A disaster was waiting to happen (Reichhardt *et al.*, 2005; Comfort, 2006).

At 6.10 am local time on 29 August 2005, hurricane 'Katrina' made landfall in south-east Louisiana. The floodwalls and levees were designed to withstand a Category 3 hurricane but stand on unconsolidated deposits and, in some cases, date back to the 1920s and 1930s. The coastal towns of Biloxi and Gulfport suffered major damage, even to engineered structures, from a storm surge at least 7.5 m above sea level, before the storm swept inland (Robertson *et al.*, 2006). Driven by strong northerly winds, the waters of Lake Pontchartrain were pushed against the flood defences by the highest-measured storm surge in North America, to a height of 5.2 m above normal. Failures occurred and flood-water flowed into the northern areas of New Orleans below sea level and reached depths of 1.5–2.0 m over about 80 per cent of the city (Figure 9.5). Slightly higher areas, formed from fossil beaches and fluvial deposits related to a previous course of the Mississippi (Metairie sediments), remained dry. The depth of flooding was directly linked to land elevation, so the lowest parts of the city, which were largely residential, suffered most; almost 80 per cent of all direct property damage was in the residential sector.

About two-thirds of the flooding was due to breaks in the levee system and one-third to over-topping, with rainfall adding to the internal water levels. Altogether 50 levees were damaged, 46 of them due to breaching and over-topping caused by a mix of under-seepage, scour erosion behind the structures and erosion along the tops of levees. For example, the Industrial Canal levee was undermined by seepage through the underlying silt and sand, whilst the 17th Street canal failure was due to over-topping and collapse. Although the severity of hurricane 'Katrina' technically exceeded the design criteria for the New Orleans flood defences, the system should have performed better than it did (Interagency Performance Evaluation Taskforce, 2006). Engineering problems included:

- over-reliance on a series of hurricane design models dating back to 1965, and the piecemeal development of the levees giving inconsistent levels of protection
- failure to take into account varying rates of ground subsidence across the area and to build structures, like flood outfalls, above the appropriate datum

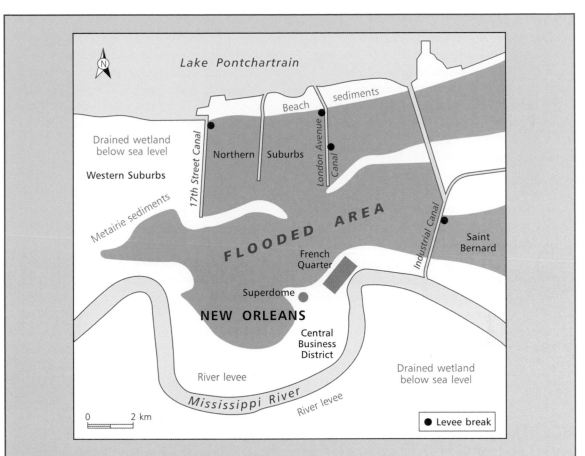

Figure 9.5 Some effects of hurricane 'Katrina' on New Orleans in August 2005. The location of the main levee breaks is indicated, along with the extent of flooding in the central urban area. Many residents found temporary shelter in the Superdome. After Waltham (2005). Reproduced from T. Waltham, The flooding of New Orleans. *Geology Today* 21 (2005): 225–31. Copyright John Wiley and Sons 2005. Reproduced with permission.

- inability of the pumping stations to cope with the demand, so that only 16 per cent of the total capacity operated during the storm
- under-estimation of the dynamic forces acting on the flood defences and the erodible soils.

Over 80 per cent of the city was flooded and at least 1,600 people died. The cost was placed at US$75 billion, making it the costliest storm in US history. Weaknesses in policy and preparedness were also evident in the disaster response, even when allowing for the scale of the emergency, which displaced 1.5 million people from their homes throughout the region, one of the largest urban evacuations in US history. As always, the poor, the elderly and the disabled suffered most. The loss of life was partially determined by depth of flooding, but more than three-quarters of deaths were suffered by people over the age of 60. Some 130,000 residents (27 per cent of the population) lacked private transport and the mandatory evacuation order was not issued until the day before the hurricane strike. Despite this, an estimated 80 per cent of the population evacuated, using extra highway lanes opened specially for the emergency, but this still left about 100,000 people – generally the most disadvantaged – within the city. Some 20,000 people were crowded into the Superdome without proper supplies of food and water for over five days before relocation to other shelters (Brodie *et al.*, 2006). Between 8 September and 14 October 2005, residents and relief workers sustained over 7,500 non-fatal injuries (Sullivent *et al.*, 2006), placing a general strain on medical services, and there was a particular lack of preparedness for dealing with the medical needs of children (Dolan and Krug, 2006). Emergency shelters were poorly equipped to deal with the needs of evacuees with disabilities and other special needs.

Hurricane 'Katrina' caused severe flooding throughout New Orleans but existing socio-economic inequalities created greater difficulties for some groups during the response and recovery periods. Low-income African American home-owners appeared especially vulnerable after the disaster and were most likely to need new jobs with a living wage and assistance with housing (Elliott and Pais, 2006). The absence of an evacuation plan for people lacking a vehicle, and for those without money or a place to go, attracted much criticism (Renne, 2006; and Litman, 2006). It is now clear that emergency planners must give attention to the needs of non-drivers, who include many people with other problems, and that buses should be made available without charge to evacuate such residents.

New Orleans faces great challenges as the city seeks to re-establish itself. Problems persist and progress is slow. Controversy exists over issues ranging from the possible restoration of the Mississippi deltaic plain (Day *et al.*, 2007) to the design standards for new levees and the renewal of the urban system at an estimated cost of US$14 billion. According to Galloway *et al.* (2009), there is a lack of resolve and federal funding for maintaining the flood-protection infrastructure for New Orleans and for restoring the coastal deltaic system. Although the initial severity of flooding showed no bias in relation to race or class, the areas of New Orleans with lower flood levels and less social vulnerability are recovering from the disaster faster than those with higher flood levels and greater vulnerability (Finch *et al.*, 2010). Above all, the need is to create opportunities that are more equal for all future residents of the city (Olshansky, 2006).

floods and landslides. Deaths were estimated at over 14,000, with 13,000 injured, 80,000 homeless and 2.5 million people temporarily dependent on aid. Material losses were put at US$6 billion, with two-thirds of the loss concentrated in the primary economic sector of agriculture, forestry and fishing. Once again, Honduras – the second-poorest country in the western hemisphere – was badly affected. In the mountains 100–150 cm of rain fell within 48 hours and created over 1 million landslips and mudflows. About 5,500 Hondurans were killed, many in the capital, Tegucigalpa, which is built on a series of floodplains and hillsides. About 60 per cent of all bridges, 25 per cent

of schools and 50 per cent of the agricultural base, mainly in the cash-crop sector of bananas and coffee, were destroyed. The economic losses reached nearly 60 per cent of the annual GDP (IFRCRCS, 1999).

Human exposure to tropical cyclones is the key impact factor. The risks of building close to the US coast were highlighted nearly 50 years ago (Burton and Kates, 1964b), but a decade later 6 million Americans were in areas exposed to hurricane storm surges that occurred at least once per century. Since then the demand for homes close to the shore has continued. Coastline counties of the USA are now home to 29 per cent

Plate 9.1 Residents along the Rio Choluteca in Tegucigalpa, capital city of Honduras, dig themselves out of flood debris created by hurricane 'Mitch' in November 1998. Many old adobe houses in the poorer neighbourhoods were destroyed. (Photo: AP Photo/Gregory Bull 9811060863)

of the population and five of its most populous cities (Wilson and Fischetti, 2010). The coastal population grew from 47 million in 1960 to 87 million in 2008, a much greater increase than that for the nation as a whole. Much of the growth took place in hurricane-prone 'Sun Belt' states, like Florida, which prove attractive to people in the over-65 age group living in mobile homes or expensive apartments close to the water's edge. Interestingly, these large-scale demographic shifts have continued, despite the experience of hurricanes. The 10 most powerful hurricanes since 1960 affected almost 51 million people living in coastline counties; if those same hurricanes had struck in 2008, the affected population would have been nearer 70 million. Figure 9.6 shows how the population of the coastline counties affected by these 10 hurricanes changed between then and 2008. For example, the counties hit by hurricane

'Donna' (1960) have since grown by 116 per cent and were some of the nation's fastest-growing areas between 1960 and 2008. The populations affected by hurricanes 'Andrew' (1992) and 'Opal' (1995) have each grown by more than 20 per cent in the decades following the storms. Only the counties affected by 'Katrina' (2005) show an overall decrease.

These trends are disturbing. Population growth combined with increasing property values is the main reason for the rise in economic losses noted for most atmospheric hazards, including tropical cyclones. However, recorded increases in US hurricane-related losses have become controversial. Figure 9.7a shows a rise in unadjusted annual economic losses from 1900 to 1995. In a pioneering study, Pielke and Landsea (1998) demonstrated that, if the data are normalized for increases in coastal population and exposed

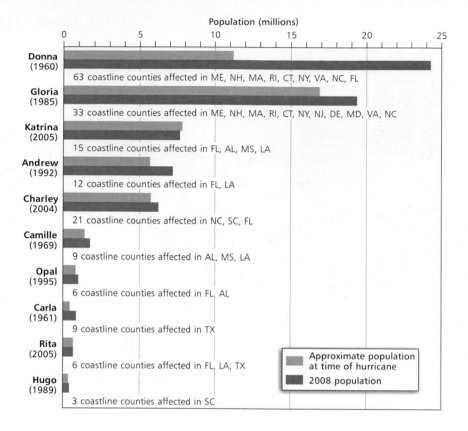

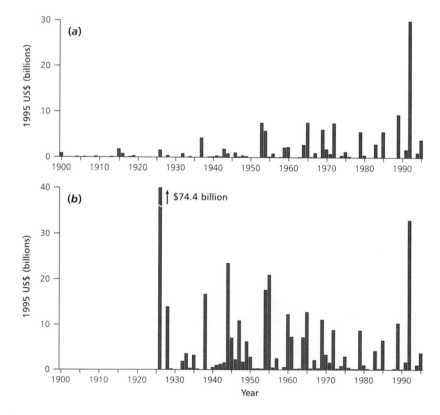

wealth, the 1970s and 1980s showed smaller damages than some earlier decades (Figure 9.7b). However, Nordhaus (2010) claimed that the ratio of hurricane damage to national GDP has continued to increase by some 1.5 per cent per year, and there is little doubt that the rush to the coast has placed more Americans, and their assets, at risk than ever before.

E SEVERE SUMMER STORMS

1 Tornadoes

A tornado is a narrow, violently rotating column of air, averaging about 100 m in diameter, that extends towards the ground. Most tornadoes are associated with 'parent' cumulonimbus clouds and are recognized by a funnel-shaped cloud that appears to hang from the cloud base above. They are highly localized storms, sometimes associated with thunder and hail, and tend to form in warm, moist air ahead of a strong cold front. This is because the contrast in air masses produces latent heating and the creation of a low-pressure area near the surface, but violent tornadoes can also be found in weakly unstable conditions in the lower atmosphere (Wesolek and Mahieu, 2011). The greatest hazard exists when the funnel cloud touches the ground and creates some of the strongest horizontal pressure gradients – and highest wind speeds – seen in nature.

A scale of tornado intensity was devised by Fujita (1973) and is shown in Table 9.4. It is believed that about one-third of all tornadoes exceed F-2 and attain wind speeds greater than 50 m s^{-1}. The forward speed of a tornado is much lower, perhaps only 5–15 m s^{-1}. Most tornadoes are of short duration and have a limited destructive path, rarely more than 0.5 km wide and 25 km long. However, in May 1917 a tornado travelled about 500 km across the Midwest of the USA and existed for over seven hours.

Over half of all tornadoes in the USA develop in the April to July period, with a marked decrease after the summer solstice. In the late winter and early spring they may be linked to strong frontal systems, but later thunder-storms develop in the Central Plains along a 'dryline' separating warm, moist air to the east from hot, dry air further west. Favoured breeding conditions are in the thunderstorms that frequently form when air near the ground flows upslope into higher ground along the front range of the Rockies; in the Texas panhandle; and in the southern High plains. In general, most occur in *Tornado Alley*, the area running from Texas through Kansas and Oklahoma and into Canada, with the maximum frequency located in central Oklahoma (Bluestein, 1999).

The United States leads the world in tornado losses. On average, over 1,000 tornadoes are recorded over land each year, resulting in around 80 deaths and more than 1,500 injuries. The 10 deadliest US tornadoes on record are listed in Table 9.5. The greatest tornado disaster recorded in the USA was the 'Tri-State Tornado' of March 1925. Losses included over 700 people dead, over 2,000 injured and damages equal to US$40 million at 1964 prices (Changnon and Semonin, 1966). The disaster resulted from a combination of physical factors, including the high ground speed plus the long track and wide path, and human

Figure 9.6 *(Facing page, above)* Population changes in coastline counties in the USA affected by the 10 most intense hurricanes since 1960. Storms are ranked according to the resident population (in millions) living in the areas affected in 2008. The graph shows the continuing movement of people to hurricane-prone coasts in the USA. After Wilson and Fischetti (2010). Reproduced with permission of the US Census Bureau.

Figure 9.7 *(Facing page, below)* Annual hurricane damage during the twentieth century in the United States (US$ millions). (a) Unadjusted values 1900–95; (b) normalized values 1925–95. After Pielke and Landsea (1998). Reproduced from R.A. Pielke and C.W. Landsea, Normalised hurricane damages in the United States 1925–95. *Weather and Forecasting* 13 (1998): 621–31, (c) American Meteorological Society.

Plate 9.2 A developing tornado threatens Caddo County, Oklahoma, in the Central Great Plains of the USA. In the second half of the twentieth century over 90 tornadoes were recorded here, including the damaging Category 5 event of 3 May 1999. (Photo: Samuel D Barricklow/Getty Images 92173219)

Table 9.4 The Fujita scale of tornado intensity

Category	Damage	Wind speed (m s⁻¹)	Typical impact
F-1	Light	18–32	Damage to trees, free-standing signs and some chimneys
F-2	Moderate	33–50	Roofs damaged, mobile homes dislodged, cars overturned
F-3	Severe	51–70	Large trees uprooted, roofs removed, mobile homes demolished, damage from flying debris
F-4	Devastating	71–92	Masonry buildings damaged, cars become airborne, extensive damage from large missiles
F-5	Disastrous	93–142	Wood-frame buildings lifted from foundations and disintegrated, cars airborne for more than 100 m
F-6 and over		>142 m	Currently thought not to exist

failings regarding a lack of warning and inadequate shelter provision. Only 3 per cent of the 900 or so tornadoes that occur in the US each year are responsible for human deaths. However, in 2011 more tornado deaths were recorded than in any year since 1936. Mortality was mainly associated with an outbreak in late April across south-eastern states that claimed 322 lives, and a storm that struck the city of Joplin, Missouri in May 2011 caused 157 deaths. This latter event was the deadliest single tornado in the country since 1947.

Tornadoes are widespread globally. For example, in Spain, over 25 per cent of all hazard-related insurance payouts are for severe wind or tornado damage (Gayà, 2011). Less frequent but more dangerous tornadoes exist in Bangladesh. A storm in May 1989 killed between 800 and 1,300 people, whilst in 1996 a tornado in the Tangail

Table 9.5 The 10 deadliest tornadoes recorded in the USA

Common name or location	Date	Number of fatalities
Tri-State tornado	18 March 1925	747
Deep South outbreak	21 March 1932	332
Great Natchez tornado	17 May 1840	317
Super outbreak	3 April 1974	310
St Louis tornado	27 May 1896	305
Palm Sunday outbreak	11 April 1965	260
Tupelo-Gainsville tornado	5 April 1936	249
Alabama–Mississippi tornado	20 April 1920	224
Dixie outbreak	24 April 1908	220
Gainsville tornado	6 April 1936	205

Source: Severe Storms Laboratory, NOAA data

Plate 9.3 Friends and neighbours help residents to find belongings and assess damage after a tornado spawned by hurricane 'Irene' destroyed many mobile homes in Columbia, North Carolina during August 2011. (Photo: Tim Burkitt, FEMA 50644)

area killed about 700 people and destroyed approximately 17,000 homes (Paul, 1997). These high death rates have been attributed to a mixture of high population densities, weak building construction, an absence of preparedness and poor medical facilities.

Most tornado damage is caused by airborne debris and building collapse. In the USA, occupants of mobile homes and road vehicles are most likely to die (Hammer and Schmidlin, 2000). The large number of survivors, often with soft tissue injuries and fractures, creates problems for local hospitals (Bohonos and Hogan, 1999). Fatalities continue to occur. Schmidlin *et al.* (1998) reported 42 people killed in Florida in February 1998 – all in mobile homes or recreational vehicles – whilst an F-5 storm in May 1999 killed 45 people and created material losses of up to US$1 billion in Oklahoma City. On the other hand, Simmons and Sutter (2005) have shown that, after allowing for changed demographic and other circumstances, recorded deaths from each F-5 tornado declined steadily during the twentieth century. By 1999, fatalities were effectively 40 per cent lower than in 1950 and 90 per cent lower than in 1900. This trend was attributed to better forecasting and warning, together with the provision of more tornado shelters.

2 Hail-storms and thunder-storms

Hail consists of ice particles falling from clouds to reach the ground. The damage potential depends on the number of particles and the surface wind speed that drives them but is also related to the size of the particles. The most destructive hailstones tend to exceed 20 mm in diameter. Large hail has been known to result in human deaths, but the main damage is to property, especially standing crops. Urban hail-storms also cause large losses in some countries, mainly attributed to damage to motor vehicles (Hohl *et al.*, 2002).

Most hail is produced when strong vertical motions are present, giving rise to cumulonimbus clouds with thunder and lightning. Hail-storms result from strong surface heating and are warm-season features. Isolated falls of hail, very often of the greatest intensity, occur in and near to mountain ranges. Few mid-latitude areas are immune from hail, but much damage occurs in the continental interiors close to mountains; hail can also be a problem at high altitude in the tropics. Most places in the United States experience only two or three hail-storms per year. However, in the lee of the central Rocky Mountains between 6 and 12 hail days are recorded each year. According to Changnon (2000), significant hail damage in the USA occurs during the most severe 5–10 per cent of all storms. In an average year, hail causes US$1.3 billion in crop losses and a further US$1–1.5 billion in property damage (at 1996 dollar prices). In Europe, hail damage can be a problem in the summer months in inland areas. Figure 9.8 shows that the highest frequency of H2-category hail in Britain, with a minimum diameter of 16 mm (the size of a typical grape) and capable of causing significant damage to fruit, standing crops and vegetation, occurs in central and eastern England, where convective activity is high in summer (Webb *et al.*, 2009).

Lightning is associated with rain, hail and the powerful up-currents of air within the clouds of summer storms. It occurs when a large positive electrical charge builds up in the upper, often frozen, layers of a cloud and a large negative charge – together with a smaller positive force – forms in the lower cloud. Since the cloud base is negatively charged, there is attraction towards the normally positive earth and the first (leader) stage of the flash brings down negative charge towards the ground. The return stroke is a positive discharge from the ground to the cloud and is seen as lightning. The extreme heating and expansion of air immediately around the lightning path sets up sound waves heard as thunder. Despite its dramatic appearance, lightning causes comparatively few deaths – perhaps about 25,000 per year worldwide – mainly to outdoor workers.

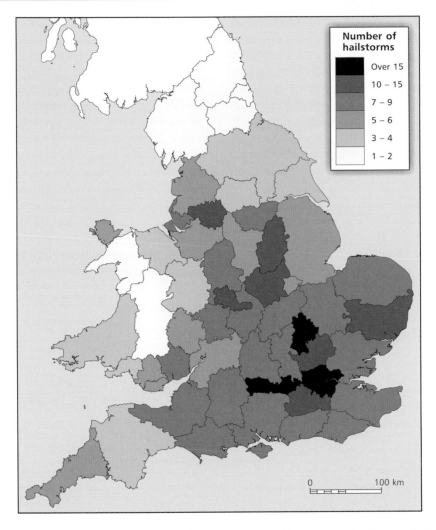

Figure 9.8 Number of hailstorms (per 1,000 km^2 per 100 years) reaching or exceeding H2 intensity for counties in England and Wales 1930–2004. These storms tend to concentrate in central-southern England during spells of high summer temperatures. After Webb *et al.* (2009). Reprinted from *Atmospheric Research* 93, J.D.C. Webb *et al.*, Severe hailstorms in Britain and Ireland: a climatological survey and hazard assessment, 587–606, (2009). Copyright, with permission from Elsevier.

F SEVERE WINTER STORMS

Extra-tropical cyclones bring strong winds during the winter season, when they may be accompanied by snow and ice hazards. An attempt will be made to distinguish between the wind-storm threat and the snow-storm threat.

1 Severe wind-storms

Severe winter wind-storms, often accompanied by heavy rain, are associated with deep mid-latitude depressions. Coastal areas are at risk because wind-driven waves erode sea defences and create dangerous storm surges, such as that of 31 January 1953, produced by a deep depression in the North Sea. The resulting northerly gale combined with a tidal surge of 2.5–3.0 m to cause exceptional coastal flooding. In the Netherlands 1,835 people were killed, 3,000 houses were destroyed and 72,000 people were evacuated. Major world cities, like Venice and London, are subject to increasing storm-surge hazard, due to long-term subsidence and rising sea levels.

These storms are often termed *extra-tropical cyclones* (ETCs). In the northern hemisphere, they are found near the Aleutian and Icelandic low pressure areas; however, the Atlantic storms are more frequent, and cover larger areas, than equivalent events in the Pacific Ocean (Lambert, 1996). Some mid-latitude cyclones develop very quickly and become *rapidly deepening depressions*.

The most favoured breeding grounds for these storms lie off the east coast of continents and correspond to areas of warm ocean currents. They tend to occur most frequently in the North Atlantic Ocean and pose the greatest winter hazard in Western Europe, notably to Britain and the near continent, which lies directly in their eastward path. A storm striking the UK on 8 December 2011 generated 74 m s^{-1} gusts over the mountains and damage costs estimated at more than £100 million. Much structural damage results from these storms, despite the fact that wind speeds rarely attain more than Category 1 or Category 2 on the Saffir-Simpson hurricane scale. One reason is that, unlike tropical cyclones, ETCs retain destructive energy as they penetrate inland.

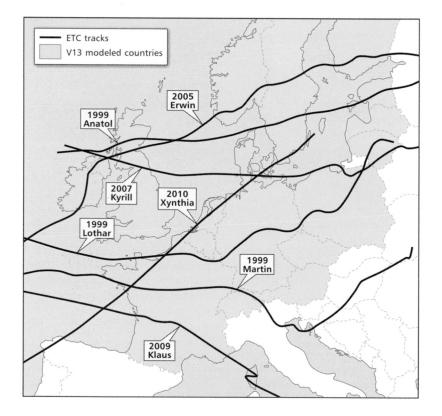

Figure 9.9 The generalized tracks of seven severe wind-storms crossing western Europe in the 1999–2010 period. Unlike hurricanes, mid-latitude cyclones can create damage well inland of coastal areas. After Kafali (2011). Reproduced from C. Kafali, *Regional Wind Vulnerability in Europe*. Report 04.2011, AIR Currents, AIR Worldwide. With permission from AIR Worldwide.

Figure 9.9 illustrates the tracks of seven major ETCs that crossed Western Europe in recent years, including the path for storm 'Kyrill', which continued to Poland and the Baltic states (Kafali, 2011).

In areas exposed to the frequent passage of winter depressions the on-going loss is high. For Britain, Buller (1986) claimed that an average of 200,000 buildings, mainly domestic properties, were damaged by wind-storms each year. In October 1987 a small depression deepened very rapidly in the Bay of Biscay and then moved over Western Europe. Heavy losses were sustained in southern England, but the storm came at night and there were only 19 direct fatalities. Casualties in other countries raised the total death toll to about 50. Forestry suffered badly; in England alone more than 15 million trees were lost. Much infrastructure damage was due to trees falling onto power lines, houses, roads and railways. In Central Europe there is evidence of a long-term increase in forestry losses due to ETCs that may be associ-

ated with a rise in winter temperatures, because most damage to woodland occurs with wind-throw when soils are unfrozen and very wet (Usbeck *et al.*, 2010).

Severe Atlantic storms appear in clusters. Between January and March 1990 four severe storms ('Daria', 'Herta', 'Vivian' and 'Wiebke') caused more damage than any previous natural disaster over Western Europe (Figure 9.10a). In these storms 230 people died and the insurance loss was more than €8 billion (Munich Re, 2002b). The storms produced high gust speeds over a wide area, mostly during daylight. Although the storms were forecast, adequate warnings advising people to seek shelter were not issued. People continued their outdoor activities, and most deaths were due to trees falling onto road vehicles. In Britain alone, an estimated 3.5 million trees were lost in this event.

In December 1999 three separate storms – 'Anatol', 2–4 December; 'Lothar', 24–27 December; 'Martin', 25–28 December – set new

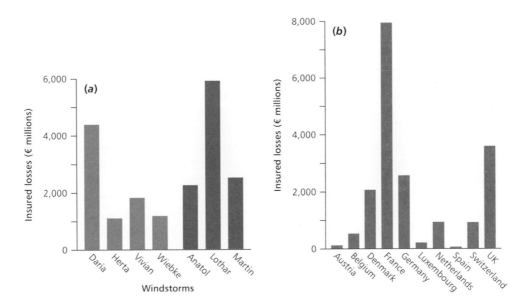

Figure 9.10 Insured losses suffered in four named European windstorms in 1990 ('Daria', 'Herta', 'Vivian' and 'Wiebke') compared to the losses in three similar events in 1999 ('Anatol', 'Lothar' and 'Martin'): (a) the losses created by individual storms; (b) the aggregate losses in individual countries. Compiled from data in Munich Re (2002b).

wind speed records and killed more than 130 people. Much of Western and Central Europe was devastated. In France 'Lothar' and 'Martin' caused many electricity pylons to collapse and 25 per cent of all transmission lines were lost, creating the greatest recorded failure of an electrical supply network in any industrialized country (Abraham *et al.*, 2000). Total insured losses in 1999 were estimated at almost €11 billion, but, after adjustment for price inflation, the real cost of insured losses was greater in 1990. Most insured losses arise from minor damage to buildings sustained over large areas.

The pattern of storm tracks largely explains the cumulative losses from all seven storms experienced by the 10 Western European countries affected (Figure 9.10b). All these storms reflected the large-scale atmospheric conditions prevailing at the time, including anomalously high sea-surface temperatures, but differed individually in terms of synoptic development and track. There may be a trend towards a greater frequency of such events in the future (see Chapter 14, p. 429).

2 Severe snow and ice-storms

Approximately 60 million persons in the United States live in urban areas of the northern states exposed to a high risk of *snow-storms*. According to Schwartz and Schmidlin (2002), the USA has about 10 blizzards per year, each affecting nearly 2.5 million people. In March 1993 a severe snow-storm occurred over the east coast of the USA and Canada. It killed over 240 people, including 48 missing at sea – around three times the combined death toll for hurricanes 'Hugo' and 'Andrew' (Brugge, 1994). The storm was initiated over the warm waters of the Gulf of Mexico and moved north along the Atlantic seaboard as a rapidly intensifying surface low. Between 1949 and 2001 there were 155 snow-storms in the USA each responsible for property losses in excess of US$1 million (Changnon and Changnon, 2005). An upward trend suggested an interactive com-

bination of a growing population, rising property assets and greater storm intensity.

The *Northeast Snowfall Impact Scale* (NESIS) was developed by Kocin and Uccellini (2004) as a relative measure of winter storms in the 13-state Northeast region of the USA. These storms have major impacts on transport and other economic activities. Using a geographical information system (GIS), NESIS takes account of the total snowfall, its geographic distribution and – different from other severe storm indices (like the Saffir-Simpson scale) – the density of the population affected. The 5-point scale was calibrated on the severity of 30 snow-storms recorded in the 1956–2000 period and captures storms described as Category 1 *notable*, Category 2 *significant*, Category 3 *major*, Category 4 *crippling* and Category 5 *extreme*. The mean NESIS value for the calibration events was 5.0 (Category 3). The greatest storm impacts are associated with heavy snowfall over areas that contain large metropolitan centres. Table 9.6 lists the top five recorded events with their NESIS score and description.

Ice (glaze) storms are also an important winter hazard in North America, especially in the Great Lakes region, where they can extend over 10,000 km². The problem arises from thick accretions of clear ice on exposed surfaces. Ice accretes on structures whenever there is liquid precipitation, or cloud droplets, and the temperature of both the air and the object surfaces is below the freezing point. Electric power transmission lines and forests are at high risk because the weight of ice may be sufficient to bring such objects down and deprive people of electrical power for many days. In January 1998 freezing rain produced ice accumulation between 40–100 mm thick on exposed surfaces in eastern Ontario, with much damage to forests and tree-related industries, such as maple sugar production (Kidon *et al.*, 2002). Most fatalities related to ice-storms are indirect. For example, over 500 lives are lost each year in the USA from carbon monoxide (CO) poisoning, many resulting from the use of domestic

Table 9.6 The five most severe winter storms affecting the north-east USA

Date	NESIS score	Category	Description
12–14 March 1993	13.20	5	Extreme
6–8 January 1996	11.78	5	Extreme
2–5 March 1960	8.77	4	Crippling
15–18 February 2003	7.50	4	Crippling
2–5 February 1961	7.06	4	Crippling

Source: National Climate Data Center, USA

generators and space heaters after the loss of electrical power in an ice-storm (Daley *et al.*, 2000).

G PROTECTION

1 Environmental control

Several countries conducted weather modification experiments during the 1950s and l960s. The outcome failed to meet expectations (Box 9.2) and no proven severe storm suppression technology is currently in routine use.

2 Hazard-resistant design

Hazard-resistant design saves lives. In very high-risk areas, special structures provide a safe refuge for people fleeing either wind speed hazards or storm surge hazards. For example, in a sample of Oklahoma City residents warned in advance of the May 1999 tornado, it was found that roughly half fled their homes – mostly for a tornado shelter – whilst the others elected to stay. None of the evacuees was injured but 30 per cent who remained in their homes were injured, with one fatality (Hammer and Schmidlin, 2002). In Bangladesh, the low-lying topography offers little escape from either high winds or storm surge. About 1,600 cyclone shelters have been built along the coast, some of which can accommodate 1,500 people, and there are also large raised mounds acting as escape platforms. Provision exists for the evacuation of up to 4 million people away from the most dangerous areas.

Natural coastlines also provide protection. Almost 25 per cent of the US coast affected by hurricanes has some storm surge mitigation in the form of man-made breakwaters, sea walls or the use of dunes and beach stabilization measures to limit coastal erosion. But in many parts of the world these defences have been removed for economic development. Mangrove forests are a good example. They are destroyed in the MDCs because they are unsightly and hinder resort construction, and in the LDCs to foster more intensive coastal activities, such as aquaculture. Coastal defences cannot provide total protection against severe storms, and computer-based techniques such as Digital Elevation Models (DEM) and Geographical Information Systems (GIS) are employed to assess the extent of coastal inundation risk (Colby *et al.*, 2000; Zerger *et al.*, 2002). Large sea waves result in the severe scouring of beaches, with the inevitable undermining of adjacent roads and buildings.

Hazard-resistant design is important for reducing property damage in wind-storms. The key to damage limitation lies in adequate, and properly enforced, building codes. A comparative study of hurricanes in Texas and North Carolina showed that nearly 70 per cent of the damage to residential property was due to poor enforcement of the building codes; 25–40 per cent of the insured losses associated with hurricane 'Andrew' were also due to non-compliance with building

Box 9.2 The dream of severe storm suppression

Hurricane modification is the most attractive of all severe storm suppression goals. The destructive power of a tropical cyclone increases rapidly with the maximum wind speed and it has been estimated that a 10 per cent reduction in wind speed would produce an approximate 30 per cent reduction in damage. Attempts at weather modification in the United States started in 1947 and culminated with Project STORMFURY, starting in 1962 (Willoughby *et al.*, 1985). The theory was that the introduction of silver iodide into the ring of clouds around the storm centre would cause existing supercooled water to freeze, thereby stimulating the release of latent heat of fusion within the clouds. It was believed that this would lower the maximum horizontal temperature and pressure gradients within the eyewall of the storm, reduce convergence and lessen the core wind speeds. Unfortunately, the computer models overestimated the amount of supercooled water available. Project STORMFURY was discontinued in 1983. Other theories for hurricane control have included spraying the ocean surface with a liquid evaporation suppressant and artificially reducing the sea-surface temperatures, either by pumping up colder water from the ocean depths or by towing icebergs from the Arctic.

Since medieval times hail suppression has been attempted in the alpine countries of Europe by firing cannon and ringing church bells when thunder clouds appear. There is no scientific basis for such methods, other than a suggestion that explosions may propagate pressure waves in the air sufficient to crack and weaken the ice making up the hailstone and thus prevent large hailstones from forming. Most hail-suppression technology has relied, like hurricane modification, on cloud seeding with ice nucleants. The theory is that the introduction of artificial ice nuclei, such as silver iodide, will introduce competition for the supercooled water droplets that hailstone embryos feed on. The expectation is that, although the total number of ice particles will increase, individual hailstones will grow to a smaller size and do less damage when they reach the ground. Hail clouds have been seeded with silver iodide by a variety of methods, including ground-based generators, over-flying aircraft which drop pyrotechnic flare devices and artillery shells.

In practice, weather modification fails to satisfy the following criteria:

- *Scientific feasibility.* This requires a more complete understanding of the microphysical processes within clouds. For example, it is thought that hurricane clouds contain too little supercooled water, and too much natural ice, for artificial nucleation to be effective even with high doses of seeding agents.
- *Statistical feasibility.* Not enough sample storms are available for treatment in any area in order to provide the statistical proof that experimental results differ from naturally occurring changes in factors such as hurricane wind speed or hail intensity.
- *Environmental feasibility.* Quite apart from the moral issue of interfering with atmospheric processes with an incomplete state of knowledge, most seeding agents or ocean temperature-reduction methods would create a pollution threat.
- *Legal feasibility.* Several hail-suppression programmes in the USA have attracted lawsuits. In some cases the complainants have alleged that their right to natural precipitation has been diminished by a reduction in rainfall, whilst in others it has been claimed that the seeding has increased storm damage.
- *Economic feasibility.* Although highly favourable estimates of the cost:benefit ratio for successful severe storm modification helped to release funds for cloud-seeding experiments in the past, the promised savings have never been realized.

codes (Mulady, 1994). Uneven code enforcement exists because of lack of funds and poor training of site inspectors. Under the EU *Eurocode Directive*, each country has its own structural building codes, generally based on the wind speed gusts of a design storm with a 1:50-year return interval. These codes reflect local building practices so that, for example, the building stock in northern parts of the UK, especially Scotland, should be more resistant to wind damage than in areas further south. When extreme winds hit regions which are less prepared, damage can be extensive. For example, by northern European standards, houses in the south of France tend to be built of relatively light masonry and stucco, with low-to-medium pitched roofs covered with tiles and weak connections between the roof coverings and the roof framing (Kafali, 2011).

The causes of wind-induced building failure are well known. Typically, shingles and other roofing materials are disturbed by wind pressures, a process which then allows rain to penetrate the building and cause additional damage. Most of these losses could be avoided if a relatively small amount of additional money were spent during construction. Better water-proofing of the roof, the use of hurricane clips (rather than staples) for fastening roof cladding and roof sheathing and the fitting of storm shutters would do a great deal to reduce damage (Ayscue, 1996). When storm proofing a property, attention should be paid to three components:

- *The roof and walls.* Houses with gabled roofs are prone to loss in high winds and can be strengthened by installing additional braces in the trusses and at the gable ends. Hurricane nails can be placed into the wall to wrap over the trusses for added resistance. Heavy roofing materials should be used. External walls of reinforced concrete can typically withstand hurricane winds and high-speed debris better than those of wood construction. Homeowners in hurricane- or tornado-prone zones should designate one 'strong room' for personal protection.

- *Windows, doors and other openings.* Most openings are susceptible to failure under wind forces or from impact by blowing debris. Such failure may then destroy the structural integrity of the building. Windows can be constructed with plastic panes or shatter-proof glass and can also be fitted with storm shutters. To prevent doors from being blown in, extra-strong bolts and pins are needed.

- *Foundations.* Mobile homes should be securely attached to concrete foundations. In coastal areas and floodplains, properties should be elevated above the expected maximum storm surge or flood level. This can be done by placing buildings on pilings made of wood, concrete or steel. The height necessary will vary over small distances, depending on the site location in relation to beach or barrier-island topography, but may be more than 5 m in some cases.

Attitudes to storm risk in the United States have changed. Following hurricane 'Hugo' in 1989, Surfside Beach became the first community in South Carolina to adopt the high-wind design standard in the Southern building code. After hurricane 'Andrew' the South Florida building code was strengthened. All new buildings should now be constructed with permanent storm shutters and protected from wind-borne debris. Exterior windows or shutters must pass a missile impact test with a 4 kg piece of timber striking at a speed of 15 m s^{-1} and shingle and tiles must be tested as a system at 49 m s^{-1}. In addition, the Florida state legislature requires that all new educational facilities in the state are designed to serve as public hurricane shelters. In March 2002, Florida adopted a building code imposing stricter standards against wind hazard, but, as so often happens, the legislation does not apply to the existing building stock and some code exemptions were agreed. In the LDCs the gradual switch from wooden to masonry buildings has helped to reduce hurricane damage, but a damage survey in

Andhra Pradesh state, south India, confirmed the need to design more buildings with hipped, rather than gabled, roofs and to make all types of roof covering more secure (Shanmugasundaram *et al.*, 2000).

In part, these developments reflect increasing knowledge about the patterns of wind-storm damage to residential property, a process which continues to foster better building techniques. Huang *et al.* (2001) collected almost 60,000 insurance claims made in South Carolina (after 'Hugo') and in Florida (after 'Andrew') and related the mean surface wind speed to the number of claims and the degree of damage for each zip (post) code. Figure 9.11a shows the aggregated claim ratio (total claims divided by total policies) and reveals that, whilst few policy-holders claim with winds less than 20 m s^{-1}, nearly all will file a claim when the speed reaches 30 m s^{-1}. Figure 9.11b graphs the wind speed against the aggregated damage ratio (amount paid by the insurer divided by total insured value) and shows that the degree of loss increases markedly with speeds more than 35 m s^{-1}. Finally, Figure 9.11c uses a long-term risk model to simulate the decline of annual losses away from the coast. The highest risks are about 2 per cent. This means that homes out on the barrier islands can, on average, expect to be damaged up to 100 per cent of their total insured value every 50 years. In comparison, only 20 km inland the risk falls to 0.2–0.3 per cent – only about one-tenth of that on the coast. Such information is useful to insurers setting policy premiums and to land planners concerned with hazard zoning.

H MITIGATION

1 Disaster aid

Tropical cyclones create major disasters in developing countries, and emergency aid alone is rarely adequate. For example, despite an injection of funds totalling over US$123 million after hurricane 'Mitch' in 1998, short-term relief covered

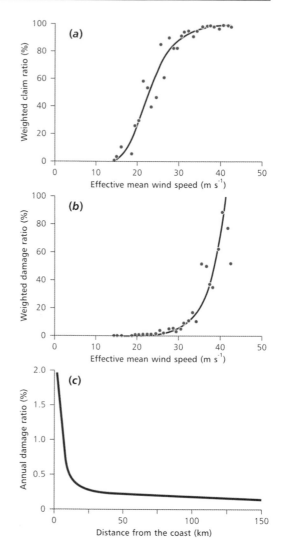

Figure 9.11 Hurricane losses to residential structures in the south-eastern USA: (a) insurance claims ratio compared to the effective mean surface wind speed; (b) building damage ratio compared to the effective mean surface wind speed; (c) expected annual damage ratio compared to distance from the coast. After Huang *et al.* (2001). Reprinted from *Reliability Engineering and System Safety* 74, Z. Huang *et al.* Long-term hurricane risk assessment and expected damage to residential structures, 239–49. Copyright (2001), with permission from Elsevier.

less than 10 per cent of the losses. Longer-term support is needed to restore damaged infrastructure. Tropical islands often suffer failures of electricity supply, due to the destruction of overhead power lines, and the lack of refrigeration poses a public health risk. After hurricane 'Gilbert' struck Jamaica in 1988, 40 per cent of the electricity transmission system, plus 60 per cent of the distribution system, was made unserviceable. It took several weeks to restore supplies (Chappelow, 1989). Aid distribution in multi-ethnic communities may be hampered by poverty, illiteracy, gender and minority status. For example, only 20 per cent of the population in Florida City (which is mainly black) applied for aid after hurricane 'Andrew', even though over 80 per cent of the homes had been destroyed. In Homestead, a mainly white area, 90 per cent of the population applied for aid and 80 per cent of such applications were successful (Peacock *et al.*, 1997).

Emergency aid is offered on the basis that appropriate donations will be welcomed, and distributed to those in need as efficiently as local conditions allow. This is not always so, especially when governments prioritize state security above human rights. Tropical cyclone 'Nargis' struck Burma (Myanmar) on 2 May 2008, to create the worst disaster recorded in that country's history. At least 138,000 fatalities resulted, placing this Category 4 storm in the top 10 deadliest cyclones of all time. Infrastructure damage was estimated at more than US$10 billion. On 6 May, Myanmar's official representative in New York requested help from the UN, by which time the international community had already started to respond. However, all relief efforts were held up by the ruling military junta. On 9 May the government agreed to accept financial aid, plus food and medicines, some of which were stock-piled – and deteriorating – in neighbouring countries, but continued to prevent the arrival of relief planes and ships. The situation eased later in the month and, eventually, help was supplied by 46 countries. Throughout, the main problem was the unwillingness of the government to issue entry visas and allow aid workers into the country (McGregor, 2010). Despite the existence of a UN principle that confirms the responsibility of the international community to protect people if their own state fails to fulfil these obligations, there is – at present – no legal right to cross national boundaries to provide emergency relief following a disaster.

2 Insurance

Tropical cyclones worldwide account for at least US$15 billion in insured losses every year. Much occurs in countries where disaster insurance is very limited. Only about 2 per cent of the losses imposed on Central America by hurricane 'Mitch' were covered by insurance. Conversely, residents of some developed countries have access to *all-risk policies* for homes, house contents and motor vehicles that cover a variety of storm-related losses. These packaged policies can make it difficult to identify the economic impact of individual weather perils on domestic property. More specialized policies are available for certain economic activities: hail insurance against crop damage, for example, is common in North America.

Eight of the ten most expensive federally declared disasters in the USA, largely driven by inappropriate coastal development, were hurricanes. Most of the structural damage to buildings was wind related. For example, insurers paid out US$2.6 billion for wind-related damage caused by hurricane 'Hugo' in 1989, whilst only 10 per cent of the total insured loss was flood related. For a few hurricanes, such as 'Opal' in Florida and Alabama in 1995, the major losses were due to storm surge, which also dominated the impact of 'Katrina'. Bush *et al.* (1996) demonstrated how storm surge property damage can be increased in the lowest-lying areas by the prior removal of dunes, as compared to the losses suffered by houses elevated on pilings or set back behind sea walls. Hurricane 'Andrew' in 1992 acted as a 'wake-up' call to the insurance industry. Some companies became reluctant to underwrite further cover in parts of Florida, whilst state catastrophe funds were used to help residents unable to buy policies on the private market. Since then, companies have become more interested in hurricane climatology, and predictions of storm activity in the forthcoming season are now routinely prepared for the insurance industry (Saunders and Lea, 2005).

According to Bouska *et al.* (2005), the direct cost of hurricane 'Katrina' to the insurance industry from wind, storm surge and flooding, excluding losses insured under the National Flood Insurance Program, was US$40–50 billion. This made the storm the most expensive disaster in American history. A breakdown of the costs of 'Katrina' shows that about half the loss was attributable to commercial property, with a further third due to claims for residential and personal property (Table 9.7). The off-shore costs were significant in this event, partly because many boats were destroyed but also because around 250 oil production platforms in the Gulf were damaged. Hurricane 'Katrina' also sparked a wind/water controversy in the industry, since flood damage due to rising water is excluded from home-owners' insurance policies because cover is available under the National Flood Insurance Program (NFIP). Florida alone accounts for almost 40 per cent of the entire NFIP portfolio.

Despite the advantages of storm-hazard insurance, there is a concern that its availability has increased the demand for coastal homes and even raised property values in some hazard zones. This is because the presence of insurance may encourage either the perception that storm events are very rare or the more cynical judgement that the certainty of compensation more than balances any personal financial risk. Since 1968, shoreline construction in the USA has been guided by the federal government through the selling of insurance to waterfront property owners through the *National Flood Insurance Program* (NFIP). Thirty years ago, home-buyers and estate agents in the Lower Florida Keys believed that the availability of flood insurance made residents more willing to

Table 9.7 Hurricane 'Katrina' estimated insured losses (US$ billion)

Category of insurance	Low estimate	High estimate
Personal property		
Residential property	14.0	17.0
Personal auto	1.0	2.0
Personal watercraft	0.2	0.3
Total	**15.2**	**19.3**
Commercial property		
Commercial property (excluding off-shore)	13.5	16.0
Business interruption (other than marine and energy)	6.0	9.0
Commercial auto	0.2	0.3
Total	**19.7**	**25.3**
Marine and energy	4.0	6.0
Liability	1.0	3.0
Other	0.0	1.0
Total all lines	**39.9**	**54.6**

Source: Bouska *et al.* (2005)

locate in flood-prone areas and also made it easier to sell at-risk property (Cross, 1985). One result was a large number of repeat claims. About 40 per cent of all claims under the NFIP have been for properties flooded at least once before. In this situation, commercial insurers either withdraw from the residential market or lobby central government for the adoption and enforcement of more stringent planning and building codes.

I ADAPTATION

1 Community preparedness

Effective public response to severe storm warnings depends on community preparedness. Most agencies in the MDCs charged with raising hazard awareness, such as FEMA in the USA and Emergency Preparedness Canada, produce leaflets and other materials that offer advice on planning for wind-storm emergencies, including the preparation of an emergency pack, the trimming of dead or rotting trees around the home, the prior choice of a shelter (such as a basement or place beneath the stairs), plus the designation of a rendezvous point where separated members of a family could meet. In the case of tornadoes, sheltering indoors within inner rooms well away from windows and getting close to the floor is important. If caught outdoors, people are advised to leave their cars and seek shelter in a ditch or other depression. Within some tornado-prone areas, such as the Midwest of the USA, substantial public buildings are clearly identified as tornado shelters. The level of storm preparedness by individuals and organizations is patchy. Most households in Florida surveyed by Baker (2011) demonstrated a capability for subsisting on their own for a three-day period after a disaster, due to adoption of the various measures shown in Figure 9.12. On the other hand, businesses surveyed in Sarasota County, Florida, showed considerable variations in their approach to storm surge risk. The highest levels of anticipation were found amongst the larger businesses and those owning their own premises; interestingly, these businesses were already located in areas less exposed to hurricanes (Howe, 2011).

Emergency planning for cyclone disasters within the LDCs has a long, and relatively successful, history. In 1980 the *Pan Caribbean Disaster Preparedness and Prevention Project* (PCDPPP) was the first regional scheme to be established. The Project concentrated on technical assistance, the training of island nationals in emergency health and water supply provisions, and the preparation of training materials. These plans were initially tested by hurricanes 'Gilbert' in 1988 and 'Hugo' in 1989. On the island of Jamaica, fewer deaths appeared to result. Thus, hurricane 'Charlie' in 1951 created 152 deaths, compared with 45 from 'Gilbert' in 1988, despite the fact that 'Gilbert's' damaging winds lasted longer and affected more of the island, which also

Table 9.8 Numbers of people killed and evacuated during tropical cyclone emergencies in Bangladesh during the 1990s

Year	Number killed	Number evacuated	Deaths as percentage of evacuees
1991	140,000	350,000	40.00
1994	133	450,000	0.03
1997 (May)	193	1,000,000	0.02
1997 (Sept.)	70	600,000	0.01
1998	3	120,000	0.0025

Source: After IFRCRCS (2002)

had a higher population in 1988 than in 1951.

Following the devastating 1970 cyclone in Bangladesh, the government intervened. Multi-storied cyclone shelters, capable of housing 500–2,500 people, were constructed, along with raised-earth platforms (*killas*) which act as refuges for 300–400 livestock. Table 9.8 shows that disaster awareness workshops and other events organized through the Cyclone Preparedness Programme (CPP) and the Cyclone Education Project have saved many thousands of lives during subsequent

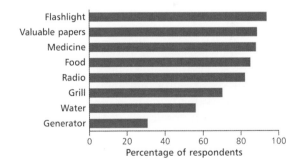

Figure 9.12 Percentage of households in Florida reporting the adoption of preparedness measures for hurricanes. These values indicated a relatively good response, as compared with levels of preparedness found for some other environmental hazards. After Baker (2011). Reprinted from Applied Geography 31, E.J. Baker, Household preparedness for the aftermath of hurricanes in Florida, 46–52. Copyright (2011), with permission from Elsevier.

storms (Southern, 2000). The hazard-prone inhabitants rely on warnings issued through Asia's largest radio network, before being advised about action at the village level by many volunteers. Paul and Rahman (2006) showed that most people have confidence in the system and an estimated 3 million people were safely evacuated before cyclone 'Sidr' made landfall in 2007.

2 Forecasting and warning

Forecasting and warning systems exist for most storm hazards and are increasingly important in preserving life. Most national weather agencies in the MDCs have systems for hurricanes, floods, tornadoes and severe thunder-storms. The warning products are distributed to a range of federal, state and local authorities and to the public via the media and the internet. Depending on the type of storm, forecasts are typically available on a variety of time-scales: long range (more than 10 days), intermediate range (3–10 days), short range (1–3 days), very short range (a few hours) and 'nowcasts' (events in progress). For example, 26 out of the total 39 National Hydro-Meteorological Services in Europe issue severe thunder-storm warnings and 8 provide tornado warnings (Rauhala and Schultz, 2009). Lead-times range from 30 minutes to 96 hours, reflecting a variety of warning philosophies for each country.

Most forecasting agencies operate a tiered release of public information based on storm 'watches' and storm 'warnings'. For example, tropical cyclone warning is likely to progress through a *watch phase*, initiated 48 hours before storm-force winds are expected to reach the coast, to a *warning phase*, when storm winds are expected within 24 hours, and finally a *flash message phase*, if any significant changes occur. Near landfall, warnings are issued hourly and contain information on both storm surge and rainfall. Some rapidly deepening depressions in the northern hemisphere are difficult to forecast accurately because the standard models used by weather forecasters under-predict the rate of deepening (Sanders and Gyakum, 1980). Local processes may be involved. The break-up of deep clouds can produce turbulent eddies that contribute to the strength of gusts at ground level, to such an extent that it results in the highest gust speeds in wind-storms.

Early storm detection and continued monitoring are necessary for effective forecasting and warning. Major improvements have followed the introduction of remote observing techniques such as geostationary and polar-orbiting satellites, automated land and sea-surface observations and Doppler radar. The data can be linked to computer graphics capable of supplying storm imagery (e.g. for wind speed and sea surge conditions) sometimes for days in advance of a disaster strike. Dynamical models of the atmosphere, using increasingly fine grid resolutions, are then combined with statistical models of the historical behaviour of similar storms to provide predictions of the track, forward speed and intensity of the developing storm with sufficient lead-times for emergency action.

Satellite sensing allows developing cyclones with well-formed 'eyes' to be located out at sea to within 30–50 km. When a cyclone is about 250 km offshore, weather radar permits a more accurate fix on the position, probably to within 10 km. In the USA, the National Hurricane Center (NHC) in Miami maintains continuous real-time monitoring of tropical cyclones and issues forecasts from 120 hour ahead down to 6-hour predictions of the central position, the extent, intensity and track. A *hurricane watch* is issued to advise a specified coastal sector that it has at least a 50 per cent chance of experiencing a tropical cyclone of hurricane force within 36 hours. Improvements in storm forecasting and warning systems have been necessary to keep pace with the growth in coastal vulnerability. A *hurricane warning* provides similar advice about a landfall expected within 18–24 hours.

Figure 9.13 shows how the average accuracy of hurricane forecasts issued by the NHC has improved in recent decades for different forecast periods. Track errors have been reduced to about 160 km (24 hours), 260 km (48 hours) and 370 km (72 hours) and wind speed errors have reduced to 9 kt (24 hours), 15 kt (48 hours) and 19 kt (72 hours). Today a three-day track forecast is as accurate as one issued for two days in the late 1980s and an intensity forecast has errors 20 per cent smaller than in the mid-1970s. Forecasting skill for hurricane 'Katrina' was generally good. The track forecasts were significantly better than the most recent 10-year average (1995–2004), with lead-times for watches and warnings

eight hours longer than average, but the intensity of the storm was under-forecast (National Weather Service, 2006). A survey of residents within 30 miles of the coast in the Miami area of Florida indicated a willingness of the public to pay for further improvements in forecast information on the timing, landfall, storm surge and wind speed characteristics of future storms (Lazo and Waldman, 2011).

Storm surge conditions are normally forecast using a variant of the SLOSH computer model (Sea, Lake and Overland Surges from Hurricanes)

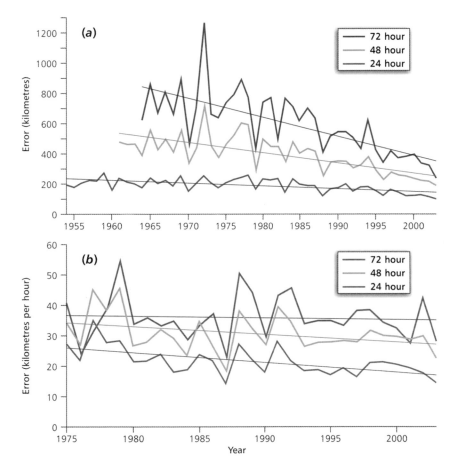

Figure 9.13 The average annual accuracy of Atlantic hurricane forecasts issued by the USA National Hurricane Center: (a) forecasts of hurricane track 1954–2005; (b) forecasts of wind intensity 1975–2003. Major improvements have occurred in the accuracy of hurricane path forecasts, but intensity forecasts have improved more slowly. Data from NOAA at www.aoml.noaa.gov/hrd/tcfaq/F6.html (accessed 10 October 2006).

that incorporates five meteorological factors: wind speed, central pressure, size (radius), forward speed and track direction of the hurricane. The calculations take into account local features: shoreline configuration, near-shore water depth (including tidal data) plus built features like roads and bridges. The SLOSH procedure is normally accurate within ±20 per cent, so that a forecast peak of 3.0 m could be expected to produce a maximum height between 2.4 and 3.6 m. For individual storms, SLOSH performance is highly dependent on the accuracy of the forecast storm track model. If the landfall prediction is in error, the surge height is unlikely to be the same as forecast at a different geographical location.

Accurate forecasting of hurricane landfall is crucial for efficient public warning and the initiation of evacuation procedures. The SLOSH model is used for defining areas subject to emergency evacuations. Generally, bulletins advising people at risk about evacuation are issued alongside forecasts, because medium-sized cities require about 12 hours to evacuate, whilst large cities, like New Orleans, Miami and Houston, need a minimum 72 hours of advance-notification time (Urbina and Wolshon, 2003). At present, hurricane warnings are issued for sections of coast averaging 560 km in length. This is partly because damaging winds extend beyond the centre of the storm, but also because of doubts about the exact landfall position. Since hurricane winds cause damage over a 190 km-wide area, about two-thirds of the coastal sector is over-warned and therefore incurs unnecessary preparation and evacuation costs (Powell, 2000). Problems attached to emergency evacuation are explored in Box 9.3.

Forecasting and warning for tornadoes operates over smaller scales of time and distance. At best, tornado watch programmes linked to Doppler radar systems can allow communities up to 2–3 hours to prepare for the event and seek shelter. In the case of the Midwest tornadoes in early May 2003, warnings issued mainly via sirens provided lead-times of 10–20 minutes only, although almost all residents who received warnings were able to take immediate shelter (Paul et al., 2003). Warnings were generally less effectively disseminated in rural areas. A more widespread use of NOAA weather radios can warn people who are asleep (even when the electricity supply fails) and would help to remedy this situation and improve overall warning capability. Given the continuing devastation from tornadoes in Bangladesh, such as the 111 people killed by a severe event in April 2004, there is a very good case for the introduction of a tornado forecasting and warning system in this country (Paul and Bhuiyan, 2004).

3 Land use planning

Much structural damage to property, either from hurricane winds or storm surge forces, occurs in foreshore dune areas or to houses within 100 m of the shore. Although restrictions on near-shore development are important for reducing losses, changes in land use fuelled by the perceived desirability of waterfront locations continue to expose people and buildings to risk. The coast of the United States is subject to a variety of federal, state and local laws. Land use planning through zoning ordinances has operated in the vulnerable Lower Florida Keys area for well over 40 years but has had comparatively little influence on residential development. Since 1975, newly constructed houses have floors built at least 2.4 m above mean sea level (the 1:100-year flood level), in order to comply with National Flood Insurance Program. Although 90 per cent of new residential construction is elevated on stilts, the protection offered has been eroded by the building of enclosed garages and recreation rooms in the space below the property. Moreover, ground-level houses remain an attractive option, especially for many elderly residents who claim they would rebuild at the same location if the property were ever destroyed.

Where damaged property has been rebuilt, stronger building codes can be applied, although set-back compliance can be limited, with shoreline

Box 9.3 Improving hurricane evacuation in the United States

Achieving efficient emergency evacuation is a recurring problem. For example, over 75 per cent of residents in Bangladesh at risk from cyclone 'Sidr' in 2007 were aware of an evacuation order but failed to leave their homes (Paul and Dutt, 2010). In the USA, hurricane warnings can trigger coastal evacuation authorized by the state Governor. Evacuation orders can be *Voluntary*, *Recommended* or *Mandatory* but even Mandatory orders can be difficult to enforce, with variable consequences. Residents in hurricane-risk areas are not unaware of the threat. Peacock *et al.* (2005) found a strong relationship between risk perception and wind-hazard zones in Florida, and most coastal residents in Texas correctly identified their own risk zone (Zhang *et al.*, 2004). As a result, maximum evacuation rates of over 90 per cent have been achieved for communities living on beaches and barrier islands. Although only 54 per cent of households threatened by hurricane 'Andrew' evacuated entirely, the proportion rose to over 70 per cent in the lowest-lying coastal zone (Peacock *et al.*, 1997).

Failures to evacuate arise because people take many factors into account when considering action, including: environmental conditions, e.g. day or night timing; social cues, e.g. neighbour behaviour; and perceived impediments e.g. availability and quality of shelters (Lindell *et al.*, 2005). Many South Carolina residents who did not comply with the 1999 mandatory evacuation order during hurricane

Plate 9.4 Highway congestion on route 37 heading north-west to San Antonio as residents evacuate from Padre Island and Corpus Christi, Texas in order to escape hurricane 'Bret' in August 1999. (Photo: Dave Gatley, FEMA 316)

'Floyd' rated household circumstances and previous experience as more important to their decision making (Dow and Cutter, 2000). Non-evacuees are less likely to respond to future evacuation orders, especially if problems existed for those who did evacuate, like the traffic congestion that delayed the return of residents to the Florida Keys after hurricane 'Georges' in 1998 (Dash and Morrow, 2000). In summary, people at risk do not always act as emergency planners wish.

Improving the evacuation process is important for the USA, given the increase in coastal population density. In 1999 3 million people fled from hurricane 'Floyd'. This created traffic gridlock, despite the use of special contra-flow lanes on the main routes, and it is clear that more sophisticated traffic management will be required in future (Urbina and Wolshon, 2003; Wolshon et al., 2005). The 'shadow' evacuation of up to 10–20 per cent of additional households who are not under direct threat but leave anyway increases pressure on the evacuation routes. In addition, the limited availability of public shelters, the long lead-times – relative to the length of hurricane warnings – necessary for communities to be evacuated and the existence of groups without their own vehicles all create problems. For example, tourists are likely to be a low-mobility group lacking experience of previous evacuations and emergency procedures.

A better understanding of evacuation behaviour is crucial. Not surprisingly, Mandatory, rather than Voluntary, orders are needed to encourage most people to evacuate, including those at high risk, but storm intensity and the perceived risk of flooding – rather than the risk from high winds – are factors (Whitehead et al., 2000). Discrepancies clearly exist between the needs of the public and the priorities of emergency managers. Special problems exist. The Lower Florida Keys are over 100 km from the closest mainland. As much as six hours before hurricane landfall, a storm surge may start to flood low points on the highways, whilst the highway network would struggle with the traffic flows if large numbers of people had to be evacuated along US Highway 1.

Hurricane 'Katrina' highlighted the need for more care provision for evacuees. When people are displaced they suffer physical and emotional stress. These problems are intensified for the disadvantaged, such as low-income groups without health insurance cover. It is important to appreciate the scale of the emergency communications and support services required. During the 2005 hurricane season, the American Red Cross opened 1,400 shelters and provided 3.8 million overnight stays, together with more than 68 million meals and snacks. Eight months after the storms more than 750,000 evacuees were still displaced from their homes throughout the USA.

structures remaining vulnerable to beach erosion (Platt et al., 2002). Subsidized shore-protection schemes, such as those undertaken by the US Army Corps of Engineers, have been criticized for encouraging development near beach areas, but a study in Florida showed no influence of these schemes on either house prices or development activity (Cordes et al., 2001). This may be because stricter land use controls linked to shore protection offset the benefits expected from reduced storm damage. In addition, some hurricane-prone communities have started to slow their population growth. According to Baker (2000), the island of Sanibel, Florida accepted restrictions on the annual number of new housing units as early as 1977, citing restrictions on emergency evacuation as the prime reason. Since then, the state has regulated residential and commercial development more firmly than in the past.

FURTHER READING

Elliott, J.R. and Pais, J. (2006) Race, class and Hurricane Katrina: social differences in human responses to disaster. *Social Science Research* 35: 295–321. A case study with much wider implications.

Lindell, M.K., Lu, J-C. and Prater, C.S. (2005) Household decision-making and evacuation in response to Hurricane Lili. *Natural Hazards Review* 6: 171–9. Another demonstration of issues facing emergency managers.

Olshansky, R.B. (2006) Planning after Hurricane Katrina. *Journal of the American Planning Association* 72: 147–54. Raises questions not always asked following a major disaster.

Pielke, R.A. Jr and Pielke, R.A. Sr (1997) *Hurricanes: Their Nature and Impacts on Society.* J. Wiley and Sons, Chichester. This remains the best introduction to tropical cyclones and their hazards.

Schwartz, R.M. and Schmidlin, T.W. (2002) Climatology of blizzards in the coterminous United States 1959–2000. *Journal of Climate* 15: 1765–72. A reminder of the disruption due to severe winter storms.

Shanmugasundaram, J., Arunachalam, S., Gomathinayagam, S., Lakshmanan, N. and Harikrishna, P. (2000) Cyclone damage to buildings and structures – a case study. *Journal of Wind Engineering and Industrial Aerodynamics* 84: 369–80. This is a careful and instructive example of practical problems.

WEB LINKS

Hurricane Insurance Information Centre www.disaster information.org/

NOAA hurricane 'Katrina' web portal www.katrina. nooa.gov/

US National Hurricane Center www.nhc.noaa.gov/

US National Oceanic and Atmospheric Administration www.noaa.gov

Environmental and Social Impacts Group, US National Center for Atmospheric Research (NCAR) www.esig.ucar.edu/sourcebook

Seasonal hurricane forecasts for the Atlantic basin www.typhoon.atmos.colostate.edu

CHAPTER TEN

Weather extremes, disease epidemics and wildfires

10

A INTRODUCTION

Major fluctuations in weather and climate create environmental hazards. Most obvious are the 'mainstream' atmospheric hazards, such as severe storms, floods and droughts. However, variations beyond the normal range interact with living organisms, including humans, to produce biologically related hazards. The most obvious examples occur when extreme spells of hot or cold weather threaten human life through physiological stress. These hazards, like others in this chapter, have a biological component. They are multifactorial in origin: the health risk from thermally related physiological stress is not determined by atmospheric processes alone. This statement applies equally well to epidemics of infectious diseases and outbreaks of wildfires. That having been said, all so-called *biophysical hazards* have much in common, including a sensitivity to weather and climate conditions and an ability to create rapid-onset threats similar to other hazards found in this book.

One important feature of biophysical hazards is a recurring pattern of inter-seasonal and/or inter-annual variation. Thus, cold stress- and heat

stress-related mortality in temperate latitudes are strongly linked to the winter and summer seasons, respectively; wildfires tend to occur at the end of a dry summer in Mediterranean climates, when vegetation burns easily and weather conditions encourage fires to spread out of control. It is true that many infectious diseases have epidemic cycles unrelated to external environmental factors. But when the pathogen develops in the environment – or in an intermediate host or vector – rather than in the human body, it comes under the influence of the prevailing weather conditions. Most viruses, bacteria and parasites are unable to complete their development if the ambient temperature is below a certain threshold, e.g. 18°C for the malaria parasite. Conversely, temperatures slightly above a given threshold will shorten the time needed for development and increase reproduction rates and the risk of a disease outbreak.

After abnormal weather events, there is often a rapid upsurge in pests and diseases with the potential to threaten human food security. An outbreak of potato blight disease (*Phytophora infestans*) in mid-nineteenth-century Ireland was triggered by unusually warm and wet weather in 1845. This led to at least 1.5 million famine-related deaths over the following three years and the emigration of about 1 million people. Like most organisms, pests produce more offspring than are required for replacement of the adults when they die, and the stability of populations is maintained through high juvenile mortality. If the environmental factors controlling juvenile mortality are eased, many species have the capacity to attain plague proportions. This is most likely to occur in arid and semi-arid areas that can be transformed by rains.

The desert locust (*Schistocerca gregaria*) responds quickly to conditions that enable it to switch from a solitary phase to a swarming phase. Rain encourages the female to lay her eggs in wet ground and provides vegetative growth to feed the immature, wingless locusts after they have hatched. Once they become adult, winged locusts migrate between areas of recent rainfall, sometimes covering thousands of kilometres in a matter of weeks. Typically, they fly at night, aided by low-level winds. About one-fifth of the land area of Africa, the Middle East and south-west Asia is at risk. A small part of a swarm, about one tonne of locusts, can consume as much food in one day as about 2,500 people; a 1 km-wide strip

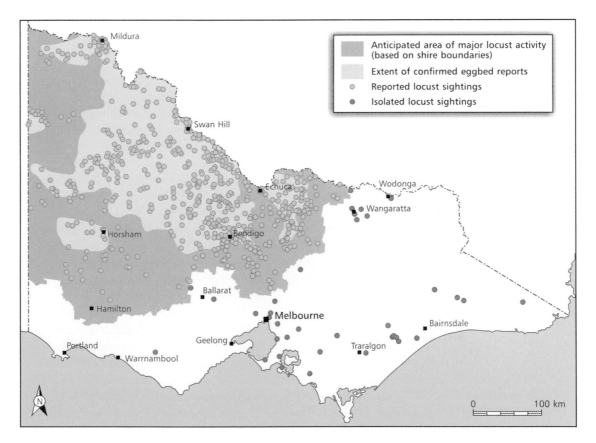

Figure 10.1 Map of reported locust activity in the State of Victoria, Australia between November 2009 and August 2010. The average time between egg-laying and the emergence of an adult locust is 6–8 weeks although high temperatures speed up the process. After Department of Primary Industries, State Government of Victoria at http://www.dpi.vic.gove. au/agriculture/ (accessed 6 February 2012). © State of Victoria Department of Primary Industries 2010.

of swarming locusts can consume 10 tonnes of crops – approximately one-third of their combined bodyweight – within 24 hours.

In 2004, locust invasions of the Sahel region were the worst for 15 years, following a very wet winter in the mountains of north-west Africa (IFRCRCS, 2005). In total, 13 West African counties were affected. Due to the loss of cereal crops and animal fodder, 9 million people were short of food. About 4 million ha of land were treated, largely unsuccessfully, with expensive airborne insecticide sprays in an attempt to control the infestation. Australia has a native species of plague locust that mainly affects parts of New South Wales, Queensland and adjacent areas of South Australia. During plague years, rural industries suffer heavy losses to pasture and crops. Following a series of drought years, warm, wet summers in 2009–10 and 2010–11 caused widespread swarms. Figure 10.1 shows the location of reported locust activity in the state of Victoria between November 2009 and August 2010. Detailed monitoring is a necessary preliminary to the use of local control methods by farmers, including the deployment of light aircraft with insecticide sprays. Locust control measures are reasonably cost-effective in developed countries. For example, Millist and Abdalla (2011) estimated that during the 2010–11 control season some A\$963 million of potential agricultural loss was averted, at an overall favourable cost-benefit ratio of 1:20 (i.e. A\$20 was saved for every dollar spent).

B EXTREME TEMPERATURE HAZARDS

The average human body is most efficient at a core temperature of 37°C. Compared with the natural variations of ambient air temperature, physiological comfort and safety can be maintained within only a relatively narrow thermal range. Irreversible deterioration and death frequently occurs if the internal body temperature falls below 26°C or rises above 40°C. The epidemiology of temperature-related death is complex. Barnett et al. (2010) concluded that, whilst relatively hot and cold temperatures do increase mortality around the world, it was difficult to identify the best temperature measure, due to wide variations in test results between age groups, seasons and cities. In Montreal, Canada, a clear link between summer mortality and daily maximum temperatures above 27°C was established by Goldberg et al. (2011), although no equivalent relationship existed for cold stress in the winter. In many cases, thermal effects are obscured by other factors such as air pollution, timelags in response and socio-economic conditions such as housing conditions, ill health in the population and other demographic indices.

1 Cold stress

Cold stress can create physiological damage in the form of hypothermia or frostbite. The effects of low temperature alone are compounded by wind and moisture, so that *windchill*, for example, is caused by the combination of low temperature and high wind speed. Outside the tropics, most temperature-related mortality is associated with outbursts of cold, Arctic air into the mid-latitudes during spells of severe winter weather. A regular pattern of excess winter deaths is found in many industrialized countries when mean daily temperatures fall below 18°C, an impact at least as severe as the more publicized loss of life due to heat stress.

Within Europe, the highest excess winter death rates are in Portugal, Spain and Ireland (rather than Scandinavia) and have been linked to poverty and poor housing conditions (Healy, 2003). The UK also fares badly in comparison to countries with colder climates. Wilkinson et al. (2001) found that deaths in England from heart attacks and strokes were 23 per cent higher during December and March than in other months. Mortality rose by 2 per cent with every 1°C fall in outdoor temperature below 19°C. For England and Wales, total excess winter mortality has been estimated at 30,000 per year. In Scotland there is a difference of 30 per cent between the summer trough of weekly death rates and the winter peak (Gemmell *et al.*, 2000). In Spain, Montero *et al.* (2010) found that mortality was linked to the duration of cold spells and high relative humidity. Such periods towards the end of winter created the greatest mortality.

Generally, the highest cold-related death rates occur during the most severe weather, which is often associated with winter anticyclones. These events also favour the trapping of high concentrations of air pollution in the cold air of the lower atmosphere, a contributory factor to excess deaths resulting from hypothermia and cold-aggravated illnesses. Older people (over 65 years) living in poorly heated houses with low energy efficiency are at greatest risk. This has been attributed to the adverse effect of low temperatures on existing underlying illness, such as influenza, coronary thrombosis and respiratory disease, but it has proved difficult to establish strong links between winter mortality and deprivation. Other deaths have been attributed to short-term outdoor exposure, rather than low temperatures indoors, a factor that increasing levels of car ownership might well reduce. Indirect winter deaths occur due to heart attacks from snow shovelling, house fires due to the use of emergency heaters and automobile accidents in ice and snow. Once again, the elderly and the poor, including the homeless, suffer the most.

The most common response to cold stress is additional clothing and improved housing but it is not always easy to maintain adequate indoor temperatures. In the MDCs, where domestic central heating systems are widely used, the rising cost of energy – plus concern about greenhouse gas emissions – makes this response impractical for poor people living in properties which have inadequate heat sources and lack good thermal insulation. Other adaptation strategies include education and outreach to vulnerable home-owners to encourage less reliance on heating systems that can create carbon monoxide risks indoors, together with guidance on safe outdoor activity during cold weather (Conlon *et al.*, 2011).

2 Heat stress

Heat stress is greatest when both atmospheric temperature and humidity are high and physical discomfort leads to mortality. The amount by which the temperature exceeds the local seasonal mean is more important than the absolute value of temperature. The threat is also high in the first heat-wave of the season, before acclimatization occurs. After several days of excessive heat, the typical mortality rate rises to two or three times the normal seasonal rate.

Extreme heat-waves are experienced frequently in the United States, where they are the number one weather-related cause of death. For example a 36 per cent increase in deaths was recorded over a five-day period in 1966 (Bridger and Helfand, 1968); in 1955, 946 excess deaths resulted in Los Angeles (more than twice the mortality recorded in the 1906 San Francisco earthquake and fire) (Oechsli and Buechly, 1970); and over 700 people died in Chicago in 1995 (Kleinberg, 2002). Even so, Ostro *et al.* (2009) claimed that existing statistics on heat-wave-related deaths, as reported in California, probably under-represent the real extent of the hazard, due to a lack of clear case definition and the multifactorial nature of such deaths. The elderly and those with existing heart disease are at greatest risk. During the month of July 1993, 213 heat-related deaths were recorded in Philadelphia, mostly at the place of residence (Johnson and Wilson, 2009). Figure 10.2 illustrates the frequency of heat-related death by age group. The mean age of the deceased was 65.7 years, with maximum values in the 60–90 years band. Other high-risk groups are the urban poor, especially those who lack domestic air conditioning or are dependent on alcohol or drugs.

Urban heat stress is an environmental hazard so far identified primarily in high-income countries. Due to global warming, the threat is expected to increase, with a possible doubling of

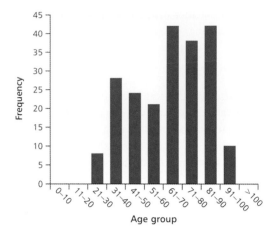

Figure 10.2 Frequency of heat-related deaths by age group in Philadelphia, USA during July 1993. Mortality was concentrated in the 60–90 age band. After Johnson and Wilson (2009). Reproduced from *Applied Geography* 29, D.P. Johnson and J.S. Wilson, The socio-spatial dynamics of extreme urban heat events: the case of heat-related deaths in Philadelphia, 419–34. Copyright (2009), with permission from Elsevier.

heat-related deaths worldwide by 2020 (Gabriel and Endlicher, 2011). Stott *et al.* (2004) claimed that man-made atmospheric warming has already more than doubled the risk of European summers as hot as that of 2003. The 2003 summer in Europe was the warmest since 1500, with temperatures around 40°C, and over 30,000 excess heat-related deaths were reported (Haines *et al.*, 2006). Half of these were in France during August (see Figure 10.3), when overall excess mortality averaged 60 per cent; parts of Paris experienced rates over 150 per cent (Poumadère *et al.*, 2005). Apart from temperature, many socio-economic factors were implicated, including age, gender, dehydration, medication, urban residence, poverty and social isolation. Most at risk were elderly people living alone in cities when many medical staff were absent for the traditional summer holiday. According to Lagadec (2004), this disaster exposed a lack of preparedness in the developed countries, and France has subsequently developed a national heat-wave plan, which helped to reduce mortality

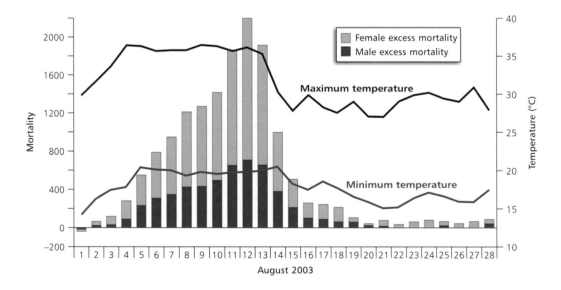

Figure 10.3 Number of excess deaths recorded in France each day during the 2003 heat-wave in relation to maximum and minimum temperatures. Between 4 and 15 August almost 15,000 deaths were directly attributed to heat stress caused by dehydration, hyperthermia and heat stroke. After Poumadère *et al.* (2005). Reproduced from M. Poumadère *et al.*, The 2003 heat wave in France: dangerous climate change here and now. *Risk Analysis* 25 (2005): 1483–94. Copyright John Wiley and Sons 2005. Reproduced with permission.

when a larger-scale, but less intense, event took place in 2006 (Fouillet, A.R. *et al.*, 2008).

Although large urban areas pose the greatest risks, Loughnan *et al.* (2010) found evidence of increased heat-stress mortality in persons aged 65 and older living in small towns in rural parts of Victoria, Australia. Metropolitan centres are especially problematic, due to intensification of heat stress by the *urban heat-island* effect and socio-economic disadvantage. There is growing awareness of the need to reduce the heat-related health inequalities in inner cities (Harlan *et al.*, 2006). This problem was highlighted for a heat-wave in the New York–New Jersey metropolitan area over 30 years ago, when up to 200 extra deaths occurred in the city core, over the number expected in the suburbs (Beechley *et al.*, 1972).

Microclimates in urban areas can be modified through the shading of houses by trees against direct solar radiation and by the use of high-albedo materials to reflect more of the incident radiation on individual properties (Solecki *et al.*, 2005). A study in Lisbon, Portugal found that a small green space, 0.24 ha in extent, lowered the temperature in the surrounding densely urbanized area significantly, an effect that was most apparent on the hottest days (Oliveira *et al.*, 2011). The planting of trees can help to reduce urban air pollution, often a contributory factor in heat stress, although inner cities rarely have space for large tracts of vegetation. More domestic air conditioning seems a likely response in the MDCs, but this is a less available option for inhabitants of Third World countries.

C THE NATURE OF DISEASE EPIDEMICS

Infectious diseases account for over one-quarter of all human deaths globally and for two-thirds of all deaths in children less than five years old. Just six infectious diseases – HIV/AIDS, tuberculosis, malaria, pneumonia, diarrhoeal infections and measles – are responsible for about 50 per cent of all premature deaths. Many of these diseases are endemic in poor countries, where epidemics are also common. Thus, outbreaks of diarrhoea occur regularly in tropical countries, especially when floods contaminate drinking water supplies or destroy sewerage systems. Such public health disasters also occur when a pathogen (virus, bacteria or parasite) triggers a disease outbreak in a human population lacking immunity. Worldwide disease outbreaks, called *pandemics*, have occurred throughout history; the Black Death pandemic during the fourteenth century most probably killed more than 50 million people (McMichael, 2001). Bacterial, viral and parasitic infections are all capable of causing public health disasters, especially in poor countries, through the transmission to humans of pathogens via insects, rodents or other *vector organisms*. Some vectors (lice, bugs, fleas and certain mosquitoes) benefit from human-aided transport. Mosquitoes, biting midges and flies are sensitive to weather conditions like temperature, humidity and rainfall (more surface water) and tend to migrate spontaneously (Lounibos, 2002).

The World Health Organization defines disease epidemics as:

> the occurrence of a number of cases of a disease, known or suspected to be of infectious or parasitic origin, that is unusually large or unexpected for the given place and time. An epidemic often evolves rapidly so that a quick response is required.

A variety of statistics-based thresholds are used to identify epidemics. According to Kuhn *et al.* (2005), any disease showing large inter-annual variability can be considered to be epidemic, even though the link with weather and climate may be weak or unquantified. An important factor is the vulnerability of the population at risk. This is influenced by factors such as malnutrition, the local level of disease immunity and previous exposure to infection. Even when the linkages between disease and atmospheric conditions are relatively strong, epidemics are likely to occur only if a significant number of non-immune people are present in the population. In most cases, weather conditions are not the most important drivers of disease outbreaks. However, some epidemics can be linked directly to conditions created by natural hazards and disasters (see Box 10.1).

Over millions of years, biological adaptation has enabled *Homo sapiens* to evolve by natural selection, a process that has helped the species to resist disease. Mass migrations and intermarriage have produced a genetic diversity that promotes disease resistance because it reduces the chance of a new organism being introduced into a community. In Africa, long-term exposure to malaria provides some immunity. The increase in medical knowledge and the spread of good practice in public health has led to greater controls on infectious diseases. During the 1960s and 1970s, with the introduction of new antibiotic drugs and vaccines, the goal of eliminating disease epidemics appeared attainable, at least for the MDCs. This mood of complacency has since been replaced by one of concern, due to the re-emergence of 'old' diseases, such as *plague* responsible for the 'Black Death' of the Middle Ages and the emergence of new ones, such as *Ebola hemorrhagic fever*, identified in Sudan and Zaire in 1976 (Noji, 2001). The first entirely new disease epidemic of the twenty-first century was the SARS (Severe Acute Respiratory Syndrome) outbreak of early 2003. The disease started in southern China during November 2002, when an animal virus jumped species to infect a human. Several hundred people died. Most countries used strict isolation measures to prevent the virus infecting the general public but the disease proved difficult to control in warm, humid areas of south-east Asia.

Good public health services are vital for *disease prevention* by means of the greater use of clean water, sanitation, safe food, prophylactic drugs and immunization, health education and mass screening. A rather different emphasis on *disease cure* is encouraged by multinational pharmaceutical companies and the privatization of government health care. This tends to create a neglect of public health services. Another cause of recent disease outbreaks is the crowding of refugees into emergency camps. In these circumstances, the risk of infection from communicable disease increases for all three modes of transmission:

- *person-to-person transmission* – e.g. measles, meningitis, tuberculosis
- *enteric (intestine) transmission* – e.g. diarrhoeal diseases, hepatitis
- *vector-borne transmission* – e.g. malaria.

In the poorest countries of Asia and Africa, endemic diseases like measles and tuberculosis have a continuing debilitating effect on people. Local populations have a reduced ability to produce food or earn a living, and become vulnerable to other hazards. It has been estimated that almost 60 per cent of deaths due to infectious diseases are among the poorest 20 per cent of the world's population, whilst only 7 per cent of victims are in the richest 20 per cent (IFRCRCS, 2000). Such imbalances are likely to remain, given that less than 5 per cent of global spending on biomedical research is allocated to the chief killer diseases in the LDCs. Much ill-health is climate related. For example, on a world scale, diarrhoea is one of the three main causes of death in children under five years of age. In sub-Saharan Africa the prevalence of outbreaks is associated with a shortage of rainfall in the dry season, when safer hygiene practices become harder to follow (Bandyopadhyay *et al.*, 2012).

The re-emergence of old diseases and the emergence of new ones can be attributed to a complex set of factors involved in global change (Molyneux, 1998; Murphy and Nathanson, 1994). These factors fall into three categories:

- *Changing environmental factors.* Changes to environmental conditions alter the ecological niches occupied by infected hosts and the vectors of existing disease and allow epidemics

Box 10.1 Diseases and disasters

Noji (1997) defined *epidemiology* as 'the quantitative study of the distribution and determinants of health-related events in human populations'. It is a specialized branch of public health medicine served by many agencies, including the Centres for Disease Control and Prevention (CDC) based in Atlanta, Georgia, plus a variety of emergency-response units worldwide. An important objective of epidemiology is to limit the potential for disease epidemics following natural disasters.

A range of communicable disease epidemics can occur after disasters (Ligon, 2006). For example, floods can increase the transmission of both *water-borne* diseases (typhoid, cholera, hepatitis A) and *vector-borne* diseases (malaria, yellow fever, dengue fever). Periodic flooding associated with ENSO events in the dry coastal area of northern Peru has been linked with malaria epidemics in the region. In practice, disease outbreaks arise from many causes, including the presence of disease in the population before the event, ecological change (e.g. the spread of surface water), damage to public utilities (e.g. the contamination of water supplies), the interruption of disease-control programmes, plus the movement of displaced people into over-crowded refugee camps. Epidemiological control involves:

- *surveillance* – on-going study of endemic disease levels, plus an associated vulnerability analysis of populations, before disaster strikes
- *assessment of disaster impacts* – short-term field surveys and other methods to monitor disease outbreaks and the nature of the emergency responses
- *evaluation* – the appraisal of the overall response to the health impacts of disaster as an aid to better planning for the future.

In the period immediately following a disaster, the epidemic risk is high. A major threat arises from the large-scale displacement of disaster victims (Watson *et al.*, 2007). Aid agencies give priority to basic post-disaster precautions, such as disease-vector control, the management of human-waste disposal, good personal hygiene and the safe preparation of food. Cholera is a major concern because of its capacity to spread rapidly within a population. Following an earthquake in January 2010, cholera was confirmed in Haiti on 21 October 2010. By March 2011, over 250,000 cases had been reported and over 4,600 people had died. A possible source of the epidemic was reported to be drinking water from the Artibonite River contaminated by sewage from a United Nations camp with deficient sanitation.

Some conditions necessary for disease epidemics often exist before disaster strikes. The main predisposing factor is poverty. Infants and children in the LDCs are several hundred times more likely to die from diarrhoea, pneumonia and measles than those in Europe or North America (Cairncross *et al.*, 1990). Poor housing, malnutrition, lack of hygiene to protect against vectors, inadequate clean water supplies and restricted access to health-care facilities all play a part. For example, in Peru the poor suffer from over 20 endemic water-borne diseases; a major cholera outbreak in the early 1990s was attributed to a combination of contaminated water supplies and poor hygiene (Witt and Reiff, 1991).

Disasters often provoke mass population movements and rehousing in refugee camps. People, frequently malnourished and with low levels of disease immunity, are crowded into temporary shelters with inadequate sanitation facilities, contaminated water supplies and a shortage of food (Morris *et al.*, 1982; Waring and Brown, 2005). After the 1991 eruption of Mount Pinatubo, Philippines, over 100,000 refugees were accommodated in 100 such camps. Migrants bring new pathogens or move into a contaminated area and catch disease because of their lower resistance. Such factors combine. For example, vector-borne diseases – like malaria and yellow fever – increase after floods in tropical areas, due to the increase in mosquito and other insect breeding sites. A loss of housing forces people to live outdoors, at greater risk from biting insects. When hurricane 'Flora' struck Haiti in 1963, 75,000 cases of malaria were reported in the next six months (Mason and Cavalie, 1965). Interacting factors included an incomplete malaria eradication programme, the washing of insecticide from houses by heavy rain, an increase of mosquito breeding in areas of standing water and a lack of shelter for the local population.

In some cases, epidemics may spread to urban areas that were previously disease free through infected people sheltering with relatives or moving around in search of food, construction materials and employment. Other sources of epidemic disease are rats, which act as reservoirs of plague and often emerge from sewers after floods, and abandoned dogs infected with rabies that are likely to bite humans following a breakdown in living conditions. The physical disruption of water supply and sewage disposal systems is most serious in areas where sanitation levels are already low. In Bangladesh four-fifths of the population rely on tube wells for drinking water and use surface sources – such as shallow ponds – for bathing, washing and cooking. After the 1991 cyclone, about 40 per cent of the tube wells were damaged, the surface sources became contaminated with sewage and salt and there was a large increase in diarrhoeal diseases (Hoque *et al.*, 1993).

to spread to new areas. These changes include urbanization, economic development, water resource development (more dams and irrigation), deforestation and climate change. All can potentially increase human exposure to insect vectors or sources of new pathogens.

- *Changing socio-economic factors.* Changes in medical practice and in human behaviour can assist the rise and the spread of both old and new diseases. Relevant factors include a trend to more cross-border travel, delays in developing new antibiotics, reduced disease surveillance in areas known to be disease prone, reduced funding for public health-care facilities. Other major factors are wars, poverty and changes in human sexual behaviour

- *Changing viral profiles.* Changes in drug resistance are a feature of several disease agents, a feature sometimes caused by the over-use of antibiotics and other drug therapies. In particular, new virus diseases appear to be emerging in both animals and humans with greater frequency, due to continuous virus evolution and genetic mutation (Box 10.2). Such changes occur when viruses replicate, because the genes may recombine and re-assort. Some viruses can recombine with genetic elements of their host cells and thus acquire new genes.

D INFECTIOUS DISEASES AND CLIMATE

Given the complexity of epidemics, it is important to identify the diseases most associated with environmental conditions, and Table 10.1 lists important communicable diseases with epidemic potential that are influenced by weather and climate. It can be seen that cholera and malaria have *important* links whilst seven other diseases have a *significant* link. Of course, weather conditions are never the sole driver of disease epidemics. In almost all cases temperature conditions are relevant; increases in temperature associated with high rainfall or humidity are the most common combinations.

Box 10.2 The emerging flaviviruses

The most significant newly emerging diseases are those of the *Flavivirus* family. This group is named from one of greatest plague diseases, yellow fever. 'Flavus' is Latin for yellow and the diseases are associated with jaundice and the yellowing of a victim's skin. They originated from a common ancestor 10,000–20,000 years ago but are now evolving effectively to fill new ecological niches (Solomon and Mallewa, 2001). More than 70 flaviviruses have been identified, of which about half cause disease in humans. The natural viral host is local wildlife and generally the disease is carried to humans by arthropod (insect) vectors – mosquitoes in the tropics and ticks in the higher latitudes. Tick-borne flaviviruses are less important for human disease than mosquito-borne viruses because tick species feed on animals in the wild and tick-borne diseases tend to be more restricted geographically. Many mosquito-borne diseases have been known for centuries but have recently upsurged, due to combinations of environmental, socio-economic and viral factors.

Dengue fever has been widespread in the tropics for over 200 years, with intermittent pandemics emerging at roughly 10–40-year intervals. It is now the most widely distributed mosquito-borne disease of humans. Together with dengue hemorraghic fever (DHF), it is caused by one of four related virus serotypes of the genus *Flavivirus*. Infection with one serotype provides no immunity against any of the other three. It is also unusual in that humans are the natural hosts for the virus. Humans are infected by *Aedes aegypti*, a domestic, day-time-biting mosquito. The population density of this insect is highly dependent on human habitation. Water-storage facilities and the availability of breeding sites around residential buildings are key factors in promoting the disease.

Emergence during the late twentieth century is attributed to poor vector control, over-crowding of refugee and urban populations and more frequent international travel. For a time, DDT sprays eliminated *Aedes aegypti* from many countries but dengue fever epidemics have increased, partly because effective mosquito control has been relaxed in areas where it is endemic. In the Pacific region, dengue viruses were reintroduced in the 1970s after an absence of over 25 years and there has been a re-emergence in Central and South America, where the geographical distribution is believed to be wider than it was before the eradication programme. For the patient, dengue fever produces a range of viral symptoms capable of developing into severe and fatal haemorrhagic disease. There are approximately 100 million cases of infection per year, with 2.5 billion people at risk (Ligon, 2004). No dengue vaccine is available.

Yellow fever has been an important tropical disease for nearly 500 years. In 1900 the mode of transmission by *Aedes aegypti* was discovered and, as early as 1908, the deliberate suppression of mosquito breeding sites had eliminated yellow fever from many urban centres. However, in 1932 the disease was found to have an independent *zoonotic* (animal-borne) transmission cycle involving monkeys. There are three types of transfer:

- *Sylvatic (jungle) yellow fever* occurs in tropical rainforests when monkeys become infected by wild mosquitoes, which pass on the virus when bitten by other mosquitoes. The infected wild mosquitoes then bite humans in the forest, such as timber workers. Disease incidence is low, due to the sparse population, but the virus can be transferred to unvaccinated inhabitants of nearby towns.

- *Intermediate yellow fever* occurs mainly in the savannahs of Africa, where semi-domestic mosquitoes infect both monkey and human hosts to create small epidemics. Infected mosquito eggs can survive several months of drought before hatching in the rainy season, so the virus is well suited to this climatic environment. Increased contact between humans and infected mosquitoes in the wet-and-dry tropics, where water projects and other developmental changes increase mosquito density, is a major cause of African outbreaks.
- *Urban yellow fever* of epidemic proportions typically occurs when migrants introduce the virus into crowded townships, where the disease spreads by domestic mosquitoes directly from person to person. In the savannah areas of Africa, water is commonly stored in large earthen pots and the consequent high rates of household breeding for *Aedes eygypti* have been implicated in several yellow fever epidemics in Senegal, Ghana, the Gambia, Côte d'Ivoire, Nigeria and Mauretania during the 1965–87 period.

An estimated population of over 500 million people, living in Africa between latitude 15°N and 10°S of the equator, is at risk of yellow fever infection, whilst the disease is endemic in nine South American countries and some Caribbean islands. There are an estimated 200,000 cases of yellow fever per year, with 30,000 deaths, but the disease is much under-reported. There is no recognized treatment for yellow fever. The most important preventive measure is a highly effective vaccine that has been available for 60 years. For adequate protection 80 per cent of the population should be vaccinated, but the immunization cover is below 40 per cent in most countries where yellow fever is endemic.

West Nile virus (WN) was not recognized until 1937, when it was clinically isolated in the West Nile district of northern Uganda (Campbell *et al.*, 2002). It is endemic in Africa, Asia, Europe and Australia and was introduced into the USA in 1999 via an outbreak in New York City. There have been several WN fever epidemics, notably that of 1973–74 in South Africa. The virus is maintained in endemic disease areas through a mosquito–bird–mosquito transmission cycle. The transfer of the disease to new areas is mainly by migratory birds. The incubation period for WN fever is typically from two to six days and, in the worst 15 per cent of cases, the development of encephalitis leads to coma. In some areas of Africa, immunity to WN virus is thought to reach 90 per cent in adults. But in Europe and North America, where the disease is likely to become more prevalent, such background immunity is almost non-existent.

1 Influenza

Influenza is one of the world's oldest, most common and most deadly diseases. 'Flu' has been known for over 2,000 years; the first well-described pandemic was in 1580. It is an acute respiratory illness caused by the influenza viruses A and B and occurs due to minor changes in the influenza viral antigenic proteins. It is thought that the viruses causing worldwide pandemics come from animals, notably swine and birds.

Influenza epidemics usually occur in winter, but pandemics can occur anytime. During the 1918–19 pandemic at least 21 million people died, more than twice the number of people killed during World War I. Lung damage was the major cause of these deaths, at a time when antibiotics were unavailable. It has been estimated that a flu pandemic today could affect 20 per cent of the world's population, of whom 30 million would be hospitalized and about 25 per cent would die (Fouchier *et al.*, 2005). In 1984, the World Health

Table 10.1 Selected infectious diseases with epidemic potential, in which climate plays at least a moderate role, ranked according to global significance

Disease	Mode of transmission	Distribution	Climate–epidemic link	Strength of climate link and comments
Influenza	Air borne	Worldwide	Decrease in temperature (winter)	Moderate A range of human-related factors more significant
Diarrhoeal diseases	Food and water borne	Worldwide	Increases in temperature and decreases in rainfall	Moderate Sanitation and human behaviour more important
Cholera	Food and water borne	Africa, Asia, Russian Federation, South America	Increases in sea and air temperature, plus El Ninō events	Important Sanitation and human behaviour also important
Malaria	Bite of female *Anopheles* mosquitoes	Epidemic in over 100 countries	Changes in temperature and rainfall	Important Local factors also relevant
Meningoccol meningitis	Air borne	Worldwide	Increases in temperature and decreases in humidity	Significant
Lymphatic filariasis	Bite of female *Culex, Anopheles* and *Mansonia* mosquitoes	Africa, India, South America, South Asia and Pacific islands	Temperature and rainfall affect distribution	Moderate
Leishmaniasis	Bite of female phlebotomine sandflies	Africa, Central Asia, Europe, India, South America	Increases in temperature and rainfall	Significant
African trycomo-pariasis	Bite of male and female tsetse flies	Sub-Saharan Africa	Changes in temperature and rainfall	Moderate Cattle density and vegetation patterns relevant
Dengue	Bite of female *Aedes* mosquitoes	Africa, Europe, South America, South-East Asia, Western Pacific	High temperature, humidity and rainfall	Significant Other factors also important
Japanese encephalitis	Bite of female *Culex* and *Aedes* mosquitoes	South-East Asia	High temperature and heavy rains	Significant Animal factors important
St Louis encephalitis	Bite of female *Culex* and *Aedes* mosquitoes	North and South America	High temperature and heavy rain	Moderate Animal factors important
Rift Valley fever	Bite of female culicine mosquitoes	Sub-Saharan Africa at end of epidemic	Heavy rains, cold weather	Significant Animal factors important
West Nile virus	Bite of female culicine mosquitoes	Africa, Central Asia, South-West Asia, Europe	High temperatures and heavy precipitation at onset	Moderate Non-climatic factors may be more important
Ross River virus	Bite of female culicine mosquitoes	Australia and Pacific islands	High temperatures and heavy precipitation at onset	Significant Host immune factors and animals important
Murray Valley fever	Bite of female *Culex* mosquitoes	Australia	Heavy rains and below-average atmospheric pressure	Significant
Yellow fever	Bite of female *Aedes* and *Haemagogus* mosquitoes	Africa, South and Central America	High temperature and heavy rain	Moderate Population factors important

Source: After Kuhn *et al.* (2005). Reproduced from K. Kuhn *et al.*, Using Climate to Predict Infectious Disease Epidemics (World Health Organization, 2005). With permission of the World Health Organization.

Notes: Global significance assessed on the calculation of Disability-Adjusted Life Years (DALYs). Strength of climate link assessed on 5-point scale: weak, moderate, significant, important, primary.

Organization (WHO) established the Global Influenza Surveillance Network, using 110 laboratories in 82 countries, as an alert system for the identification of new viruses (Kitler *et al.*, 2002).

Influenza spreads rapidly in seasonal epidemics. Most people recover from the infection, but it can cause fatal complications like pneumonia. It has been estimated that epidemics worldwide result in 3–5 million cases of severe illness and 250,000–500,000 deaths in an average year. In the MDCs, children under two years of age, adults aged 65 or older or anyone with underlying health issues, such as chronic heart or lung disease, are most at risk. It is a serious public health problem that takes an economic toll through lost productivity in the workforce and by straining health-care services. Much less is known about influenza outbreaks in the LDCs.

2 Malaria

This is the world's major vector-borne parasitic disease and is endemic in over 100 tropical and subtropical countries (Figure 10.4). These areas possess three necessary elements – a suitable climate, infected humans and mosquitoes. The cause of malaria is a single-cell parasite called *Plasmodium*. Humans act as the only vertebrate host and the disease is transmitted from person to person by the bite of the female *Anopheles* mosquito, which uses the blood to nurture her eggs. The disease is widely under-reported, especially in Africa. It is estimated that about 3.3 billion people – half of the world's population – are exposed to *Plasmodium falciparum*, the most life-threatening form of the disease. Every year at least about 250 million clinical cases incidents of malaria, and about 1 million malaria-related deaths, are reported.

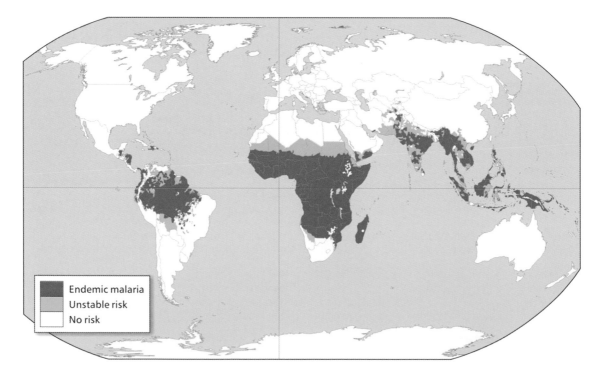

Figure 10.4 The spatial distribution of *P. falciparum* malaria endemicity in 2007. There are high levels of endemicity in Africa, where there is an urgent need for control measures, combined with opportunities to eliminate the disease elsewhere. After Hay *et al.* (2009).

About 90 per cent of malaria-related deaths occur in Africa, mainly in sub-Saharan Africa, where the resources needed to undertake long-term anti-malarial campaigns are lacking. The Middle East and Asia are also malaria prone. In South Africa, 30 per cent of cases occur in patients under 15 years of age, although the majority of infections are in the economically active age group of 15–50 years (Govere *et al.*, 2001). Economic growth in countries with high rates of malaria transmission has historically been lower than that in many malaria-free areas. It has been estimated that, in some African countries, malaria is responsible for a 'growth penalty' amounting to as much as 1.3 per cent of GDP each year (Roll Back Malaria Partnership, 2011). Malaria generally attacks the most vulnerable people in a community and about three-quarters of infections are in children. A study within a poor urban area of Kolkata, India found a high incidence of the disease and that low-income and illiterate household members living in a building not built of brick were at the greatest risk of malaria (Sur *et al.*, 2006).

Epidemics of malaria are major disasters. It has been estimated that epidemic malaria is responsible for over 100,000 deaths each year in Africa alone. Epidemics are caused by many factors, including an influx of non-immune people into endemic areas, a break in normal malaria control measures and an increase in associated diseases or malnutrition. Environmental changes, like deforestation, irrigation or flooding, also play a part (Guintran *et al.*, 2006). Cities with unplanned and expanding peripheries, which invade the surrounding malaria-infested rural areas, are at high risk. For example, Khartoum, Sudan suffered an epidemic after extensive floods in 1988, but most epidemics are linked to a rise in vector numbers, due to shifts in weather and climate conditions, especially spells of increased temperature and/or rainfall.

Key weather influences are:

- *Temperature.* At ambient temperatures below 20°C the malaria parasite has a limited life cycle, although it can exist in small, warmer locations such as houses. As the temperature rises from 21°C to 27°C it thrives because the time required for development progressively decreases.
- *Rainfall. Anopheline* mosquitoes require bodies of surface water for breeding. Adequate rainfall is essential, but survival rates depend on the season and on rainfall intensity, e.g. heavy rainfall can wash away the breeding larvae.
- *Relative humidity.* Atmospheric humidity is normally high in the wet season and during spells of rainfall. If the relative humidity falls below 60 per cent, the life of the mosquito is shortened and transmission of the parasites to humans is limited.

In Africa, weather-related epidemics of malaria afflict mostly highland or semi-arid areas and occur in 2–7 year cycles linked to climate variability (Pascual *et al.*, 2008). Similar links have been established in other regions, such as Sri Lanka (Briët *et al.*, 2008) and Thailand, where outbreaks of malaria are closely associated with seasonal rains, although farming practices and population movements are involved too (Wiwanitkit, 2006; Childs *et al.*, 2006). In Venezuela and elsewhere outbreaks of malaria have been linked to El Niño Southern Oscillation (ENSO) fluctuations.

3 Cholera

Cholera is an acute intestinal infection caused by the bacterium *Vibrio cholerae*. Ecologically speaking, this bacterium is part of the flora found in brackish water and estuaries in tropical areas and is spread to humans via contaminated food and water. During the nineteenth century cholera spread throughout the world from its original source in the Ganges delta of India and, so far, there have been seven pandemics. The disease is endemic in tropical latitudes (Figure 10.5). Up to 6 million cases of infection are reported annually. About 10 per cent of cases lead to severe

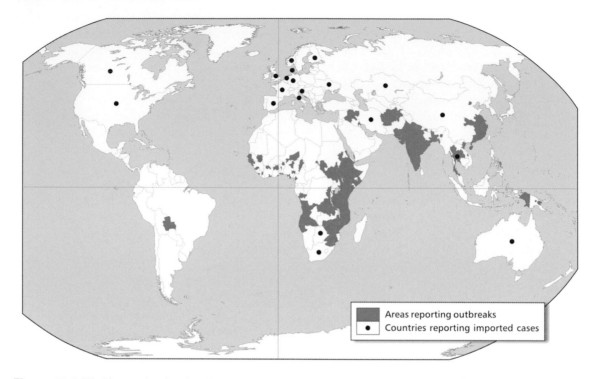

Figure 10.5 World map showing the distribution of areas reporting cases and outbreaks of cholera 2007–9. The delta regions of the Ganges and Brahmaputra rivers are the native habitat for cholera, which is endemic in the Indian sub-continent. After World Health Organization at http://gamapserver.who.int/mapLibrary/Files/Maps/Global_CholeraCases 0709_20091008.png (accessed 7 July 2011). Reproduced with permission.

dehydration that, without treatment, can lead to death. As a result, there are 100,000–120,000 deaths each year. Most fatalities occur in people with low immunity, such as children and those living with HIV.

Cholera has an impressive ability to spread rapidly into new areas and to become epidemic. This is because *Vibrio cholerae* is so common in the aquatic environments of the tropics and has a very short incubation period of between two hours and five days. Consequently, there is a rapid onset of the early symptoms of diarrhoea and vomiting. Like most infectious diseases, it is particularly dangerous when newly introduced into an unprepared area, where the fatality rates can rise to 50 per cent of infections. For example, in 1970 and 1991 the disease struck West Africa and Latin America respectively, although both

regions had not suffered major outbreaks for around 100 years. It is now endemic in both continents.

Many epidemics are multifactorial. Inadequate environmental management and poor public health facilities are root causes. A prime issue is the absence of sanitation in the peripheral urban slums of the LDCs. The growing number of vulnerable people living under these conditions in many world cities is mainly responsible for the relentless rise in cholera cases. The lack of clean water and good sanitation in 'temporary' over-crowded refugee camps is also a factor (see also Box 10.1). Rural areas are not immune. In north-east India, inadequate anti-bacterial treatment, plus poor hygiene and sanitation, were responsible for an outbreak in 2002 (Phukan *et al.*, 2004), whilst urban epidemics in Peru during 1991 were

Plate 10.1 Children wade across a stream flowing through the Kroo Bay slum area of Freetown, Sierra Leone. The stagnant water, choked with rubbish, is used by many residents for defecation. Floods occur in the rainy season and the stream is a source of cholera and malarial mosquitos. (Photo: Panos/Aubrey Wade AWA00236SRL)

associated with the microbial contamination of water supplies and ineffective chlorination procedures (Tickner and Gouveia-Vigeant, 2005).

4 Ross River virus

Ross River virus (RRV) is an example of a regionally restricted disease, currently limited to Australia and the Pacific Islands. It is transmitted by *Aedes* and *Culex* mosquitoes and is the most common mosquito-borne viral disease in Australia. More than 5,000 cases are reported annually.

About 30 per cent of all infected persons develop flu-like symptoms, including a rash and a fever. This occurs within 3 to 11 days following infection. RRV outbreaks tend to occur in rural areas of southern Australia near intensive irrigation schemes with standing surface-water bodies or salt marshes. Outbreaks peak in the summer and autumn months, especially during spells of warm, humid weather with heavy rainfall. In southern Australia such weather episodes are often related to El-Niño cycles (see Chapter 14, pages 413–14). RRV has been an officially notifiable disease in Australia since 1991, but no specific treatment is available.

E DISEASE HAZARD REDUCTION

As with other environmental hazards, reducing the risks from infectious diseases depends on replacing public apathy and emergency responses with a strategy aimed at long-term control and prevention. For example, the WHO-sponsored Roll Back Malaria campaign aims to reach near-zero deaths from malaria by 2015, in pursuit of UN Millennium Development Goals (Roll Back Malaria Partnership, 2011).

1 Protection

Where an effective vaccine is available, immunization is the best approach to infectious disease. For example, a good vaccine exists for yellow fever. This provides immunity within one week for 95 per cent of people vaccinated. The immunity lasts at least 10 years and has few side-effects. Following the success of global immunization programmes to eradicate smallpox in the 1960s and 1970s, the WHO reorganized

its emergency division in 1993 with a view to providing more effective responses to epidemics. Mass vaccinations, covering 80 per cent of the population, would save millions of lives each year from common diseases like tuberculosis, measles, whooping cough, tetanus and diphtheria.

However, the cost is high and the support structure of health-care clinics is often missing in poor countries. Routine issues are exemplified by Zimbabwe, where political instability and economic collapse have led to a decline in general health care and lapses in the treatment of six vaccine-preventable diseases. The immunization service here is faced with many difficulties, including a lack of gas for cold-chain equipment, limited supplies of fuel and transport and inadequate staff levels (Chadambuka *et al.*, 2012). New viruses create special problems anywhere. Although the

Plate 10.2 A man uses a 'fogging' technique to spray a village in northern Thailand with insecticide against mosquitoes. Such aerial spraying has had only limited success in the control of malaria. (Photo: Panos/William Daniels WDA00040THA)

delivery of influenza vaccine in the USA is well funded and highly organized, it is estimated that a major pandemic due to a new virus could affect 200 million people, with up to 300,000 fatalities, and require an unprecedented emergency response.

Vector control is the other key element in disease prevention. For example, special attempts are made to suppress mosquitoes by pesticide applications at the beginning of a disease outbreak, before vaccination can take effect. For dengue fever, the absence of vaccines or any other cure means that preventive measures are the only option. Vector control has had little success (Brightmer and Fantato, 1998). Such responses could be improved by more detailed maps showing mosquito breeding sites and preparedness campaigns designed to increase local awareness of such sites. Although liquid pesticides can be effective in controlling diseases like malaria, the longer-term ecological effects of large-scale applications are not well understood.

2 Mitigation

Surveillance is critical for the prompt recognition and control of an emerging epidemic, especially in areas where there is poor vaccination cover. The longer an epidemic remains undetected and uncontrolled, the higher the morbidity and mortality. Cholera was the first disease for which surveillance and reporting was employed on a large scale. If a disease is common – like cholera – it will be easily diagnosed and appropriate responses will quickly follow. Other diseases, like dengue fever, are more difficult to identify. They can spread quickly into areas with high concentrations of *Aedes aegypti* mosquitoes and produce new virus strains and serotypes against which local resistance is limited. The use of remotely sensed data in association with terrain analysis is now providing more timely information on vector breeding habitats for malaria and other diseases.

Apart from hindering an effective response to infections, poor surveillance leads to the under-reporting of many diseases, such as yellow fever. Once a disease outbreak is suspected, there is a need for rapid access to laboratory testing and diagnostic facilities. Guintran *et al.* (2006) described the requirements for early detection of malaria epidemics in Africa, which include improved data collection, the use of early indications in the population and a better definition of epidemic thresholds. Here the overall aim is to progress from surveillance to early warning.

Treatment can be effective in dealing with the clinical symptoms of some infectious diseases. In the case of malaria, it has been suggested that infants and children in Africa should be treated even before they show any symptoms (Vogel, 2005). Simple treatments can be effective. For example, the dehydration and fever associated with yellow fever can be treated with oral rehydration salts and paracetomol, although any bacterial infection will need antibiotics. Unfortunately, many patients die before reaching hospital – often through a lack of local transport. In many urban areas there is a shortage of medical provision and, in parts of East Africa, over 50 per cent of hospital beds are occupied by AIDS victims.

Reliable hospital therapies are not available for all diseases. For example, malaria parasites quickly become resistant to drugs. Resistance to chloroquine is high, particularly in south-east Asia, and there is a need for new anti-malarial drugs. In other instances, such as cholera, oral vaccine may be available, but in such small quantities that it is used for individual travellers rather than the public as a whole. Some countries lack the clinical expertise and infrastructure necessary to administer therapies in an optimal way. Because public health facilities are under threat in so many countries, there is a need for international partnerships capable of taking a longer-term view. For example, Roll Back Malaria is a scheme initiated in 1998 by the WHO,

UNDP, UNICEF and the World Bank to involve governments, NGOs and the private sector in a partnership to reduce the global impact of this disease.

Forecasts and warning

Early warning systems exist for many health hazards. For example, following the heat-waves in summer 2003, a number of European countries introduced heat-health warning systems (Matzarakis *et al.*, 2011). In 2000 the WHO initiated the *Global Outbreak and Response Network* to identify unusual agents and pathogens, in order to improve international rapid responses. The prospect of effective early warning systems (EWS) for disease epidemics based on seasonal climate forecasts, or other environmental indices, provides one of the most hopeful ways forward (Kuhn *et al.*, 2005). This is because forecasts of an epidemic many months in advance would – in theory – justify increased surveillance and preparedness in the areas at risk. Thus, cholera epidemics may be predictable by monitoring the seasonal abundance of zooplankton in aquatic habitats using remotely sensed vegetation images (Colwell, 1996).

Long-term datasets in Bangladesh that merge climatic parameters with cholera outbreaks have indicated a link between cholera and the El Niño Southern Oscillation, an association that has become more apparent with intensified ENSO signals in recent decades. However, disease-surveillance and disease-control measures would need to be improved before a warning system could be successfully introduced using real-time climate data. A similar situation exists with regard to malaria. In a pioneer work, Bouma and van der Kaay (1996) found a statistical relationship between historic malaria epidemics on the Indian subcontinent, sea-surface temperatures and ENSO episodes. Since then, monitoring and early detection has improved and some countries have begun to trial EWS using climatic information. However, the work remains predominantly research led (Thomson *et al.*, 2006).

Education

In the longer term, people at risk must become better informed about infectious diseases. Ideally, public education should be reinforced by primary health care, including local facilities – like pharmacies and reference laboratories – combined with good community health practice taught to emergency health managers and local officials through regional workshops. In many areas, parallel improvements in domestic water supply and sanitation are required to combat diseases like cholera that spread through contaminated food and water.

Some educational responses are simpler, cheaper and quicker to implement at the community level than others. An understanding of the importance of better personal hygiene, the effective maintenance of latrines, the introduction of safe water supplies and suitable methods for rubbish disposal would help. The safe disposal of human waste, clean water and food supplies are crucial steps against cholera epidemics, especially when linked to basic household tasks such as the washing of hands before preparing food and the thorough cooking of food before it is eaten. The 1998 floods in Bangladesh caused major diarrhoea epidemics directly associated with low socio-economic status, poor water handling and inadequate household sanitation (Kunii *et al.*, 2002). Over three-quarters of people surveyed believed that water collected from tube wells and rivers was contaminated, yet only 1 per cent treated the water by boiling and a further 7 per cent used chlorination.

F WILDFIRE HAZARDS

Wildfire is a generic term for uncontrolled fires fuelled by natural vegetation. In Australia and North America the terms *bushfire* and *brushfire* are used, respectively, for such fires. Apart from Antarctica, no continent is free from the various combinations of ignition source, fuel and weather conditions necessary for wildfire hazard. In the past, wildfires were started naturally by lightning strikes in unpopulated areas. Today they are primarily due to human actions, especially when high temperatures and drought follow a period of active vegetation growth. A seasonal pattern is found in most areas with a Mediterranean-type climate, where most rain falls in the winter and vegetation is dry during the annual summer drought. Some countries – like Australia, France and Greece – are popular holiday destinations with an added risk of fire starts by tourists. The dry season in tropical climates has rather similar seasonal biophysical conditions, whilst large continental interiors – like those of the USA or Eurasia – experience dry air for much of the year, with the potential for a long fire season.

The world's greatest wildfire disaster to date occurred in October 1871, when about 1.7×10^6 ha of land in Wisconsin and Michigan, USA was burned. About 1,500 lives were lost. The fires began on the night that an urban fire in Chicago killed 250 people and were preceded by a 14-week drought across the Midwest. Many small fire outbreaks were not considered a threat until strong winds whipped up the flames to create a widespread, uncontrollable blaze. Today, in the industrialized nations, the wildfire risk to lives and property is highest on the rural–urban fringe – often termed the *Wildland–Urban Interface* (WUI) – where housing blocks are intermixed with natural vegetation. The life-style attractions of a semi-rural environment have encouraged the expansion of low-density suburbs in Sydney, Melbourne and Adelaide in Australia, plus the Los Angeles and San Francisco Bay communities in the USA, into the surrounding bush-land. In recent decades the risk has increased, due to the large-scale use of fire to clear forested land for development. This has become an international issue in parts of south-east Asia, often because of the resulting smoke pollution. For example, in 1994 over 3 million people were affected in Indonesia. In Greece, biomass burning extended to over 12 per cent of the forested area of the

country during 2007; 67 people died and dense smoke plumes travelled thousands of kilometres to the south (Kaskaoutis *et al.*, 2011).

Australia is the most fire-prone country in the world. Fires caused by lightning strikes have been characteristic for at least 100 million years and the native vegetation is adapted to regular burning. About 2,000 wildfires occur each year, many now started illegally, with some extending to over 100,000 ha. The most hazardous feature is the speed with which they spread. According to Mercer (1971), Australian bushfires can engulf up to 400 ha of forest in 30 minutes, as compared with as little as 0.5 ha over the same period in the slower-burning coniferous forests of the northern hemisphere. The lack of large surface water sources in inland Australia limits the options for fire-fighting. During 1974–75, an estimated 15 per cent of the continent burned, although this was largely in remote, arid land and little economic loss ensued. On the other hand, the 'Black Saturday' fire of February 2009 claimed 180 lives in the state of Victoria. Under similar conditions of 'fire weather' – temperatures of 40°C combined with wind speeds over 20 m s^{-1} – the 'Ash Wednesday' fires in Victoria and South Australia during February 1983 caused 76 deaths, 8,000 people were made homeless and economic losses were put at A\$200 million (Bardsley *et al.*, 1983).

Rural wildfires damage ecosystems and create indirect losses. After a major event, timber and forage resources may be destroyed, animal habitats disrupted, soil nutrient stores depleted and amenity value reduced for many years. If the burned-out areas consist of steep canyons, debris flows, rill erosion and floods are likely to follow. Such fires also adversely influence timber production, outdoor recreation, water supplies and other natural assets, and land restoration is expensive. A major threat exists in the dry, inland part of the western USA, where over 15 million ha of forests are at risk. Since 1990, over 90 per cent of all large (>400 ha) forest fires, and over 95 per cent of the area burned in the USA, have been in this region. In 1988 nearly 300,000 ha of the Yellowstone National Park was burned out, despite the efforts of more than 9,000 fire-fighters, raising important policy issues about fire management in areas with an important heritage status (Romme and Despain, 1989).

G THE NATURE OF WILDFIRES

1 Human influences

Wildfires are multifactorial events. The most obvious risk factor is increased human exposure, due to the spread of people into areas of predominantly natural vegetation. The population of Australia has grown fivefold in the last 100 years, most people now living in suburban areas. Handmer (1999) described a wildfire that affected Sydney in January 1994, when 4 deaths occurred and 200 houses were destroyed, despite the efforts of over 20,000 fire-fighters mobilized from all over Australia. In Canberra, the Australian capital, a series of semi-natural ridges, used for open-space recreation and nature conservation, run through the city, which, in many suburbs, backs directly onto rural areas without any transitional land uses (Lucas-Smith and McRae, 1993). During January 2003, bushfires in the Canberra suburbs killed four people, injured 300 others, destroyed 500 homes and forced more than 2,000 residents to evacuate their homes. Total costs were placed at A$400 million.

In the USA during the early 1990s, over one-quarter of the fires attended by public fire departments were in timber, brush and grass areas housing rural communities of less than 2,500 people (Rose, 1994). It was estimated that these rural residents were almost twice as likely to die in a fire as people living in larger communities of 10,000–100,000 people (Karter, 1992). The chief risk is in California, where 8×10^6 ha of highly flammable brush land has been developed to create an urban/wildland 'intermix' (Hazard Mitigation Team, 1994). At this time (1994), five fires creating the greatest loss to buildings in California had occurred within the previous five years. In 1991 a wildfire in the East Bay Hills area of San Francisco killed 25 people, injured more than 150 and made over 5,000 homeless (Platt, 1999). With estimated losses of US$1.5 billion, it was the third-most costly urban fire in US history. The fire started under classic conditions of high temperatures, low air humidity and strong winds and spread rapidly, aided by a dry vegetation cover. Fire-fighters were hampered by congested access roads, plus a loss of water pressure, and some 60 years of urban development in this area was destroyed, with only the building foundations remaining.

Recent work has challenged some perceptions surrounding the societal context of bushfire risk in Australia (McAneney et al., 2009; Haynes et al., 2010). For example, despite the high public profile accorded to bushfires in Australia, the normalized insurance losses suffered in the most costly fire disaster, in 1983, were easily outstripped by damages in the 1989 Newcastle earthquake, the 1974 Darwin cyclone and the 1974 Brisbane floods. In addition, wildfire losses fail to show the increasing time trend that might be expected due to enhanced human exposure in the WUI. Figure 10.6, whilst indicating associations between fatalities and dwellings destroyed by fire in individual events, shows no clear upward trend for either loss category. Rather, the message from this diagram is that wildfire disasters are relatively rare, random events that take place as a result of extreme weather conditions. Even more surprisingly, it was calculated that the chance of a randomly selected house in the WUI being destroyed by a bushfire was over six times lower than the probability of a structural house fire. Interestingly, it was concluded that, if the relative risk of bushfire were to be perceived rationally by people living in the rural–urban fringe, it might further reduce the incentive for home-owners to protect themselves and their properties from wildfire.

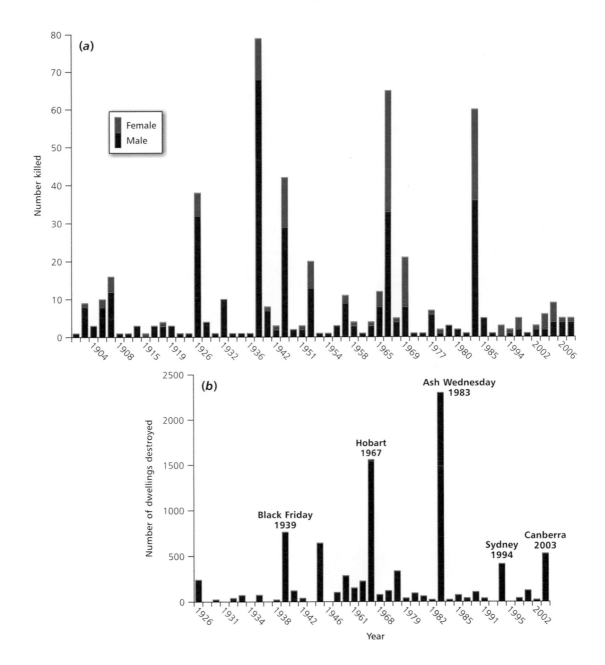

Figure 10.6 Deaths and damages caused by bushfires in Australia. (a) Annual number of male and female deaths 1900–2008. After Haynes *et al.* (2010). Reprinted from *Environmental Science & Policy*, 13(3), K. Haynes *et al.*, Australian bushfire fatalities 1900–2008: exploring trends in relation to the 'Prepare, Stay and Defend or Leave Early' policy, 185–94. Copyright (2010) with permission from Elsevier. (b) Annual number of dwellings lost to bushfires 1926–2003. After McAneney *et al.* (2009). Reprinted from *Journal of Environmental Management* 90(8), J. McAneney *et al.*, 100-years of Australian bushfire losses: is the risk significant and is it increasing?, 2819–22. Copyright (2009) with permission from Elsevier.

2 Fire ignition

Lightning strikes remain the chief cause of wildfires in very remote areas. The origin of many other fires is unclear but most are due to human actions. Figure 10.7 compares the causes of wildfire ignition on public land in the state of Victoria, Australia (a high-risk area averaging 600 fires per year) and in Bages County, Catalonia, Spain (a typical western Mediterranean area of rural depopulation averaging 15 fires per year). In both cases, natural causes (lightning) are small; arson may be higher than indicated in Bages County, due to the high percentage of unknown causes. Accidental sources are a mix of agricultural and recreational activities.

Deliberate fire-raising is a widespread problem that has attracted the attention of criminologists (Willis, 2005). In California one-quarter of all wildfires are believed to be due to arson but only 10 per cent of police investigations lead to an arrest. In Australia, about 13 per cent of all bushfires for which there is a recorded cause are logged as deliberate but a further one-third are suspicious, making it possible that around half the total may be deliberately lit.

3 Availability of fuel

The amount and moisture status of vegetation influences both the intensity of a bushfire (heat-energy output) and the rate of spread. Thus, grassland fires rarely produce the intensity of burn, and the degree of threat, associated with forest trees and mature shrubby vegetation. Apart from its quantity, the moisture content of fuel is important. This depends largely on the weather and climate. Climatic conditions ensure that there is a marked seasonal procession of risk in most countries. Figure 10.8 illustrates the normal pattern for Australia, where the risk is driven by the sequence of rains. According to Cunningham (1984), south-east Australia – where the peak danger period occurs in summer and autumn – is the most hazardous wildfire region on Earth. This is because many forests are dominated by the genus *Eucalyptus*. Most Australian forests accumulate a great deal of litter on the forest floor, mainly from bark shedding, after a number of fire-free years. Apart from creating a source of fuel, bark shedding creates a special problem of rapid fire-spread known as 'spotting'. This occurs when ignited fuel is blown ahead of an advancing fire front by strong winds to create 'spot' fires. Australian eucalypts have the longest spotting distances in the world. The reason is bark shedding by the *stringybark* and *candlebark* species that

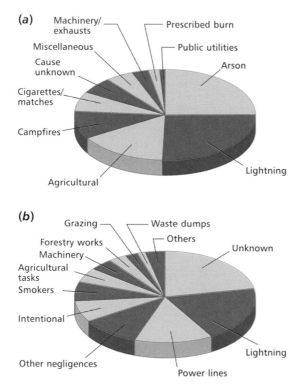

Figure 10.7 Sources of wildfire ignition in two different regions. (a) The State of Victoria, Australia. (b) Bages County, Catalonia, Spain. Adapted from State of Victoria at http://www.nre.vic.gov (accessed 29 January 2003) and after Badia *et al.* (2002). Reprinted from *Global Environmental Change* 4, Badia *et al.*, Causality and management of forest fires in Mediterranean environments, 23–32. Copyright (2002), with permission from Elsevier.

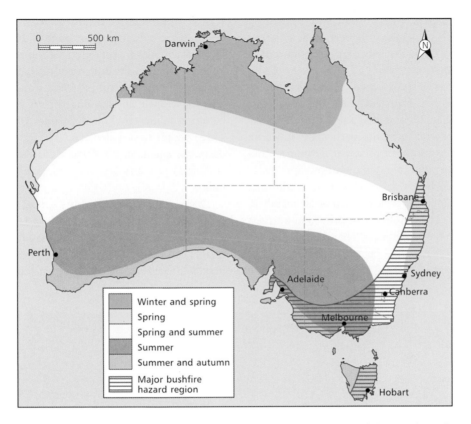

Figure 10.8 Seasonal patterns of bushfire activity in Australia. The centre of the continent is sparsely vegetated and populated, so the main hazards are in parts of South Australia, New South Wales, Queensland and Tasmania. Adapted from *Bushfires in Australia* by R.H. Luke and A.G. McArthur, Commonwealth Scientific and Industrial Research Organization, AGPS, Canberra, 1978.

produce loose, fibrous tapers easily torn loose by strong winds and convection currents. Spotting distances of 30 km or more have been authenticated, at least twice the distance recorded in the deciduous hardwood and coniferous forest fires of North America. In addition, eucalyptus trees contain volatile waxes and oils in their leaves that release a high heat output when burned and greatly increase the flammability of the vegetation (Chapman, 1999). At ambient fire temperatures of around 2,000°C these oils can create a spontaneous gas explosion.

4 Prevailing fire weather

Weather conditions are crucial for the ignition and development of wildfires. Drought periods provide a drying effect on vegetation and may also have atmospheric conditions suitable for 'dry lightning' storms, when no appreciable rain falls. Such storms are most active during unstable weather conditions in the summer months and ignite 60 per cent of all fires on public land in the western states of the USA. In the summer of 2000, 122,000 wildfires were started in this area and burned out 3.2×10^6 ha (Rorig and Ferguson, 2002). There is evidence that increased spring and summer temperatures in the western USA since the mid-1980s have led to a longer wildfire season and, in turn, to more frequent large wildfires (events >400 ha) that burn for longer durations (Westerling *et al.*, 2006).

Brotak (1980) compared extreme fire hazard situations in the eastern USA and south-east Australia, and found that most fire outbreaks occur near weather fronts, particularly in warm, dry conditions ahead of a well-developed cold front with unstable temperature lapse rates and strong winds at low levels. In California, easterly *Santa Ana winds*, which occur mainly in September and October – the driest and warmest months in the San Francisco Bay Area – create an extreme hazard in the fall season. Strong north-easterly Santa Ana-type winds developed in late

July 1977 and led to a disastrous wildfire which began in the hills and advanced to within a mile of the downtown area of the city of Santa Barbara. Over 230 homes were destroyed (Graham, 1977). In October and November 1993, 21 major wild-fires developed in six Southern Californian counties, fanned by hot, dry Santa Ana winds. Three people were killed, 1,171 structures were destroyed and some 80,000 ha were burned. The combined property loss was estimated at US$1 billion.

Once ignited, the rate of fire spread is closely related to the surface wind strength and direction. This is because the burning fire front advances by firstly heating – and then igniting – vegetation in its path through a combined process of convection

Plate 10.3 A fire truck moves away from an advancing bushfire in the Bunyip State Forest near Tonimbuk, west of Melbourne, Australia, on 7 February 2009 (Black Saturday). These fires were the deadliest in Australian history and claimed 173 lives. (Photo: AP Photo/01_1789 0754 accessed via www.boston.com)

and radiation. Most wildfire damage, including loss of life, occurs during a relatively short period of time – usually a few hours – compared with the total duration of the fire. These high-loss episodes are associated with extreme fire-risk weather, often involving high winds that shift in direction and cause the fire to accelerate in unexpected directions. Fire acceleration is greatly aided by the topography. For a fire driven upslope, wind and slope acting together increase the propagating heat

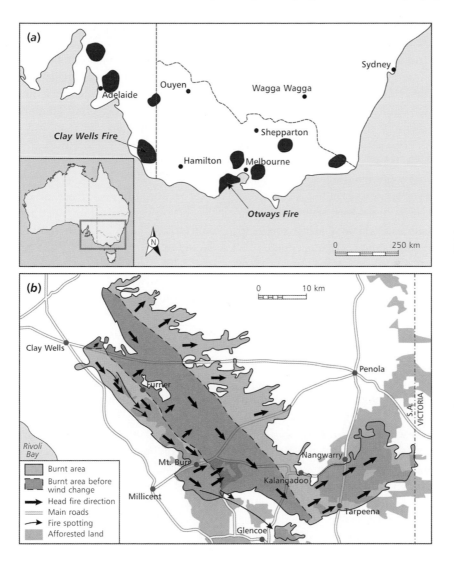

Figure 10.9 The 'Ash Wednesday' bushfires of 16 February 1983 in south-eastern Australia: (a) the location of major fires; (b) the progress of the Clay Wells fire, South Australia. This fire originated with a long, narrow shape created by strong, dry, north-westerly winds ahead of a cold front. The change in shape was due to the later onset of a south-westerly wind. Various spot fires can also be seen. After Keeves and Douglas (1983). Reprinted from A. Keeves and D.R. Douglas, Forest fires in South Australia on 16 February 1983 and consequent future forest management aims. *Australian Forestry* 46(3) (1983): 148–64. With permission from the Institute of Foresters of Australia.

flux by exposing the vegetation ahead of the fire to additional convective and radiant heat.

The combined effect of wind and slope is to position the advancing flames at an acute angle so that, once the slope exceeds 15–20°, the flame front is effectively a sheet moving parallel to the slope. Data from experimental fires in eucalypt and grassland areas in Australia have shown that the rate of forward progress of a fire on level ground doubles on a 10° slope and increases nearly four times when travelling up a 20° slope (Luke and McArthur, 1978). It is these conditions which pose a serious risk to fire-fighters as well as residents. The 1994 South Canyon fire in Colorado, USA killed 14 fire-fighters when a change in wind conditions caused flames to leap across the canyon floor and ignite a steep slope of Gambel oak trees immediately below the men. Within seconds, flames up to 90 m high spread up the slope at a speed impossible to out-run.

The combined effects of fuel and weather conditions were evident in the Ash Wednesday bushfires that raged across south-east Australia in February 1983 (Figure 10.9a). These fires included the largest conflagration experienced in the Forest Reserves of South Australia (Keeves and Douglas, 1983). The area had been in drought for the previous six months and the fires were all ignited between 11.00–16.30 hr, when air temperatures and solar radiation were high, relative humidity was low and the winds were strong and gusty. The first outbreak (the Narraweena fire) started at about 12.10 hr in grassland and, within four hours, travelled 65 km south-east through intensively managed agricultural land before veering with a change in wind direction. The parallel Clay Wells fire began at 13.30 hr in roadside grassland and quickly spread to native forest and adjacent pine plantations where large quantities of fuel created crown fires and allowed the development of spot fires down-wind (Figure 10.9b). By 16.00 h the wind had changed from north-west to west-south-west, and increased in speed from 30–60 km h^{-1} to 50–80 km h^{-1}, with gusts over 100 km h^{-1}, before dying down several hours later. In total, fire damaged about 30 per cent of the area planted with conifers in the forests of South Australia.

H WILDFIRE HAZARD REDUCTION

Wildfires present complex problems arising from the interaction of physical, biological and social factors in different landscape settings. Consequently, a range of practical solutions has to be employed (Gill, 2005).

1 Protection

After wildfire disasters, there are demands for stricter fire ban legislation. Such measures are difficult to enforce, although *Total Fire Bans* on days of extreme fire danger are needed and are used. They usually apply to a particular weather forecast district and last for 24 hours, during which period no fire may be lit in the open. Over-use of Total Fire Bans can increase the risk of a future wildfire, due to the growing supply of fuel that has been allowed to build up over time. The recognition of this relationship has led to the increasing use of low-intensity fires ('controlled burns'). The purpose of controlled burns is to consume existing fuel in relatively small, supervised fires that reduce the risk of major outbreaks later. This may be a cost-effective policy for genuine wildland areas but is less useful in the mixed landscape of the rural–urban fringe where farmland, forest plantations and suburban gardens coexist. Prescribed burning is labour intensive and can lead to uncontrolled fires, cause air pollution and have controversial effects on local ecosystems, such as a reduction in the diversity of flora. A simulation study of prescribed burning around Sydney, south-east Australia, showed that very frequent levels of burning were needed to improve fire safety significantly (Bradstock *et al.*, 1998). Such levels are difficult to achieve because of the high costs in steep terrain and the lack of sufficient dry days in some winters. Due to these difficulties, there is an increasing view that excess fuel should be removed by mechanical means – including commercial timber harvesting – as well as controlled burning.

2 Mitigation

Disaster aid

Aid provides limited assistance when disaster strikes. The 1983 'Ash Wednesday' fires in Victoria and South Australia raised a total fund of some A$12 million, which was channelled through an appeal fund administered by the Department of Community Welfare (Healey *et al.*, 1985). About three-quarters of this sum originated within Australia itself, including federal funds released under the National Disaster Relief Arrangements. A large part of the federal assistance was in the form of interest-free repayable loans, rather than direct grants. Similar responses followed the 2009 'Black Saturday' disaster, estimated to cost A$4.4 billion.

Disaster appeals inevitably raise questions about compensation for both insured and un-insured residents. Current fire insurance tends to rely on the private sector. For example, about two-thirds of all home-owners affected by the 1991 East Bay Hills fire in California had replacement-cost insurance cover. This was a major factor in defraying the costs of the federal government and ensured that recovery and rebuilding went ahead quickly (Platt, 1999). However, policy premiums are rarely calculated according to individual risk. At best, the standard residential policy considers only the presence or absence of adequate fire-fighting services. In future, there is more scope for premiums to reflect community responses to the enforcement of fire-safe building codes and vegetation management. For example, roofing materials are a major risk factor and premium reductions could easily be offered for fire-resistant roofing materials.

3 Adaptation

Community preparedness

Preparedness, including identifying vulnerable areas and the early detection and suppression of wildfires, is a vital element in disaster reduction. Geospatial technology is increasingly used to prepare wildfire risk maps which typically combine the three key variables of *fuel* (type, moisture level, size, shape), *topography* (elevation, slope, aspect) and *weather* (temperature, precipitation, relative humidity, wind speed), as shown by Lein and Stump (2009). Attempts have also been made to assess residents' perceptions of risk mitigation and their willingness to pay for mapped information (Martin *et al.*, 2009; Mozumber *et al.*, 2009).

Rural fire-fighting groups are the first line of defence. These bodies are composed of volunteers and are often taken for granted by state and federal governments. For example, in the USA the value of rural fire-fighting services to the nation has been estimated to exceed US$36 billion each year, but the fire-fighters feel they neither influence policy nor obtain the resources needed to work effectively (Rural Fire Protection in America, 1994). In Australia there are over 200,000 volunteers but, as in North America, the number is declining rapidly due, to socio-economic factors, including demographic trends that reduce the proportion of the population between 25 and 45 years of age (McLennan and Birch, 2005). Rural fire services require costly training and specialist equipment. Because piped water supplies are not always available, fire teams need methods to deliver and use water more efficiently. This implies dedicated items such as tankers for transporting water or access to aircraft. There is also a need for more general tools such as earth-moving plant to construct access tracks and firebreaks.

In the United States, major wildfires cross local government boundaries and affect land managed by private landowners and state and federal agencies. A comprehensive fuel modification plan should be agreed to reduce fire intensity, including prescribed burns and vegetation thinning. It is also necessary to have an overall view of fire-fighting requirements, including water supply and equipment. This approach was tried in California after the Oakland–Berkeley Hills firestorm of 1991, when the cities of Oakland and Berkeley formed a consortium with other 'inter-mix' land-owners to develop a coordinated hazard reduction plan. Similar bushfire management committees, representative of local interest groups, exist in Australia.

There are many reasons why preparedness for wildfires is often low. In Edmonton, Canada, households hold a variety of views on the effectiveness of fire reduction and rarely implement the full range of measures available (McGee, 2005). From a study in California, Collins (2005) concluded that residents were reluctant to remove vegetation from around their property because they attached a high amenity value to their semi-natural environment. Others, such as those living in areas lacking basic community services (roads,

piped water) and those who did not own their properties, lacked the incentives and the financial means to make hazard adjustments. In parts of Victoria, Australia, residents recognized the risk but relied on fire-fighters for protection and made few preparations of their own (Beringer, 2000). Awareness of fire hazard tends to grow with residence time in the area. Figure 10.10 shows that the deployment of self-reliant protection measures can increase fourfold with residence periods of 25 years or longer.

Forecasting and warning

This option plays a limited role. For example, in Australia a *fire season* may be declared by emergency agencies during which certain restrictions on outdoor fires apply. Daily *fire danger ratings* are issued by the Bureau of Meteorology throughout the season. *Fire-weather warnings* are given on days with a forecast of extreme fire risk, when a total fire ban is likely to be imposed. Comparatively little is known about the accuracy and the effectiveness of these warnings.

In populated areas, look-out towers may be sufficient for early fire detection. In more remote regions, regular surveys by aircraft or other remote sensing means may be necessary. During dry weather, plants reduce the amount of evapotranspiration from their leaves, with a consequent increase in the surface temperature of large vegetation stands, such as forests. These temperature changes can be detected on satellite images and the derivation of an appropriate 'vegetation stress index' can be used as an indication of where wildfire outbreaks are most likely to occur (Patel, 1995). It is then possible to intensify ground surveillance in these areas and to exclude the public until the fire risk starts to fall. In the western USA, most fires are caused by summer lightning strikes, so investment in automated lightning detection systems, with options for follow-up aerial survey, is prudent. Florida, on the other hand, has a year-round fire season, due to the fuel types and weather patterns, and most of the fires are human-caused. Here, early fire detection is

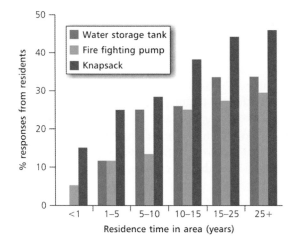

Figure 10.10 Relationship between length of residence in the North Warrendyte area, Victoria, Australia and the ownership of fire-fighting equipment. This area is exposed to high bushfire risk. After Beringer (2000). Reprinted from *Fire Safety Journal* 35, J. Beringer, Community fire safety at the urban/rural interface, 1–23. Copyright (2000), with permission from Elsevier.

based on a fixed-position, passive infra-red system (Greene, 1994). This uses sets of computer-controlled infra-red sensors, weather monitors and video cameras located at remote observation points to scan the horizon and mountain sides for thermal variations.

Early fire detection is important because risk reduction options decline rapidly once an outbreak occurs. In Australia, Handmer and Tibbits (2005) showed that late evacuation from homes, as opposed to staying inside, resulted in high fatality rates. More recent data reinforced the dangers of late evacuation and illustrated a gendered difference in bushfire deaths; men are mostly killed outside, protecting assets, whilst women and children died sheltering in the house or attempting to flee (Haynes *et al.*, 2010). Attention has turned to self-help measures, as residents are encouraged to develop a household strategy for voluntary evacuation well before a fire outbreak. This has been termed 'Prepare, Stay and Defend or Leave Early' (SDLE). The national policy has been renamed 'Prepare, Act, Survive' (Tolhurst, 2010). The choice between either early evacuation or sheltering in domestic properties remains difficult (Cova et al., 2009). Such decisions can be aided by the use of a dynamic GIS to activate *evacuation trigger buffers*, which are pre-established boundaries around a community that trigger an evacuation order if a fire crosses the line (Pultar *et al.*, 2009; Larsen *et al.*, 2011).

Land use planning

Some wildfire threats exist because local governments have not adequately factored the hazard into the development control system (Buxton *et al.*, 2010). Following the 2009 bushfire disaster in Australia, new policy measures were proposed for Victoria, including more stringent building codes designed to resist fire temperatures and greater control on future developments. In turn, residents need to comply with safety requirements and take some responsibility for their own security.

Improved land planning and public education would aid hazard reduction. The basic tool is a large-scale, local map showing severe wildfire hazard areas that can be used to steer development. For example, wide *fire-breaks* are an integral part of rural land planning either to exclude fire or to isolate crops, timber plantations or other high-risk assets. Within high-risk areas, detailed landscaping is necessary, including:

- cluster development – individual homes or apartments built in small groups (saving land for community open space)
- low overall housing density – individual residential lots at least 0.5 ha
- minimum spaces between buildings (approx. 10 m) and clusters
- access lanes wide enough for fire-fighting equipment
- all properties with a minimum set-back from the natural bush (approx. 30 m)
- mature trees and large shrubs on edge of bush pruned to avoid direct spread of fire from one tree to another
- intermediate area managed by clearing all dead vegetation and planting grass or small, low fuel-volume vegetation.

More thought should be given to limiting occupancy rates, and to raising the capability of road networks for emergency evacuation, when fire-prone areas are approved for new residential development (Cova, 2005). It is important that neighbouring local authorities work together on land and vegetation management and develop a common database for land planning in the WUI. Public fire-prevention education is also important. A persuasive approach can be reinforced in a variety of ways. For example, the provision of barbecue places set up by local authorities in safe clearings alongside roads tends to discourage indiscriminate fire lighting. An increased understanding of the benefits of prescribed burns could help officials to obtain the cooperation of landowners for fuel management practices. Finally, when areas have been burned out, consideration should be given to the government acquisition of land for public open space and the rebuilding of properties at lower densities on larger plots.

Communities threatened by wildfire often have little hazard awareness and there is a key role for residents. Home-owners build with highly flammable materials, such as weatherboard or wood-shingle roofs, retain thick vegetation too close to their property and disregard the adequacy of fire-fighting equipment. This may happen despite the existence of legislation designed to prevent such actions, such as that in California, which requires property owners in State Responsibility Areas to remove flammable vegetation for a distance of at least 10 m from a structure or to the property line, whichever is closer.

Other safety steps include:

- planting low-growing, fire-resistant plants near to the house
- installing a fire-resistant roof and using such material on surrounding decks and walkways
- removing dead vegetation nearby and branches from nearby trees
- ensuring the property is accessible to fire-fighters
- stacking logs and other flammable items at least 10 m from the house
- compliance with any fire-safety inspections and subsequent recommendations.

FURTHER READING

Beringer, J. (2000) Community fire safety at the urban/rural interface: the bushfire risk. *Fire Safety Journal* 35: 1–23. Gives a clear account of wildfire issues in Australia.

Noji, E.K. (ed.) (1997) *The Public Health Consequences of Disaster.* Oxford University Press, New York. This remains a useful source of reference.

Kaskaoutis, D.G. *et al.* (2011) Satellite monitoring of the biomass-burning aerosols during the wildfires of August 2007 in Greece: climate implications. *Atmospheric Environment* 45: 716–26. A case study of the wildfire threat in the Mediterranean parts of Europe.

Kuhn, K., Campbell-Lendrum, D., Haines, A. and Cox, J. (2005) *Using Climate to Predict Infectious Disease Epidemics.* World Health Organization, Geneva. An important treatment of this subject.

Lagadec, P. (2004) Understanding the French 2003 heat wave experience: beyond the heat, a multi-layered challenge. *Journal of Contingencies and Crisis Management* 12: 160–9. A thought-provoking insight into just one of the likely consequences of climate change.

Ligon, B.L. (2006) Infectious diseases that pose specific challenges after natural disasters: a review. *Seminars in Pediatric Infectious Diseases* 17: 36–45. Demonstrates the link between disaster and disease.

WEB LINKS

Centers for Disease Control and Prevention, USA www.cdc.gov

Pan American Health Organization www.paho.org/disasters

History of bushfires in the Australian Capital Territory www.esb.act.gov.au/firebreak/actbushfire.html

Food and Agriculture Organization Locust Watch Group www.fao.org/ag/locusts/en/info/index.html

Fire Weather Information Center USA www.noaa.gov/fireweather/

Roll Back Malaria Partnership www.rbm.who.int/

World Health Organization www.who.int/en/

Wildfire Impact Reduction Center, USA www.westernwildfire.org/

CHAPTER ELEVEN

Hydrological hazards

Floods

A FLOOD HAZARDS

Floods are a common environmental hazard, due to the widespread distribution of river floodplains and low-lying coasts that have proved attractive for human settlement. Over 3,000 flood disasters were recorded in the EM-DAT database between 1990 and 2010. They were responsible for more than 200,000 deaths and adversely affected nearly 3 billion people, mostly through homelessness. For example, a total of 12 million people were displaced by floods during 2008 in India and 2010 in Pakistan. At present, 800 million people live in flood-prone areas; 70 million are exposed to flooding every year (Peduzzi *et al.*, 2011). Floods coexist with other hazards and can be difficult to classify; they can be the consequence of storms and tsunamis but may cause landslides and disease epidemics.

About 90 per cent of all flood-related deaths and 50 per cent of the economic damage occurs in Asia, notably in China and Bangladesh. Since 1900 the five most disastrous river floods worldwide have occurred in China, causing almost 6.5 million deaths and affecting 900 million people, mostly through homelessness. In Bangladesh over 500,000 people have been killed by coastal and river floods since 1970. Damaging floods are not confined to the LDCs. They are the most frequent environmental disaster in Europe and Table 11.1 lists five recent events. All occurred in the summer half of the year and display low fatality rates combined with high damage costs. Similarly, the Midwest floods of 1993 in the United States, which affected over 15 per cent of the country, created total losses of US$15–20 billion although there were fewer than 50 deaths (US Department of Commerce, 1994). In 2005, coastal flooding caused by hurricane 'Katrina' led to the costliest natural disaster in US history and took around 1,600 lives.

Floods pose various threats, depending on the depth and velocity of the water, the duration of the flood and the quality of the water (sediment load, salts content, presence of sewage, chemicals, etc.). Some approximate damage thresholds for depth and velocity are shown in Figure 11.1. People and cars can be washed away in fast-flowing water 0.5 m deep. Buildings, and other bankside obstructions, create turbulent scour effects and the foundations of houses can start to fail at velocities of 2 m s^{-1}. The stresses on

Table 11.1 Five flood disasters in Europe 2000–10

Year	Country	Month	Deaths	Number of people affected	Total economic losses (thousand US$)
2002	Germany	August	27	330,108	11,600,000
2002	Czech Rep	August	18	200,000	2,400,000
2005	Romania	July	24	14,669	800
2005	Romania	September	10	30,800	No data
2007	United Kingdom	July	7	340,000	4,000,000

Source: Compiled from EM-DAT

structures like bridges increase when high flows contain large-calibre debris such as rocks or ice blocks. The collapse of sewerage systems and storage facilities for products like oil or chemicals produces pollution hazards. In November 1994 over 100 people were killed in Durunqa, Egypt, when floods destroyed a petroleum storage facility and carried burning oil into the town in a na-tech flood disaster.

The threats to human life are drowning and asphyxiation. Flood depth and mortality are closely related and males are more at risk than females. Post-flood morbidity is most likely in low-income countries with poor infrastructure (Ahern *et al.*, 2005). Faecal-borne intestinal diseases (cholera, typhoid) break out where sanitation standards are low or when sewerage systems are damaged. In tropical countries the increased incidence of vector-borne diseases, such as typhoid or malaria, can double the endemic mortality rate. Rodent-borne disease (leptospirosis) can also emerge. In the MDCs, survivors tend to suffer from mental health problems (anxiety, depression, post-traumatic stress disorder). Eighteen months after the 1972 flood disaster at Buffalo Creek, West Virginia, over 90 per cent of survivors had related mental disorders (Newman, 1976). In a study of 771 fatalities from hurricane 'Katrina', two-thirds of the deaths could be attributed to direct physical impacts, mainly experienced by elderly people (Jonkman *et al.*, 2009). Other deaths were recorded outside the flooded areas of New Orleans, or in hospitals or shelters within the flood zone, as a result of the inadequate public health arrangements after the disaster.

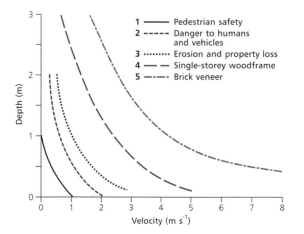

Figure 11.1 Flood hazard thresholds as a function of depth and velocity of water flow. Curve 1 indicates approximate risk to pedestrians; Curve 2 danger of death by drowning; Curve 3 threat of bank erosion and the loss of shanty-type housing; Curves 4 and 5 the destruction of more permanent residential buildings. Adapted from D.I. Smith (2000) and other sources.

Direct structural damage to property, especially in wealthy urban areas of the LDCs, is the major cause of tangible flood losses. There are also longer-term, secondary losses, due to reduced house values when flood-prone property is resold (Tobin and Montz, 1997). Crops, livestock and the agricultural infrastructure suffer major

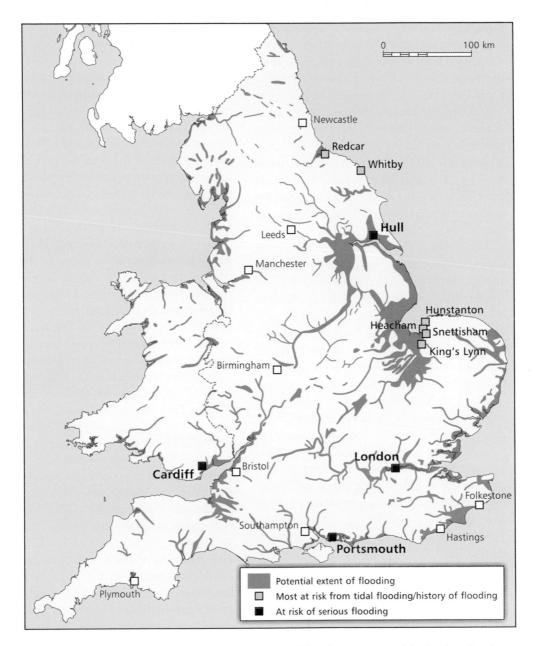

Figure 11.2 Areas of potential flooding in England and Wales and the urban centres at risk of serious flooding. Coastal flooding poses the greatest threat, especially in low-lying parts of eastern and southern England. After UK Environment Agency at www. environment-agency.gov.uk/mediacentre (accessed 19 September 2011). Contains Environment Agency information © Environment Agency and database right.

damage in intensively farmed rural areas. In India, for example, almost 75 per cent of overall flood damage has been attributed to crop losses, whilst riverbank erosion of farmland and villages in Bangladesh destroys crops and renders up to 1 million people landless and homeless every year (Zaman, 1991). However, there is a growing economic threat in the mega-cities of the LDCs. Ranger *et al.* (2011) estimated that the total losses from a 1:100-year flood in Mumbai, India could triple from US$700 million at present to US$2,305 million by the 2080s unless remedial actions take place.

More than any other environmental hazard, floods bring benefits as well as losses (Smith and Ward, 1998). The seasonal 'flood pulse' is a vital element in maintaining a diverse range of habitats in wetland ecosystems. After the initial physical and ecological disturbance due to major floods, there is a burst of biological productivity. Floods maintain the fertility of soils by depositing layers of silt and flushing salts from the surface layers. Although silt-laden flood-waters regularly cover only a small area of Bangladesh, the alluvium enriches the phosphorous and potash content of the soil. Along low-lying coasts and estuaries, periodic inundation helps to maintain salt-marshes and mudflats, often rich in wildlife, as well as more specialized vegetation such as man-grove forest.

Most traditional societies are well adapted to the flood pulse. Floods provide water for irrigation and for fisheries, which are a major source of protein. Flood-retreat agriculture, where the moist soil left after flood recession is planted with food crops, is widely undertaken in the tropics. The seasonal inundation of large floodplains in semi-arid West Africa is of crucial ecological and economic importance and supports a larger agricultural output than that obtained from formal irrigation systems (Adams, 1993). In a normal year, floods bring all these benefits; it is only the rare, high-magnitude events that create disaster.

B FLOOD-PRONE ENVIRONMENTS

The nature and scale of flood risk varies greatly. In most countries rivers are the greatest hazard, as in the United States, where river flooding accounts for about two-thirds of all federally declared disasters. Conversely, rivers in the UK represent only one-third of the total flood risk. This is because storm rainfall maxima in the UK are low when measured against world extremes and produce less-aggressive rivers. In addition, virtually all buildings are constructed of brick or stone and are not easily damaged. However, sea flooding is a serious threat caused by the coastal configuration of eastern and southern England, combined with long-term land subsidence, rising sea levels and under-investment in coastal flood protection over many years (Figure 11.2). In February 1953 over 300 people died in eastern England when the sea walls were over-topped.

Wide differences exist in the proportion of the national populations exposed to flood risk (Parker, 2000; Blanchard-Boehm *et al.*, 2001):

- France 3.5 per cent
- United Kingdom 4.8 per cent
- United States 12 per cent
- Netherlands 50 per cent
- Vietnam 70 per cent
- Bangladesh 80 per cent.

Exposure is related to high rural population densities, especially in the LDCs, and to the location of urban areas along rivers and coasts in more industrialized societies. In China, vast alluvial river plains contain half the total population and the problems are vast (Box 11.1). In countries like Bangladesh and Vietnam there is a combined threat from river, delta and sea floods. Although New Zealand has a low population density, nearly 70 per cent of the towns and cities with populations in excess of 20,000 have a river-flood problem (Ericksen, 1986). In densely populated countries, the issues are greater. About

Box 11.1 Flood hazards on the Yangtze River, China

China has a long record of flood disasters. Attempts to protect cities date back over 4,000 years (Wu, 1989). Most human settlement and cultivable land is on the alluvial plains of seven great rivers prone to flooding. The largest of these is the Ch'ang Chiang (Yangtze) – third-longest river in the world – which flows for 6,300 km towards the Pacific Ocean through an area populated with more than 75 million people. This river killed more than 300,000 people during the twentieth century. In the middle reaches, two huge connected depressions – the Dongting lake (covering an area the size of Luxembourg) and the smaller Poyang lake – provide flood storage and help to protect downstream areas. In an average year, the Yangtze carries 500×10^6 tonnes of sediment, mostly in the flood season. Under natural conditions, most rivers in the Yangzte basin would change course frequently in response to the progressive rise of the land, due to silt deposition.

The catchment area is subject to summer monsoon rains and tropical cyclones. Records over the last 2,000 years show that damaging floods occur, on average, once in every 10 years and, over the last 500 years, the variability of floods and droughts can be linked to ENSO episodes (Jiang *et al.*, 2006). In 1998, 32 million ha of land were flooded, over 3,000 people were killed or injured and more than 200 million people were affected with the direct property damage estimated at US$20 billion.

Flood management includes:

- *Levees.* The earliest flood levees date back to 345 AD and are designed to control routine floods with a 10–20-year return period (Zhang, 2004). There are now about 3,600 km of main river levees and 30,000 km of tributary levees protecting farmland, oilfields and cities. But, due to silt deposition in the constricted river channel, the level of the Yangtze in high flood can attain 10 m above the land behind the levees, which themselves reach 16 m in height in places. Many levees are old and weak, and subject to breaching.
- *Lakes.* Many surface depressions are used to regulate large floods that over-top the levees. Lake Dongting provides detention storage but starts to break its banks when full and threaten 667,000 ha of densely populated farmland and over 10 million people, many in cities such as Yueyang and Wuhan. Major lake failures are more frequent, due to silt deposition and land reclamation for economic development. The capacity of the lake has been reduced by nearly 80 per cent since 1950. In August 2002, the Dongting lake reached a record high level of 35.9 m, a state of emergency was declared and more than 80,000 people were mobilized to strengthen the levees.
- *Dams.* China has built about half of all the world's 45,000 large dams, mostly in the 1950s and 1960s. This tradition ensured the construction of the Three Gorges Dam (TGD) near Chingquing, designed to produce hydro-electricity and control floods in the middle and lower reaches of the Yangtze. The TGD stores up to 39 billion m^3 of water, held in a reservoir 600 km long behind a dam 175 m high and nearly 2 km long. It is the largest hydropower station and dam in the world. Water impounded by the dam has flooded 13 cities, 140 towns and over 1,300 villages. At an estimated cost of US$22.5 billion, it is probably the most expensive single structure – and one of the most controversial – ever built. Social costs have been high. Hwang *et al.* (2007) recorded high levels of stress and depression in residents subjected to involuntary migration, especially for poor rural people who had never previously moved in their lives.

The reservoir began to fill on 1 June 2003. The TGD project began operating immediately but was not declared complete until 2009. During the dry season between December and March the dam discharges water for agricultural and industrial purposes down-river. Before the onset of the flood season in June, storage is reduced to store flood flows equivalent to 22 billion m^3. This should reduce the peak flow of the 100-year flood in the downstream Jinjiang section from 86,000 m^3 s^{-1} to less than 60,000 m^3 s^{-1}, a discharge within the safe capacity of this part of the river. As an example, inflows to the dam during July 2010 peaked at 70,000 m^3 s^{-1}. This exceeded the maximum discharge recorded during the 1998 floods but the regulated outflow remained at about 40,000 m^3 s^{-1}.

Future success of the TGD project is uncertain. Sediment deposition behind the dam is likely to limit its efficiency and many alleged negative aspects include landslides, losses of biodiversity and wildlife and the destruction of over 300 heritage sites. Legislation to encourage afforestation, combined with controls on illegal logging, now exists but is unlikely to be fully implemented or have much influence on peak flows. Early experience indicates that the TGD has had significant effects on the flow of the Yangtze river below the dam, although the effects decay downstream, due to inflows from downstream tributaries (Guo *et al.*, 2012*).* Complex interactions between the river and Poyang lake have been found. One positive consequence may be the partial mitigation of flood risks in the lake basin during the July–September wet season. On-going monitoring of erosion and sediment deposition and efficient flow management will be crucial in optimizing the flood-control benefits of the TGD (Fang *et al.*, 2012). Emergency flood evacuation could involve up to 1 million people with their livestock and possessions. The flood problems on the Yangzte river remain daunting.

Source: Material adapted from review papers and other documents made available by the World Commission on Dams website http://www.dams.org (accessed 17 March 2002) and the Chinese Embassy website www.chinese-embassy.org.uk (accessed 3 December 2006).

2.4 million people and one in six residential properties in England are at flood risk, despite the existence of more than 25,000 miles of structural flood defences (Anonymous, 2009).

The most vulnerable landscape settings for floods are:

Low-lying parts of major floodplains

When left in their natural state, these setting suffer frequent inundation. As a result, such flood zones in the MDCs are typically afforded protection by engineering works and are subject to relatively strict planning controls. Within the LDCs the risk of disaster is much greater. In Bangladesh over 110 million people live with limited protection on the floodplain of southern Asia's most flood-prone river system – the Ganges–Brahmaputra–Megna. This river basin extends over more than 1,750,000 km^2 and, in an average year, receives about four times the annual rainfall of the Mississippi basin in the USA. As shown in Figure 11.3, the complex deltaic terrain of Bangladesh experiences several types of flooding. Half of the country is less than 12.5 m above mean sea level and seasonal floods regularly cover 20 per cent of the total land area. In very high flood years, up to two-thirds of the country may be inundated at any one time. The 1988 floods affected 46 per cent of the land area and killed an estimated 1,500 people; in 1998 over 1,000 people were killed and direct damages reached US$2–3 billion, the highest recorded to that date (Mirza *et al.*, 2001).

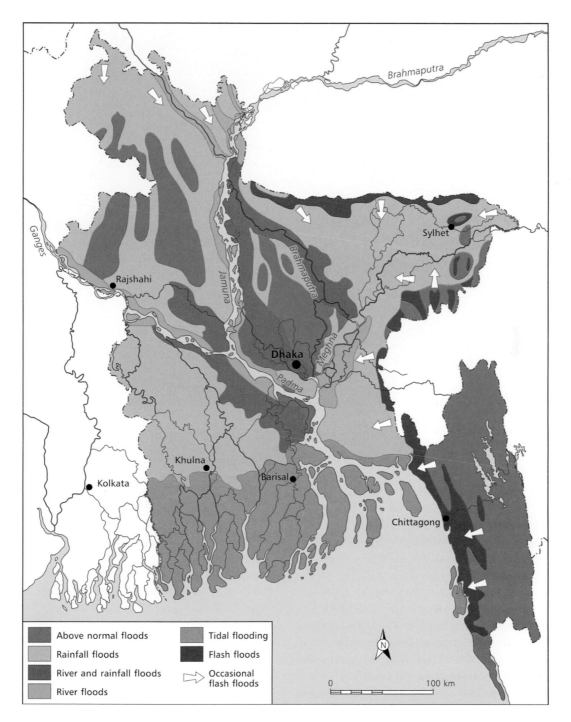

Figure 11.3 Types of flooding in Bangladesh. Some areas are affected by more than one type of flood. The highest risk from normal seasonal flooding is along river courses and at the edge of the delta. After Brammer (2000). Reproduced with permission from H. Brammer, Flood hazard vulnerability and flood disasters in Bangladesh. In D.J. Parker (ed.) *Floods*, vol. 1 (Routledge, 2000), 100–15.

Plate 11.1 A woman carries a water jug whilst wading through floods in search of clean drinking water in Shyamnagar Upzila, Satkhira District, Bangladesh. Many thousands of people were made homeless when cyclone 'Aila' struck Bangladesh in May 2009, creating tidal surges and extensive floods. (Photo: Panos/G.M.B. Akash AKA03267BAN)

Low-lying coasts and deltas

These are essentially estuarine areas exposed to high risk from a combination of river floods and high tides, as in the case of the lower Thames in London, England. These areas can be submerged by a mixture of fresh and marine water when peak river flows caused by inland flooding are prevented from reaching the sea, perhaps as a result of high-tide conditions. More direct marine flooding occurs when salt water is driven onshore by wind-generated waves, and storm surges are responsible for most of the worldwide loss of life from coastal floods (see Chapter 9). Much rarer marine invasions result from tsunami waves created by earthquakes (see Chapter 6).

Coasts and deltas are often densely populated. In Vietnam all low-lying areas have been intensively exploited by the rural population for wet-rice cultivation, especially in the deltas of the Red River in the north and the Mekong River in the south (Department of Humanitarian Affairs, 1994). As a result, the distribution of population and the distribution of flood-prone land coincide almost exactly (Figure 11.4). Many urban areas are also at risk. At the start of the twenty-first century, 17 out of the 25 cities with populations in excess of 9 million were on the coast (Timmerman and White, 1997). These cities, often surrounded by populous rural areas, tend to be in countries lacking effective coastal-zone management and

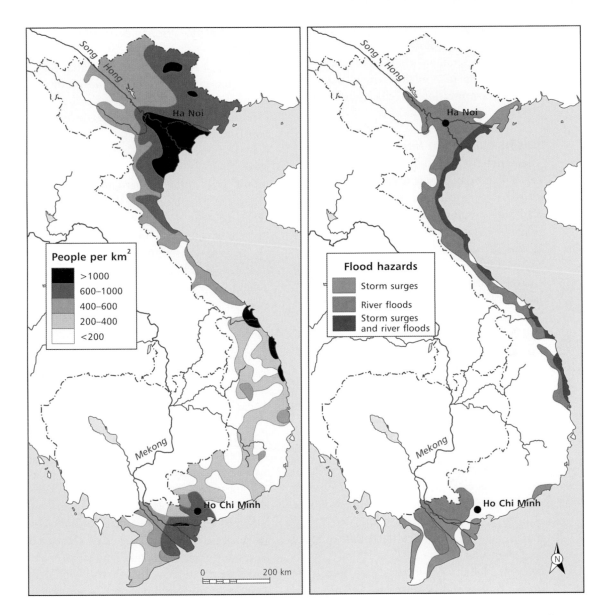

Figure 11.4 Human vulnerability to flooding in Vietnam: (a) distribution of population; (b) areas subject to different to types of flooding. Alluvial deltas and coastal plains, intensively settled for wet-rice cultivation, are prone to river floods and storm surge. About 70 per cent of the population is at risk from flooding. Reproduced from K. Smith and R. Ward, *Floods: Physical Processes and Human Impacts* (John Wiley and Sons, 1998). Copyright John Wiley and Sons 2009. Reproduced with permission.

planning controls. Mumbai, India has the highest exposure to coastal flooding of any city in the world. Nearly 3 million people are at risk, a figure projected to rise to 11 million, with little prospect of effective protection or relocation (Hanson, 2011).

Basins subject to flash floods

These are at risk from short, high-intensity rainfall events that occur locally over small – often un-gauged – drainage basins. They are mainly associated with combinations of steep topography, limited vegetation cover and severe convective storms in arid and semi-arid zones. However, narrow valleys and heavily developed urban settings elsewhere are also likely to be flood prone. It has been suggested that 90 per cent of the lives lost through drowning in tropical countries result from intense rainfall on small, steep catchments upstream of poorly drained urban areas. The city of Kuala Lumpur, Malaysia lies at the foot of a relatively steep, fan-shaped basin with almost perfect hydrological conditions for generating flash floods.

Flash floods are a hazard in Europe in drainage basins less than 500 km² in extent, although damaging events in the Mediterranean region can develop over catchments twice this size (Gaume *et al.*, 2009). The 1999 Aude flood in France claimed 35 lives and damages estimated at €3.3 billion; the 1962 Barcelona flood (Spain) killed over 100 people and the 2002 Gard flood (France) cost €1.2 billion. At present, flash floods are poorly understood because physical processes are often inferred from post-event analysis rather than real-time monitoring. Hazard warning times are either short or non-existent. In the Big Thompson Canyon, Colorado flood of July 1976, 139 people were drowned after a thunder-storm produced 300 mm of rain in fewer than six hours. Many of the dead were tourists with little awareness of the dangers and the need to escape quickly from the canyon floor.

Areas below unsafe or inadequate dams

These areas have great disaster potential. According to the International Commission on Large Dams, there are 45,000 dams worldwide that exceed 15 m in height. About three-quarters were built before 1980; many are poorly maintained. In the USA alone more than 2,000 communities are at risk from dams believed to be unsafe and there is little opportunity for warning and evacuation. When the foundations of the Malpasset dam, France failed in 1959, 421 people died. Even dams that are structurally sound may be over-topped by surges of water induced by earth movements. In 1963 a landslide created a major flood surge behind the Vaiont dam in Italy. Although the structure held, the subsequent wave of water killed 3,000 people downstream. When a dam burst in 1972 in the coal mining valley of Buffalo Creek, West Virginia there was no warning and 125 people were killed and 4,000–5,000 became homeless. Few countries have prepared inundation maps or made emergency plans for such events.

Low-lying inland shorelines

These extend for thousands of kilometres and involve much property, as around the Great Lakes and the Great Salt Lake in North America. Fluctuating lake levels from river inputs are the main problem. Lake levels tend to rise to damaging heights only after a period of wet years but the erosion of barrier islands, sand dunes or bluffs removes any natural protection from wind-driven wave attack on buildings and other shoreline facilities.

Alluvial fans

These environments present a special type of flash flood threat, especially in semi-arid areas. About 15–25 per cent of the arid American West is covered by alluvial fans that can provide attractive development sites, due to their commanding views and good local drainage (FEMA, 1989). The flood hazard is under-estimated because of the prevailing dry conditions, which lead to long intervals between successive floods, and the absence of well-defined surface watercourses. The braided drainage channels meander unpredictably across the steep slopes, often ignored by the local population until flash floods reach velocities of 5–10 m s^{-1} and contain high sediment loads.

C THE NATURE OF FLOODS

All floods represent the temporary state existing when a body of water rises to inundate land not normally submerged. A river flood occurs when water over-tops the banks – either natural or artificial – of a river; a marine flood occurs when the sea invades low-lying coastal zones. Both become hazards when life and property are threatened. For a hydrologist, flood magnitude is best expressed in terms of peak river flow (*discharge*), whilst the hazard potential relates more to the maximum height (*stage*) that the water reaches. The hazard potential reflects a mix of physical and human factors. Smith and Ward (1998) distinguished between primary physical causes (mainly climatological forces) and secondary physical conditions (catchment-specific factors).

1 Physical factors: river floods

Climatic conditions

Figure 11.5 indicates that some floods may be a product of other environmental hazards, but climatic extremes, notably *excessive rainfalls*, are the principal cause of flood disasters. Such conditions vary from semi-predictable annual rains over wide geographic areas, which give rise to the wet-season floods of tropical areas, to almost random convectional storms over small drainage basins.

South-west Asia is prone to widespread damaging floods in the monsoon season. For example, 70 per cent of India's average annual rainfall comes during 100 days during the summer. In 2010, Pakistan suffered exceptional floods after heavy monsoon rains reached the Suliaman mountain range in the north-west of the country on 22 July. Khyber Pakhtunkhwa Province (formerly North-West Frontier Province) received 9,000 mm of rain in one week – about 10 times the yearly average (OCHA, 2010). Local flash floods and landslides contributed to significant flooding on all the major rivers downstream. Within a few weeks 160,000 km^2 – about 20 per cent of the national territory – was under water (Figure 11.6). Some 1,980 people were killed, 30,000 were rescued and 18.1 million were affected out of a population totalling 180 million. Some 14 million people became in urgent need of humanitarian assistance. About 80 per cent of residents in the flooded areas depend on agriculture for their livelihood and 2.4×10^6 ha of cultivable land were devastated. In addition, more than 1.6 million homes were damaged or destroyed, along with 430 community health facilities and 10,000 schools (United Nations in Pakistan, 2011). Given that, in 2010, Pakistan was ranked 125 out of 169 countries on the HDI scale, and almost one-quarter of the population is malnourished, full recovery will take a long time.

Excess rain across large drainage basins is also associated with tropical cyclones (mainly late summer in the subtropics) or major low pressure systems (mainly winter in the mid-latitudes). In February 2000 heavy rains fell over much of southern Africa. These regular rains are due to activity in the inter-tropical convergence zone and strong convection in hot moist air during the wet season, but the 2000 floods were supplemented by rain from two tropical cyclones. The main problems were in the low coastal plains of Mozambique, the most indebted country in the world relative to income. On the Limpopo River, the water reached levels not exceeded in the last 150 years and an area almost the size of the Netherlands and Belgium combined was submerged (IFRCRCS, 2002). Some 700 people were killed, 450,000 made homeless, 544,000 displaced and a further 4.5 million otherwise affected. In total, 10 per cent of all cultivated land was destroyed,

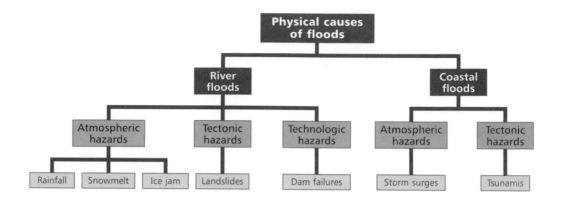

Figure 11.5 Causes of floods in relation to other environmental hazards. High-rainfall episodes are most important but the diagram also illustrates the multi-causal nature of flooding.

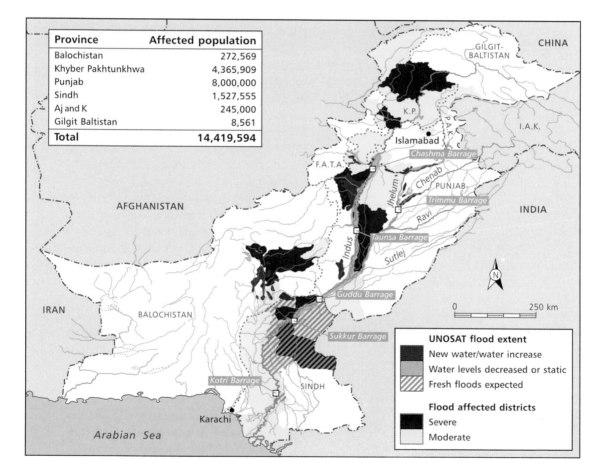

Figure 11.6 Map showing extensive flooding over Pakistan in August 2010. At this time over 14 million people were affected and by the end of the disaster the total had risen to more than 20 million. This figure is far larger than that for any other disaster in recent years. After OCHA (2010).

including one-third of the staple maize crop and 80 per cent of cattle. Direct costs were estimated at US$273 million. Since 2000, the government has enacted a large resettlement programme for almost 60,000 families, which helped to minimize losses when flooding reoccurred in 2006–7.

High-intensity rainfall

High-intensity rainfall occurs in more localized storms. If strong convectional cells are located over small drainage basins, catastrophic flash floods can result. This is especially so during the summer season in large continental interiors, although the most hazardous flood regime of Mediterranean Europe occurs in autumn (Marchi *et al.*, 2010). Catchment morphology is important because orographic effects intensify precipitation and promote the concentration of peak stream-flows, with great damage potential. In June 1972, Rapid City, South Dakota was devastated by a flash flood and the 238 deaths remain the highest loss of life from a single flood in the United States.

Snow and ice

Snow and ice are responsible for many flood disasters in higher latitudes. Melting snow in late spring and early summer routinely causes extensive flooding in the continental interiors of both North America and Asia. The most dangerous conditions arise from rain falling onto melting snow. This combination occurred in the Romanian floods of May 1970, when the Transylvanian basin was devastated by heavy rain from a deep depression plus snowmelt from the Carpathian Mountains. In April 1997 exceptional snowmelt produced a flood with an estimated return period of 1:100–200 years on the Red River of the North at Grand Forks, USA. The damage was estimated at over US$2 billion, making it the most costly flood to date, on a per capita basis, for a major metropolitan area in the US. Melt-water floods are compounded by ice-jam flooding, which occurs when floating ice, resulting from the spring break-up, causes the temporary damming of a river. The ice lodges at bridges and other constrictions in the channel, or at shallows where the channel freezes solid. The largest ice masses can destroy buildings and shear off trees above the water level; near lake shorelines, pressure ridges in the ice can dislodge houses from their foundations.

Drainage basin conditions

These conditions intensify the flood response to the precipitation input over a river basin. Some factors are permanent (the hydraulic geometry of the basin), whilst others (the effect of seasonally frozen soils in reducing infiltration) are temporary. Together with precipitation characteristics, they determine key features of a flood such as the speed of onset, the flow velocity, the peak flow and the duration of the event. Nepal is a mountainous country much influenced by flood-intensifying conditions. Small drainage basins, steep deforested slopes and melt-water from snow and glaciers create flash floods and landslides in the monsoon season between June and September. A special hazard is glacial-lake outburst floods, resulting from breached glacial moraines. These have created peak discharges well in excess of $2,000 \text{ m}^3 \text{ s}^{-1}$ that extend for tens of kilometres down the valleys, destroying villages and covering cultivated land with debris (Cenderelli and Wohl, 2001). In the 2000 monsoon season, over 500 people were killed, plus a further 250,000 affected. For a time, over 30,000 people were isolated in remote upland areas.

Flood-intensifying conditions can also arise from human actions, notably changes in land use. Some changes may be deliberate, such as agricultural land drainage designed to remove surface water from productive areas, but inadvertent land use changes are often more important.

Urbanization

Urbanization increases the magnitude and frequency of floods in several ways:

- Figure 11.7 is a schematic illustration of how the spread of impermeable urban surfaces,

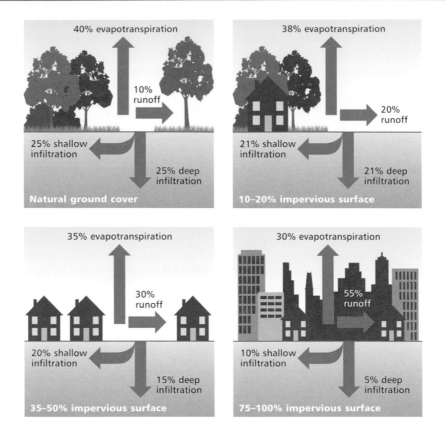

Figure 11.7 Schematic illustration of the influence of urbanization on the hydrological cycle. There is a progressive increase in surface run-off from 10 per cent to 50 per cent of the total rainfall, due to the spread of impervious surfaces. After Federal Interagency Stream Restoration Working Group (2001), USDA–Natural Resources Conservation Service, http://www.nrcs.usda.gov/Internet/FSE_DOCUMENTS/stelprdb1044574.pdf US government material not subject to copyright protection in the United States.

such as roofs and roads, changes the local water cycle so that a higher proportion of storm rainfall is translated into surface run-off (Anonymous, 2008). Small flood peaks may be increased up to 10 times and the 1:100-year event may be doubled in size by introducing a 30 per cent paving cover into an undisturbed rural area.

• Hydraulically smooth urban surfaces, serviced with dense networks of drains and underground sewers, deliver water more rapidly to the river channel than before. This affects flood onset, perhaps reducing the lag period between storm rainfall and peak flow by half. Overall,

this means that, downstream of urbanized areas, higher flood peaks occur which arrive more quickly – with less opportunity for hazard warning – than under pre-urban conditions (Figure 11.8).

• Natural river channels are often constricted by bridge supports or riverside facilities, thus reducing the carrying capacity so that high flows over-top the banks more frequently. For example, successive navigation works on the Mississippi river have reduced the capacity of the natural channel by one-third since 1837 (Belt, 1975). As a result, the flood of 1973 was viewed as a rare 1:200-year event in terms of peak water level, although the flow volume had an average recurrence interval of only 30 years.

- Insufficient storm-water drainage capacity following building development is a major cause of intra-urban flooding. The typical design capacity may be for storms with return periods as low as 1:10 to 1:20 years. This problem is increased where old, poorly maintained sewerage systems exist – as in parts of the UK – and storm water is frequently surcharged onto low-lying urban areas. It is estimated that US$300 billion will be needed over a 20-year period to upgrade sewerage infrastructure to an acceptable standard across the USA.

Deforestation

This can cause greater flood run-off from valley slopes, together with a reduction in channel capacity due to increased sediment deposition. In small drainage basins, fourfold increases in peak flows have been recorded, together with suspended sediment concentrations as much as 100 times greater than in rivers draining undisturbed forested land. The 1966 flood that claimed 33 lives and damaged 1,400 works of art and 300,000 rare books in the city of Florence, Italy was partially attributed to long-term deforestation in the upper Arno basin.

On the other hand, direct cause-and-effect relationships between forest cutting in the headwaters and major floods far downstream are hard to find for large river basins. Hamilton (1987) conceded that forest cutting, followed by abusive agricultural practices, in the Himalayas may aggravate flooding but cautioned against misunderstanding natural processes. Despite long-term hydrological records, no statistically reliable increase in flooding has been found in the plains and delta areas of the Ganges–Brahmaputra river system (Ives and Messerli, 1989). Mirza *et al.* (2001) concluded that the high monsoon rains in the Himalayas, combined with very steep slopes, ensure rapid run-off and high sedimentation rates, irrespective of the vegetation cover. Any recorded increases in flood damage were attrib-

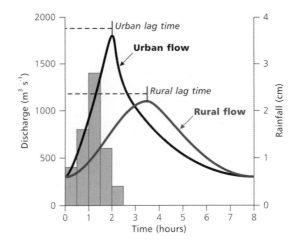

Figure 11.8 Idealized flood hydrographs from rural and urban areas, showing the reduced time-lag and higher peak flow associated with urbanization. This increases flood risk because warning time is reduced and the peak flow is more difficult to contain within the river channel.

uted to population growth and an expansion in the agricultural area.

2 Physical factors: coastal floods

Hazardous flooding of coasts and estuaries occurs when the sea surface is raised above the normal levels created by waves and tides. Such increases in height result from both short-term factors and much longer-term processes.

Short-term factors

These include storm surges driven by onshore hurricane-force winds (see Box 9.1 Hurricane 'Katrina',) and tsunami waves created by sea floor earthquakes (see Box 1.1 Tōhoku 2011). In addition, meteorological and hydrological conditions can combine with coastal configuration to create floods. For example, the semi-enclosed, low-lying coast of the North Sea is exposed to northerly gales that force wind-driven water to pile up towards the south, where the sea narrows. This situation led to the storm surge disaster of 31 January–1 February 1953. As shown in

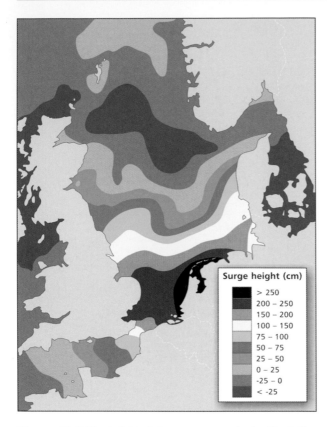

Figure 11.9 The height of the storm surge in the North Sea on 31 January 1953. Strong northerly winds caused the sea surface to pile up in the constricted area between the United Kingdom and the Netherlands. Low-lying land in coastal areas of south-east England and the Netherlands was flooded. After Deltawerken Online at http://www. deltawerken.com/Climatic-circumstances/483/html (accessed 4 August 2011).

Figure 11.9, the height of the surge varied between 1 m and 2 m along the east coast of England but was greater on the Dutch coast, where it reached 4.55 m above NAP (Normal Amsterdam Water Level) in some inlets. Much land was flooded (850 km^2 in England and 1,600 km^2 in the Netherlands) and over 2,000 lives were lost across the UK, Belgium and the Netherlands.

Even greater losses would have been experienced if the event had coincided with river floods or high spring tides. In January 1928 the river Thames, England produced a high flood peak caused by heavy rain and snowmelt. The passage of the flood crest was impeded by a spring tide enhanced by on-shore winds and the water level reached 1.8 m above the expected height, resulting in extensive flooding in the estuary.

Long-term factors

Any relative rise in sea level along low-lying coasts raises the frequency with which sea defences are over-topped. During the last 100 years, there has been a *eustatic* (worldwide) increase in sea level of 0.10–0.20 m. This is due to a combination of the thermal expansion of sea-water and the melting of ice-caps after the end of the last ice-age, a process accelerated by global warming. In addition, some coastal areas have experienced an *isostatic* (local) increase in sea level, due to a lowering of the land surface. For example, the south-east corner of England is slowly sinking as the north-west of Britain rises in response to the removal of the mass of ice that accumulated there more than 10,000 years ago. The city of Venice is sinking into the Adriatic, due to local land subsidence brought about by over-extraction of ground-water. In the lowest-lying coastal zones, the increased volume of water in the ocean basins and local subsidence have resulted in a net rate of sea-level rise of about 0.3 m per century. As a consequence, natural shore defences, such as salt marshes, beaches and dune systems, have suffered increased erosion and many of the 300 barrier islands along the coast of the United States have been driven landward by on-shore storm winds.

2 Human factors

In many countries, significant floodplain invasion did not occur until the late nineteenth century but then expanded rapidly. By 1975 more than half of the floodplain land in the USA was developed and urban areas were spreading onto floodplains at the rate of 2 per cent per year. Rapid City, South Dakota is a typical case. The initial site was laid out south of the floodplain but there was progressive floodplain invasion from 1940 onwards (Rahn, 1984). By 1972, the year of the flash flood disaster,

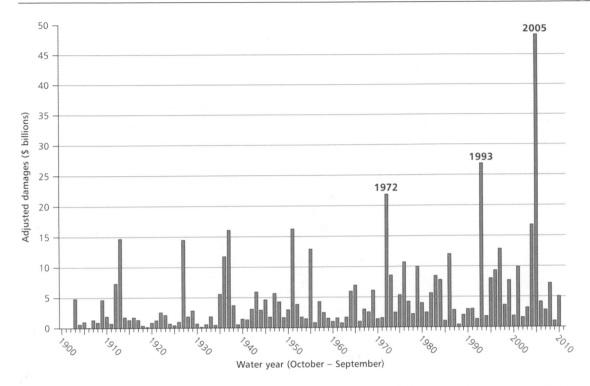

Figure 11.10 Estimated annual losses from river floods in the USA 1904–2010. The data are for water years (1 October to 30 September) adjusted for inflation and do not include coastal flooding. There is a general upward trend, most marked in high-loss years such as 1972 (hurricane 'Agnes'), 1993 (Midwest floods) and 2005 (hurricanes 'Katrina' and 'Rita'). Compiled from data at the US National Weather Service, Hydrologic Information Center.

the entire floodplain within the city limits had been urbanized. Mainly as a result of floodplain invasion, average annual river flood damages in the USA grew fourfold during the twentieth century – from around US$1 billion to US$3.5 billion – even when adjustment is made for cost inflation (Figure 11.10). Similar trends are evident for coastal cities. Over 20 per cent of the world's population lives within 30 km of the sea and these populations are growing at twice the overall global rate (Nicholls, 1998).

Floodplain invasion is driven by countless individual decisions rooted in a belief that the locational benefits outweigh the risks. An appreciation of these attitudes is as important as flood hydrology in understanding flood hazard. Floodplain development is not necessarily irrational. A net economic gain exists if the additional benefits derived from locating on the floodplain (i.e. the benefits over and above those available at the next-best flood-free site) outweigh the average annual flood losses. Unfortunately, it is almost impossible to assess costs and benefits accurately at both local and national levels. For example, a major flood can easily wipe out benefits accumulated over previous years. What is more certain is that, once floodplains become urbanized, there follows a demand from the local community for flood protection that may well lead to even greater losses.

The levee effect

Engineered flood embankments and levees, funded by central government to protect local development, have been the prime response to flood risk. The structural approach is likely to

Plate 11.2 Flooding along the Phra Pinkalao road, Bangkok, in November 2011. More than 500 deaths occurred and over one-third of the city's districts were subject to evacuation orders, largely due to overflowing of the Chao Phraya River draining to the Gulf of Thailand. (Photo: © thai-on.)

prove counter-productive when urban flood-control works are erroneously perceived to render a floodplain completely safe for investment and development. It may well create the 'levee effect', whereby newly erected flood defences increase the demand for building on low-lying areas now thought to be fully protected against future flooding. Land values start to rise. If new building development follows, more property is placed at risk, so generating claims for even greater levels of protection. Many studies have shown that, despite the widespread use of structural controls in the USA, floodplain invasion and flood losses have continued to increase (Montz and Gruntfest, 1986).

The pressure on flood-prone land – both protected and unprotected – is most difficult to control in communities experiencing high economic growth and lacking alternative, risk-free sites for development. Neal and Parker (1988) investigated the case of Datchet, a town of 6,000 people on the Thames floodplain, England. Despite the absence of flood protection works, the planning control system failed to prevent the construction of 425 new houses on the floodplain in the decade 1974–83. In England and Wales, proposed new development is subject to planning approval based on a *Flood Risk Assessment* (FRA) which may be linked to the construction of flood defences (White and Howe, 2002). Even so, the demand for development land continues to grow (Figure 11.11). Applications to build on floodplain land increased from 8 per cent of all development applications in 1996–97 to

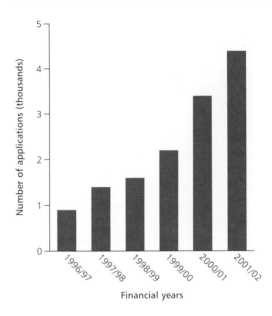

Figure 11.11 The number of planning applications for residential and non-residential development on floodplain land in England between financial years 1996/97 and 2001/2. After Pottier *et al.* (2005). Reprinted from *Applied Geography* 25, N. Pottier *et al.* Land use and flood protection: contrasting approaches and outcomes in France and in England and Wales,1–27. Copyright (2005), with permission from Elsevier.

13 per cent in 2001–2 and the number of housing units proposed, though not actually built, rose almost sixfold during the period (Pottier *et al.*, 2005).

The Environment Agency (EA), advises local authorities in England and Wales about the flood risk attached to planning applications. About two-thirds of all EA objections to applications by developers arise from absent or inadequate FRAs. Most objections are sustained, leading to withdrawal or modification of many proposals, but some appeals by developers are successful. Central government plans for further urban expansion in southern England (Thames Gateway, the M11 Corridor and the South Midlands) could add over 100,00 new homes to the local floodplains. In Greater London, about 1 million people in over 500,000 homes are already located

on floodplains. Surprisingly, the levee effect can remain strong even after a damaging flood. The 1993 Midwest floods created up to US$16 billion in damages and the loss of 7,700 properties, but this event was soon followed by a rush for new development on the floodplains. In the St Louis metropolitan region alone, 28,000 new homes were built and nearly 27 km^2 of commercial and industrial land were developed – amounting to US$2.2 billion in new investment – on land that was under water in 1993 (Pinter, 2005).

The circular link between flood control works and floodplain encroachment can be explained by three factors:

- The more intensive the floodplain development, and the greater the existing investment, the greater are the local economic benefits perceived to result from flood-control structures. Flood protection schemes can then be justified on cost-benefit grounds.
- The cost-benefit ratio turns in favour of new building when land gains a perceived high level of protection from risk and can be freed for development. The higher land values in the 'protected' area make further floodplain invasion more likely.
- The process exists because the real costs are not borne by those gaining the benefits. Most flood defence is financed by central government, in a search for national economic efficiency, whilst planning authorities pursue more local development goals. Since private investment in the floodplain is protected by public money, it is perfectly rational for an individual or a company to locate there and transfer any hazard-related costs elsewhere.

Flood impacts appear to be increasing but it is difficult to identify all the factors. Climate change may well become an increasingly important driver in the future, but most observers believe that present-day increases in flood losses are due to a combination of better flood monitoring and more intensive land use. As already seen, urbanization itself transforms hydrological systems and creates a growing risk through continued floodplain invasion and rising property values (Mitchell, 2003; Hall *et al.*, 2003). Most countries have found it difficult to reverse these trends. Canada – faced with mounting flood losses in insurance claims and disaster relief – introduced a comprehensive Flood Disaster Reduction Program in 1971 (Shrubsole, 2000). The aim was to decrease reliance on structural schemes in favour of a strategy based on floodplain mapping and public education. Administration of the Program fell between two federal agencies (Environment Canada and Emergency Preparedness Canada) and was closed because of budget cuts in 1999 (de Loë and Wojtanowski, 2001).

D PROTECTION

The hydrological and engineering criteria necessary to control floods are well understood and structural intervention is used against floods more than any other environmental hazard. There are two main approaches. *Embankments* and *sea walls* are used to protect urban areas, high-quality farmland and other assets in low-lying areas. *Storage dams* are used to retain flood-waters upstream so that downstream flows can be regulated within the carrying capacity of the river channel. Engineering works have proved effective in limiting the flood risk but there is concern about certain side-effects of large-scale control works. Other forms of flood protection include flood abatement and flood proofing.

1 Flood embankments

Embankments and walls, sometimes called dykes, stop-banks, are the most favoured form of river control (Starosolszky, 1994). These are linear structures built parallel to a river channel or shoreline to contain high-water levels. In general, embankments tend to be used for river floods, and walls for coastal floods. Embankments, or levees, are essentially compacted mounds of earth raised above the original ground level to a height equivalent to that of the design flood (Figure 11.12a). Most floodplains and river deltas are characterized by mixed alluvial sediments and it is important that structures have sufficiently deep foundations to withstand long-term under-seepage. Thus, levees often have a clay core to reduce the risk of water seepage. Floodwalls perform the same task but usually have a concrete cut-off below ground to prevent seepage undermining the structure (Figure 11.12b). The height of control structures is determined by the predicted level of the *design flood* (often with a return interval 1:100 years), plus an additional element of *freeboard* for uncertainty. This safety factor, often about 0.5 m, reflects caution regarding the design flood estimate, weather conditions at the time of the predicted event (strong winds and waves) and longer-term concerns such as local ground subsidence and climate change. For example, structures newly built today may be raised by 20 per cent above previous design levels to cater for future conditions.

Earth embankments are relatively cheap to construct. In China, many dykes built since 1949 protect large alluvial plains from floods with a 10–20-year return period. Over 4,500 km of the Mississippi river, USA is embanked in this way. Major cities, such as New Orleans, lie below river level and rely on such structures, as during the Midwest floods of 1993. Selecting the right design criteria and location is crucial. Most federal levees are designed for 1:100 to 1:500-year return intervals, whilst other levees are designed to withstand smaller floods with recurrence intervals of 50 years or less. In the UK, where the 1:100-year flood is the normal standard, flood defences provide protection for about 145,000 households. It is reckoned that most new structures should produce a 1:8 cost-benefit ratio, reducing flood losses by £8 for every £1 spent (Anonymous, 2009). Some observers argue that flood protection should not be based entirely on econometric principles but should include some measure of the risks to people (Jonkman *et al.*, 2008).

Defence works fail for a variety of reasons. Over-topping is the main cause, followed by structural failure causing a breach and undermining by sub-surface seepage. Therefore, there is a need for on-going maintenance (see Box 9.1). In the UK, two-thirds of the flood protection budget is spent on maintenance and improvement. These costs may be raised by poor design, inadequate construction or damage due to the erosive force of major floods. In 1993 the Mississippi levees rarely failed until the river stage reached a metre or more above the design level. When flood banks are breached, local losses arise but the natural floodplain storage at that point normally helps to limit flood stages downstream. Figure 11.13 shows the effects along a 150 km stretch of the Mississippi during July 1993. The levees above Keithsburg held, resulting in a smooth variation in river stage within the channel, whilst multiple levee failures upstream of Hannibal led to water spilling onto the floodplain and sudden drops in the river level at these sites. In contrast, during hurricane 'Katrina', levees in New Orleans failed before over-topping, due to construction defects and poor maintenance. For at least one New Orleans levee, it appears that the metal sheet-piling used to anchor the levee into the sub-soil simply did not penetrate far enough into the ground (Kintisch, 2005).

Embankments are often supplemented by *river channel enlargement*. Dredging of the river Arno at Florence, Italy after disastrous flooding in November 1966 lowered the river bed near two of the old bridges by one metre. This was designed to raise the capacity of the channel from 2,900 $m^3 s^{-1}$ to 3,200 $m^3 s^{-1}$. Natural river channels can be made straight and smooth to increase the flow velocity and carry flood-waters downstream more quickly. In addition, flood-relief channels can be built to provide extra overspill storage or to divert water around an area of urban development.

All these schemes intrude visually on the landscape. They also isolate the river from its alluvial plain, with negative consequences for the riparian ecosystem. Over large deltaic areas of

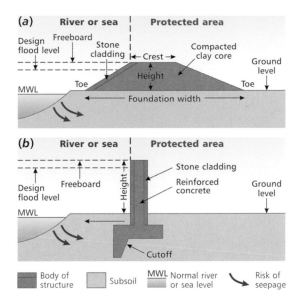

Figure 11.12 Engineered flood defence measures in common use to protect land and development against river and marine floods: (a) cross-section through an earthen flood levee or embankment; (b) cross-section through a typical flood wall. Each structure is raised to a height to contain a flood stage with a given return period.

Bangladesh and Vietnam the use of embankments to protect agricultural land has closed tidal channels and reduced surface drainage efficiency in the monsoon season. Higher water levels and faster flows in the rivers and canals then increase bank erosion and the risk of embankment collapse (Choudhury *et al.*, 2004; Le *et al.*, 2006). The destruction of wetland habitat and increased river levels due to large-scale upstream river straightening have sometimes transferred flood risks to downstream communities (Birkmann, 2011). Subsequent river restoration projects seek to achieve a better functioning of the 'river corridor' as a whole (Mitsch and Day, 2006; Bechtol and Laurian, 2005).

2 Flood-control dams

Flood-control dams have existed for well over 2,000 years. They provide temporary storage of

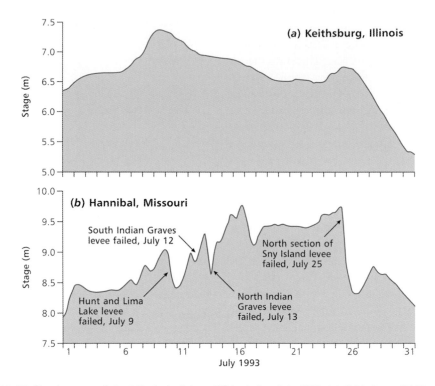

Figure 11.13 Flood stages of the Mississippi river, USA, during July 1993: (a) Keithsburg, (b) Hannibal. The levees upstream of Keithsburg held and the river stage shows a smooth transition in water levels. Just upstream of Hannibal levee failures produced sudden falls in river levels, due to widespread floodplain inundation and water storage. After Bhowmik (ed.) (1994). Reproduced from Flood stages of the Mississippi River July 1993, from *The 1993 Flood on the Mississippi River in Illinois* by N.H. Bhowmik et al. (1994) Illinois Water Survey. Reproduced with permission.

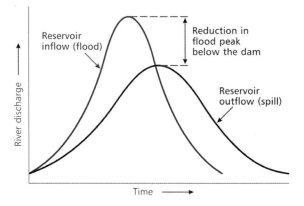

Figure 11.14 Idealized flood hydrographs for water inflowing and discharging from a reservoir. The reservoir storage regulates the downstream flow so that the peak flood discharge is reduced.

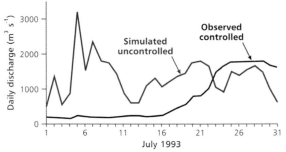

Figure 11.15 Simulated flood discharges on the upper Mississippi river, USA, during July 1993, assuming the absence of reservoirs. Without reservoir storage, the Big Blue River near Manhattan, Kansas, would have quickly over-topped the federal levee and downstream flooding would have been more severe. After Perry (1994).

Table 11.2 Estimate of the reduction in flood losses due to levees and dams on the Mississippi and Missouri rivers during the 1993 floods

River basin	Dams	Levees	Reduction (US$ billion)
Mississippi	3.6	3.9	8.0
Missouri	7.4	4.1	11.5
Total reduction	11.0	8.0	19.1

Source: After US Army Corps of Engineers (quoted in Green *et al.*, 2000)

water so that downstream flood peaks are lowered (Figure 11.14). Most large dams are multipurpose but, globally, around 10 per cent have flood alleviation functions. About 50 per cent of Japan's population live in flood-prone urban areas, many of which are protected by dams. Well-designed and safely operated dams are highly effective. For example, the 66 flood reservoirs in the upper Mississippi and Missouri basins worked well in combination with the levee system during the 1993 flood (Table 11.2). Flood discharges were reduced by 30–70 per cent, despite the fact that the inflow behind some dams was several times their total storage capacity (US Department of Commerce, 1994). The maximum benefit was on the Big Blue River, within the Kansas River basin, where Tuttle Creek Lake withheld a daily mean flow of 3,029 m^3 s^{-1} on 5 July (Figure 11.15), thus greatly reducing the peak, which would have caused far more damage than the 1,700 m^3 s^{-1} controlled release later in the month (Perry, 1994).

Large dams have been judged a mixed blessing (World Commission on Dams, 2000). They are expensive to construct and can be vulnerable to earthquake damage and rapid siltation. In some countries, like Bangladesh, the annual floods are simply too large to retain in storage reservoirs. When very large dams have been built, such as the Aswan High Dam, capable of storing 1.5 times the mean annual flow of the Nile, flood protection has to be balanced against the loss of fertility to floodplain soils due to natural silt deposition.

Where dams are multipurpose, conflicts in water management arise between retaining water (for power generation) and releasing water (to create flood storage). In some cases, dams have led to the loss of forest land, wildlife and aquatic diversity. It is estimated that more that 40 million people from poor, indigenous communities have been displaced and resettled by dam construction.

3 Coastal flooding

Coastal floods create special problems. Economic development tends to 'harden' the shoreline through the construction of sea walls and related infrastructure. This process will interfere with near-shore processes and block natural shoreline retreat. One consequence may be increased beach erosion elsewhere, leading to a reduction in sand supply and available recreational space. Generally, coastal flooding is best addressed by avoidance rather than confrontation. Set-back strategies – or 'managed retreat' – are increasingly preferred, unless key functions and large population clusters need protection. For example, beach replenishment is a method of placing sand on an eroding or limited-width beach, in order to extend it seaward and keep floods at bay (Daniel, 2001). It has been successful for storm mitigation and has benefits for wildlife habitat and the tourism industry. However, it may not pass conventional cost-benefit tests and there can be problems in obtaining environmentally sustainable sand supplies (Jones and Mangun, 2001; Nordstrom *et al.*, 2002).

4 Flood abatement

The integrated management of soil, vegetation and drainage processes in order to reduce floods is a well-established objective: New Zealand set up a Soil Conservation and Rivers Control Council as early as 1941. But the practical results are often inconclusive, except for comparatively small catchment areas. Even then at least half of the catchment area has to be managed by reforestation or reseeding of sparsely vegetated areas (to increase evaporative losses); mechanical land treatment of slopes, such as contour ploughing or terracing (to reduce the run-off coefficient); comprehensive protection of vegetation from wildfires, over-grazing, clear-cutting of forest land or any other uses likely to increase flood discharges and sediment loads. For extensive drainage basins, it would take decades of reforestation and soil conservation to have an appreciable effect.

In brief, headwater forests will not prevent floods or sedimentation in the lower reaches of major river basins, nor will they significantly reduce flood losses generated by intense storms over smaller basins. Peak flows on minor rivers can be reduced by the clearance of sediment and other debris from headwater streams, construction of small water- and sediment-holding areas (farm ponds) and the preservation of natural water detention zones such as sloughs, swamps and other wetland environments. Within urban areas limited water storage can be achieved by the grading of building plots and the creation of detention ponds and parkland.

5 Flood proofing

Individual buildings at risk from floods can be constructed or retro-fitted to become more hazard resistant. Several methods exist:

- *elevation and set-back* – elevation involves raising the habitable parts of a property above flood level, on stilts or on landfill (Figure 11.16). The design-flood level is the minimum elevation of the underside of floors in habitable buildings. Other measures, such as placing the building back from any water body and the flood proofing of basement spaces, may be specified in local planning regulations.
- *wet flood proofing* – making uninhabited parts of a property (e.g. a cellar) resistant to damage but allowing water to enter during floods
- *dry flood proofing* – sealing a property (doors and windows) to prevent flood-water from entering
- *floodwalls* – building a floodwall around a property to hold back the water
- *relocation* – moving the entire house, if timber-framed, to higher ground
- *demolition* – demolishing a flood-damaged property and either rebuilding more securely on the same site or rebuilding at a safer site.

The most common permanent response consists of raising the living spaces above the likely flood level, guided by floodplain zoning maps and local ordinances. Emergency measures may be activated by flood warnings. These include the blocking up of entrances, the use of movable shields to seal doors and windows and the laying

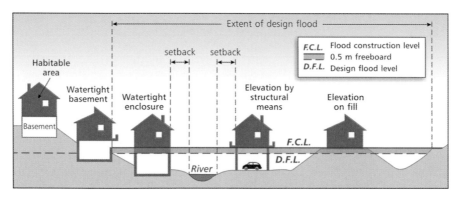

Figure 11.16 Schematic of flood-proofed residential buildings on a river floodplain. Habitable areas are raised above the flood construction level (FCL). The FCL allows 0.5 m of freeboard above the predicted maximum height of the design flood, e.g. the 1:100 year event. Adapted from Rapanos *et al.* (1981).

of sandbags to keep water out of the property. Further low-cost measures include removing damageable goods to higher levels and pre-flood greasing of mechanical equipment.

E MITIGATION

1 Disaster aid

Recovery from major flood disasters in developing nations is beyond local capacity alone. An international financial package, supported by bodies such as the World Bank, with contributions from the national government and locally based NGOs, is often required. Past misuse of funds and poorly distributed aid has prompted donors to channel more and more assistance through the NGOs. However, following a damaging flood in 1998, the Bangladeshi government significantly improved the delivery of aid to flood victims, in an attempt to raise its political profile within the country (Paul, 2003). For LDCs it is the rural sector – agriculture – that bears most of the direct

losses. It is also true that reconstruction costs often outweigh the direct damages. After exceptional floods over Mozambique in February and March 2000, over 12 per cent of the cultivated land was inundated and reconstruction costs were 60 per cent higher than those needed for direct damages.

The true scale of humanitarian need in disasters is not immediately apparent, so the aid process evolves through time. The 2010 Pakistan flood disaster began with the arrival of heavy monsoon rains on 22 July. By 10 August about US$156 million of aid had been pledged in cash or in kind. On 11 August the UN and the Pakistani government launched a joint *FLASH appeal* for assistance. FLASH appeals aim to finance and coordinate the humanitarian response in the first 3–6 months of a new emergency. They are often issued within one week of the start of the event and revised within 30 days. The initial plea was for nearly US$460 million for relief in seven priority sectors (Table 11.3); food security, shelter and WASH (water, sanitation and hygiene) accounted

Table 11.3 Aid requirements and potential beneficiaries identified in the main appeal for humanitarian aid following the 2010 floods in Pakistan

Sector	Requirements US$	%	Targeted beneficiaries
Food security	156,250,000	34	Food assistance for up to 6 million flood-affected people (US$5.7 million for surviving livestock)
Health	56,200,000	12	14 million potential beneficiaries, especially children under five and women of child-bearing age
Shelter and non-food items	105,000,000	23	Initial target 300,000 families with damaged or destroyed houses
Water, sanitation and hygiene	110,500,000	24	Approximately 6 million people, including 3 million children
Logistics, emergency telecommunications and coordination	15,624,000	3	Aid partners and wider humanitarian community in Pakistan
Nutrition	14,150,847	3	1.35 million people, including children under five and pregnant and lactating women
Protection	2,000,000	<1	500,000 individuals
Total	**459,724,847**	**100**	

Source: Adapted from OCHA (2010)

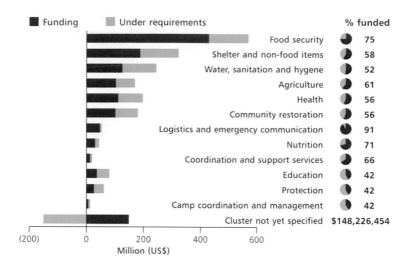

Figure 11.17 Humanitarian funding levels at 22 July 2011 following the Pakistan floods emergency in 2010. Large sums needed for emergency relief had been raised, so that about 70 per cent of the funding needs had been met. After United Nations in Pakistan (2011).

Plate 11.3 Flooding in the centre of the city of York, England during autumn 2000. The river Ouse burst its banks and raised water levels 5–6 m above summer flows in certain areas. About 40 people had to be rescued from their homes. Total costs were estimated at £3 million. (Photo: Panos/Trygve Sorvaag TSO00010UK)

for 80 per cent of the total. On 5 November 2010 a scaled-up appeal was made for US$1.96 billion.

At the time, this was the largest-ever UN appeal for a single natural disaster. Food security was the main requirement because 80 per cent of the people in the flood-affected areas depend on agriculture for their livelihood. An estimated 1.3 million ha of standing crops had been destroyed and 1.2 million farm animals had drowned. Extensive water-logging of soils, silt deposition and damage to irrigation structures created a challenge for farmers hoping to recover land in time for planting wheat, the staple crop, in September and October 2010. One year later, about 70 per cent of the appeal target had been achieved, albeit with considerable variations in unmet requirements, as shown in a detailed allocation list (Figure 11.17). High demands for aid in agriculture are typical. After the 1988 floods in Sudan, almost one-quarter of all reconstruction costs were incurred in this sector.

2 Insurance

Accurate flood damage assessments are important as a basis for loss-reducing decision making, although the optimum use of the information is not always clear (Merz *et al.*, 2010). In the UK and Germany, householders buy flood insurance as part of a packaged policy covering buildings and their contents, and flood losses are subsidized by the market as a whole. But buildings insurance is mandatory only during the life of a mortgage. Many householders – notably tenants, pensioners and those in lower socio-economic groups – either fail to take out insurance or are under-insured. These people are unlikely to recover well after a flood, and their situation prompts frequent policy discussions between government, the insurance industry and property owners.

Governments are unwilling to use tax revenues to recompense uninsured households, but floods in Germany during August 2000 showed that government compensation and public donations continued to be important methods of loss recovery. This does not encourage preparedness for future events and loss-reduction measures by either house owners or the insurance companies. Only 14 per cent of insurers surveyed by Thieken *et al.* (2006) rewarded policy holders who voluntarily reduced their exposure to floods and only 25–35 per cent gave householders advice on flood-loss reduction measures. In the face of rising losses, attitudes are changing. Insurance companies are now exploring financial incentives, such as lower premiums, to those adopting damage-mitigation measures. In a study of precautionary measures taken against floods by private home-owners in Germany, it was found that whilst large investments – like building a sealed cellar – were rarely justified financially, smaller measures – such as protecting an oil tank – were worthwhile (Kreibich *et al.*, 2011). One way forward would be to make low-cost measures mandatory through local building codes in flood-prone areas, in order to maintain the insurability of flood losses in high-risk areas.

The UK insurance industry seeks to increase public awareness of flooding and shed some of the risk, especially due to the possibility of increased losses in the future (Treby *et al.*, 2006). About 5 million people, living in some 2 million homes, are at risk from floods which cost the insurance industry about £1.5 billion per year. As a general rule, commercial insurers in the UK will not cover new-build property with an annual flood risk greater than 0.5 per cent. For a time, there was a loose agreement between central government and the industry that, as long as 'adequate' investment in flood defences was maintained, home-owners and small businesses would be offered cover. However, floods in 2007 cost insurers £3 billion (Box 11.2) and the industry no longer guarantees to provide affordable insurance for new developments built against the advice of the Environment Agency from 1 January 2009 onwards. This restriction is expected to extend to all households and businesses from 2013 onwards (Anonymous, 2009). In due course, flood insurance may be unavailable to many householders.

Box 11.2 Floods in England: the summer of 2007

Most floods in England are associated with deep Atlantic storms in winter. During the summer months localized floods arise from thunder-storms but relatively dry weather tends to prevail, due to northward extensions of the Azores high-pressure cell. In June and July 2007, the Azores high remained weak and a southerly displacement of the jet stream brought Atlantic depressions, embedded in warm, humid, south-westerly air, across southern Britain. The early summer from May to July was the wettest since records began in 1766 and rainfall over England and Wales was more than double the average amount.

Widespread flooding occurred over central England and Wales. More than 11,000 homes were flooded in East Yorkshire, over 8,000 of them in the city of Hull, where approximately 90 per cent of the houses are below sea level. The prime cause was *intra-urban flooding* as storm water and rising ground-water levels flooded the drainage system beyond the capacity of the elderly pumping system. Most properties were flooded to a depth of less than 0.5 m, but about 1,200 people were evacuated for several months to temporary accommodation, mainly in caravans. The schools were disproportionally affected and many were forced to close. Local residents were generally unprepared and many failed to receive a flood warning.

Different problems occurred near Tewksbury, due to combined flood peaks at the confluence of Severn and Avon rivers. Upton-on-Severn was inundated because de-mountable flood barriers, designed for the town but stored 20 miles away, arrived two days late, due to motorway congestion. Local flash floods trapped people overnight in vehicles, emergency shelters were opened and RAF helicopters were deployed to evacuate hundreds of people in one of the largest-ever peace-time rescue missions. Nearly 10,000 homes were flooded. In Gloucester 15,000 people were without electricity, and 140,000 were without a mains water supply for several days when a water-treatment works became disabled. Firemen were drafted in from other areas to help with pumping-out operations and the army transported water bowsers and bottled water into town.

In total it is estimated that over 1 million people were affected in some way and losses were placed at £5 billion.

The 2007 floods exposed several weaknesses in preparedness:

- *Lack of warning.* The localized intra-urban flooding that characterized much of northern England is difficult to forecast, but residents frequently complained about the lack of warning. The Environment Agency admitted that only 30 per cent of all flood-prone houses in England and Wales had signed up for the telephone flood-warning service. All flood-prone householders should be informed of the risks and be given detailed advice on appropriate methods of preparedness and flood proofing.
- *Lack of investment.* Spending on flood defences in 2007 amounted to £600 million, a cut of £14 million from the previous year. Some commentators have contrasted this situation with the profits achieved by the privatized water companies; in 2007 the Severn Trent water authority realized profits of some £300 million. During 2005–6 these authorities were supposed to spend £4.3 billion on improving infrastructure but only about £3.4 billion was invested, despite above-inflation increases in domestic water bills.
- *Lack of flood-protection standards.* There were no national flood-defence standards designed to protect key sites and infrastructure, such as schools and hospitals, despite the fact that over half

the water- and sewage-pumping stations and 14 per cent of the electricity infrastructure in the country are located in flood-risk areas. It is becoming clear that the aspiration to protect urban areas against the 1:100 flood may be unobtainable. For example, since the Thames flood barrier started operating in 1983, the estimated annual flood risk in the area has been doubled, from 1:2,000 to 1:1,000.

- *Lack of sustainable planning.* In the 2007 floods, many urban drainage systems failed in areas remote from river courses. Apart from the presence of outdated sewerage and water-pumping systems, many householders have paved over their front gardens to provide car parking, in response to increased traffic congestion. This has sometimes happened in contravention of planning rules and can lead to the local accumulation of excess surface water. With the prospect of more housing development on floodplains, water-proof ground floors and the introduction of domestic services such as electricity supply at a higher level in the home may become normal requirements.

Flooding is the number one cause of damages from natural disasters in the USA and the *National Flood Insurance Program* (NFIP) was conceived as a tool for better floodplain management. The Program was introduced in 1968 because of rising flood losses and a growing reluctance by the industry to continue selling cover for some areas. The scheme is a partnership between the federal government, state and local governments and the insurance industry – administered by the Federal Emergency Management Agency (FEMA) – to provide financial assistance to flood victims and to discourage unwise occupancy of flood-prone areas. Since the NFIP was introduced, cover for flood damage due to rising water has been excluded from home-owners' private insurance policies.

The first step in the NFIP is the publication of a *Flood Hazard Boundary Map* that outlines the approximate area at risk from either river or coastal flooding. In order to join the NFIP, a community must agree to adopt certain minimum land use controls within this area during the so-called 'Emergency Program'. In return, flood insurance is made available at subsidized rates. FEMA then supplies more detailed maps to define the 1:100-year floodplain and the floodway – the area within which the 100-year flow can be contained without raising the water surface at any point by more than 0.3 m (see Figure11.18). The 100-year flood is adopted in recognition of the benefits, as well as the costs, associated with flood-plain development, although a few communities also regulate development in the 1:500-year floodplain. Designated floodplains now cover an area almost the size of California and contain about 10 per cent of the nation's households.

At this point, the community must join the 'Regular Program' and use more stringent land use controls, such as prohibiting further development in the floodway and elevating residential development in the rest of the floodplain (floodway fringe) to at least the 1:100-year flood level. The designated flood area is divided into risk categories on the basis of a large-scale *Flood Insurance Rate Map* (FIRM) so that insurance ratings can be applied to individual properties. All new property holders within the 1:100-year floodplain must then buy insurance at commercial rates, although reductions are available for properties erected before the FIRM map was produced. NFIP operating costs and damage claims are paid by insurance premiums rather than the tax-payer. When premiums fail to cover costs, the deficit is plugged by loans from the US Treasury.

Initially, few local governments adopted the scheme. In 1973 the Flood Disaster Protection Act

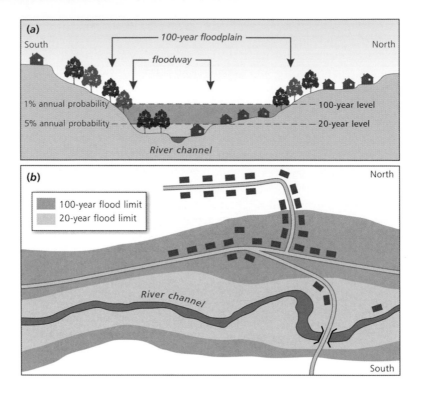

Figure 11.18 Schematic representation of the river flood hazard: (a) river stage for different return periods in relation to land use across a floodplain; (b) map of the floodplain indicating land-planning zones.

was passed to encourage more participation by denying non-compliant local authorities various federal grants-in-aid and making property owners ineligible for flood insurance or federal flood relief. Since then the number of policies taken out has reached a total of around 4.5 million. The NFIP has achieved some success:

- Nearly 20,000 flood-prone communities across the USA have adopted floodplain regulations and zoning.
- Low-cost insurance has been made available and 35–40 per cent of all properties insured benefit from subsidized rates.
- Building construction has improved so that new, flood-resistant homes suffer about 80 per cent less flood damage than other properties.

- Annual flood-damage costs have been reduced by nearly US$1 billion, with associated savings in disaster assistance.

However, Burby (2001) and others have criticized the NFIP because:

- Flood hazards are ill-defined. For example, the exclusion of localized storm-water drainage flooding means that premiums for individual properties do not always match the risk.
- The scheme has not stopped unwise development. Floodplain invasion has increased by over 50 per cent during the NFIP's existence, especially in coastal areas.
- The aim of spreading risk more widely has also failed because only about 25 per cent of individuals exposed to floods purchase insurance.

In a study of two cities in Nevada, Blanchard-Boehm *et al.* (2001) attributed the poor uptake of insurance to perceptions that flood risks were low, insurance was not good value for money and government assistance would always be available. Fewer than one-third of householders indicated a willingness to purchase insurance. Even if insurance were made compulsory, many would not be eligible, either because their houses were built before the publication of FIRMs or because, with an ageing population, increasing numbers of property owners have paid off their mortgages. After four decades, the NFIP is scheduled for a major overhaul.

A major task, also relevant to preparedness in general, will be to correct a tendency of the public to under-estimate flood risks. Work in Switzerland by Siegrist and Gutscher (2008) showed that residents not previously exposed to floods underestimated the negative consequences – especially the emotional uncertainties and insecurities – as compared to experienced victims. In another study, some residents in a flood-prone area of North Carolina were unable to make correct assessments of their risk from updated FIRM maps and consistently downgraded the threat (Horney *et al.*, 2010).

F ADAPTATION

1 Preparedness

Many countries rely on voluntary organizations and the armed forces to combat flood disasters. This can be viewed as a low-cost option, although emergency responses to UK flooding in autumn 2000 accounted for about 15 per cent of all flood losses (Penning-Rowsell and Wilson, 2006). Preparedness is an important tool in the LDCs. During the Mozambique floods of 2000, almost 500,000 people were either displaced from their homes or trapped in flood-isolated areas. Thousands of overseas aid workers arrived in the country and a high-profile search-and-rescue operation

Table 11.4 Numbers of flood victims rescued by air and boat in the Mozambique floods of 2000

Operator	Air	Boat*
Mozambique military		17,612
Mozambique Red Cross		4,483
Local fire service, private boats		7,000
South African military	14,391	
Malawian military	1,873	
French military	79	
Air Service (international NGO)	208	
Totals	**16,551**	**29,095**

Source: After IFRCRCS (2002)

Note: *Many of the boats used were donated by international agencies.

was organized. However, most survivors were rescued by boats manned by Mozambicans themselves (Table 11.4)

There are 30,000 Red Cross-trained flood volunteers in Bangladesh charged with a wide range of tasks from raising hazard awareness, health and hygiene education and first aid techniques through to emergency response skills that include warning of villages through loud hailers and the evacuation of people to refuges and higher ground. Other adaptations are more indigenous. For example, on the *chars* – the vulnerable silt islands in Bangladeshi rivers and offshore – ordinary life is highly flexible and flood responses include:

- quickly moving livestock and possession away from flood or erosion threats
- sometimes dismantling thatched houses and moving them by boat to a temporary site on higher ground
- using reeds to stabilize new silt deposits, ready for cultivation
- planting rice in moveable seed beds for transplanting when flood-waters recede
- seeking marriage partners on other *chars* to secure escape routes and refuges.

2 Forecasting and warning

Advances in hydrometeorology and flood hydrology have enabled the modelling of storm rainfall and flood flows to high levels of accuracy. Automatic rainfall and river gauges, linked to satellite and radar sequences, permit real-time data handling. Meso-scale computer models supply the forecast information used for Numerical Weather Prediction (NWP) and Quantitative Precipitation Forecasting (QPF) so that estimates of storm precipitation can be fed into hydrological models. In turn, these produce forecasts of the height and timing of flow levels moving through a drainage basin. Flood waves can also be tracked by satellites in real time, a useful feature in a country like Bangladesh, where about 90 per cent of river flow originates outside the national territory. Generally speaking, flood-forecasting and warning schemes are most effective for large rivers with long lead-times, such as the Danube in Europe and the Mississippi in the USA (Werner et al., 2005). There is a constant search for new forecasting technologies and longer lead-times (Golding, 2009,) plus greater integration of schemes across wider geographical areas (Werner et al., 2009).

The most urgent need is for better forecasting of flash floods. These warnings are not always accurate or timely because standard prediction methods are overridden by local factors, such as soil type, soil moisture levels or the degree of urbanization (Montz and Gruntfest, 2002). In the UK, a network of weather radars is used for flood-warning purposes but more than half of all dwellings at risk in England and Wales have less than six hours' flood lead-time. Hydro-meteorological monitoring rarely takes place on the fine spatial scale required to forecast flood waves that travel at 3 m s^{-1} and can threaten these communities. Thus, Javelle et al., 2010 suggest a combination of radar rainfall data and estimates of antecedent soil moisture to improve flash-flood warning for un-gauged catchments in Europe. A similar problem exists for forecasting more localized surface-water flooding in the UK, where 4 million urban properties are at risk. The development of *Extreme Rainfall Alerts* (ERAs) may offer some scope for improvement if the warnings are reliable and tailored to the needs of emergency responders (Parker et al., 2011).

Flood-forecasting and warning schemes under-perform when either the forecast component or the communication link with decision makers fails. In the 1997 Grand Forks, USA flood disaster both elements under-performed (Todhunter, 2011). Flood peaks were much higher than forecast and uncertainties in the forecast were not properly communicated to emergency responders. It is likely that flood forecasts and warnings could reduce damages by up to one-third on the floodplains of large rivers but such estimates assume optimum loss-reducing actions by the emergency services and the general public. In practice, some people will not receive a warning, despite the claim by the UK Environment Agency that more than 70 per cent of at-risk properties in England would have the offer of a *Floodline Warnings Direct* (FWD) alert by 2011 (Anonymous, 2009). When warnings are received the response rate may be poor. There is a multitude of influences on the public's behavioural response to flood information, including a lack of understanding, mistrust of authority and inadequate knowledge regarding appropriate mitigating action (Parker et al., 2009). Flash flood victims are unlikely to be warned. For example, 60 per cent of those who survived the Big Thompson Canyon, Colorado flash flood in 1976 received no official warning.

Flood forecasting and warning in the LDCs is hampered by less access to technology and poor communication systems. In 2003 the World Meteorological Organization stated that little more than one-third of its members could run NWP models and apply them to flood forecasting. Since then a programme for improving meteorological and hydrological forecasting for floods has been introduced, but weaknesses remain, such as:

- limited access to radar and satellite data
- meteorological forecasts qualitative not quantitative
- meteorological and hydrological services not integrated
- fragmented and non-standard data archives
- shortage of qualified personnel
- lack of a lead agency responsible for flood warnings
- flood warnings not focused on those most at risk.

High rates of illiteracy and limited response capabilities remain problems. When floods struck the Limpopo river catchment in 2000, a survey of two communities in Mozambique showed that official warnings failed to reach nearly 60 per cent of households and residents relied entirely on relatives and friends instead (Brouwer and Nhassengo, 2006).

3 Land use planning

The concept of *Integrated Flood Management* implies that a balance is struck between the growing social and economic needs of society, including urbanization, and keeping the level of flood risk at an acceptable level (Anonymous, 2007b). In practice, this means that the catchment-wide management of land and water resources has a role to play, ensuring that land use practices do not alter flood hydrology adversely. The task is to maximize the net benefits that arise from the occupation of floodplains, including intangible ecological and social gains, whilst minimizing flood losses. The greatest challenge is to limit further floodplain invasion which fails to meet the need for sustainable development.

The relatively high frequency of flood events usually provides sufficient information for accurate flood risk mapping, including information on water depth, flow velocity and flood duration. According to Marco (1994), detailed flood maps were first introduced in the USA, but many countries now have large-scale maps of flood risk

down to the level of individual properties. Figure 11.19 is a map of the floodplain in Northam, a small town on the river Avon about 100 km north-east of Perth, Western Australia. It shows the limits of the 100-year flood and the designated floodway. On well-monitored floodplains, it may be possible to designate different risk categories for certain locations. For example, the UK Environment Agency identifies three threat levels:

- Low – <0.5 per cent (1:200 chance in any given year)
- Moderate – 0.5–1.3 per cent (1:200–1:75 chance in any given year
- Significant – >1.3 per cent (1:75 chance in any given year).

However, there are calls for more societal-based risk mapping in densely populated urban areas, to avoid under-counting the more vulnerable members of society who may require extra attention with preparedness and during emergency relief periods (Maantay and Maroko, 2009).

Urban communities adopt controls on land use to protect flood-prone infrastructure and limit new development. Flood risk maps are required to develop policies like those embedded into the US National Flood Insurance Program (Burby, 2000). In the UK, regional 'structure plans' are first approved by central government before the detailed development of land is devolved to local planning authorities. These bodies have the power to refuse 'planning permission' on land zoned liable to flooding, usually on advice from the Environment Agency, but refusal decisions can be overturned on appeal. London's floodplain alone accounts for about 16 hospitals, 200 schools and 500,000 properties. New housing developments continue to appear in such areas, despite the increased rejection of applications, and the pressure of competition for land continues to grow, especially for floodplain land in south-east England. Until recently, low rates of population growth helped to manage demand, but this situation changed, due to a population influx and

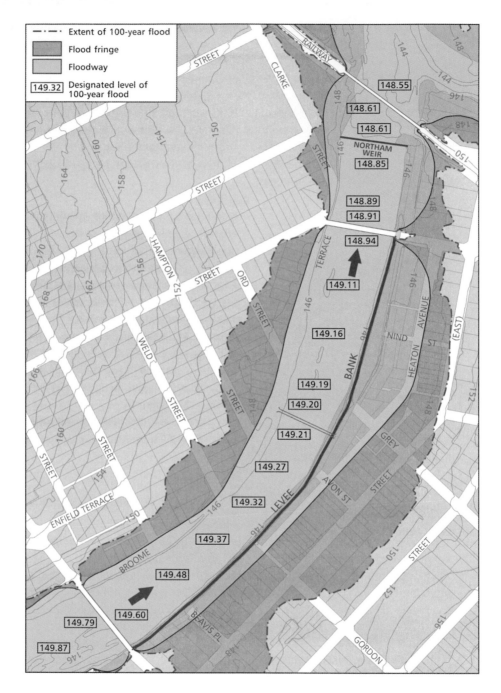

Figure 11.19 Floodplain map for the Avon River at Northam, Western Australia showing the floodway, flood fringe and the predicted extent of the 1:100 year event. Maps at this scale are necessary to identify risk at the plot level and provide a basis for detailed responses such as the rating of insurance premiums for individual households. After Water and Rivers Commission (2000). Reproduced from *Water Facts 14: Floodplain Management*. Water and Rivers Commission, July 2000, Government of State of Western Australia. With permission of the Department of Water.

socio-economic changes, including the trend to more fragmented families and single occupancy.

When urban areas suffer from either a single major disaster or from small-scale repeated loss, there can be a call for public funds to purchase flood-prone land, thus enabling property buy-outs and the relocation of residents to safer sites nearby. In practice, this happens rarely and, even then, only on small scales. A typical example is the Cedar Grove mobile home park in King County, Washington State, which was severely flooded by the Cedar Creek three times in the 1990s. As a result, the local government acquired an 8 ha site in 2008 in order to move 41 residents to higher ground, at a cost of almost US$7 million, under a plan that includes financial assistance and counselling benefits for the residents. In many cases, the task is simply too large. The commercial centre of Hat Yai in southern Thailand, the fourth-largest city in the country, was flooded in 1988, 2000 and 2010. Proposed relocation was dismissed as an expensive over-reaction. Improved management of the canals and drainage system in this low-lying area, aimed at facilitating the removal of flood-waters from the streets within one or two days, was deemed a more economically efficient option. In many LDCs, dangerous flood-prone land is illegally occupied by squatter families. When the authorities attempt to intervene, citing safety issues, the residents are sceptical and may resist any forced removal.

The usual motive for relocation is public safety, but other benefits can result, such as the creation of parkland, the preservation of wetland habitats and the improvement of waterfront access. To be successful, such schemes have to be voluntary and offer incentives. In Australia, the authorities buy houses at an independently derived market price, and relocation offers families an opportunity to better themselves (Handmer, 1987). Buy-outs are seen as a cost-effective measure because, in return for the one-off purchase cost, the property becomes ineligible for any future disaster aid. In the USA, FEMA's Hazard Mitigation Program enables buy-outs especially for high-risk

properties that make repeated loss claims. Nearly 50,000 repeatedly flooded homes exist, including more than 5,000 where the owners have received insurance payments exceeding the value of their property. Although accounting for only 1 per cent

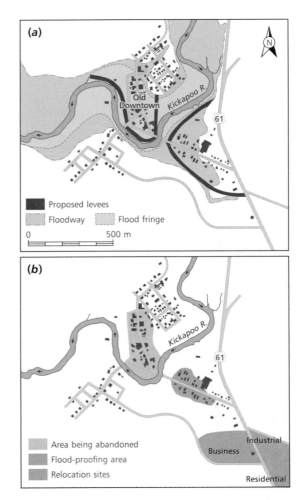

Figure 11.20 Adjustment to the flood hazard at Soldiers Grove, Wisconsin, USA: (a) the floodway and flood fringe together with the location of two proposed levees; (b) the areas eventually flood proofed , the areas abandoned and the relocation sites. After David and Mayer (1984). Reproduced from *Journal of the American Planning Association* 50 (1984), Comparing costs of alternative flood hazard mitigation plans, E. David and J. Mayer, 22–35. Copyright © American Planning Association, Chicago, IL reprinted by permission of (Taylor & Francis Ltd, http://www.tandfonline.com) on behalf of American Planning Association, Chicago, IL.

of all NFIP properties, they account for over 25 per cent of claim costs. More Federal funds are now reserved for land acquisition, relocation and similar hazard responses, but complex decisions are required of flood loss victims. Kick *et al.* (2011) found that acceptance of FEMA relocation offers by US flood victims depended on a mix of factors, including financial concerns, perception of future risk, attachment to home and community and the degree of trust placed in local flood management officials.

Early examples of relocation include the small town of Soldiers Grove, Wisconsin, USA after several floods in the 1970s (David and Mayer, 1984). The Army Corps of Engineers proposed to build two levees, in conjunction with an upstream dam, to protect the central business district (Figure 11.20a), but the residents argued that relocation of this area would yield greater benefits, not least the compensation payments for businesses to build improved premises. This scheme involved public acquisition, evacuation and demolition of all structures in the floodway, together with flood proofing of properties in the flood fringe (Figure 11.20b). Following the Midwest floods in 1993, more than 10,000 homes and businesses were relocated away from valley bottoms. About half of all relocations were in Illinois and Missouri and cost US$66 million, but these

properties had previously received US$191 million in flood insurance payouts. Along the Mississippi river, the small community of Valmeyer, Illinois was moved from the floodplain to a new site on higher ground (Table 11.5).

Occasionally, single floods can trigger relocation. On 10 January 2011, 160 mm of rain fell in 36 hours in the Toowoomba area of Queensland, Australia and a flash flood swept through the Lockyer Valley, killing 21 people. In Grantham, a settlement of 360 people about 100 km west of Brisbane, more than 130 houses were damaged. On 24 March the designation of a reconstruction area led to the Lockyer Valley Community Recovery Plan to relocate Grantham on a 935-acre site owned by the Regional Council, on ridge ground above the floodplain. This was seen as a voluntary land-swap initiative, where equivalently sized new-for-old building lots would be allocated on a ballot system to create better residential services, including town water supply and sewerage, roads, footpaths and parkland (Figure 11.21). The possible addition, over time, of a new community centre, show grounds and a school was envisaged; flooded valley land was to be retained as memorial gardens.

In future, land planning will include a greater element of the 'living with floods' approach, which recognizes the costs and benefits of flood-prone

Table 11.5 Fact-file on the post-1993 flood relocation of Valmeyer, Illinois

Before the flood	During the flood	Flood responses	After the flood
Population 900		All people flooded eligible for relocation grants	Population 1,000, although half former residents moved elsewhere
350 houses and other buildings	90% of structures substantially damaged	Temporary accommodation provided in FEMA trailers	Full rebuilding took more than 10 years
Site protected by levees built after flood in 1947	Levees over-topped. Floodwaters in town from August to October	Community decision to relocate. Federal and state funds purchase 200 ha about 3 km away and 150 m higher	New site flood free
25 businesses	Commercial sector badly hit		Commercial sector rebuilt, with growth potential

Note: This table and other information on Valmeyer kindly provided by Graham Tobin and Burrell Montz (personal communication).

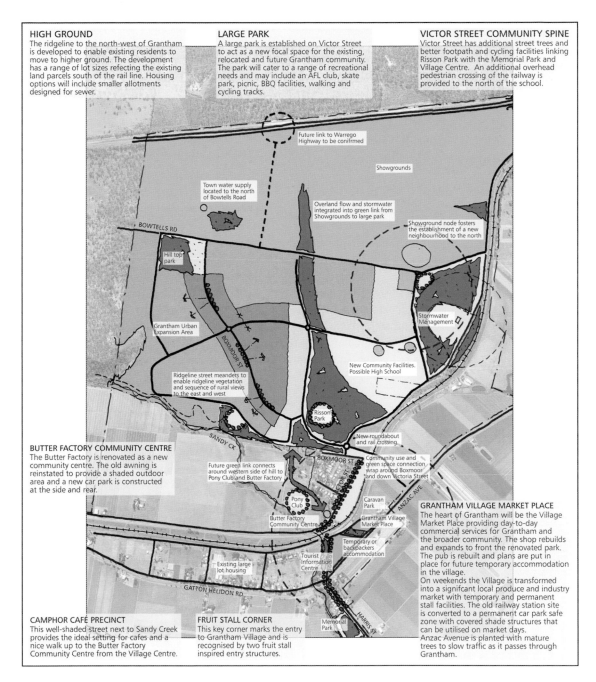

HIGH GROUND
The ridgeline to the north-west of Grantham is developed to enable existing residents to move to higher ground. The development has a range of lot sizes reflecting the existing land parcels south of the rail line. Housing options will include smaller allotments designed for sewer.

LARGE PARK
A large park is established on Victor Street to act as a new focal space for the existing, relocated and future Grantham community. The park will cater to a range of recreational needs and may include an AFL club, skate park, picnic, BBQ facilities, walking and cycling tracks.

VICTOR STREET COMMUNITY SPINE
Victor Street has additional street trees and better footpath and cycling facilities linking Risson Park with the Memorial Park and Village Centre. An additional overhead pedestrian crossing of the railway is provided to the north of the school.

Future link to Warrego Highway to be conifrmed

Showgrounds

Town water supply located to the north of Bowtells Road

Overland flow and stormwater integrated into green link from Showgrounds to large park

Showground node fosters the establishment of a new neighbourhood to the north

BOWTELLS RD

Hill top park

Stormwater Management

Grantham Urban Expansion Area

Ridgeline street meanders to enable ridgeline vegetation and sequence of rural views to the east and west

New Community Facilities. Possible High School

Risson Park

New roundabout and rail crossing

SANDY CK

BUTTER FACTORY COMMUNITY CENTRE
The Butter Factory is renovated as a new community centre. The old awning is reinstated to provide a shaded outdoor area and a new car park is constructed at the side and rear.

BOXMOOR ST

Future green link connects around western side of hill to Pony Club and Butter Factory

Community use and green space connection wrap around Boxmoor and down Victoria Street

Pony Club

Caravan Park

GRANTHAM VILLAGE MARKET PLACE
The heart of Grantham will be the Village Market Place providing day-to-day commercial services for Grantham and the broader community. The shop rebuilds and expands to front the renovated park. The pub is rebuilt and plans are put in place for future temporary accommodation in the village.
On weekends the Village is transformed into a signifcant local produce and industry market with temporary and permanent stall facilities. The old railway station site is converted to a permanent car park safe zone with covered shade structures that can be utilised on market days.
Anzac Avenue is planted with mature trees to slow traffic as it passes through Grantham.

Butter Factory Community Centre

Grantham Village Market Place

Temporary or backpackers accommodation

Tourist Information Centre

Existing large lot housing

GATTON HELIDON RD

Memorial Park

CAMPHOR CAFÉ PRECINCT
This well-shaded street next to Sandy Creek provides the ideal setting for cafes and a nice walk up to the Butter Factory Community Centre from the Village Centre.

FRUIT STALL CORNER
This key corner marks the entry to Grantham Village and is recognised by two fruit stall inspired entry structures.

Figure 11.21 The draft reconstruction and relocation plan for the town of Grantham, Queensland, Australia after severe flooding by Lockyer Creek in January 2011. After Lockyer Valley Regional Council at www.lockyervalley.qld.gov.au

land. For example, the proposal for increased reliance on embankments along the Brahmaputra and Ganges rivers, together with defences to protect Dhaka and 80 other towns, following disastrous flooding in Bangladesh in 1988, has not been adopted in full. This is because there are alternative strategies that place more reliance on traditional and sustainable flood responses. Such self-help strategies, which fit in with present land use practices and reduce the ecological impacts of engineering schemes, are likely to assume increasing importance. Within the MDCs, there is a similar movement towards 'multi-objective river corridor management', which seeks to improve floodplain development so that these areas are better equipped to cope with the various demands placed upon them (Kusler and Larson, 1993). Various types of managed retreat and realignment will become more prominent in the future (Ledoux *et al.*, 2005). The traditional defence of floodplains and coasts is looking increasingly unsustainable in the light of climate change and potential societal risk.

FURTHER READING

Gaume, E. *et al.* (2009) A compilation of data on European flash floods. *Journal of Hydrology* 367: 70–8. Captures the nature of a major flood threat that remains difficult to forecast and to regulate.

Merz, B., Kreibich, H., Schwarze, R. and Thieken, A. (2010) Assessment of economic flood damage. *Natural Hazards and Earth System Sciences* 10: 1697–724. A good illustration of both the importance and the difficulty of measuring flood costs.

Parker, D.J. (ed.) (2000) *Floods*, vols 1 and 2. Routledge, London and New York. The most detailed and reliable general reference source.

Pinter, N. (2005) One step forward, two steps back on US floodplains. *Science* 308: 207–8. Clearly demonstrates the difficulty of changing perceptions of flood risk even after a major event.

Todhunter, P.E. (2011) *Caveant admonitus* (Let the forewarned beware): the 1997 Grand Forks (USA) flood disaster. *Disaster Prevention and Management* 20: 125–39. A cautionary tale of what can go wrong with flood forecasting and warning.

White, I. and Howe, J. (2002) Flooding and the role of planning in England and Wales: a critical review. *Journal of Environmental Planning and Management* 45: 735–45. Points the way towards better land use as a means of risk reduction.

WEB LINKS

World Commission on Dams www.dams.org

Association of British Insurers www.abi.org.uk/floodinfo/

UK Environment Agency www.environment-agency.gov.uk/regions/thames

United States National Flood Insurance Program www.fema.gov./nfip

Flood Hazard Research Centre, Middlesex University www.fhrc.mdx.ac.uk/

UK Meteorological Office www.metoffice.gov.uk/corporate/pressoffice/anniversary/floods1953 html

CHAPTER TWELVE

Hydrological hazards

Droughts

A DROUGHT HAZARDS

Drought is different from almost all other environmental hazards. It is often called a 'creeping' hazard because droughts develop slowly. They also have a prolonged duration, sometimes over several years. Unlike earthquakes or floods, droughts are not confined to a particular tectonic or topographic setting. Single events can extend over regions that are subcontinental in scale and cover several countries at one time. The human impact of drought varies across the world more than any other hazard. National wealth is the main criterion. No fatalities result from drought in the industrialized countries, but in some developing nations low rainfall can lead to famine-related deaths. But, even in the most drought-prone countries, famine is often part of a 'complex emergency' arising from a combination of factors. Food supplies are most typically disrupted under conditions of war, extreme poverty, weak governance and land degradation as well as rainfall deficiency. For these reasons, the causes and consequences of drought hazards are difficult to assess.

Put simply, drought is an exceptional dry period, but such periods are hard to recognize, especially in the early stages. A simple definition is 'any unusual dry period which results in a shortage of water'. This indicates that precipitation deficiency is the 'trigger' but that the impacts are the most important characteristics. It is the shortage of *useful water* – in the soil, in rivers or reservoirs – that creates the hazard. Drought is dependent not only on climate extremes, but also on the hydrological processes that determine what happens to the precipitation and on the societal consequences of water shortages. For example, in the western USA there is a timelag between the accumulation of the winter snowpack in the Rockies and the availability of melt-water for crop irrigation the following summer. Therefore, drought disasters are better understood in terms of the impacts on natural resources and human activities, such as water supplies, agricultural production and food availability, rather than on the basis of rainfall statistics alone.

Specifically, drought hazards are difficult to define because:

- Many indices have been developed for identifying and measuring aspects of drought such as onset, severity, duration and geographical scale (Heim, 2002; Keyantash and

Dracup, 2002). One outcome is that an overall consensus on drought events can be hard to find. Similar issues apply to famine assessment.

- Vulnerability to drought depends on both human and physical factors, including geology and soil types, water storage facilities, crop types and access to early warning information. For example, areas dependent on rivers for irrigation or urban water supplies may be far downstream from the headwaters where precipitation has failed.
- Unlike earthquakes or floods, drought does not destroy infrastructure and the impacts are less visible. In the LDCs, where drought strikes hardest, it is difficult to translate agricultural losses of crops and livestock due to water shortage into measures that adequately describe a rise in poverty or further delays to human and economic development.

In summary, drought should be regarded as a *process* which develops over time, with increasingly severe effects, rather than an *event*. Above all, the link between drought and famine is unclear, as illustrated by changes made to drought entries in the EM-DAT record. Because drought and famine are often multi-year and multi-country events, over-counting occurs if each year and each country is separately recorded. Following a review of the 807 droughts previously recorded by CRED between 1900 and 2004, some entries were merged to better represent the duration and areal extent of the hazard (Below *et al.*, 2007). Although the number of droughts fell by 56 per cent, this revision led to an increase of 20 per cent in the number of drought-related deaths and a rise of 35 per cent in economic losses. Of the 76 famines which were recorded, 68 (almost 90 per cent) were classified as drought-related events; the remaining famine entries were described as 'complex disasters'. It is believed that more than half of all deaths associated with natural hazards are due to drought and that only floods are responsible for a greater number of people adversely affected by disaster.

Droughts are not confined to areas of low rainfall, any more than floods are confined to areas of high rainfall. Drought is an integral part of natural climatic variations and influences both rural and urban areas. It is important to view water shortages in terms of *resource need* rather than absolute rainfall amounts. In other words, drought and aridity are not the same. This is because humans normally adapt their activities to the expected moisture environment: a yearly rainfall of 200 mm might be tolerable for a semi-arid sheep farmer but disastrous for a wheat farmer accustomed to an average 500 mm of rain per year. What might be termed *hazard droughts* are significant events that create major socio-economic impacts. They are unlikely to result from a single dry year within the normal range of precipitation variability, but rather from a period with consecutive years of below-average rainfall (Figure 12.1). Rainfall patterns create uncertainty in several ways:

- *variability* – from season to season and year to year
- *trends* – towards wetter or drier conditions over several years
- *persistence* – wetness or dryness grouped over a period of years.

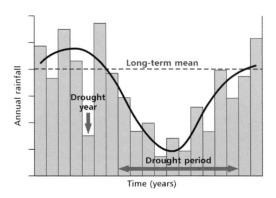

Figure 12.1 The development of a drought regime due to a persistent lack of rainfall over several years. Most drought hazards build up during a period with below-average precipitation.

Drought hazards tend to be most prominent in those semi-arid regions where there is a strong dependence, either in terms of export crops and national GDP or for subsistence purposes, on rain-fed dry-land agriculture. There are two reasons for this. First, a low mean annual precipitation is statistically associated with a high variability of rainfall from season to season and from year to year. It is this lack of *rainfall reliability* (rather than low absolute amounts) that makes water supplies uncertain. The historical record is replete with examples of over-optimistic agricultural expansion into marginal areas during wetter climatic phases, followed by retreat in the drier years later. Second, the duration of drought is longer in the drier lands. In wetter climates, a rainfall deficit is likely to persist for a few months only. For example, the 1975–76 drought over north-west Europe lasted only 16 months, whereas the late twentieth-century drought in the African Sahel was due to dry conditions that lasted for over 15 years and led to widespread famine in the mid-1980s. Some regions may be moving towards increased aridity now. According to Sousa *et al.* (2011), such a trend exists in the Mediterranean area; in 2007–9 parts of the Middle East experienced the most intense drought episode since 1940, with significantly depressed grain yields in Syria, Iraq and Iran (Trigo *et al.*, 2010).

Short-term crisis management has been the main human response to drought (Wilhite and Easterling, 1987). Emergency methods tend to focus on government intervention, like the distribution of food aid or water rationing. Longer-term adjustments favour increasing the supply of water to meet anticipated demands, e.g. building more storage reservoirs or extending irrigation schemes. In turn, this may lead to a false sense of confidence, a rise in water demands and increased risk during the next dry spell. Much less attention has been paid to improving efficiency in water use and to managing the demand for water as well as the supply. A more sustainable response to drought, including urban water recycling, more efficient irrigation practices and the use of drought-resistant crops is clearly required.

B TYPES OF DROUGHT

It is common to recognize four categories of drought hazard (Figure 12.2). Each type has different criteria for definition and assessment, although most drought indices use rainfall values, either singly or in combination with other hydro-meteorological factors (Mishra and Singh, 2010). The impacts of drought increase with the duration of the dry spell.

1 Meteorological drought

This type poses the least risks, with few hazard impacts, and is defined on statistical criteria relating to shortfalls of precipitation alone. As stated, precipitation deficiency is not necessarily hazardous because the links between precipitation and the useful water necessary to meet demands are indirect. Rainfall itself does not supply water to plants: the soil does this. Equally, rainfall does not supply water for irrigation or domestic use: rivers and ground-water do this.

The concept of meteorological drought has led to many rainfall-based definitions. The simplest method is to define drought on the minimum duration of a rain-free period, the length of which has differed from 6 days (Bali), 30 days (southern Canada) up to 2 years (Libya). As early as 1887 the UK Meteorological Office distinguished between an *absolute drought*, a period of at least 15 consecutive days with less than 0.2 mm of rain on any one day, and a *partial drought*, defined as at least 29 consecutive days when the mean daily rainfall failed to exceed 0.2 mm.

Other definitions depend on the rainfall totals that fall within a stated per centile value below the long-term average, usually during the main crop growing season or a calendar year. These definitions are of limited value unless they recognize that the impact of any rainfall deficiency is likely to vary through the period in question.

Types of drought and their characteristics		
	Major features	*Major impacts*
Meteorological drought	**Rainfall deficit** Low precipitation High temperatures Strong winds Increased solar radiation Reduced snow cover	*Loss of soil moisture* *Supply of irrigation water declines*
Hydrological drought	**Streamflow deficit** Reduced infiltration Low soil moisture Little percolation and ground-water recharge	*Reduced storage in lakes and reservoirs* *Less water for urban supply and power* *generation – restrictions* *Poorer water quality* *Threats to wetlands and wildlife habitats*
Agricultural drought	**Soil moisture deficit** Low evapotranspiration Plant water stress Reduced biomass Fall in ground-water levels	*Poor yields from rain-fed crops* *Irrigation systems start to fail* *Pasture and livestock productivity declines* *Rural industries affected* *Some government aid required*
Farming drought	**Food deficit** Loss of natural vegetation Increased risk of wildfires Wind-blown soil erosion Desertification	*Widespread failure of agricultural systems* *Food shortages on seasonal scale* *Rural economy collapses* *Rural–urban migration* *Increased malnutrition and related mortality* *Humanitarian crisis* *International aid required*

(Left vertical arrow labelled: **Drought duration and severity** *pointing downward)*

Figure 12.2 A classification of drought types based on defining components and hazard impacts. Disaster potential increases with the severity and duration of drought. Rainfall deficit alone may not produce a hazard.

The Australian Bureau of Meteorology, with its interest in agricultural drought, has used such a period-specific rainfall system, declaring a drought if the rainfall in a given area fails to exceed 10 per cent of all previous totals for the same period of the year and if the situation persists for at least three months.

The *Standardized Precipitation Index* (SPI) is a better rainfall-based measure. It uses the long-term rainfall record for a given location and period and estimates the probability of an observed precipitation deficit occurring over a given time period. The advantage is that, because the SPI can be calculated for a variety of time-scales – commonly 1 to 36 months – droughts of different lengths and severities can be assessed. It can also be related to a range of hydrological parameters, such as stream-flow and ground-water level, applied to various agricultural needs and can also be adapted for spatial–temporal analysis.

2 Hydrological drought

This occurs when stream-flows and/or ground-water levels are sufficiently reduced to threaten water resources. Hydrological drought tends to be measured by relating the reduced availability of water to the demands necessary for various uses. These demands vary greatly. For example, some approaches estimate the moisture deficit within the soil, in an attempt to assess the availability of water for plants and crops (Hunt *et al.*, 2009). The *Palmer Drought Severity Index* (PDSI) is the most widely used regional index and is based on a soil-moisture budgeting system that analyses precipitation and temperature for a given area over a period of months or years (Palmer, 1965). Drought is defined in terms of available moisture relative to the norm. The severity is assessed as a function of the length of period of abnormal moisture deficiency, as well as the magnitude of this deficiency. The PDSI produces a single hydrological measure for drought effects on soil moisture, ground-water and stream-flow and, like the SPI, can be used to rank the severity of drought episodes (Table 12.1). In some ways the PDSI is superior to the SPI, but the numerical values cannot be related directly to highly specific hazard impacts, e.g. reduced yields of different crop types (Alley, 1984; Guttman, 1997). Direct hydrological

measurements, such as stream-flow and reservoir storage, can also help to determine drought conditions relevant to water resources.

Hydrological drought is mainly linked to urban water supplies and the MDCs, although it applies elsewhere. For example, in the rural areas of north-eastern Brazil there are no permanent rivers and water supplies depend on seasonal rainfall stored in shallow reservoirs and ponds prone to high rates of evaporation. After two or three years with below-average rains, these storages dry up. Drought here gives rural dwellers less access than usual to clean water supplies; isolated communities have to rely on the distribution of water by road tankers, with negative consequences for community health.

Hydrological drought can be managed through legislation that specifies the maximum amount of water to be abstracted from a given source during dry periods. For example, the 95 per cent

Table 12.1 Drought severity ranked by return period according to Standardized Precipitation Index and Palmer Drought Severity Indices and typical impacts

Drought severity	Return period (years)	Typical impacts	SPI	PDSI
Minor	3 to 4	Short-term dryness; slower than normal crop and pasture growth; some lingering water deficits after the event.	−0.5 to −0.7	−1.0 to −1.9
Moderate	5 to 9	Reduced yields for crops and pastures; streams, wells and reservoirs at low levels; high fire risk, voluntary restrictions on water use publicized.	−0.8 to −1.2	−2.0 to −2.9
Severe	10 to 17	Appreciable crop and pasture losses; stock levels fall; fire risk very high; isolated urban and rural water shortages common; mandatory water restrictions imposed.	−1.3 to −1.5	−3.0 to −3.9
Extreme	18 to 43	Major crop and pasture losses; death and emergency sales of stock; extreme risk of wildfires; widespread water shortages combined with severe limitations on use.	−1.6 to −1.9	−4.0 to −4.9
Exceptional	>44	Complete failure of arable and pastoral production; collapse of rural economy; major wildfire outbreaks; severe water shortages with emergency arrangements in action for rural and urban supply systems; government aid required.	< −2.0	−5.0 or less

Source: Modified from Department of Space and Climate Physics and AON Benfield Hazard Research Centre, University College London, at http:// drought.mssl.ucl.ac.uk/class.html (accessed 20 August 2011). Courtesy of the UCL Global Drought Monitor.

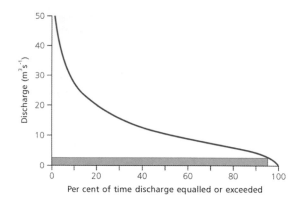

Figure 12.3 An idealized flow duration curve for a river. The definition of dry-weather discharge is based on the 95 per cent exceedance level. At this point special water conservation measures are likely to be introduced.

value on the flow duration curve, shown in Figure 12.3, is often used as the appropriate minimum discharge for a river. Legal restrictions on water abstraction will be enforced to maintain flows at or above this level. During the mid-1970s, drought was widespread in north-western Europe. In the winter of 1975–76 the recharge of ground-water into the aquifers of England and Wales was less than 30 per cent of average. As a result, many rivers were at low levels, abstractions were reduced and water rationing was imposed in some areas. The United States drought of 1988 was the most severe in the Mississippi basin since 1936. By July 1988 barge traffic was drastically reduced on the Ohio and Mississippi rivers. The reduced river flows also caused hydropower generation

Box 12.1 Drought in Australia

Drought is a recurrent feature in Australia. The most costly impacts are on agricultural productivity, especially in the south-eastern parts of the country, where most of Australia's 50,000 farming families live. Although less than 4 per cent of Australia's GDP comes directly from agriculture, the economic consequences are severe because 80 per cent of all agricultural products are exported, and account about half the value of all exported goods. Drought affects rain-fed crops, like wheat and barley, and halts the growth of pasture so that stock levels of cattle and sheep fall sharply. During the 1979–83 dry period over half the nation's farms, housing about 60 per cent of the livestock, were affected. In 1982–83, the yearly cash surplus on Australian farms fell from an average of A$21,700 down to A$12,200 (Purtill, 1983). The Federal government accepts drought as an integral feature of the climate and provides emergency funds for rural communities in times of 'exceptional circumstances'. The long-term impacts of drought include the loss of agricultural jobs, the erosion of investment capital for rural industries, damage to timber stocks due to bushfires, the degradation of natural vegetation and the wind erosion of soils.

The Australian Bureau of Meteorology recognizes two chief types of drought based on rainfall criteria:

- *Serious deficiency* – rainfall totals within the lowest 10 per cent of values on record for at least three months
- *Severe deficiency* – rainfall totals within the lowest 5 per cent of values on record for at least three months.

Australia experiences low, and highly variable, rainfall totals because the climate is dominated by the subtropical high-pressure belt of the southern hemisphere. The worst droughts occur after a prolonged spell of below-average rainfall but it is unusual for more than 30 per cent of the country to be affected at any one time (Chapman, 1999). Droughts differ greatly; some are intense and short lived, some last for years, some are localized whilst others cover large areas (Table 12.2 and Figure 12.4). Australian droughts are closely linked to negative phases in the El Niño Southern Oscillation (ENSO) phenomenon (Allan *et al.*, 1996). Under these large-scale atmospheric influences, relatively cool sea-surface temperatures prevail off northern Australia and tend to produce low rainfall over eastern and northern Australia (see Figure 12.4c).

Table 12.2 Major droughts and their impact in Australia

Period	General characteristics	Economic losses
1895–1903	The 'Federation drought' followed several years of low rainfall, especially in Queensland; an early ENSO-related event.	Devastating stock losses – 50% reduction in sheep and 40% in cattle stock. Wheat crop almost totally destroyed.
1913–16	Most severe in 1914, an ENSO year. This drought spread over most of the country.	Cattle transported in attempt to find better pasture; 19 million sheep and 2 million cattle lost. Bushfires in Victoria.
1937–45	The 'World War II droughts' mainly affected eastern Australia; 1940 and 1941 were ENSO years.	Loss of almost 30 million sheep between 1942 and 1945. Wheat yields lowest since 1914. Some large rivers almost dried up.
1963–68	Widespread drought, mainly in central Australia but also affected the eastern states 1965–68	1967–68 40% drop in wheat yields, loss of 20 million sheep, decrease in farm income A$ 300–500 million.
1982–83	Severe short-term drought due to strong ENSO event in 1982; very extensive across eastern half of Australia	Total economic cost over A$ 3,000 million, mainly due to impact on wheat yields and sheep stocks.
1991–95	An extended ENSO drought, the longest recorded to date; mostly in central and southern Queensland plus northern New South Wales	Productivity of rural industries down 10% and an estimated A$ 5 billion loss to national economy Nearly A$ 600 million provided in drought relief.
2002–8	The 'Big Dry' drought. Began in 2002, the fourth Driest year on record. Warmer and drier on average during 2004 and 2005 in many south-eastern areas. The rainfall deficits continued into 2007.	GDP down 1% in 2002–3, some 70,000 jobs lost in the rural sector. The federal government spent A$ 740 million in aid, during 2002–5 estimated farm output down by 20%.

Since the 1970s, there has been a shift in rainfall patterns, with the sparsely populated areas of the north becoming wetter and the eastern areas, where most people live, becoming drier. Australia experienced two closely spaced El Niño events (2002–3 and 2006–7) with no interspersed wet period (Nowak, 2007). During the past decade most years have also been warmer than average, with 2005 the warmest so far recorded, thereby increasing evaporation losses. More recently, Australia suffered the 'Big Dry', a long spell of unusually dry and warm conditions that began in 2002. This was the first drought in Australian history to be firmly associated with climate change by some observers and assessed as a possible 1:000-year event. Despite spatial and seasonal fluctuations in rainfall, the drought spread nationwide, affecting more than half the farmlands and being particularly severe in the agriculturally sensitive areas of eastern and southern Australia. By the end of 2006, more than

90 per cent of New South Wales was in drought. The Murray–Darling River Basin, which accounts for over 40 per cent of the nation's agricultural output, had already experienced the lowest-recorded flows during a consecutive four-year period since records began in the 1890s.

Farm production fell by 20 per cent. Crop yields for wheat and barley were down about 60 per cent, with the wool harvest set to be the lowest in 20 years, thereby reducing national economic growth by an estimated 0.5 per cent. Income in the agricultural sector was expected to fall by 70 per cent in 2006, taking A$7 billion out of the economy. In late 2006, global wheat prices reached a 10-year high, partly due to forecasts of low Australian yields in 2006–7, and increased livestock slaughtering occurred. The economic multiplier effect means that, as agricultural production declines, so does the demand for transport and other services. In October 2006 the federal government introduced an extra package of drought relief for areas already in 'exceptional circumstances' and to provide financial and counselling services for the worst-affected farms and rural enterprises. The government was expected to spend more than A$2 billion on welfare payments to about 50,000 farming families in the country.

The drought affected much more than agriculture. Despite the fact that most of Australia's cities are served by massive water supply schemes designed to withstand multi-year episodes of low run-off, most reservoirs had fallen below half their capacity by late 2006 and many towns and cities in the southern half of the country were subject to water restrictions. The Murray–Darling system

Figure 12.4 Examples of differing areal extent and temporal duration of droughts in Australia during the second half of the twentieth century: (a) localized drought; (b) short and intense drought; (c) prolonged drought. After Bureau of Meteorology, Australia, at www.bom.gov.au/climate/drought/ Copyright Bureau of Meteorology, Australia, reproduced with permission.

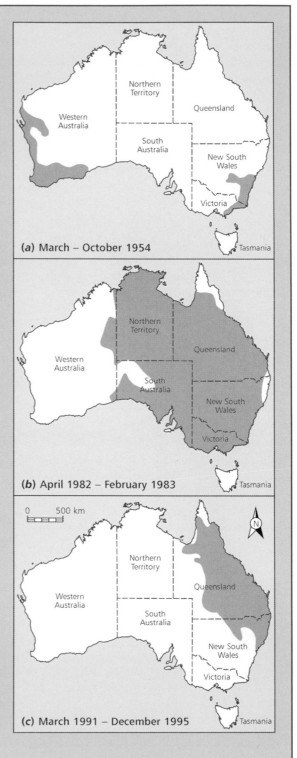

suffered water stress, with reduced allocations for irrigation, a decline in the Red Gum trees along the river, fish kills due to low flows and a build-up of salt on some floodplains and wetlands. Adelaide in South Australia is vulnerable because the city draws 40 per cent of its drinking water from the river Murray. *Per capita* consumption of water in Australia is high. In 2006 Perth began operating a water desalination plant, partly powered by electricity from wind energy, with Sydney and Melbourne expected to follow. Reliance on such supply-side 'drought-proofing' schemes has been severely criticized (Isler *et al.*, 2010). They are likely to be insufficient for future urban and industrial needs unless demand management measures – recycling waste water and higher water charges – are also used to curb demand. In 2007 all the major cities were declared 'in drought', highlighting key questions about the utilization of the nation's water resources.

to fall 25–40 per cent below average over large areas of the USA, with significant losses in company revenue (Wilhite and Vanyarkho, 2000).

3 Agricultural drought

Agricultural drought is important because of the implications for food production. This applies to the MDCs with a dependence on agricultural output for their economic well-being (see Box 12.1). Drought is probably the third-most costly environmental hazard in the USA and also impacts heavily on the LDCs where subsistence agriculture supports most of the population. For example, during the drought years of 1992 and 1994 in Malawi, agricultural sector output fell 25 per cent and 30 per cent, respectively, below normal. All farmers, whether arable or pastoral, rely on the water available for plant growth in the soil, and an agricultural drought exists when soil moisture is insufficient to maintain average crop and grass yields.

Ideally, the severity of agricultural drought should be based on direct soil-moisture measurements, but indirect assessments are usually made through water-balance calculations like the PDSI. Regional water stress on plants can be monitored from space through the *Vegetation Condition Index* (VCI), described by Liu and Kogan (1996). Unfortunately, indices like the VCI and the PDSI cannot be directly linked to drought impacts on

farm production because each crop responds differently to heat and moisture stress. Consequently, attempts have been made to derive a *Crop-Specific Drought Index* (CSDI), as pioneered by Meyer (1993) for corn.

Prosperous countries have the means to avert the worst effects of drought but are not immune from risk. Drought accounted for over 40 per cent of the estimated US$349 billion cost of all weather-related disasters in the USA between 1980 and 2003 (Ross and Lott, 2003). Droughts are a long-term feature of North America and occur on the Great Plains about every 20 years. During droughts in the 1890s and 1910s there were deaths due to malnutrition. A turning-point was reached in the 'Dust Bowl' years of the 1930s, with spells when almost two-thirds of the USA was in drought (Figure 12.5). The impact of this drought was amplified by poor farming techniques and prompted massive injections of aid, greater control of soil erosion and improved irrigation practices. Together with more sustainable farm management and crop insurance, these measures ensured that the 1950s drought had less severe impacts. Even so, problems of US drought management remain (Pulwarty *et al.*, 2007).

The main consequence of agricultural drought is reduced crop and animal production. When fodder is inadequate, mass slaughter of livestock follows. It may take up to five years for animal stocking levels to recover. In 1988 the USA

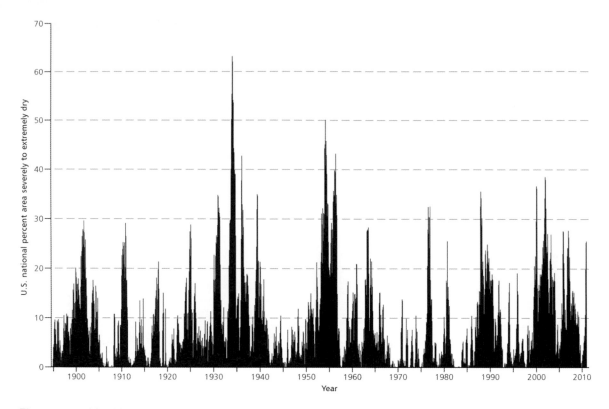

Figure 12.5 Monthly percentage area of the United States in severe and extreme drought from January 1895 to August 2009. This indicates that drought is a normal part of climate, with more severe episodes recorded in the 1930s, 1950s and more recently in 1988 and around 2000. After National Climatic Data Center, NOAA at www.ncdc.noaa.gov/ (accessed 6 February 2012). US government material not subject to copyright protection in the United States.

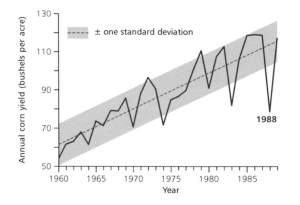

Figure 12.6 Annual corn yields in the USA 1960–89 showing the effect of the 1988 drought. Yields in 1988 were more than 30 per cent below trend, the largest annual drop recorded since the 1930s Dust Bowl years. After Donald (1988). US government material not subject to copyright protection in the United States.

experienced a costly agricultural drought over the Midwest. The 1988 corn yield was 31 per cent below the progressive upward trend that is driven by improved technology, the largest drop since the mid-1930s (Figure 12.6). More than one-third of the American corn crop was destroyed, at a loss of US$4.7 billion (Donald, 1988). Agricultural drought on this scale disrupts international trade in food, and world grain stocks fell to a 63-day supply, the lowest since the mid-1970s. At the farm level, severe drought disrupts normal activities and causes a diversion of capital from farm development to drought-reducing strategies, a fall in cash liquidity and a rise in debt.

In the poorest countries, drought disrupts subsistence food supply and increases seasonal hunger. This happened during the 1990–92 drought in southern Africa. In general, the harvest yield was 30–80 per cent below normal and 86 million people were affected over an area of almost 7×10^6 km². Although comparatively little loss of human life ensued, there was severe hardship. In Zimbabwe the volume of agricultural production fell by one-third and contributed only 8 per cent to GDP, compared to 16 per cent in normal years. By November 1992 half the population had registered for drought relief. Conditions in Zambia were typical. In some administrative districts yields of maize were down by 40–100 per cent and 2 million rural people were affected (IFRCRCS, 1994). According to Kajoba (1992), part of the grain shortfall was due to the culti-vation of hybrid maize under imported fertilizer regimes, rather than a reliance on more traditional drought-resistant crops like sorghum, millets and cassava. Communities in remote areas suffered badly. The drought closed schools and led to a decline in tourism as wildlife camps were deserted. Due to low water levels, the Kariba, Kafue and Victoria Falls hydropower stations worked at 30 per cent capacity and the government imposed daily power breaks.

4 Famine drought

Famine disasters have been recorded for at least 6,000 years and have only recently been eliminated – it is hoped – from the MDCs (Dando, 1980). Examples of excess mortality due to famine include Ireland 1845–49 (1.0 to 1.25 million); Soviet Union 1932–34 (5.0 million); Bengal 1943–46 (3.0 million) and China 1958–61 (16.5 to 29.5 million). In each case, mass starvation followed poor harvests. However, there was often enough food within the country to prevent deaths on this scale if the relevant government had acted quickly enough to organize food distri-bution to those in need (Jowett, 1989). In other words, drought rarely acts alone to create mass starvation and humanitarian disasters. Pingali

et al. (2005) reported that, whilst the number of food emergencies worldwide notified by the Food and Agriculture Organization (FAO) of the UN doubled from an average 15 per year during the 1980s to more than 30 per year in the early 2000s, over 50 per cent of the emergencies were attributable to human failures.

The interrelationships between drought and famine are complex because drought is a hydro-meteorological extreme, whereas famine is a cultural phenomenon. Drought results from a lower-than-expected amount of precipitation. Famine results from an acute food shortage. Both phenomena create problems of definition and measurement and represent processes rather than events. For example, the early stages of a famine disaster are difficult to distinguish from chronic, lesser states of hunger, such as malnutrition, and other forms of food scarcity, like seasonal short-falls in harvest yields. Famine-related deaths typically arise in the context of a 'complex emer-gency' which includes meteorological factors but also involves many other factors (White, 2005). Emergency relief efforts, such as early warning and food aid donations, are difficult to implement if there is uncertainty and disagreement, perhaps between field workers and aid donors, about the nature of a food crisis.

Most definitions of famine try to capture the situation where a food shortage lasts long enough to cause widespread suffering, severe malnutri-tion and death from starvation, particularly amongst the most vulnerable groups in an area or community (Howe and Devereux, 2004). Famine reflects a failure in food security and denies access to enough food for an active and healthy life. Across the globe probably 2 billion people lack food security. Many have a chronically inadequate diet, due to the household's inability to produce or buy enough food. Whenever food access falls below these levels, famine becomes a threat.

The United Nations recognizes five stages in the decline of food access within a region or a country:

Plate 12.1 The carcass of a dead animal lies near the over-crowded tents of the UNHCR Dadaab refugee camp, in the North Eastern Province of Kenya during August 2011. The most severe drought for 60 years led to severe food shortages and a humanitarian emergency across much of East Africa. (Photo: Panos/Sven Torfinn STO05040KEN)

- Food secure
- Moderate/borderline food insecure
- Acute food and livelihood crisis
- Humanitarian emergency
- Famine/catastrophe.

Since 2008, the United Nations has defined famine as:

acute malnutrition of more than 30 per cent of all children, two deaths per 10,000 people a day, access to less than 2,100 kilocalories and 4 litres of water each day plus complete loss of income or assets.

Associated markers include one-fifth of households with extreme food shortages, large-scale displacement of population and outbreaks of civil strife. Conflict is a recurrent feature because it leads to direct destruction of life and property, the abandonment of productive land and disruption of trade and other economic activity. Such features were evident during 2011, when many refugees from Somalia crossed the border into Kenya to escape hunger due to the worst drought in 60 years and an on-going civil war.

Malnutrition is always a contributory component of famine. It has been described as the most widespread disease in the world because at least one-third of the population of the LDCs is malnourished. There is a general under-registration of population in the LDCs and many countries lack reliable data on the causes of death. It has been said that where there are statistics,

Box 12.2 Drought and famine in the Horn of Africa: the case of Ethiopia

The Horn of Africa, a peninsula in the north-east of the continent, contains the nations of Somalia, Djibouti, Eritrea and Ethiopia, along with adjacent parts of Kenya and Sudan. It has been an arid region for at least 4,000 years and is regularly scourged by drought, as well as floods and protracted civil wars. Factors that increase vulnerability to drought include poor governance, low productivity, stagnant levels of technology and the inappropriate use of human and natural resources. Malnutrition among pre-school children is endemic, being higher amongst pastoralists than in agricultural communities, and increasing in drought years (Chotard *et al.*, 2010). When weather and socio-economic conditions deteriorate, all countries have a familiar downward spiral into high-risk coping strategies such as selling livestock and household goods, followed by migration to find pasture, water and food (Glantz, 1988). The failure of rains in late 2010, one of the driest spells in many pastoral areas since 1950–51, created the most recent food crisis, affecting some 12 million people. In 2011 the UN declared famine in parts of Somalia when up to 3,500 refugees per day were crossing the border to the Dadaab emergency camp in Kenya. The EU allocated €600 million in aid; Oxfam called for £50 million, its largest-ever appeal for Africa.

During recent decades, Ethiopia has been subject to more drought than any other part of the Horn. Between 1970 and 1996, 25 droughts/food shortages/famines were recorded, killing more than 1.2 million people and affecting over 60 million (Ferris-Morris, 2003). Chronic food insecurity is experienced by 5–10 per cent of the population, but in the 1984–85 drought more than one-fifth were at risk of starvation after widespread rainfall deficits over several years. Excess mortality was estimated at 700,000. Following three consecutive low rainfall years, a similar crisis emerged in 1999–2000, characterized by market failure, a ban imposed on livestock trading and a border conflict. Ethiopia is ill-equipped to cope with these disasters. Of the 66 million population, 85 per cent live in the countryside; half the national GDP and 90 per cent of exports depend on agriculture. Yet only 12 per cent of the land is arable, mostly dependent on rain-fed production. Less than two per cent is in permanent crops, often grown for export and exposed to fluctuations in world commodity prices. Cattle herding has been the main traditional activity, occupying over 60 per cent of the land area, but drought and mismanagement have degraded the rangeland vegetation to favour of camels and small ruminants, with an attendant rise in poverty (Kassahun *et a.*, 2008). Over half the people live on less than US$1 per day and Ethiopia is the annual recipient of economic aid valued at almost US$100 million.

It was previously believed that rainfall deficiencies in Ethiopia were driven by the moist summer monsoonal flow of air over Africa. It is now known that the key rainy period from June through to September (the *Kiremt* season) is largely dependent on the El Niño Southern Oscillation (ENSO) cycle, with sea-surface temperature anomalies in the Atlantic and Indian oceans of lesser significance (Korecha and Barnston, 2007). *Kiremt* rains result principally from localized convective activity over the Ethiopian highlands of western and west-central Ethiopia. The north-eastern area and the south-eastern pastoral lowlands are normally relatively dry at this time of year (Figure 12.7a). Figure 12.7b shows that this season provides about half the country with over 60 per cent of its annual rainfall; the resulting harvest supplies over 90 per cent of the nation's food. Unfortunately, these rains are unreliable. Ethiopia's farmers are well aware of this and adopt what strategies are available to them in unfavourable conditions. For example, agriculturalists in the north of the country have shifted towards more drought-resistant crops and a shorter growing season (Meze-Hausken, 2004), combined with more supplementary irrigation during September, taken from surface run-off water captured in farm ponds in July and August, in an attempt to prolong the growing period (Araya and Stroosnijder, 2011).

In a year with good rains, Ethiopia can produce enough food for its needs, although households still require the financial means to access supplies. Food emergencies are often complex, as in the 1999–2000 drought, when meagre *belg* rains during spring 1999 led to crop losses in the north-east highlands and cattle-herd reductions of up to 80 per cent in the south and south-east were coincident with a border dispute with Eritrea. By January 2000, an estimated 7.7 million people were affected and by July 2000 a humanitarian crisis existed over much of the country. Ethiopia is in urgent need of alternative, sustainable coping mechanisms to combat drought.

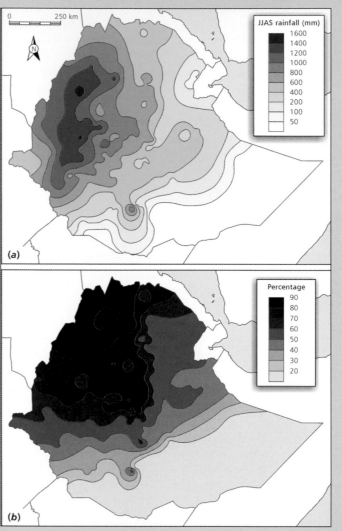

Figure 12.7 Rainfall patterns over Ethiopia: (a) total summer (JJAS) amounts over Ethiopia 1971–2000; (b) percentage of 1971–2000 mean annual rainfall contributed by JJAS rains. Summer rainfall over Ethiopia varies greatly in amount and timing. Both factors are crucial for food production and water management. After Korecha and Barnston (2007). Reproduced from D. Korecha and A.G. Barnston, Predictability of June–September rainfall in Ethiopia. *Monthly Weather Review* 135 (2007): 628–50. Copyright American Meteorological Society, reproduced with permission.

there is no malnutrition, and, more significantly, where there is malnutrition, there are no statistics. Given these limitations, it is impossible to produce a reliable estimate of the average annual number of people who die or are affected by drought-related famine.

Most famine-related deaths during recent years have occurred in the semi-arid areas of sub-Saharan Africa. In February 1985 the United Nations estimated that 150 million people living in 20 African countries were affected, of which 30 million were in urgent need of food aid. An estimated 10 million of these people abandoned their homes in search of food and water and up to 250,000 people died. There were also huge losses of cattle and sheep. One area especially prone to drought and famine is the Horn of Africa (Box 12.2). Asian countries, such as India, are also affected by drought but some have sought to avoid famine through national self-sufficiency in the production of food grains (Mathur and Jayal, 1992). In South America, the semi-arid area of north-east Brazil suffered frequent droughts in the twentieth century (1915, 1919, 1934, 1983 and 1994). Children younger than five years of age constitute almost 20 per cent of the population and endemic malnutrition, especially within marginalized groups, greatly increased their vulnerability to drought periods.

Today, famine strikes mainly at the LDCs with semi-arid areas of subsistence or near-subsistence agriculture when the rain-fed crops fail. However, evidence from the Darfur region, Sudan during the African drought of 1984–85 challenged the common concept of mass starvation arising directly from crop failure. According to de Waal (1989), the great majority of people survived in Darfur, despite a doubling of the overall mortality rate. The excess mortality was heavily concentrated on children and the elderly, with people between 10 and 50 years old accounting for fewer than 10 per cent of the excess deaths. Many deaths were caused by the transmission of disease (measles, diarrhoea and malaria) arising from the crowding of refugees into centres where water supplies were poor and health care was inadequate. This interpretation casts doubt on the conventional indicators of famine drought, such as mass starvation, and on the assumed efficacy of conventional disaster reduction strategies, such as the supply of food aid.

Although scarcity of food is a leading factor in famine drought, its effects are compounded by underlying and long-term socio-economic and health-related problems, such as limited access to potable water, a lack of modern sanitation and inadequate health care, especially for the very young. Another feature of severe droughts is that they undermine rural stability by encouraging out-migration. After the 1985 drought in northeast Brazil, almost 1 million people – mainly men – abandoned their small farms in search of work, creating a wave of rural–urban migration that added to the *favellas* or shanty towns surrounding every Brazilian city.

C CAUSES OF DROUGHT HAZARDS

1 Physical factors

Drought originates from anomalies within the general atmospheric circulation, but the climate processes that cause drought are not completely understood. More research is required into *climate dynamics*, a broad field of science that examines the entire coupled system of atmosphere, oceans, cryosphere, biomass and land surface as interacting components that produce the global climate. Much of this work is geared to the analytical and numerical modelling necessary to explain – and ultimately predict – aspects of climatic variability on all scales of space and time (Goosse *et al.*, 2008). Some variations depend on external forcing factors, such as solar activity or volcanic eruptions, whilst others, like the El Niño Southern Oscillation (ENSO) and the North Atlantic Oscillation (NAO) occur due to internal actions and reactions within the system (see also Chapter 14). Feedbacks within the system and lag effects, due to inertia in the oceans and ice sheets, apply, and small changes in the forcing factors can lead to important consequences for the climate if a significant threshold is crossed.

Climate dynamics stresses the role of *tele-connections*. These are links between climatic anomalies occurring at long distances apart, especially interactions between the atmosphere and the oceans. For example, it is known that sea-surface temperature anomalies (SSTAs) played a role in the late twentieth-century Sahel drought. Direct sea–air interactions influence the flux of both sensible heat and moisture at the ocean–atmosphere interface. One example would be the initiation of drought by negative (relatively cold) SSTAs leading to descending air and anti-cyclonic weather. It is possible that the starting-point for the 1975–76 drought over north-west Europe was very low sea-surface temperatures over the Atlantic Ocean north of 40°N which caused near-surface stability in the atmosphere and led to blocking anticyclones. The case for oceanic forcing of drought is best established for El Niño

Southern Oscillation (ENSO) events (see Chapter 14, pp. 427–9). According to Dilley and Heyman (1995), worldwide drought disasters double during the second year of an El Niño episode, as compared with all other years. This happened in 1982–83, when droughts in Africa, Australia, India, north-east Brazil and the United States coincided with a major El Niño phase. Over smaller areas, other climate drivers, like the NAO, may not fully capture the detailed aspects of drought, due to weather types and synoptic-scale processes (Vicente-Serrano and López-Moreno, 2006).

The African Sahel is drought sensitive because it occupies a transitional climatic zone between the Sahara to the north and tropical rain forests to the south. Under 'normal' conditions the mean annual rainfall varies from about 100 mm (hyper arid) on the Saharan edge to 800 mm (dry sub-humid) along the southern margins. But mean values are highly misleading. These rainfall values are characterized by high variability on all climatic time-scales – seasonal, year to year and decadal.

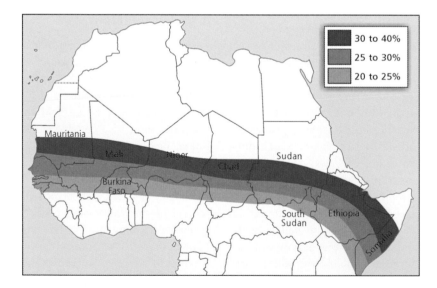

Figure 12.8 Countries of the Sahel region prone to drought. The coloured areas show the average annual departures from normal rainfall. When rainfall totals are low and variability is high, drought is likely to be a recurrent feature. Source: USAID. US government material not subject to copyright protection in the United States.

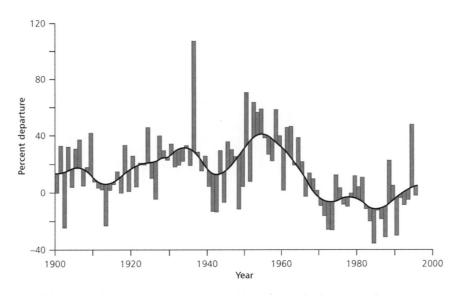

Figure 12.9 Sahelian rainfall during the rainy season (June–September) expressed as a percentage of the 1961–90 mean. The downturn in the late 1960s was a major factor in the famine disasters of the later twentieth century. After M. Hulme at http://cru.uea.ac.uk (accessed on 31 May 2003), used with permission.

More than 80 per cent of the annual rainfall is expected in the summer months of July, August and September. The year-to-year variability, expressed by the coefficient of variation, is high, ranging from approximately 20 to 40 per cent, and leads to low reliability of the rainy season (Figure 12.8). This pattern has been disrupted by prolonged rainfall anomalies such as the relatively wet periods in 1905–9 and 1950–69. From the mid-1960s, there was a pronounced decline in annual rainfall (Figure 12.9). Overall, the Sahel experienced a period of some 30 years with below-average rainfall conditions, broken only by widespread rains in the 1994 wet season.

This was the most striking trend in precipitation in the world during the twentieth century (Hulme, 2001). The agricultural impact of the drought was made worse by good rains in the 1950s and 1960s, which encouraged rain-fed cropping into marginal lands and larger herd sizes. Reviews concluded that the drought was driven mainly by a rise in sea-surface temperatures in the southern hemisphere and the Indian Ocean which produced changes in the atmospheric circulation

(Brooks, 2004). Broadly speaking, it is believed that the tropical rain belts were attracted to the relative warmth of the southern hemisphere, so that the Sahel, on the northern extension of these rain belts during the Northern summer, effectively lost its rainy season. Further work has confirmed that the Sahel region is highly sensitive to the variability of SSTs in all tropical ocean basins, including the remote Pacific. This is because unusually warm low-latitude waters favour deep convection over the oceans that, in turn, weakens the summer monsoon convergence and creates drought from Senegal to Ethiopia (Giannini *et al.*, 2003).

Following a wide geographical survey, Peel *et al.* (2002) concluded that the variability of annual precipitation was appreciably higher in ENSO-affected climatic zones than in other areas. Teleconnections involving SSTs and rainfall have been established for several parts of the world. Across eastern Australia (Queensland, New South Wales, Victoria and Tasmania), annual spring and summer rainfall is strongly linked to the Southern Oscillation Index (SOI), with a correlation

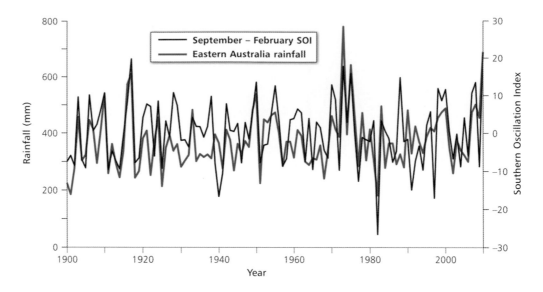

Figure 12.10 Rainfall in eastern Australia during the September–February period in relation to the Southern Oscillation Index (SOI). The SOI measures the strength of La Niña and El Niño events, a normal part of climate, which are strongly correlated with summer rainfall in eastern Australia and other parts of the world. After Nicholls (2011). Used with permission.

coefficient of 0.66 (Nicholls, 2011). Figure 12.10 shows a clear relationship between low and negative SOI values and individual drought years such as 1940, 1982 and 1997. Whilst ENSO-based models can reproduce year-to-year variability quite well, they have been less successful in handling multi-decadal drying trends. McCabe *et al.* (2004), for example, demonstrated that SSTAs over the North Atlantic Ocean, as well as the Pacific, contribute to drought incidence in the United States. Together, these influences accounted for more than half of the spatial and temporal variation in multi-decadal drought in this region.

Another complex climatic link exists between ocean surfaces and the strength of the Indian summer monsoon system. Whilst ENSO alone accounts for about 30 per cent of the inter-annual variability in monsoon rainfall over India, un-expected drought years – like 2002 – have occurred. The cause has been attributed to inter-actions between ENSO and weather conditions over the more local Indian Ocean, where deep

convection in the atmosphere over the equatorial marine surface varies annually between the eastern and western parts of the ocean, to create the Indian Ocean Oscillation (Gadgil *et al.*, 2003). This feature, termed the *Indian Ocean dipole* (IOD), modulates the ENSO–Indian monsoon teleconnection on decadal time-scales. Strong ENSO events are associated with a weak monsoon air-flow and potential drought hazards over India. Conversely, in the absence of any ENSO tendency, the Indian Ocean effect is strong and positive IOD years (pIOD) produce above-average monsoon rainfall. When El Niño events coincide with pIOD conditions, the drought-forming subsidence of ENSO is counteracted so that a near-average monsoon season results (Ummenhofer *et al.*, 2011). Figure 12.11 shows annual rainfall anomalies over the core monsoon region of India for the summer monsoon season (JJAS) from 1877 to 2006 in relation to ENSO, pIOD and joint ENSO–pIOD events. More than 85 per cent of ENSO years experienced anomalously low rainfall, some deficits exceeding

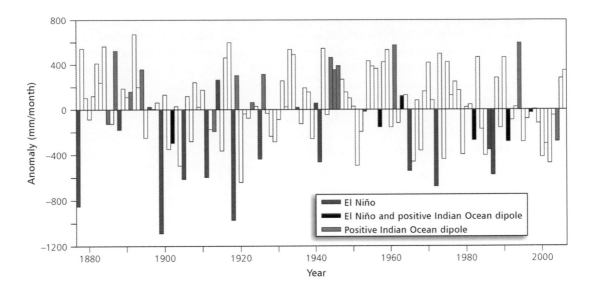

Figure 12.11 Time-series of rainfall anomalies (mm per month) during June–September 1877–2006 for three interacting categories of climatic conditions over the core Indian monsoon region. Anomalies are shown for El Niño events (blue), El Niño co-occurring with positive Indian Ocean dipole (pIOD) events (purple) and positive Indian Ocean dipole events (orange). After Ummenhofer *et al.* (2011). Reproduced from C.M. Ummenhofer *et al.*, 2011 Multi-decadal modulation of the El Niño–Indian monsoon relationship by Indian Ocean variability. *Environmental Research Letters* 6: 1–8. With permission from IOP Publishing Ltd.

800 mm per month; pIOD years had above-average rainfall, whilst co-occurring ENSO/pIOD events led to near-average conditions.

Droughts are initiated by large-scale atmospheric processes, but the dryness may be prolonged by *regional forcing*. This is due to localized feedback mechanisms between the land surface and the lower atmosphere. Excessively dry ground helps to maintain the rainless atmospheric state because a higher proportion of the incoming solar radiation is used to heat the ground and the air, compared to normal conditions when more energy would be used in evaporation. Where prolonged dryness has reduced the vegetation cover, changed the surface albedo and created greater dustiness, it is possible that drought may become almost self-perpetuating. The theory is attractive for the Sahel, where a lack of rain, combined with pressure on land resources, has produced environmental deterioration and even desertification. Other regional feedbacks include the increase in aerosols due to large-scale biomass burning, such as occurred in Indonesia (Field *et al.*, 2009).

Similar mechanisms have been found elsewhere. The 1988 North American drought was linked to an ENSO episode which led to a northward displacement of the Inter-tropical Convergence Zone south-east of Hawaii and the eventual appearance of a strong anticyclone at upper levels over the American Midwest (Trenberth *et al.* 1988). In fact, most droughts in the USA can be explained by primary forcing and anomalous SSTAs, but the exceptional 'Dust Bowl' drought of the 1930s was probably amplified by regional factors. Once dry conditions are established, the subsiding air, combined with the drying-out of open water surfaces and the upper soil layers, reduces the relative humidity in the lower atmosphere. Potentially, this limits the ability of local weather disturbances to create rainfall. Cook *et al.* (2007) and Cook *et al.* (2011) simulated climatic conditions at the time using general circulation models (GCMs) and found that anomalous sea-surface temperatures alone were insufficient to account for the severity of the 1930s drought. However, when land-surface factors – de-vegetation and dust – were incorporated into the model, together with feedbacks between the atmosphere and the land, the rainfall anomalies and other drought features were represented more accurately.

2 Human factors

Human factors do not cause drought but they do influence the impact of drought hazards. Major drought disasters are concentrated in the semi-arid, developing countries prone to 'complex emergencies'. This is well illustrated in Africa, a continent where two-thirds of the area is dry land. At the height of the 1986 drought, 185 million people were at risk of famine and disease (Dinar and Keck, 2000). During the 1991–92 drought spell in southern Africa there was a 6.7 million tonne deficit of cereal supplies, affecting more than 20 million people; in the 1999–2000 Ethiopian crisis about 10 million people were in need of food assistance. Drought impacts in Ethiopia were increased by socio-economic conditions due to rural destitution, growing environmental degradation, a war with neighbouring Eritrea and conflicts between humanitarian and political objectives (Hammond and Maxwell, 2002).

Poverty is a key factor. Sub-Saharan Africa contains over two-thirds of the world's poorest countries, where people struggle to cope with fluctuations in their physical and societal environments. Drought impacts differ greatly; some people prosper, some migrate to refugee camps and others die. Worst hit are the landless and jobless, especially women and children in the rural areas who lack the means to ensure their own food security. Factors such as the declining terms of trade for primary agricultural products, market protection by the industrialized countries, extreme commodity price fluctuations on international markets and the need to service enormous overseas debts have all restricted the ability of African governments to address complex internal problems.

In the Sahel, rural population densities have increased, due to a doubling of the population every 20–30 years. Despite the importance of agriculture – accounting for more than 40 per cent of GDP in some countries – population growth has outstripped food production. In certain areas, the progressive conversion of natural ecosystems into farmland has led to desertification through the over-cultivation of croplands, shortening of fallow seasons, over-grazing of rangelands, mismanagement of irrigated cropland and deforestation. About 90 per cent of pasture land and 85 per cent of the cropped lands in nations close to the Sahara have been affected. Deforestation is an important catalyst of land exhaustion and soil erosion, partly driven by the fact that more than 90 per cent of cooking and other energy needs are met by wood.

The reliance on rain-fed agriculture throughout sub-Saharan Africa creates vulnerability. In such a low-technology system the management options during drought is limited to the selection of a particular crop type for sedentary cultivators and reduced stocking rates for pastoralists. That having been said, the traditional pattern of agricultural land use in the Sahel was well adapted to the uncertain rainfall conditions. Generally speaking, the drier northern zone was used for livestock, whilst the more humid southern Sahel relied on rain-fed crops. This system permitted a degree of flexible interdependence. The pastoralists followed the rains by either seasonal migration (transhumance) or the practice of full nomadism, whilst the cultivators grew a variety of drought-resistant subsistence crops, including sorghum and millet, to reduce the risk of failure. Long fallow periods were used to rest the land for perhaps as much as five years after cropping in order to maintain the fertility of the soil. In the absence of a cash economy, a barter system operated between nomads and sedentary farmers, leading to the exchange of meat and cereals.

This system has been eroded. Population growth, with the need for more food supplies, has placed more pressure on the land. One consequence has been soil erosion as cultivation spread into the drier areas formerly used for livestock. In turn, the rangelands have been over-grazed, with degradation of the resource base. This de-vegetation has been described as a 'hidden' cause of drought in southern Africa (Msangi, 2004). The need of national governments for export earnings and foreign exchange produced a trend towards cash crops, which competed for land with basic grains and reduced the fallowing system. Subsistence crops have been discouraged, to the extent that farm produce prices have consistently declined in real value over many years. At the same time, the build-up of food reserves has been seriously neglected, under pressure from international banks wanting loan repayments. In addition, a lack of government investment to improve the productivity of rain-fed agriculture and a failure to organize credit facilities for poor farmers have also tended to undermine the stability of the rural base.

The pastoral herdsmen of Africa are at particular risk. National governments have legislated against nomadism and attempts have been made to settle the herdsmen. In northern Kenya, for example, the Catholic Church has been influential in settling pastoralists in mission towns (Fratkin,

Plate 12.2 Pastoralists collecting water during March 2006 from a hole dug in a dry river bed in the Somali region of Ethiopia. Shallow groundwater reserves just below the ground surface are exploited wherever possible in order to supply goats and camels as well as humans during dry seasons and drought years. (Photo: Panos/Dieter Telemans DTE01214ETH)

1992), although livestock remain important in this dry area. People still depend on their animals for subsistence and trade but mobility and resilience have been limited. Typically, foreign aid has been ear-marked for sedentary agriculture rather than pastoralism. The traditional system of animal accumulation during years with good rains was not appreciated by governments, who taxed animals in the belief that the herdsmen should sell cattle to reduce pressure on land. Increasingly strict game preservation laws have been introduced which restrict the possibility of hunting for meat during drought. Other forms of employment, such as caravan trading, have gone, due to the enforcement of international boundaries and customs duties, plus competition from motor transport. In short, the traditional strategies evolved by pastoralists are no longer adequate. New thinking is overdue, for example to minimize the conflicts between pastoralists and crop farmers when the former need access to watercourses and forage for livestock during drought periods (Orindi *et al.*, 2007).

These are serious problems, but varying perspectives exist on the famine drought of the 1980s. Some observers have optimism for the future in that, despite the social upheavals and the loss of life, many of the traditional adaptive strategies of the people worked well (Mortimore and Adams, 2001). In a review paper, Batterbury and Warren (2001) stressed the continuing flexibility of Sahelian ecosystems and suggested that some of the resource limitations might be alleviated through migration, asset sales, cash-crop production and the generation of non-farm income.

D PROTECTION

1 Environmental control

In theory, the artificial stimulation of rainfall by cloud seeding could reduce the drought hazard. But experiments have shown that this technique can apply only to clouds with natural precipitation potential. These do not exist in viable numbers during drought conditions and there is little practical scope for this option, although experiments continue.

The development of additional water supplies is not necessarily a solution. Every year about 5,000 ha of new land come under irrigation in the Sahel, but this is balanced by about the same area going out of use through waterlogging or soil salinity. The drilling of new boreholes in dry areas is an example of how aid and technology, without proper local management, can actually increase disaster. Along the southern edge of the Sahara desert new tube wells were constructed to provide water-points so that further rangeland could be opened up. Without the imposition of effective controls, the borehole sites provided an attractive focus for many cattle and humans. The water encouraged the growth of herds beyond the available feed, until the new areas were stripped and the cattle died. Other inappropriate uses of irrigation water exist. Some of the new supplies have been used to irrigate export crops, such as pineapples, and rice grown for the urban elites. Both these crops are highly consumptive of water and have done nothing to alleviate food shortages in the rural areas.

2 Hazard-resistant design

The standard defence against hydrological drought is the use of dams and pipelines for the artificial storage and transfer of water supplies. The emphasis on these engineering solutions is symbolized by the global spread of large dams. Regulated rivers smooth out the seasonal variations in river flow and are able to provide enhanced dry-weather flows for water abstraction purposes. Figure 12.12 shows the half-monthly regulated flows for the river Blithe, England during the 1976 drought, at a point downstream from a regulating reservoir. When the observed flows are compared with the modelled 'naturalized' flows occurring in the absence of a reservoir, it can be seen that storage was able to contain river discharges within a narrow band right through the year. During the winter months, reservoir storage retained flood peaks and reduced average flows. Between May and September the regulated river discharge was enhanced above the natural flow regime. The river never fell below the designated minimum acceptable flow of $0.263 \text{ m}^3 \text{ s}^{-1}$, despite the severity of the drought conditions, which would have reduced natural flows below this level for three months.

Reservoirs have been used extensively to maintain urban water supplies. The greatest buffering against drought exists for those areas with a sufficient margin between the supply capacity of the system and the maximum use. Figure 12.13 illustrates how water-supply reservoirs smooth out seasonal variations in river flow to manage supplies in order to meet demand. The annual surplus available for supply will be eroded if storage is not increased to cope with the progressive rise in demand, due either to population increase or greater water use by individual households, typical of most areas. During a drought period water shortages, with attendant restrictions on use, are likely. A drought exists when the managed supply available for

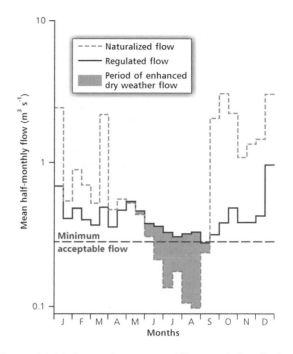

Figure 12.12 Reservoir storage and flow regulation alleviate hydrological drought of 1976 on the river Blithe, England. In the absence of a reservoir, the natural flow would have fallen below the minimum acceptable level for more than three months during the summer. After D.J. Gilvear, personal communication. Reproduced with permission.

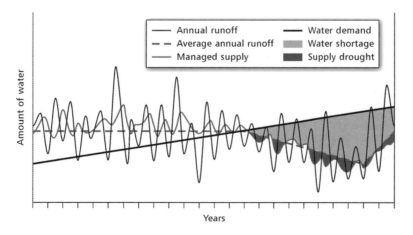

Figure 12.13 Idealized emergence of a water supply drought. Water demand rises progressively over time. Without any increase in supply, relatively minor reductions in run-off create a water shortage. When the managed supply falls below average annual or seasonal run-off, an urban water-supply drought exists.

distribution falls below the average annual run-off, leading to an interrupted service and the possible need to import extra supplies from elsewhere.

The response to water shortages depends on the severity of the conditions. For example, the responsible bodies in England and Wales – Environment Agency (EA), water companies, local authorities and government – cooperate along the following lines:

- *Stage One.* EA and water companies conserve water by encouraging all consumers to reduce demands and by in-house operational measures, such as lowering the water pressure in the distribution system.
- *Stage Two.* Water companies introduce restrictions on hosepipes for domestic garden use and car washing. The EA uses powers to ban

agricultural spray irrigation. Water conservation publicity is stepped up by the EA, water companies and local authorities.
- *Stage Three.* Legally enforceable *Drought Orders* or *Drought Permits* are granted, either by the EA or by central government, to restrict or ban water abstraction for agriculture and other specified uses, whilst allowing water companies to abstract increased amounts of surface and ground-water
- *Stage Four.* Water companies cut off domestic supplies at preadvertised times of day, standpipes are installed in urban areas and tankers are used to import supplies into critical areas.

Many reservoir-based urban water supply systems are designed to provide a predetermined minimum supply during roughly 98 per cent of

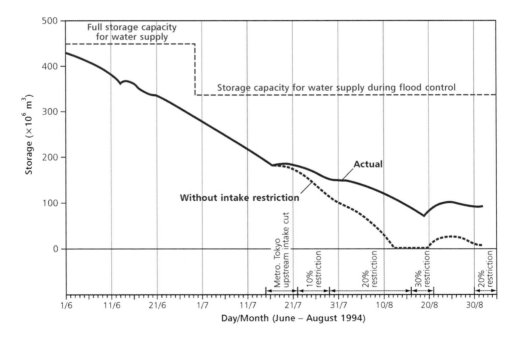

Figure 12.14 Changes in water storage in reservoirs along the upper river Tone, Japan, showing the effect of intake restrictions during the summer drought of 1994. After Dr Toshikatsu Omachi (1997). Originally published in IDI Water Series No. 1, *Drought Conciliation and Water Rights: Japanese Experience* (Infrastructure Development Institute, Japan, March 1997). Reproduced with permission from the Infrastructure Development Institute, Japan.

the time (2 per cent probability of failure), although minor shortages may be accepted more frequently. With an element of over-design and careful crisis management, it is possible for these systems to perform well during droughts of a magnitude beyond the 1:100-year drought event. Figure 12.14 shows the operation of domestic water restrictions on the river Tone, Japan which helped to maintain supplies during the summer drought of 1994 (Omachi, 1997). Without these actions, the content of the reservoirs would have declined more quickly and storage would have been exhausted by 12 August 1994.

E MITIGATION

1 Disaster aid

In financial – as well as humanitarian – terms, food aid is the most important response by the international community to all disasters (Leader, 2000). For some LDCs, food aid is almost synonymous with drought relief. The 1991–92 El Niño-related drought in southern and eastern Africa threatened 30 million people and generated a major international aid effort. From April 1992 to June 1993 roughly five times more food and relief goods were shipped into southern Africa than were delivered to the Horn of Africa during the 1984–85 famine (IFRCRCS, 1994). By August enough food was arriving to prevent famine, although there were distribution problems due to congested railheads and poor road transport. A 'Programme to Prevent Malnutrition' (PPM) was established to coordinate activities between the agencies representing each of the geographic areas targeted for food assistance. This structure gave over 50 NGOs more direct access to nearly 250,000 tonnes of maize for distribution to about 2 million people.

Despite its life-saving role in emergencies, food aid is controversial, and was described as a 'blunt instrument' by de Waal (1989). This is because it can be diverted from the needy to more prosperous elite groups and because it is based on the western view of famine as a mass starvation event. For example, Kelly and Buchanan-Smith (1994) argued that, in the absence of many excess deaths due to starvation (i.e. a substantial 'body count'), donors are unwilling to contribute fully to relief. Whilst large-scale distribution of food appears to be sensible, famine-related deaths are often age-specific and dependent on factors like the presence of disease. As always, it is desirable to prioritize aid to those most vulnerable but it is difficult, in practice, to provide selective assistance at the household level (Kelly, 1992). For example, targeting food according to anthropometric criteria, such as weight-for-height indices, has sometimes led to the deliberate under-feeding of children to ensure that the household qualifies for rations. In the short term, disaster relief can be better deployed only when those in most need have been identified and transportation methods are improved. Above all, it is most difficult to optimize food aid in complex emergencies where civil unrest, war and political interference obscure basic humanitarian aims (Ojaba *et al.*, 2002).

Longer-term drought aid has not always been invested wisely. Comparatively little money has been spent on agriculture and forestry or on field action at a local level. More aid should be directed to small farmers so that the rural sector can be stabilized again. Part of the difficulty is that students selected for overseas training come from the urban elites. After their return, the temptation is to remain in the cities. So the transfer of agricultural technology is from city to city, rather than into the rural areas where the food must be grown. More attention must be paid to sustainable development in the rural areas. In the short term, this might involve the provision of food aid via work programmes, as in Zambia, where free maize is distributed to those without food or cash resources. Recipients are required to participate in self-help projects like repairing roads, digging pit latrines, drilling boreholes and wells and constructing dip-tanks for cattle. A similar cash-for-work scheme has operated in north-eastern Brazil. The programme is meant to guarantee a small

Plate 12.3 A mother from Somalia (right) nurses her 18-month-old malnourished son whilst another woman feeds her child, at the stabilization centre at the Dadaab camp, Kenya, July 2011. During June 2011 about 30,000 refugees fled here to escape hunger and conflict, making Dadaab the largest refugee camp in the world. (Photo: Panos/Sanjit Das SDA01588KEN)

salary, and in 1993 an estimated 2 million people, with another 4–6 million dependants, were employed in this way.

Emergency drought relief has been a priority for governments in the MDCs too. In a comparison of drought policy in the USA and Australia some years ago, Wilhite (1986) showed that actions had a loss-sharing character, dependent on loans and grants, and that most drought mitigation has occurred in a crisis-management framework similar to that for emergency overseas aid. In severe droughts, governments are the only bodies able to intervene at the scale required and the costs can be high. For example, the total cost of drought relief programmes during the 1974–77 drought in the USA was US$7–8 billion.

The rising cost of drought relief in the industrialized countries has moved policy away from emergency subsidies funded by the tax-payer and towards long-term self-reliance on the part of rural communities. In 1989 the Australian government removed drought from the terms of reference of the Natural Disaster Relief Arrangements (O'Meagher *et al.*, 2000). The *National Drought Policy* recognized that drought is an integral part of the Australian climate and should be factored into all agricultural decisions. However, during the 1990s this policy became confused in the public mind with farm poverty and continued government acceptance of 'severe droughts' enabled 'exceptional' relief payments to continue (Botterill, 2003). Central-government responsibility for drought assistance has also declined in New Zealand, with progressive tightening of the definition of a drought eligible for support (Haylock and Ericksen, 2000). In 1996 the definition was restricted to a 1:50-year event (2 per cent annual probability of exceedance). The present aim is to devolve drought response to rural communities, within a more sustainable approach to natural resources, but it is unclear how national policy will engage with long-term drought management.

2 Insurance

At the present time, drought insurance is very limited, especially in the LDCs, where it is most needed. The private sector has shown little interest, but pilot initiatives sponsored by the World Bank and others suggest that some form of weather-based insurance may have a future role (Hazell and Hess, 2010). Barriers to progress include a perception of high risk by potential insurers, too few weather stations for monitoring purposes, the inability of farmers to fund policies and the absence of intermediaries to link farmers with insurers for trial schemes. Attitudes to insurance will also depend on the views of people in rural areas about the long-term significance of drought for their lives, whether they live in developing countries such as Iran, or in parts of Australia (Karpisheh et al., 2010; Raphael et al., 2009).

F ADAPTATION

1 Community preparedness

According to Wilhite (2002), preparedness is the key to drought hazard reduction. Arguably, it has met with some success in the traditional societies that have evolved 'coping' strategies for food insecurity. To cope with a 'normal' drought, nomadic people in the Sahel adopted the practice of herd diversification, involving camels, cattle, sheep and goats, all with different grazing habits, water requirements and breeding cycles, which helped to spread any risk of pasture failure. During years with abundant rainfall, the tribes increased their herds for food storage and as an insurance against drought. When drought occurred, people migrated to find good pasture and, in the most severe episodes, could either eat or sell off surplus livestock. Informal systems of communal loss sharing allowed the transfer of gifts or loans of any spare animals available to those in greatest distress and various fall-back activities, such as gazelle hunting or caravan trading, were intensified as temporary measures to help survive the drought. In a similar way, villagers in rural Mali adjusted to lower harvests by diversifying their income sources from non-agricultural activities (Cekan, 1992).

Under severe drought conditions, rural people have to do more. They often start by eating less, in an attempt to conserve food stocks. Agricultural adjustments include crop replacement (drought-resistant crops preferred at the normal planting time), gap-filling in fields (where germination of an earlier crop has been poor) and resowing or irrigating crops. When food stocks are exhausted, they turn to a range of wild 'famine' foods that are not part of the normal diet because of their low nutritional value. In Zambia, for example, this includes eating honey mixed with soil, wild fruits and wild roots, some of which are poisonous unless boiled for several hours before eating. The selling of livestock is usually underway by this stage, although a case has been made for external intervention to ensure that de-stocking takes place early in the drought cycle to prevent ecological damage to the grazing land (Morton and Barton, 2002). Unfortunately, de-stocking accelerates the fall in the value of livestock at a time when the cost of grain is rising. This produces a disadvantageous change in the terms of trade for nomadic people. Poor pastoralists have to sell a larger proportion of their animals in order to buy food than do the wealthy. During a severe drought, therefore, many of the poor are squeezed out of the pastoral economy and forced to settle in towns to live on famine relief or from wages paid to herders or labourers (Haug, 2002). Without food and other resources, rural dwellers routinely turn to local wage labour for support, rather than work on their own unproductive land.

For some households, outstanding debts and other favours can be called in. Cash or food entitlements may be borrowed from more prosperous relatives, neighbours or other support groups in a non-agricultural response strategy. Without access to external help, there is little option but to resort to the trading of valuables, such as jewellery, or other capital assets, such as radios, bicycles or firearms, which can be sold to buy grain. As incomes decline, health conditions also deteriorate. This deterioration is exacerbated by poor nutrition and growing competition for declining, and increasingly polluted, water supplies. Wherever possible, villagers poach wild game in order to survive. Table 12.3 shows that during a severe drought in 1994–95 that affected over 10 per cent of Bangladesh, households adopted a variety of non-agricultural adjustments. Over half of those questioned sold livestock and over 70 per cent of the respondents either sold or mortgaged land (Paul, 1995).

Ultimately, the family starts to break up. Some children may be sent to distant relatives out of the famine zone and male members may seek work in the towns. This can lead to large-scale migration that may be permanent if families lose their land rights because of moving. In a study of rural households that had migrated from famine-affected communities in northern Darfur, Sudan, it was found that asset wealth did not enhance famine resistance, as some of the earliest migrations were undertaken by 'wealthy' families (Pyle, 1992). As with other hazards, prior experience enhances the chances of survival, but some traditional responses are now less available than in the past. For example, livestock raiding by pastoralists has been a means of rebuilding herds destroyed by drought, but this activity has been disrupted in parts of Kenya by external raiders (Hendrickson et al., 1998).

More research on staple grains and better dryland farming techniques, such as terracing, strip cropping and soil erosion control, is needed. It may even be more productive to support nomadic pastoralism rather than irrigation schemes. New attitudes are required in order to change funding priorities. Improving the physical infrastructure in areas at risk by the provision of better roads will not only reduce short-term vulnerability by helping the distribution of emergency food aid but will also allow the optimum location of new facilities, such as well-equipped health clinics, which will lead to longer-term resilience. Above all, there is a need to release local initiatives in order to produce more self-reliance in the people and less dependence on famine aid.

Famine drought is unknown today in the MDCs, and also within urban areas in most countries. The short-term adjustments used by water authorities during hydrological droughts are aimed mainly at the domestic consumer and are a nuisance rather than a hazard. They include both supply–management and demand–management practices. Supply-management methods tend to concentrate on the more flexible use of available supplies and storage, as shown above. This can be achieved by switching water abstraction between surface and ground sources and by water transfers between different supply authority areas to ease the greatest shortages. Temporary

Table 12.3 Adoption of non-agricultural adjustments to drought by households in Bangladesh

Adjustment	Number of households	%
Sold livestock	166	55
Sold land	112	37
Mortgaged land	106	35
Mortgaged livestock	2	1
Sold possessions	26	9
Family members migrated	1	0

Source: After Paul (1995)

Note: 265 households were surveyed. Multiple responses are possible.

HYDROLOGICAL HAZARDS: DROUGHTS **365**

engineering, such as the laying of emergency pipelines, may allow importation of water from more distant sources. Other technical measures include reducing the water pressure in the main supply pipes and repairing as many leaking pipes as possible in the distribution system. When all else fails, water can be rationed in the worst-hit areas by rota cuts. Attempts to manage (reduce) consumer demand normally include a mix of legal measures and public appeals to conserve water. Special legislation may be introduced, such as the Drought Act rushed through the British Parliament in August 1976 to prohibit non-essential domestic uses of water like washing cars or watering gardens. Combined with 'save water' publicity campaigns, these measures can cut residential demand for short periods by up to one-third.

But crisis management is no substitute for longer-term planning and water conservation in urban areas. Where hydrological drought is more common, as in Adelaide, South Australia, the management of water demand has been a central plank of policy for many years. During the summer months, when rainfall is almost entirely absent, as much as 80 per cent of the water consumed within the metropolitan area has been used to irrigate gardens. As part of an overall conservation strategy, domestic use can be reduced through a combination of financial measures (seasonal peak pricing), technical measures (curbs on inefficient water-using appliances and advice on watering methods) and social measures (persuading people to grow native plants in their gardens rather than more water-demanding European varieties).

2 Predictions, forecasts and warnings

To be effective, drought forecasts should be available many months ahead in order to aid decisions on crop planting and water supply management. The best prospects lie with climate-based models that couple the atmosphere and the oceans in order to predict events like the El Niño Southern Oscillation (ENSO) which affect food production in many countries. For example, Selvaraju (2003) stated that during warm ENSO phases total food grain production (rice and wheat) over India declined by up to 15 per cent, with annual losses of around US$773 million. But the science is not entirely reliable. According to Ropelewski and Folland (2000), some skill exists in seasonal rainfall prediction but most precipitation estimates are for broad regions, averaged over several months, and often lack the precision needed by individual decision makers.

Various drought monitoring and forecasting agencies exist. One example is the intergovernmental *Climate Prediction and Applications Centre* (ICPAC) set up in Nairobi in 2003 with responsibility for 10 countries in the Greater Horn of Africa. ICPAC provides climate monitoring and prediction services for early warning and mitigation of extreme climate events by issuing regular regional advisories, including 10-day, monthly and seasonal bulletins. Climate Outlook Forums are held before the start of major rainfall seasons to provide consensus outlooks and identify drought strategies. Other bodies are attached to academic institutions, like the US National Drought Mitigation Center at the University of Nebraska and the Global Drought Monitor maintained at the University of London.

Future disaster reduction requires a proactive approach that integrates monitoring and forecasting of significant shortfalls in seasonal rainfall with efficient food security management, including the distribution of food aid (Tadesse *et al.*, 2008). This process has become increasingly sophisticated, due to improved understanding of climate variability and its implications for agriculture, food supplies and the welfare of rural communities. Methodologies differ in detail but a common sequence of information requirements and actions in the build-up to a drought disaster might be:

seasonal climate outlook

↓

agro-meteorological monitoring

↓

pre-harvest crop assessments

↓

emergency appeal

↓

delivery of aid

Crop-yield models can be incorporated into climate forecasts. In addition, developments in satellite remote sensing have led to much more accurate agricultural monitoring of cropped areas, crop development and estimates of production. For example the *Normalized Difference Vegetation Index* (NDVI), described by Gouveia *et al.* (2009), can be used as an early indicator of both crop and pasture failure. A similar measure of vegetation health was found to be highly correlated with drought events between 1981 and 2009 over large parts of Africa (Rojas *et al.*, 2011).

Two important global warning systems – the UN-sponsored *Global Information and Early Warning System* (GIEWS) and the USA-sponsored *Famine Early Warning System* (FEWS NET) – operate in order to anticipate crop failure and food shortage (Table 12.4). These systems were established after severe food emergencies in the 1970s and 1980s. They rely on multi-agency support and focus on large-scale monitoring and forecasting in preparation for intervention at a more local level. The primary data come from satellites that produce near-real-time images on a regular 10-day basis. For example, GIEWS operations rely on the Africa Real Time Environmental Monitoring Information System (ARTEMIS). The European METEOSAT satellite monitors cloud types to produce proxy rainfall estimates for Africa. These data are then linked to Advanced Very High Resolution Radiometer (AVHRR) information on the status of the vegetation cover at a spatial resolution of 8 km, via the Normalized Difference Vegetation Images (NDVI) available from NOAA's polar-orbiting satellites. Rainfall and vegetation estimates are then processed into maps of current and forecast weather conditions for staple food crops and pasture land. Regular reports on rainfall, food production and famine vulnerability are published and, when

Table 12.4 Global monitoring and warning for drought and food shortages

Organizations	Global information and early warning systems (GIEWS)	Famine early warning system (FEWS NET)
Origins	Started 1975 by the Food and Agriculture Organization (FAO) of the United Nations; HQ Rome, Italy	Started 1986 as the FEW program by USAID, an agency of the US federal government; renamed FEWS NET in 2000; HQ Washington, DC
Objectives	To improve food security in 22 drought-prone African countries and improve planning to reduce famine vulnerability	Monitoring food supply and demand in all countries, with emphasis on 80 low-income, food-deficit nations
Cooperating agencies	Mainly other UN and FAO bodies – including WFP, UNDP, EU, OCHA	Chemionics International Consultancy, NASA, NOAA
Routine operations	Regular reports on global and regional crop production, demand for staple foodstuffs, reserve stock levels, agricultural trade	Monthly bulletins from surveillance for 22 African host countries to determine level of food alert – Watch, Warning or Emergency
Emergency operations	FAO HQ issues Special Alerts for areas where crops or food supplies are threatened to activate decision makers and aid donors	FEWS/Chemionics head office issues warnings to decision makers based on advice from field-based staff

danger threatens, local offices undertake rapid-assessment surveys to clarify the situation on the ground.

In summary, the agricultural monitoring process integrates field observations, crop models and remotely sensed data to assess harvest production in a particular year and the likelihood that food shortages requiring humanitarian intervention will occur. If this process can be streamlined, the needs assessment window could be brought forward to extend the lead-time for decision making in the vulnerable areas (Haile, 2005). Figure 12.15a shows the current situation for sub-Saharan countries. At present, the agro-meteorological monitoring phase extends from February to October or November. If required, emergency appeals are launched in January, which allows only a limited time for raising disaster

funds before the onset of seasonal hunger. It is desirable to link the monitoring and assessment process more closely to humanitarian decision making. If the appeal process could begin as early as October, more opportunity time would be available for farmers and others to respond to the oncoming drought (Figure 12.15b). Other key improvements would include taking greater account of the needs of pastoralists and making early warnings of drought more user friendly so that indigenous knowledge and experience could be more easily incorporated into the system.

Large-scale surveillance, however successful, cannot detect food security issues at local levels, when the prompt detection of failing supplies is necessary to trigger swift reactions from donors. After the famine droughts of the mid-1980s, several sub-Saharan countries – notably Chad and

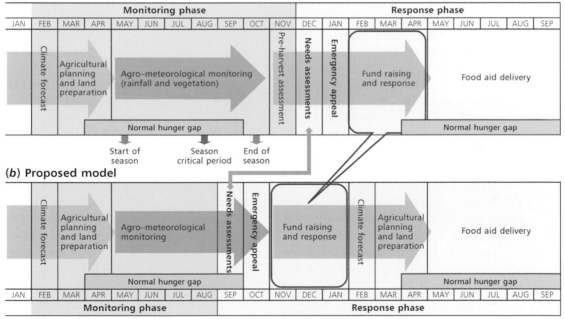

Figure 12.15 Drought response and food security in sub-Saharan Africa: (a) current arrangements for drought monitoring, fund raising and food-aid delivery; (b) proposed model for earlier response and preparedness. Decision making will be improved by bringing the whole humanitarian process forward by about four months. After Haile (2005). Reproduced from M. Haile, Weather patterns, food security and humanitarian response in sub-Saharan Africa. *Philosophical Transactions of the Royal Society B – Biological Sciences* 360 (2005): 2169–82. With permission from the Royal Society.

Mali – set up comprehensive food and nutrition monitoring systems (Autier, *et al.* 1989). Regional expertise and indigenous support are vital, as shown by the establishment of the *Southern African Development Community (SADC) Remote Sensing Unit* in Harare, Zimbabwe and the *AGRHYMET Regional Centre* in Niamey, Niger. Created in 1974, this is a specialized institute sponsored by nine sub-Saharan states for improving food supplies and natural resource management in the Sahel. It provides food security assessments and NDVI product enhancement but also trains local staff in agro-meteorological and hydrological monitoring, statistics and data compilation and dissemination designed to predict food shortages.

The difference between regional food availability and household-level access to food means that an early indication of the downward spiral into famine is dependent on local nutritional field surveys. These surveys measure body conditions such as height for age, weight for age and weight for height, to identify those, like pre-school-age children, with the greatest needs. Other reasonably reliable famine precursors are rising grain prices combined with falling livestock prices and wages as the economic balance shifts from assets and services, like jewellery and labour, to food, which rises in value both absolutely and relatively. However, the range of risk assessment methods used can produce discrepancies in the assessment of need and the distribution of food aid. Despite these uncertainties, it is most unfortunate that an early humanitarian response to famine drought seems to be increasingly dependent on clear evidence of excess mortality, without which many potential donors have become unwilling to act.

3 Land use planning

Drought increases pressure on land resources. Overgrazing, poor cropping methods, deforestation and improper soil conservation techniques may not create drought but they do amplify drought-related disaster. There is a need, therefore, for sustainable agricultural practices that involve the optimum conservation and use of both land and water resources in a long-term strategy (Gupta *et al.*, 2011).

Small-scale farming in marginal drylands is so dependent on seasonal rains that it is desirable to spread risk by engaging in forms of mixed agriculture that include livestock and tree crops as well as annual arable produce. In parts of Asia an integrated pattern of sugar cane and livestock farming is found. This is complemented with cattle rearing combined with vegetable and maize cultivation in the uplands. Where irrigation is possible, the development of improved crop varieties and fertilizer application can increase output, but these options are less relevant for rain-fed subsistence farming. Short-term responses are frequently required, not least in the regions dependent on monsoon rains. For example, if the seasonal rains are weaker than expected at the time of planting, possible mid-season adaptations include thinning of field crops and the spreading of green material as a soil to reduce water loss. When the monsoon is late, but produces adequate rainfall, the selection of short-duration crops such as pulses and oilseed compensates for the reduced length of the growing season. In years when rainfall is inadequate overall, leguminous crops and oilseeds tend to provide the best yields. At all times, it is necessary to control weeds. These compete with crops for moisture and nutrients, thereby reducing yields, especially in dry periods.

Sustained dry-land farming is dependent on soil conservation measures against water and wind erosion. A grass or legume cover is an effective control against water erosion, as are strip cropping and contour cultivation, which retard the flow of water downhill. Wind erosion can be reduced by maintaining a trash cover at the soil surface, plus the use of crop rotations and shelterbelts – where possible – to lower the wind velocity at the soil surface. Local watersheds provide the most suitable land use unit for managing land and water in drought-prone regions where the top priority is to conserve rainwater for crops, livestock and people. In many LDCs, maintaining the quantity and quality of water in the village pond is critical because this acts as a common-property, multi-purpose resource.

The basis of successful water management is the retention of surplus resources in the wet season for later use in the growing season. An obvious method is to divert intermittent surface flows into farm and village ponds, but high evaporation rates, combined with sedimentation and pollution, limit the survival time of these resources in the dry season. The artificial recharge of ground-water is often a better method. This is achieved by delaying and spreading the downslope movement of water to promote greater infiltration into the soil and shallow aquifers. Several methods exist. Following pioneering research in the 1930s, contour ridging and ploughing is extensively used, sometimes combined with earthen bunds at the terrace edge, to retain the maximum amount of water on a slope. These techniques also minimize the threat of soil erosion. In limited areas with intermittent concentrated flows of water, like small ravines or vegetated ditches, *check dams* are useful. The lowest-cost structures are made of rock, vegetation or sandbags and form barriers that reduce water velocities and pond-back temporary high flows to increase infiltration into adjacent aquifers and wells (Figure 12.16). They cannot be used for permanent streams and are rarely employed for channels draining more than 5 ha, unless constructed from concrete. More permanent check dams have been installed in parts of India. The state of Uttar Pradesh is estimated to have lost about 1.5×10^6 ha of productive land, due to erosion of light alluvial soil in ravines. Small concrete structures can arrest up to 80 per cent of the seasonal run-off in some ravines and have raised the local water table by several metres.

Even in the wetter areas of the MDCs, land use policies to combat drought are necessary because rural areas rarely have the large water storages and the options for reducing consumer demand that are available in the cities (Campbell *et al.*, 2011). Therefore, drought strategies should prepare the agricultural sector to withstand shortfalls of precipitation. This involves the careful management of surpluses in good years, adoption of appropriate stocking rates so that the pasture is not easily exhausted, the build-up of a reserve of fodder and the improvement of on-farm water supplies. The installation of an irrigation system may offer some security but the amount of storage may not provide complete drought proofing. Pigram (1986) cited heavy losses sustained by irrigators of rice and cotton in New South Wales, Australia during the latter stages of the 1979–83 drought, when water allocations were suspended in the middle of the irrigation season. Flexible decision making is always needed and resilience will be strengthened by a greater diversity of cropping patterns and income sources in drought-prone areas. For example, further scope exists for the development of more drought-resistant crops, and for crops with varying production cycles, that make it easier for rural communities to exist when the rains fail.

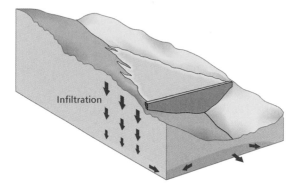

Figure 12.16 The use of check dams across intermittent water courses. Small quantities of water can be stored for domestic supply or plot irrigation, whilst infiltration aids local soil moisture conditions around the dam.

FURTHER READING

Below, R., Grover-Kopec, E. and Dilley, M. (2007) Documenting drought-related disasters. *The Journal of Environment and Development* 16: 328–44. A serious attempt to unravel the true significance of drought for disaster databases.

Botterill, L.C. (2003) Uncertain climate: the recent history of drought policy in Australia. *Australian Journal of Politics and History* 49: 61–74. Illustrates the problems that all governments face when coping with the societal consequences of climatic variability.

Haile, M. (2005) Weather patterns, food security and humanitarian response in sub-Saharan Africa. *Philosophical Transactions of the Royal Society B – Biological Sciences* 360: 2169–82. Shows exciting possibilities for reducing drought-related famine in the region of greatest need.

Mishra, A.K. and Singh, V.P. (2010) A review of drought concepts. *Journal of Hydrology* 391: 202–16. A science-based benchmark paper that forms an excellent reference source.

Vicente-Serrano, S.M. *et al.* (2012) Challenges for drought mitigation in Africa: the potential use of geospatial data and drought information systems. *Applied Geography* 34: 471–86. An impressive multi-authored contribution that looks beyond the conventional reactive responses to drought hazards.

Wilhite, D.A. (ed.) (2000) *Drought: A Global Assessment*, vols 1 and 2. Routledge, London and New York. The most comprehensive survey of drought hazards made in recent years.

WEB LINKS

US Drought Information Center www.drought.noaa.gov/

Australian Bureau of Meteorology www.bom.gov.au/climate/

National Integrated Drought Information Center, USA www.drought.gov/

CHAPTER THIRTEEN

Technological hazards

A INTRODUCTION

Technological hazards are usually seen as 'man-made accidents' triggered by human action (or inaction) rather than by natural processes. The root cause is human fallibility in decision making that permits failures within complex systems, due to a combination of technical and operational defects (Chapman, 2005; Shaluf *et al.*, 2003). However, the environment has a role to play, not least with respect to the growing significance of na-tech disasters where the trigger event – by definition – is an extreme natural event. As with other hazards, the release of excess energy or harmful substances has to be sufficient to threaten human life, material assets or the environment on a disastrous scale. This means that the damaging effects have to travel outwards from the accident site along natural pathways to reach a wider area. For example, when dangerous materials are released from an industrial site the resulting pollution can be transferred over large distances by the prevailing environmental conditions such as wind speed and direction and the flow of local rivers.

The term 'technology' has been applied in ways that range from a single toxic chemical to an entire industry, like nuclear power. The term 'environmental hazard' is also contentious. Sometimes health risks from long-term exposure to chemical pollutants or low-level hazardous waste have been included (Cutter, 1993). Arguably, there is too little awareness of how the use on-going of technology can routinely compromise health and safety. One example is the added risk of death or injury to road users when adverse weather conditions, like snow or high winds, make driving conditions difficult. The problem is that whilst the cumulative impacts of low-level pollution and road accidents are significant, the definable losses in individual events very rarely meet the disaster thresholds adopted in this book. Some observers have drawn parallels between technological hazards and terrorism and warfare. But terrorism and warfare are examples of the harmful use of technology. As such, they are deliberate acts of violence against others – like crime – and are not 'accidental'. The only link between warfare and technological hazards would occur if a dangerous technology developed for military purposes went out of control. This could happen, for example, from the accidental release of toxic material during the manufacture of weapons of mass destruction.

There are obvious differences between natural and technological hazards, but important similarities too. Just as natural hazards often represent what – in a less extreme form – would be a resource, so technology creates benefits as well as risks. The construction of a river dam may produce water supply and hydropower but also carries the risk of a flood disaster due to structural failure. The true balance between risks and benefits is not always apparent. When the internal combustion engine was first introduced, it was impossible to foresee either the extent of our present dependence on the invention, or that the global total of deaths from road accidents would

Table 13.1 Some examples of early technological accidents

Structures (fire)

1666	Fire of London, England	13,200 houses burned down
1772	Zaragoza theatre, Spain	27 dead
1863	Santiago church, Chile	2,000 dead
1871	Chicago fire, USA	250–300 dead, 18,000 houses burned
1881	Vienna theatre, Austria	850 dead

Structures (collapse)
Dam

1802	Puentes, Spain	608 dead
1964	Dale Dyke, England	250 dead
1889	South Fork, USA	>2,000 dead

Building

1885	Palais de Justice, Thiers, France	30 dead

Bridge

1879	Tay bridge, Scotland	75 dead

Public transport
Air

1785	hot air balloon, France	2 dead
1913	German airship LZ-18	28 dead

Sea

1912	*Titanic*, Atlantic Ocean	1,500 dead

Rail

1842	Versailles to Paris, France	>60 dead
1903	Paris Metro, France	84 dead
1914	Quintinshill junction, Scotland	227 dead

Industrial accident

1769	San Nazzarro, Italy (gunpowder explosion)	3,000 dead
1858	London docks, England (boiler explosion)	2,000 dead
1906	Courrières, France (coal-mine explosion)	1,099 dead
1907	Pittsburgh steelworks, USA (explosion)	>59 dead
1917	Halifax harbour, Canada (cargo explosion)	>1,200 dead

average around 250,000 every year. Similarly, although technology is the cause of some environmental problems, it can help to clean up pollution from accidental spills.

Not surprisingly, there is no universally agreed definition of a technological hazard. In this book, technological hazards are defined as:

> *accidental failures of design or management relating to large-scale structures, transport systems or industrial processes that that may cause death, injury, property loss or environmental damage on a community scale.*

These are not new threats. Nash (1976) showed that river dams and other built structures have failed, with disastrous consequences, since antiquity. Table 13.1 lists various disasters that occurred before the end of World War I, organized into the three categories mentioned in the definition above:

- *Large-scale structures* – public buildings, bridges, dams. In this case, the risk is usually defined as the probability of structural failure during the lifetime of the structure.
- *Transport systems* – road, air, sea, rail travel. In this case, risk is usually defined as the probability of death or injury per distance travelled.
- *Industrial processes* – manufacturing, power production, storage and transport of hazardous materials. In this case, risk is usually defined as the probability of death or injury per person per number of hours exposed to the materials.

B THE SCALE AND NATURE OF THE HAZARD

In the period 1900–2011, 7,244 technological disasters were archived by CRED. They were responsible for a total of 348,506 deaths worldwide. These figures compare with 6,350 events and more than 2 million deaths recorded for natural disasters in just 30 years between 1974 and 2003. Globally, technological disasters have an average annual frequency about one-third less than natural disasters and kill only one-tenth of the people who die in natural disasters. Clearly, they are less frequent and less deadly than natural disasters; the number of people affected and the amount of economic damage is also low. On the other hand, technological disasters can be the source of major environmental pollution such as that following the Deepwater Horizon drilling accident in the Gulf of Mexico. During April 2010, an oil rig operated by BP exploded and subsequently sank, with the loss of 11 crew members. The sea-floor well was not successfully capped for almost three months and created the largest offshore oil spill in US history. BP was obliged to set aside a US$20 billion fund to compensate coastal residents and the many businesses serving the seafood industry.

The relative impact of technological hazards can also be shown by a comparison of major disasters. In the period 1900–2011, the greatest mortality attributed to a single technology-related event was 4,386 deaths caused by a 1987 marine transport accident in the Philippines. This is insignificant when placed alongside the 3.7 million people killed by floods in China during 1931 and the 3 million people killed by drought in China in 1928. Smets (1987) claimed that, apart from three industrial disasters involving the concentrated release of toxic substances, no instance of accidental pollution had – at that time – *directly* caused more than 50 deaths anywhere in the world. Fritzsche (1992) showed that the fatality rate from man-made disasters in the late twentieth century was about the same in highly developed regions (Europe and North America) as in the rest of the world (Table 13.2). At around 0.01 fatalities per 100,000 people per year, the overall rate is low when compared with the mortality associated with natural disasters. Yet the much larger part of the global population living outside Europe and North America means that the absolute loss of life from technological hazards –

Plate 13.1 Supply boats attempt to extinguish fires on the Deepwater Horizon oil rig located off the south-east tip of Louisiana in the Gulf of Mexico on 21 April 2010. The fires were started the previous day by an explosion when 126 personnel were on board, and the rig sank into the sea on 22 April. (Photo: AP Photo/Gerald Herbert 100421038603)

Table 13.2 Annual death toll, averaged over the 1970–85 period, due to natural (N) and man-made (M) disasters for the world, North America and Europe

Populations and fatalities	World		North America		Europe	
Population (millions)	4,264		245		477	
Cause of death	N	M	N	M	N	M
Fatalities per year	88,900	5,500	220	310	450	540
Fatality rate per 100,000 per year	2.1	0.13	0.09	0.13	0.09	0.11

Source: After Fritzsche (1992)

as with natural hazards – is concentrated in the LDCs.

The EM-DAT archive places technological disasters into three groups: industrial accidents, miscellaneous accidents and transport accidents. On a continental basis, the largest number of events and technology-related deaths between 1900 and 2011 occurred in Asia, which accounted for 45 per cent of all disasters and 46 per cent of all deaths (Table 13.3). In every continent, transport-related accidents were the most frequent type of technological disaster, responsible for almost two-thirds of all events and 60 per cent of all deaths during the period in question.

The number of technological disasters and the related mortality has increased over time

Table 13.3 Technological disasters 1900–2011 by continent and disaster type

Continent	Disaster types			Totals
	Industrial accidents/ chemical spills	**Miscellaneous accidents/structural collapse**	**Transport accidents**	
Africa				
No. of events	113	192	1,364	1,669
No. of deaths	6,218	5,169	50,685	62,072
Americas				
No. of events	194	234	835	1,263
No. of deaths	7,653	15,110	35,200	57,963
Asia				
No. of events	757	601	1,901	3,259
No. of deaths	29,256	33,621	96,549	159,426
Europe				
No. of events	212	186	601	999
No. of deaths	8,909	39,366	19,131	67,406
Oceania				
No. of events	6	11	37	54
No. of deaths	51	115	1,527	1,693

Source: Compiled from EM-DAT database

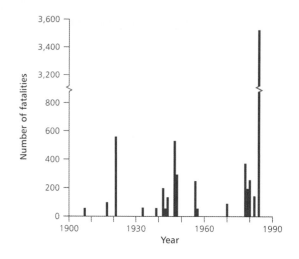

Figure 13.1 Annual number of deaths 1900–84 from industrial accidents causing more than 50 fatalities. The 1984 Bhopal gas-leak accident was an exceptional event. Compiled from data presented in Lagadec (1987).

(see Figures 2.3 and 2.5). From an early study by Lagadec (1987) it can be seen that industrial accidents causing more than 50 deaths to workers and third parties were rare in the first half of the twentieth century (Figure 13.1). It was not until 1948 that more than one such accident occurred in any one year, and not until 1957 that the first incident occurred outside the industrialized world (Europe, the USA, the Soviet bloc and Japan). However, in 1984 three industrial accidents collectively caused around 3,500 deaths:

- *Cubatao, Brazil, 25 February* – petroleum spillage and fire in a shanty town built illegally on the industrial company's land – 500 deaths
- *Mexico City, Mexico, 19 November* – multiple explosions of liquefied petroleum gas in an industrial site in a heavily populated poor area – at least 452 deaths, 31,000 homeless, 300,000 evacuated
- *Bhopal, India, 2–3 December* – release of toxic gas from an urban factory – well over 2,000 immediate deaths, 34,000 eye defects, 200,000 people voluntarily migrated. This remains the world's deadliest industrial accident to date.

These events marked a watershed. Technological hazard was no longer confined to the MDCs and, as shown by the Mexico City disaster, the *'domino' disaster* had arrived. These are event 'chains' that occur when an accident in one industrial unit causes one or more knock-on accidents nearby. A common sequence would start with a loss-of-containment accident, perhaps a leak of flammable gas, which reaches nearby systems and causes a further loss of containment, due to ignition and explosion. According to Khan and Abbasi (2001), domino accidents result from increasing congestion within industrial complexes, coupled with large concentrations of population around such plants. These characteristics are particularly common in the LDCs. The Bhopal disaster sprang from some of these factors, as did the refinery accident at Vishakhapatam, India, in September 1997 that claimed 60 lives.

After the mid-1980s there was a steady increase in the annual number of technological disasters. This trend was not fully reflected in the death toll, possibly due to the growing effectiveness of health and safety legislation. Of course, any apparent trend in disaster data is complicated by the very high impacts recorded for the most severe events. The importance of major technological disasters can be understood by reference to Table 13.4, which lists the 10 deadliest events recorded for each EM-DAT accident category. Causal agents are also interesting. For example, in addition to the 1987 Philippines ferry disaster, several other disasters taking over 500 lives each have involved seriously over-loaded ferries in the LDCs. Major industrial disasters have more varied causes, although explosion, as at a dynamite factory in Cali, Colombia (1956), when at least 2,700 were killed, and the release of toxic gas, as at Bhopal, India (1984) are important. As might be expected, miscellaneous accidents have the widest range of causes.

Some of the most severe events (Japan, 1923; USA, 1906) were classic 'na-tech' disasters involving urban fires caused by earthquakes. Lindell and

Table 13.4 The 10 deadliest transport, industry and miscellaneous accidents 1900–2011

Transport			Industry			Miscellaneous		
Country	Year	Deaths	Country	Year	Deaths	Country	Year	Deaths
Philippines	1987	4,386	Colombia	1956	2,700	Japan	1923	3,800
Haiti	1993	1,800	India	1984	2,500	Turkey	1954	2,000
Canada	1917	1,600	China	1942	1,549	China	1949	1,700
UK	1912	1,500	France	1906	1,099	Japan	1934	1,500
Senegal	2002	1,200	Nigeria	1998	1,082	Saudi Arabia	1990	1,426
Japan	1954	1,172	Iraq	1989	700	India	1979	1,335
China, PR	1948	1,100	Soviet Union	1989	607	Iraq	2005	1,199
Egypt	2006	1,028	Germany	1921	600	USA	1906	1,188
Canada	1914	1,014	USA	1947	561	Nigeria	2002	1,000
USA	1904	1,000	Brazil	1984	508	Guyana	1978	900

Source: After CRED database

Note: All events listed have at least one of the following criteria: 10 or more people reported killed, 100 people reported affected, a call for international assistance, a declaration of a state of emergency.

Perry (1996) voiced general concern about seismic activity and technological accidents in the USA but, until recently, na-tech accidents have attracted little attention. This situation has started to change, not least as a result of the 2011 Fukushima disaster in Japan (see Box 1.1). Earthquakes continue to pose risks, but most na-tech accidents are triggered by floods and lightning strikes. According to Cozzani *et al.* (2010), na-tech events account for up to 4 per cent of all reported technological accidents, although they are not always properly recorded. For example, in an analysis of 134 flood-related na-tech cases, economic-loss data were available for only six events. Significantly, it was not until the first decade of this century that attempts were made to introduce a formal quantitative assessment of this type of hazard, by Antonioni *et al.* (2009). Much of the research so far has stressed the impacts of extreme natural events on industrial facilities and the related spills of oil, chemicals or other dangerous materials (Young *et al.* 2004).

Flood-induced technological accidents are complicated by interactions between hazardous substances and water and the environmental pollution caused by contaminated flood-water. For example, the release of over 30,000 m³ of oil from storage tanks damaged by hurricane 'Katrina' – a volume approaching the 1989 *Exxon Valdez* petroleum spill in Alaska – created much soil pollution in the local area. The fact that 'Katrina' struck a densely industrialized zone of New Orleans resulted in over 200 onshore releases of hazardous chemicals, petroleum or natural gas (Santella *et al.*, 2010). In addition to direct discharges from petroleum tanks, other hazardous-materials releases were attributed to flaring events due to start-up and shut-down operations at oil facilities, other equipment damage and loss of containment from chemical storages.

Lightning-induced na-tech accidents are no less complicated. Some authorities regard these as the most common na-tech events. For example, from a study of two databases, Rasmussen (1995) concluded that over 60 per cent of na-tech incidents at storage and processing facilities were initiated by lightning. Apart from the damage to structures by a direct strike, and the disruption of control systems and electrical circuitry, other equipment categories are at risk. From a subset of

190 accident records, Renni *et al.* (2010) found that over 90 per cent of lightning-related na-tech events were either in oil and gas facilities or in the petrochemical sector. Storage tanks were the equipment category most often damaged and were associated with toxic releases, mainly oil, diesel and gasoline and tank fires.

Subsequent research is likely to confirm the growing importance of na-tech hazards in other areas, such as the effects of volcanic ash on air transport, the threat to shipping from icebergs and the damage to power stations and other key infrastructure arising from major floods and from earthquakes.

C AN OUTLINE OF THEORY

Many accounts of technological accidents dwell on case-study detail but, as with natural hazards, there are at least two opposing schools of thought about the root causes of such disasters (Sagan, 1993). These views can be summarized as:

The High Reliability School

This view admits the potential for human error in complex, dangerous technologies, and the possibility of accidents, but argues that properly designed and managed organizations can compensate for such errors. The argument is that:

- High-risk industries always seek failure-free performance and give top priority to reliability and safety. All personnel are subject to constant on-the-job training and are well-informed about the potential dangers.
- Complex organizations have in-built redundancy. Therefore, duplication and overlap of components and procedures provides a back-up system and a fail-safe environment if problems arise.
- In large organizations there is a culture of local decision making so that delegated authority can lead to swift accident-preventing decisions.

The Normal Accidents School

Conversely, this view believes that serious technological accidents are 'normal', much as natural disasters are a normal part of Earth processes (Perrow, 1999). In particular, the failure to control minor breakdowns or recurrent disruptions suggests that more serious accidents are likely to follow. Key features of this school are:

- Safety and reliability are not undisputed priorities. They compete with other objectives, such as increasing demands on performance and the need for profitability. Built-in redundancy simply increases the complexity of the technology and may encourage complacency within the organization just when the system becomes more difficult to understand.
- Competitive pressures for innovation can produce design faults in new equipment. At the same time, routine maintenance of older components may be overlooked and lead to eventual failure. These risks are increased whenever dangerous technologies spread into remote or hostile geographical environments.
- Constant training and local decision making cannot eliminate operational failure. Operators often work unsocial shift patterns in relative isolation, a pattern associated with boredom and the temptation for substance abuse. The substitution of computer-control is not a guaranteed response, due to the potential for hardware failure and defects in software.

There are plausible elements to both viewpoints. It is true that – so far – the nightmare scenario of an accidental nuclear war has not yet materialized. It is also possible that the rigorous formal checks in place continue to make this unlikely. On the other hand, defective design and inadequate management have already created near-accidents with nuclear weapons systems. According to Dumas (1999), the public record for nuclear weapons-related accidents, which totalled 89 worldwide from 1950 to 1994, was incomplete because of military and political secrecy about such matters, especially in some totalitarian countries.

The evidence presented in this book favours the Normal Accidents School. Technological disasters can occur anywhere – in both MDCs and LDCs – so long as the responsible bodies fail to prioritize safety measures and governments do not enforce compliance with health and safety regulations. The balance between technological risk and health and safety is constantly changing. Technical and legislative developments that foster occupational and public safety have to be set against factors and trends that increase risks, as illustrated in the following section.

D TECHNOLOGICAL HAZARDS IN PRACTICE

1 Some factors that increase risks

The rise of the modern chemical and petro-chemical industry created a suite of entirely new technologies. This industry, like some other processing operations, has tended to group on large sites near to significant concentrations of population. Over 30 years ago, a study of Canvey Island – a major chemical and oil-refining complex on the north shore of the river Thames about 40 km downstream from London, England – revealed that the quantities of flammable and toxic materials either in process, store or transport represented a serious threat to public safety (Health and Safety Executive, 1978). The hazards included fire, explosion, missiles and the spread of toxic gases. It was concluded that the existing industrial installations possessed a quantifiable risk of killing up to 18,000 people.

Glickman *et al.* (1992) found that major industrial accidents tended to occur at refineries and manufacturing plants or during the transportation of dangerous materials. These accidents were linked to the nature and scale of industrial activity; harmful energy was released in either *mechanical impact form* (dam burst, waste tip slippage, vehicle deceleration) or *chemical impact*

form (explosion, fire). The most hazardous materials were seen as high-level radioactive materials, explosives and a limited number of gases and liquids that are poisonous when inhaled or ingested. Chemicals are most hazardous if they are flammable, explosive, corrosive or toxic in low concentrations. In order to constitute a community-scale risk, such substances must be present in large quantities and be stored or transported in an insecure manner. Toxic materials are hazardous if they are transferred to the affected population by severe air pollution in the form of a 'toxic cloud'. An important feature of severe pollution episodes is that the adverse effects, both on the human body and on the environment, can often outlast the impacts associated with natural disasters.

The relentless rise in demand for energy has created risks, including those associated with the nuclear industry. The exhaustion of easily won fossil fuel sources has pushed the exploitation of hydrocarbon deposits into increasingly hostile physical environments. Offshore oil and gas development in physical environments like Alaska and the North Sea has proved challenging, and some large drilling platforms have suffered major accidents. For example, the 1988 Piper Alpha platform disaster in the North Sea claimed 167 lives. According to Paté-Cornell (1993), this was a largely self-inflicted disaster caused by design and management errors that ranged from insufficient protection of the structure against intense fires to poor communication regarding equipment that had been turned off for repair. Explosions and fires on offshore installations make rapid response by *evacuation, escape and rescue* (EER) operations vitally important. Human error is very possible in such emergencies and more needs to be learned about the most effective command and control arrangements and likely human behaviour in these specialized circumstances, in order to reduce fatalities in the future (Skogdalen *et al.*, 2012).

Another factor has been the increased transportation of dangerous materials, including radioactive waste. Incidents involving hazardous materials transport in the USA between 1971 and 1991 claimed a total of 375 lives and cost an estimated US$205 million (Cutter and Ji, 1997). Most incidents resulted from road transportation, although the injury rate per incident was greatest for water transport accidents. Similar disasters have been recorded elsewhere. For example, in 1978 more than 200 people were killed and 120 injured when a road tanker containing liquid propane gas (LPG) exploded near a camp site in Spain. Another gas disaster involving explosions and flash fires occurred at Viareggio, Italy in 2009, when a train of LPG tanker cars was derailed (Brambilla and Manca, 2010). In this disaster 14 people died instantly and a further 1,100 were evacuated. Infrastructure damage was estimated at €32 million. In November 1979, a freight train carrying a mix of hazardous materials including propane and chlorine was derailed in Mississauga, near Toronto, Canada. Although no lives were lost, the incident created a week-long emergency and almost 250,000 people had to be evacuated from local homes and hospitals. Most of the loss from these accidents was suffered within minutes of the event. No time was available to activate emergency plans, indicating the need to implement protective measures well in advance of such disasters.

Many of the risks in the LDCs are due to new technologies transferred from more technologically advanced nations into countries with different social cultures and industrial practices. The global spread of multinational corporations has introduced high-tech manufacturing and processing techniques into countries lacking the safeguards needed to handle the associated risks, as illustrated by the catastrophe at Bhopal, India, in 1984 (Box 13.1). Much of this technology is planted in rapidly urbanizing cities, where the infrastructure may be poorly designed, with few controls on land use development. Regulatory frameworks are often weak because the legal system, in terms of both relevant new legislation and the enforcement of controls, fails to keep pace with the speed of innovation. These weaknesses

can also apply to relatively low-level technology. In October 1968, a four-storey building nearing completion in the Malaysian capital of Kuala Lumpur collapsed, killing 7 persons and injuring 11 others (Aini *et al.*, 2005). In the general absence of qualified construction engineers and adequate site supervision, the collapse was attributed to serious under-design of the building, combined with basic failings that led to the poor quality of the reinforced concrete work.

2 Some factors that improve safety

Many changes have taken place that help to reduce technological risks (Lagadec, 1982). For example, in the case of fire hazard, whole urban areas rarely burn down as in the past, because of improved fire regulations and more efficient fire-fighting services. During the twentieth century, improvements in engineering design and a growing awareness of health and safety issues, reinforced by government legislation, have made all large structures, including large dams, much safer than in the past. In the case of public transport, individual cars, ships, trains and aircraft are all built to higher safety standards than a few decades ago.

Dam failure is an example of a low-risk, high-impact hazard; it does not occur often but the consequences can be catastrophic. In August 1975, the Banqiao dam on the upper Ruhe river in Henan province, China failed after heavy rainfall and contributed to floods that inundated over 1×10^6 ha of land and killed some 20,000 people. In 1993 the Gouhou dam in Qinghai province suffered structural failure and a further 1,200 lives were lost in floods. Within Europe, the 1959 failure of the high gravity-arch Maupassant dam in southern France led to more than 450 deaths in the town of Fréjus. This dam collapsed only 5 years after completion, and illustrates the fact that about 70 per cent of all dam failures occur within 10 years of construction.

The rate of dam failure is much lower than in the past. Figure 13.2 illustrates the improving safety record during the first 20 years of service for

Box 13.1 The 1984 gas disaster at Bhopal, India

Methyl isocyanate (MIC) is a fairly common industrial chemical used in the production of pesticides but has qualities that make it hazardous (Lewis, 1990). First, it is extremely volatile and vaporizes easily. Since MIC can boil at a temperature as low as 38°C, it has to be kept cool. Second, MIC is active chemically and reacts violently with water. Third, MIC is highly toxic, perhaps 100 times more lethal than cyanide gas and more dangerous than phosgene, a poison gas used in World War I. Fourthly, MIC is heavier than air and, when released, stays near ground level.

During the early morning of 3 December 1984, over 40 t of MIC gas leaked from a pesticide factory in the industrial town of Bhopal, India within a two-hour period and created the world's worst industrial disaster in a city of over 1 million people (Hazarika, 1988). The chemical was stored in an underground tank that became contaminated with water and produced a chemical reaction, followed by a rise in gas pressure and subsequent emissions into the atmosphere. The Bhopal factory was built within 5 km of the city centre by Union Carbide, a multinational company based in the USA. A dense cloud of gas drifted over an area with a radius of some 7 km. It is believed that over 3,000 people may have been killed by cyanide-related poisoning, with a further 300,000 injured or harmed by genetic defects passed on to following generations. In fact, a total of 600,000 injury claims and 15,000 death claims were ultimately filed with the Indian government.

Most fatalities were in the poor neighbourhoods located in low-lying parts of the city, including a shanty town of some 12,000 people near the gates of the factory. Most of the victims were the very young and the very old, although pregnant women suffered badly too. The disaster was severe because of the large numbers of people inhaling the gas and the lack of any emergency planning. There was no local knowledge of the nature of the chemicals in the factory, no adequate warning and only limited means of evacuation. The company provided no information about the medical treatment required by the victims and key resources such as oxygen (needed to treat respiratory problems) were in short supply

An investigation revealed that safety devices failed through a combination of faulty engineering and poor maintenance practice, although the company claimed that the cause was sabotage. A contributory factor was that the air-conditioning system, normally in use to keep the MIC cool, was shut down at the time of the accident. Safety procedures were clearly inadequate. For example, the plant lacked the computerized warning and fail-safe system used by the company in the USA.

The plant was unprofitable at the time of the accident and, because cut-backs had been made in maintenance, blame was attached to the local Indian management. Over the following two years, the parent company slimmed down, partly by distributing assets to shareholders and creditors, who were mainly banks. This strategy was deemed necessary in order to fend off a hostile take-over bid but it also served to off-load assets that were not then exposed to compensation claims. At the same time, the US legal system overturned precedent and opposed compensation claims for such an overseas liability, on the grounds that it would unfairly tax the US courts. So the responsibility was passed back to the Indian government.

The Indian government made itself the sole representative of the victims and filed compensation claims against the company both in the USA and in India. In 1989 Union Carbide made a final out-of-court compensation payment of US$470 million. This compares unfavourably with the US$5 billion awarded in the USA after the *Exxon Valdez* oil spill. Special compensation courts typically awarded sums of £500 for injury and £2,000 for death. In the meantime, the Indian government distributed relief at about £4 per month for each family affected. But the Indian government failed to organize efficient legal or medical aid for the victims. As a result, victims found it difficult to have their cases brought to court without resorting to bribes or paying private lawyers. Many medicines that should have been supplied free to patients were obtainable only on the black market. Families sometimes had to spend double their monthly government allowance on medicines. Ten years after the event, it was estimated that fewer than one-quarter of the total claims had been settled and that less than 10 per cent of the damages paid by Union Carbide had reached the victims.

The Bhopal accident led to a greatly increased awareness of the hazards associated with large chemical plants. The main Union Carbide plant in West Virginia was quickly closed and about US$5 million was spent on technical improvements at other Union Carbide sites in the USA (Cutter, 1993). The Bhopal accident left a legacy of improved regulation in the chemical industry, although safety innovation is more apparent in the MDCs, such as Canada (Lacoursiere, 2005), than in the developing countries. There is now much greater awareness of safety issues, better regulatory consultation between the chemical process industry and governments and more preparedness for emergencies.

Plate 13.2 The Atal Ayub colony at Bhopal, Madhya Pradesh, India in 1992. People continue to live in the shadow of the now defunct and abandoned Union Carbide pesticide plant, responsible for the worst industrial disaster on record. (Photo: Panos/Rod Johnson RJH00053IND)

As an example, the term 'process-safety' was used around 500 times per year as a key-word in science and engineering journals at the time of Bhopal but usage rose to an annual frequency of over 2,500 by 2004 (Mannan *et al.*, 2005).

Despite these advances, Bhopal is a reminder of how an inability to learn from comparatively minor events can lead to disaster. According to Gupta (2002), at least six serious accidents occurred at the Bhopal plant in the four years before the tragedy. More significantly, it is claimed that there had been nearly 60 leaks of MIC at Union Carbide's West Virginia plant between 1980 and 1984, most of which were unreported. Clearly, in 1984 a Bhopal-style accident could have happened anywhere in the world. The question today is – how far has that situation changed?

dams constructed up to 1950. More recently, the average failure rate has fallen below 0.5 per cent. It should also be remembered that most dams worldwide are small. Most failures can be linked to the type of dam involved. The most common dams are older *fill-type* structures, built of compacted earth or rock. They are vulnerable both to over-topping by floods and to inadequate foundations that fail to prevent sub-surface erosion. Most of the world's highest and largest-capacity dams were built within the last 25 years and have proved controversial on ecological and socio-economic grounds rather than on account of safety standards. No major failures of *concrete-gravity* dams have been recorded in recent times, although doubts have been expressed about the Three Gorges dam on the upper Yangtze River, China. This is built on granite bedrock and designed so that its weight resists the pressure from the stored water, but concerns have been expressed about the possibility of induced earthquake activity in the area, due to the great weight of the stored water, and the possible risk from large landslides.

Some technologies have been rendered less dangerous by the improved scrutiny of industrial processes and accidents through international organizations like the UN and the OECD. In many industries, increasingly strict health and safety regulations have been introduced and harmonized through cooperation between individual countries. For example, UNEP has the *APELL*

Programme (Awareness and Preparedness for Emergencies at Local Level), intended to improve responsiveness of local communities. These programmes sponsor manuals and guidelines for hazard reduction in high-risk industries dealing with hazardous materials and specific activities, such as mining (Emery, 2005). Individual countries enact their own legislation to improve industrial safety.

One result has been more consistent definitions for 'major accidents' which can then be notified to the appropriate regulatory authority

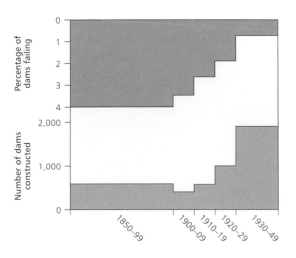

Figure 13.2 Inverse relationship between the percentage failure rate for all dams and the number of dams constructed between 1850 and 1950. Compiled from data presented in Lagadec (1987).

Table 13.5 Simplified minimum criteria for the mandatory notification by an EU member state of a 'major accident' to the European Commission

Substances involved	Any fire, explosion or accidental discharge of a 'dangerous substance' involving 5% of the 'qualifying amount' specified
Injury to persons and damage to buildings	Involves a 'dangerous substance' and: – one death, – six persons injured on-site and in hospital for 24 hr – one person off-site hospitalized for 24 hr – dwellings off-site damaged and unusable because of the accident – specified levels of evacuation of people and interruption of basic services
Immediate damage to the environment	– 0.5 ha of terrestrial habitat protected by legislation – 10 ha of more widespread habitat, including agricultural land – 10 km of river or canal – 1 ha of a lake or pond – 2 ha of a delta – 2 ha of a coastline or open sea – 1 ha of aquifer or underground water
Damage to property	– Damage on-site of €2 million – Damage off-site of €0.5 million
Cross-border damage	Any accident directly involving a 'dangerous substance' giving rise to effects outside the territory of the member state concerned

Source: After Kirchsteiger (1999)

Notes:

- Official notification is required for any accident which meets at least one of the consequences detailed above.
- Accidents or 'near misses' which member states regard as being of technical interest for preventing major accidents and limiting their consequences, and which do not meet the quantitative criteria above, should also be notified to the Commission.
- In the period 1984–98, 312 major accidents were notified to the Commission.

(Kirchsteiger, 1999). For example, within the European Union, the Seveso II Directive of 1997 provided a quantitative scale of minimum criteria for the notification of an accident to the European Commission (Table 13.5), based on the experience of the chemical process industry. Since 1984, all reported accidents have been archived in the *MARS database (Major Accident Reporting System)* in order to record and exchange data on accidents on industrial sites falling under the EU Directive. Other examples include the *MHIDAS system* (Major Hazard Incident DAta Service) supervised by the Health and Safety Executive in the UK, whilst the *NRC* (National Response Center) holds the federal archive on oil and chemical spills in the USA.

Attempts have been made to improve the safety of hazardous-materials transport using better design and engineering for rail vehicles and their loads. Other approaches include routing road vehicles away from populated areas, but there is often a conflict between the routes that minimize accident risks and those offering the lowest operating costs. Route restrictions can include prohibiting the use of specific roads, tunnels or bridges for the transport of certain materials, regulations that require advance warning of hazardous shipments, special speed limits on permitted routes and curfews to control the hours when certain routes and facilities can be used for hazardous materials transport. Such arrangements have the potential to create friction, due to

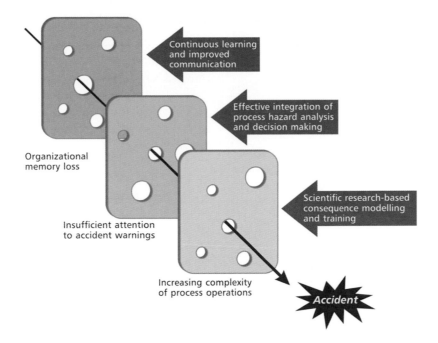

Figure 13.3 Safety challenges and the organizational responses required in the process industry, as illustrated by the Swiss Cheese disaster model. After Qi *et al.* (2012). reprinted from *Process Safety and Environmental Protection* 90, R. Qi *et al.*, Challenges and needs for process safety in the new millennium, 91–100, Copyright (2012), with permission from Elsevier.

the imposition by a national government of specific routes and regulations on state and local authorities without providing any support that would enable the directives to be met.

Although progress has been made since the 1984 Bhopal disaster to make safety an integral function for all process industries, accidents still happen, due to failures to identify best practices and implement adequate risk reduction programmes. Qi *et al.* (2011) summarized present-day challenges under three categories:

- failure to learn from past incidents in order to improve process design, procedures, site training etc.
- insufficient attention paid to leading indicators of risk
- growing complexity of operations, combined with inadequate communication.

This situation is illustrated using Reason's Swiss Cheese model (see also page 51) in Figure 13.3, along with the respective research needs for a safer future. The basic message is to learn from previous mistakes and reduce risk to the lowest practicable level, although organizational lapses may prevent this from happening (Pidgeon and O'Leary, 2000). For the most dangerous technologies, this may not be enough, especially if the residual risk is judged to be intolerably high. If perfection cannot be achieved, at what point can continuation of the activity be justified?

E PERCEPTION: THE TRANSPORT AND NUCLEAR INDUSTRIES

Turner (1994) estimated that up to 80 per cent of large-scale accidents can be explained by social, administrative or managerial failings. Given such a clear association between technological disasters and people, risk perception and human behaviour – whether by technology operators or the general public – assumes considerable importance. For example, operator error was the key factor in an accident at Ladbroke Grove, London when two trains collided causing 31 deaths and injuring over 400 others. In this case, it is believed the driver ignored three in-cab warnings, overrode three automatic applications of the emergency brake and continued through a signal set at danger (Stanton and Walker, 2011). Similar combinations of dangerous human factors have been noted in other modes of transport but remain difficult to explain or predict.

Most people perceive technological risks as tolerable when they are outweighed by the benefits (see Chapter 4, page 73). In the medical field certain treatments require patients to take chemical substances in prescribed measures, or expose themselves to ionizing radiation through X-rays, knowing that larger doses can be extremely harmful. Such public acceptance is critical, not least because, when compared to natural hazards, there is usually less statistical evidence for a probability-based assessment of technological risks. Problems arise when public perception of the risks and benefits of a technology differs from that of scientists and technical experts. Gardner and Gould (1989) found that lay people emphasize so-called 'dread risks' beyond statistical experience. Such risk amplification occurred at Henderson, Nevada in 1988, when explosions at an industrial site killed two people and injured some 300 more (Olurominiyi *et al.*, 2004). This incident alerted residents about other latent – but unconnected – safety issues in the community, ranging from the confidentiality laws to the political responsibility for land use.

Amplified risk perception can be demonstrated by comparing a relatively large *actual risk* industry (transport) with a relatively large *perceived risk* industry (nuclear power).

1 The transport industry

The long-term rise in transport-related deaths is mainly a function of the rise in the distance travelled per year and the growing size of vehicles (Yagar, 1984). Increase in business travel, together with the greater amounts of leisure time, has led to more mobility. Car ownership is widespread. Air travel is as commonplace today as rail travel was for the previous generation. Therefore, the total exposure to transport-related risks has grown. Also, many passenger vehicles carry more passengers, so that when an accident occurs it creates more victims. This feature was illustrated in 1987 when the Zeebrugge ferry accident killed 167 people, and again in 1994 when the sinking of the ferry ship *Estonia* claimed a further 800 lives in Europe. As a result, most passenger carriers routinely undertake risk assessment exercises (van Dorp *et al.*, 2001).

Most forms of transport are getting safer. Table 13.6 shows that, with the exception of rail travel, which reflects the importance of two major accidents in the period concerned, the risk of death per passenger-distance travelled in the UK fell during the late twentieth century (Cox *et al.*, 1992). Air travel is particularly safe. According to Lewis (1990), despite the media attention paid to air crashes, the average risk in the USA is one fatality per billion passenger miles. The public tends to trust commercial airlines. Barnett *et al.* (1992) investigated the public response in the USA to the Sioux City disaster of 1989. This was the third DC-10 crash caused by the loss of hydraulic power and killed 112 out of 282 passengers on board. Despite adverse publicity, within two months bookings recovered to about 90 per cent of the level expected in the absence of the incident. This seems to be a demonstration of the 'willingness-to-pay' principle in safety management, which lets the market decide what constitutes 'acceptable' risk (McDaniels *et al.*, 1992).

Road travel is more dangerous. Traffic accidents claimed over 30 million lives worldwide during the twentieth century. In the USA, road

Table 13.6 Deaths per 10^9 kilometres travelled in the UK

Years	1967–71	1972–76	1986–90
Railway passengers	0.65	0.45	1.1
Passengers on scheduled UK airlines	2.3	1.4	0.23
Bus or coach drivers and passengers	1.2	1.2	0.45
Car or taxi drivers and passengers	9.0	7.5	4.4
Two-wheeled motor vehicle passengers	375.0	359.0	104.0
Pedal cyclists	88.0	85.0	50.0
Pedestrians*	110.0	105.0	70.0

Source: After Cox *et al.* (1992)

Note: *Assuming travel at 8.7 km per person per week

collisions account for about half of all accidental deaths. In Japan, traffic accidents account for 0.01 per cent of all deaths, compared with a death rate of only 0.00025 per cent for natural disasters (Mizutani and Nakano, 1989). In the UK the average driver faces a risk of about 8 in 100,000 per year of being killed in a car accident and a 100 in 100,000 risk of being seriously injured. The risks to other persons are greater and are strongly age dependent. For example, driving accidents account for about three-quarters of all accidents in the 16–19 age group, and drivers age 21 years or under are responsible for about one-quarter of all road deaths. These are the highest technology-related risks routinely faced by the public, but the personal convenience of car travel is widely perceived to justify the threat. Indeed, the spread of car ownership is such a token of human progress that over 70 per cent of all road deaths now occur in the LDCs, where the annual cost of traffic accidents rivals the amount of international aid received by these countries.

Public perception does not fit the facts. Although road deaths in private vehicles dwarf public transport deaths in most countries, the latter attract most attention. This pattern probably stems from the larger group deaths associated with public transport accidents and also from the extra opportunity for blaming large corporations, in an age when litigation is a growth industry. Generally speaking, investment in highway safety is a 'good

buy'. Risk reduction has been achieved at relatively low cost through improvements in car design, more use of motorways and legislation such as the compulsory wearing of seat belts and stricter enforcement of drink–driving laws. This is not to say that other highway risks are not emerging. In parts of the MDCs the number of vehicle miles travelled by large trucks is increasing at a faster rate than that of other vehicles. Increasing competition for road space between commercial vehicles and cars is likely to create more multi-vehicle collisions.

2 The nuclear industry

There are about 500 nuclear power plants either operating or under construction around the world, with major clusters in Europe, the USA and Japan. About 25 per cent of the existing plants are over 20 years old. Large nuclear power stations have the capability to cause many deaths, extreme societal disruption and long-lasting pollution. Because of this, nuclear plants are rarely sited near urban areas and the industry is highly regulated. For example, the *Nuclear Regulatory Commission* (NRC) supervises the use of nuclear power for civilian purposes in the USA; in the UK the *Office for Nuclear Regulation* (ONR) is responsible for authorizing use in both the civilian and defence sectors. On a world scale, these responsibilities fall to the *International Atomic Energy Agency*

Table 13.7 The International Nuclear Event Scale

Level of event	Criteria		
	Off-site impact	**On-site impact**	**Defence-in-depth degradation**
Major accident	Major release, widespread health and environmental effects		
Serious accident	Significant release, full implementation of local emergency plans		
Accident with off-site risks	Limited release, partial implementation of local emergency plans	Severe core damage	
Accident mainly in installation	Minor release, public exposure of the order of prescribed limits	Partial core damage, acute health effects on workers	
Serious incident	Very small release, public exposure at a fraction of prescribed limits	Major contamination, over-exposure of workers	Near accident, loss of defence-in-depth provisions
Incident			Incidents with potential safety consequences
Anomaly			Deviations from authorized functional domains
Below scale			No safety significance

Source: After International Atomic Energy Authority (personal communication)

Note: Nuclear events leading to off-site impact are rare and none has occurred in Europe since the Chernobyl 'major accident' of 1986.

(IAEA), which in 1992 formalized a nuclear event scale that defines various severities of incidents leading up to major accidents (Table 13.7). The world's worst nuclear accident to date occurred during the night of 25–26 April 1986, at Chernobyl about 130 km north of the city of Kiev, in the Republic of Belarus, with many fatalities and trans-continental pollution (Box 13.2). There are many European nuclear reactors located either on national boundaries, such as rivers, or within 25 km of a national border, so that the risk of another trans-boundary incident remains.

Practical problems in managing the industry arise from the extreme public dread of the risks from nuclear plants and toxic waste sites.

Opposition to this technology is driven by a concern that the risks are very great (Slovic *et al.*, 1991). In a wide-ranging study in Japan and the United States, Hinman *et al.* (1993) found that people in both countries dreaded nuclear waste and nuclear accidents at a level exceeding their fear of crime or AIDS. Hazardous-waste sites have been cited as the most worrying environmental problem in polls of public opinion conducted in the USA (Dunlap and Scarce, 1991). Yet, according to Lewis (1990), the risk from a properly constructed nuclear waste repository is 'as negligible as it is possible to imagine … [and] a non-risk'.

Nuclear power threatens people with a dose of ionizing radiation beyond the background levels

Box 13.2 The 1986 nuclear disaster at Chernobyl, Belarus

The Chernobyl disaster in April 1986 occurred at a nuclear reactor constructed to a flawed design that was operated by poorly trained personnel. The immediate cause of the accident was an unauthorized experiment conducted by workers at the plant to determine the length of time that mechanical inertia would keep a steam turbine freewheeling, and the amount of electricity it would produce, before the diesel generators needed to be switched on. During the experiment, the routine supply of steam from the reactor was turned off and the power level was allowed to drop below 20 per cent, well within the unstable zone for this type of water-cooled, graphite-moderated reactor design.

During the experiment, the reactor was not shut down and a number of the built-in safety devices were deliberately overridden. In this situation vast quantities of steam and chemical reactions built up sufficient pressure to create an explosion which blew the 1,000 t protective slab off the top of the reactor vessel. The resulting steam explosions and fires ejected at least 5 per cent of the radioactive reactor core into the atmosphere, where it was carried downwind. Lumps of radioactive material were ejected from the reactor and deposited within 1 km of the plant, where they started other fires. The main plume of radioactive dust and gas sent into the atmosphere was rich in fission products and contained iodine–131 and caesium–137, both of which can be readily absorbed by living tissue.

Immediate efforts were made to control the release of radioactive material. A major limitation was that water could not be used on the burning graphite reactor core because this would have created further clouds of radioactive steam. Instead the fire had to be starved of oxygen by the dumping many tonnes of material (lead, boron, dolomite, clay and sand) from over-flying helicopters. Two plant workers died on the night of the accident and a further 28 were killed during the next few weeks by acute radiation poisoning. A further 200 people were exposed to over 2,000 times the normal annual dose from background levels of radiation. Eventually some 135,000 people were evacuated from within a 30 km-radius exclusion zone around the plant and the nearby town of Pripyat was abandoned.

In the two weeks following the accident, the radioactive plume circulated over much of north-western Europe. Away from Chernobyl itself, the greatest depositions of radioactive material occurred in areas affected by rain, which flushed much of the particulate material out of the atmosphere. These areas included Scandinavia, Austria, Germany, Poland, the UK and Ireland. Some of the heaviest fall-out was experienced in the Lapland province of Sweden, where it affected the grazing land of reindeer, contaminated the meat and dealt the Lapp culture a great blow. More widely, the immediate consequence was a general contamination of the food chain, and restrictions on the sale of vegetables, milk and meat were imposed. Some countries also issued a ban on grazing cattle out of doors and warnings to avoid contact with rainwater.

It has proved difficult to assess the long-term health consequences from potentially fatal cancers attributable to the Chernobyl accident. The 50,000 soldiers who fought to control the fire on the reactor roof clearly suffered the greatest exposure to radiation, followed by the 500,000 workers who subsequently cleaned up the site. Others subjected to high doses include some of the total 400,000 people who were relocated. Over 15 years later, it was estimated that 2 million people in Belarus were affected with various health disorders, including a marked drop in the human birth rate (IFRCRCS, 2000). Rahu (2003) was more cautious and claimed that the only direct public-health evidence of radiation exposure was 1,800 cases of childhood thyroid cancer recorded in the 1990–98 period, but did acknowledge many cases of psychological illness attributable to factors like fear of radiation, relocation and economic hardship.

As a result of political change, there is now more openness. Since 1993 a UN-appointed Co-ordinator of International Cooperation has acted as a catalyst between organizations and member states in addressing issues such as better medical provision for children with thyroid cancer, the establishment of socio-psychological rehabilitation centres, the creation of an economic development zone in the affected area and the restoration of contaminated land to safe agricultural use.

At present, Chernobyl Unit 4 is enclosed within a large concrete shield, hastily erected in October 1986, and holds around 200 t of highly radioactive material. This shield has a limited life expectancy. The internationally financed Chernobyl Shelter Fund exists to place a new safe-confinement structure over the entire site, and complete other projects, by 2015. In July 2010 the Belarus government published plans to repopulate the 'contaminated areas' over a similar time-scale. Restrictions on settlement have been relaxed in over 200 villages and plans for new infrastructure (roads, housing and schools) have been announced. The aim is to return the land to productive agriculture and forestry, with tree planning mainly on land with high concentrations of radionuclides.

Chernobyl acted as a 'wake-up' call for the nuclear industry. Not only are all Soviet-designed reactors now safer, but nuclear plants and their operating procedures have been improved worldwide through cooperation between engineers from Eastern and Western Europe.

For more updated details see: *www.world-nuclear.org/info/chernobyl/inf07.html*

---

emitted via the Earth and its atmosphere. High-level nuclear wastes are products from reactors, including spent or used fuel, with a radioactive half-life (the period taken for half the atoms to disintegrate) of more than 1,000 years. Inter-mediate-level waste has a shorter half-life but exists in larger quantities. The usual solution to the disposal of nuclear waste has been to store it for years in pools of water near to the power plant so that the temperature falls and some radio-activity decays. The material is then taken to a permanent storage site elsewhere, via public highways.

Waste transport is itself is a highly contentious 'dread' issue. In a study of radioactive waste transport through Oregon, MacGregor *et al.* (1994) found no reduction in public concern with distance away from the transport corridor. But the permanent storage risks arouse greater anxiety. By the year 2000, the USA had some 40,000 t of spent nuclear fuel stored at about 70 sites awaiting disposal. Sharp differences between public opinion and the technical community were highlighted when, in 2002, the US Congress designated Yucca Mountain, Nevada as the sole repository site for the nation's high-level nuclear waste. Yucca Mountain, a long ridge of volcanic ash 1,500 m high, was scheduled to open in 2017 to receive deliveries of waste from across the USA for many years. Once inside the storage tunnels, it was planned up to 70,000 t of spent fuel would be placed in titanium-covered tubes and monitored for 300 years before the mountain was sealed. The plan was always controversial. It was opposed by environmentalists and state citizens, some of them 160 km distant in Las Vegas (Flynn *et al.* 1993a and 1993b), whilst others have argued that storing nuclear waste above ground for the next 100 years, prior to burial in a permanent repository, would be US$10,000–50,000 million cheaper than the Yucca Mountain project (Keeney and von Winterfeldt, 1994). The plan was opposed by the current federal government and funding was withdrawn in September 2011.

F PROTECTION

Technological hazards offer more potential for suppressing events at source than exists for geophysical hazards. Some observers have gone so far as to advocate a policy of *inherent safety*, also called primary prevention, which effectively means the elimination of technological hazard (Hansson, 2010). Examples include the substitution of dangerous substances, such as inflammable materials, by less dangerous alternatives. The technical scope for this is limited and many technological accidents have roots in a combination of faulty engineering and human weakness. Since the latter includes common human flaws such as greed and carelessness, against which there are few reliable defences, it is the engineering route that offers the better chance of success.

Risk-free design and construction for all types of technology are not attainable; it would simply be too expensive to build against all possibilities of failure. But industrial plant design has all too often been changed after, rather than before, an accident. The risks at Chernobyl would have been less had the reactor been surrounded by a protective shield. The risks at Bhopal would have been reduced, although not eliminated, if the factory had been equipped with an effective gas exhaust facility, including a very high chimney that would have pierced the nocturnal inversion layer and dispersed the toxic material through a much larger volume of air. There is a special need to ensure that, when multinational corporations operate within the LDCs, the safety standards in the subsidiary plants at least match those at the parent site.

The mitigation of frost hazards on highways is an example of pre-emptive hazard reduction. Icing on road surfaces increases the risk of deaths and vehicle damage from skidding accidents. Spells of low temperature can usually be forecast with some accuracy and salt is an effective de-

icing agent down to temperatures of −21.0°C. However, it must be spread sparingly to reduce the financial and environmental costs. The most efficient use of road salt occurs when it is spread as an *anti-icing* agent at low application rates of about 10 g m². This is sufficient to prevent the formation of a thin film of ice. When ice has already formed, salt has to be used as a *de-icing* agent at application rates which are some five times higher per unit area. Advances in technology and highway meteorology have enabled road engineers to monitor and forecast localized road temperatures. Together with the use of automatic ice-detection sensors, this has led to economies in winter salt usage amounting to 20 per cent or more in Europe and elsewhere (Perry and Symons, 1991).

G MITIGATION

Loss-sharing arrangements differ between natural and technological hazards. Explicit allocation of blame is likely after technological accidents if they can be attributed to human decisions and actions. The attachment of blame is encouraged by public pressure for corporate manslaughter charges to be brought following cases of organizational failure. Because of perceived corporate negligence in technological disasters, the victims tend to attract less public support than those suffering from natural disasters. Compensation funds are sought from those deemed responsible and the role of international disaster appeals is much reduced.

One example is the 1987 *Herald of Free Enterprise* incident, when 193 people died in the North Sea after a ferry capsized due to a failure to close the bow doors. This was the worst maritime disaster for a British-registered ship in peacetime since the 1912 sinking of the *Titanic*, but the subsequent court case stalled halfway through. It was not until 1994 that 400 years of legal history were swept away, with the UK's first successful case of corporate manslaughter, albeit against a small company organizing outdoor activities held responsible for the deaths of four school children in a canoeing accident. Since then, other cases have been proved.

1 Compensation

Compensation is a less spontaneous form of loss sharing than disaster aid. Indeed, it is often resisted by the donor and has to be legally enforced. Where litigation is involved, the final settlement may include a punitive element that goes beyond the recovery of costs. Persons in the MDCs now seek compensation for actual and perceived harm, including emotional distress, caused by industrial emissions (Baram, 1987). In the USA, the legal system allows people either injured or at risk to bring 'toxic tort' actions against industry and secure high monetary damages. Governments also sue industry for large sums in order to clean up hazardous waste sites. Some of the firms involved are multinationals, so the repercussions of such actions on profits and jobs may be worldwide.

Although legal compensation can provide high monetary returns, it is not always the best method of loss sharing. Apart from the costs faced by industrial plants that lose liability suits, it is not always effective for disaster victims. Litigation can delay settlements for years. In some cases, the plaintiff may die or the company may go out of business before compensation can be paid. In India the lack of formal documentation held by many Bhopal victims, combined with inertia and inefficiency, delayed the settlement of many claims for more than 10 years after the accident. Clearly, it would be better to have compensation schemes that discharge sums quickly, in order to help victims whilst also safeguarding the financial future of responsible companies producing or using hazardous substances.

Such compensation schemes are unlikely to exist without government intervention. Theoretically, a government could establish a national technological disaster fund or could require industries – like chemical processing – to set aside monies financed by a levy imposed on the product. Neither of these arrangements is wholly satisfactory, since they devolve the cost onto innocent third parties – the tax-payer or the safely run industrial plant. There are instances where a contribution from the tax-payer is appropriate. It might be deemed equitable to compensate from general taxation a community that assumes local risks, perhaps from a nuclear power plant, on behalf of the nation as a whole. But, ideally, direct government intervention should be motivated by the principle of *making the hazard maker pay*. This suggests that government involvement might be best directed to ensuring that individual plants carry full insurance cover against civil liability for death, injury and environmental damage arising from their industrial activities and emissions.

2 Insurance

Insurance can be used to spread the financial risks associated with technological hazards. Within the MDCs, many people are likely to be insured through personal life and accident policies that cover 'all-risks'. Cover is normally available for exposure to hazardous substances, although such policies invariably exclude exposure to radiation. Property insurance similarly tends to cover all-risks and include hazardous materials, unless they are specifically excluded. On the other hand, personal insurance has practical disadvantages. For example, it may be difficult to establish a link between liability and damage, especially in more delayed-effect cases involving toxins such as carcinogens. This is particularly true when the damage results from long-term, low-level leakage rather than from a single, rapid-onset accident.

Insurance can also be taken out by industry. With increasing demands for corporate safety, there is a need for industrial plants to have full insurance cover against civil liability for human injury or environmental pollution. A proactive partnership between industry and insurers could encourage a more responsible attitude towards hazard reduction. But, as with other types of hazard insurance, the difficulty lies in setting realistic premiums for industry to pay. Unless the premiums are fully economic, industry will not take technological hazards seriously and insurance companies may either fail commercially or withdraw from this market. Equally, unless premiums are weighted according to the actual risks involved at the level of individual plants, insurance will amount to little more than an unfair tax on the safe, well-managed sites within that industry.

H ADAPTATION

1 Preparedness

Comprehensive management of risks in the high-hazard sectors of industry within the UK began with the Health and Safety at Work Act 1974 (Health and Safety Executive, 2004). This legislation established two bodies: the *Health and Safety Commission* (HSC), responsible for new laws, setting of standards, research and information and the *Health and Safety Executive* (HSE), responsible for enforcement of the regulations (in association with Local Authorities) and advice to the HSC. During the first 30 years, several major fatal accidents occurred (Table 13.8). Some of these – the Piper Alpha accident in the energy field and the King's Cross and Clapham rail disasters in the transport field – led to the transfer of greater responsibilities to the HSE.

For high-hazard sites, mainly those operated by the chemical industry, two pieces of legislation have proved significant:

- *Control of Industrial Major Accident Hazards Regulations (CIMAH).* This 1984 Act was designed to implement, within the UK, the European Community 'Seveso Directive'. This Directive resulted from the accidental release of dioxin at Seveso, Italy in 1976, which caused widespread contamination. Industrial-scale manufacturers and storekeepers of specified dangerous substances were required to identify high-hazard sites, to provide evidence to show

Table 13.8 Major technological accidents causing deaths in the UK during the twentieth century

1974	Flixborough chemical explosion	28 deaths
1979	Golbourne colliery explosion	10 deaths
1984	Abbeystead water pumping station explosion	16 deaths
1985	Putney domestic gas explosion	8 deaths
1986	Clapham rail crash	35 deaths
1987	King's Cross underground station fire	31 deaths
1988	Piper Alpha offshore oil explosion and fire	167 deaths
1999	Ladbroke Grove rail crash	31 deaths

that adequate control measures were in place to prevent major accidents and also to limit the effects of any accidents that might occur (Welsh, 1994).

- *Control of Major Accident Hazards Regulations (COMAH).* This 1999 Act replaced the CIMAH regulations by implementing new EC legislation known as the 'Seveso II Directive'. The Act applies to the chemical industry but also includes some other high-risk storage activities and nuclear sites. It aims to mitigate the potential harm from dangerous substances such as chlorine, liquefied petroleum gas and explosives. The regulations are enforced by the HSE in association with the Environment Agency in England and Wales and the Scottish Environment Protection Agency in Scotland. The land use planning aspects of the Directive are covered by separate planning laws under the Office of the Deputy Prime Minister, the Scottish Executive and the National Assembly for Wales. Minor amendments were made to the Act in 2005.

A pre-planned, preventive approach to technological hazards is essential. In the UK, legal efforts to enhance safety are normally required to be carried out so far as reasonably practicable (see page 89). Attention has also been given to harmonizing control measures within the

European Community. Amendments to the 1982 'Seveso Directive' require that premises storing or using more than certain specified quantities of very hazardous substances are designated 'major hazard sites'. The operators of such sites must prepare emergency plans and communicate the dangers to the public nearby. In the USA the *Chemical Emergency Preparedness Program* (CEPP) has been developed by the Environmental Protection Agency to lower threats from the chemical industry. One focus has been the preparation of a list of acutely toxic chemicals that might endanger public health in the event of an accidental air-borne release. Emergency response planning for hazardous industrial sites is less advanced in the LDCs. Some progress is being made, for example in India (Ramabrahmam and Swaminathan, 2000). But such planning is rarely a priority and inaction by both national and local government authorities is possible (de Souza, 2000).

Although preparedness may be high at individual sites, the level of readiness for chemical emergencies at the community level is often low. Little attention has been given to the needs of the most vulnerable groups during chemical emergencies. Phillips *et al.* (2005) conducted a random 10 per cent sample survey of the estimated 31,000 households situated within the 13–16 km radius Immediate Response Zone surrounding an army depot holding stockpiled chemical weapons in Alabama, USA. Emphasis was given to those in the lowest income quartile, a group containing 43 per cent of all households and believed to require special assistance in the event of a chemical emergency. Despite being more concerned than other residents about the threat of a chemical accident (Figure 13.4a), this group was less well informed about the provisions of the federal Chemical Stockpile Emergency Preparedness Program (CSEPP), designed in part to aid such people (Figure 13.4b). Lower-income households were also less willing to participate in preparedness classes (Figure 13.4c) and had less access to their own transport when faced with a warning to evacuate (Figure 13.4d).

New initiatives are required. For example, attempts to raise preparedness are likely to be ineffective if nothing is done to address the poverty of vulnerable population groups. Another priority is better, more scientific training for the local emergency responders – such as the police, fire and medical services. In addition, effective public response depends on more 'freedom of information' with regard to industrial hazards. Despite legislative strides in this direction by certain countries, like the USA, important restrictions are applied when commercial competitiveness or terrorist activity might be involved.

As far as off-site safety is concerned, the best emergency response plans are in the nuclear power industry. Following the 1979 accident at the Three Mile Island nuclear plant in Pennsylvania, all reactors in the USA were required to produce emergency response plans that met criteria laid down by the Federal Emergency Management Agency and the Nuclear Regulatory Commission (NRC). Formal approval of these plans is a condition for granting and maintaining operating licences for commercial nuclear power plants. The plans normally include three protective measures in any radiological emergency – *in-door shelter* (to protect against the short-term release of radionuclides); *medical treatment* (use of potassium iodide as a thyroid blocking agent) and *evacuation* (to remove the population from exposure to the pollution plume). These measures should be seen within the context of two standardized *Emergency Planning Zones* (EPZs). The first EPZ (the plume exposure pathway) extends over an approximate 10 miles (16 km) radius of the plant, downwind from the plant. It represents the area within which whole-body exposure and particle inhalation might be expected to occur. The second EPZ (the ingestion exposure pathway) extends to approximately 50 miles (80 km) from the plant where the hazard would be, largely due to contamination of water supplies and crops.

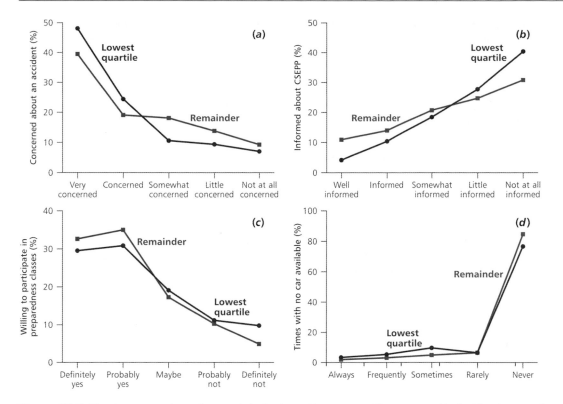

Figure 13.4 Disaster preparedness for people in the lowest income quartile, compared to that for the rest of the population, living near a chemical weapons depot in Alabama: (a) percentage concerned about an accident; (b) percentage informed about the appropriate federal emergency program; (c) percentage willing to attend preparedness classes; (d) number of occasions when the groups lack access to personal transport. Adapted from data presented in Phillips *et al.* (2005).

2 Forecasting and warning

Most technological accidents, such as industrial explosions, provide little scope for forecasting and warning, although anomalous heat emissions from the Chernobyl accident in 1986 were detected after the event by satellite imagery (Givri, 1995). For certain types of structural failure and the release of toxic materials from industrial plants, a warning to the local population is possible by sirens, or other audible means, but the limited timescale between the initiating event and the hazard often precludes preventive action. Where longer lead-times are available, warnings would be beneficial. Smith, D.I. (1989) quoted studies of warning systems for major dam failures following flash floods in the United States. In the cases where more than 90 minutes' warning had been possible, the average loss of life was as little as 2 people per 10,000 residents. When the local community received either less than 90 minutes' warning, or no warning at all, the average number of lives lost rose to the equivalent of 250 per 10,000 residents.

US federal guidelines for warning effectiveness in a nuclear emergency are:

• the capability to disseminate messages to the population inside the 10-mile zone within 15 minutes

- assurance of direct coverage of 100 per cent of the population within a 5-mile (8 km) zone around the plant
- arrangements to ensure 100 per cent coverage within 45 minutes to all who live within the 10-mile radius and who may not have heard or received the initial warning.

These plans have been criticized, especially regarding the expectations for evacuation. Cutter (1984) saw little evidence that the public will follow evacuation procedures, including the time-frame specified and the identified routes, as laid down by the authorities. An early survey of the population affected by the emergency response plan for the Diablo Canyon nuclear power plant on the southern Californian coast revealed that only one-third of the households had any familiarity with the plan and fewer than 6 per cent claimed that they had information about the action to take during an emergency (Belletto de Pujo, 1985). In terms of response, only about half of the households questioned said that they would follow emergency instructions from the authorities, despite the fact that 40 per cent perceived the risk of a major accident at this power plant to be either high or very high.

It is sometimes asserted that there has been a failure to learn from evacuation responses after natural disasters, although it could be misleading to transfer that experience directly to technological hazards. The fear of nuclear radiation is so great that during the Three Mile Island incident, 196,000 people engaged in a 'shadow' type of evacuation, despite the fact that no formal evacuation order was issued. As a proportion of the local population, this is a much greater response than can be expected when people are ordered to evacuate after natural disasters.

3 Land use planning

The purpose of land use planning is to resolve the conflicts, and reduce the risks, associated with the location of dangerous facilities. High-hazard installations will almost always be unwanted by most of the local population. The least acceptable facilities tend to be nuclear waste and toxic chemical disposal sites, together with chemical processing plants, nuclear power plants and fuel storage depots. Major industrial accidents tend to result from a planning decision which locates a dangerous technology in an inappropriate place, often combined with a failure to control subsequent intensive land uses from invading the area around the site.

It is rarely possible to place other industrial activities, housing and people sufficiently far away from high-hazard sites to guarantee zero risk. At the simplest level, land use planning aims to separate densely populated areas from very high-risk facilities and their transport routes, and to reduce any residual threat through the use of buffer zones. It is self-evident that chemical plants should not be located near schools, hospitals or densely populated areas, but there is no universally valid rule that determines exactly where development should take place. Ideally, all vulnerability issues, such as the level of preparedness and the demographic characteristics of surrounding communities, should be considered. But the spatial context of technological risk appears to be poorly understood. As a result, risk is poorly integrated into land planning, even in the LDCs, not least because of vested financial interests in the urban land market (Walker *et al.*, 2000). Better planning will require more understanding of the strengths and weaknesses of the different approaches used for technological risk assessment and improved public information and acceptance.

In the UK, the Flixborough chemical plant disaster of 1974, which killed 29 people and injured more than 100 others, was a policy watershed. After that event, the Health and Safety Executive (HSE) reviewed the location of major industrial risks, with a view to integrating high-hazard sites with other land uses. The current system requires any site containing hazardous materials above specified quantities to obtain special consents from the responsible authority, normally the *local authority planning department* (LPA). The LPA is advised by the HSE when dealing with such planning applications. The HSE may recommend imposing conditions on any consent in order to limit risks, e.g. restricting the amount and nature of substances stored on site or requiring tanker delivery rather than on-site storage. This system is under fresh scrutiny following the Buncefield accident (Box 13.3 and Figure 13.5).

To aid planning decisions around hazardous plants governed by the COMAH regulations, the HSE normally produces a map for the *Consultation Distance* (CD) surrounding the site. All planning applications for new development within this zone have then to be referred by the LPA for consultation with the HSE. Typical maps have three risk contours showing the probabilities of any person receiving a 'dangerous dose' of chemicals, or any other specified level of harm, in any one year. A 'dangerous dose' is defined as one likely to cause some fatalities (at least 1 per cent), a substantial need for medical attention (including hospital treatment) and severe distress to a typical house resident within the Consultation Distance. The chance of receiving a dangerous dose increases with proximity to the site, and the risk is usually expressed in 'chances per million per annum' (cpm).

Figure 13.6 shows three contours representing the levels of individual risk of a dangerous dose rising from 0.3 cpm in the Outer Zone to 10 cpm within the Inner Zone. In principle, these risk levels can be combined with socio-economic data and information on other land uses to moderate the advice the HSE offers to the LPA. However, the concept of 'dangerous dose' does not provide a consistent measure of societal risk that would be obtained, for example, by estimating the individual risk of fatality or serious injury within the Consultation Distance. Importantly, this approach has not linked individual planning decisions sufficiently closely to the Control of Major Accident Hazards Regulations (COMAH).

Too often, planning approval has been given to development on a case-by-case approach, which does not prioritize either on-going review or the control of cumulative risk. This has allowed other land uses to develop in proximity to high-hazard sites.

In Europe and elsewhere there has been a trend towards harmonizing quantitative risk-assessment methods to create a wider picture of the threats to society and the environment around a high-hazard site. This approach depends on laying down *risk acceptance criteria*, i.e. a definition of the level of societal risk that is tolerable and capable of enforcement (Duijm, 2009). It involves the determination of a specified *safety distance*, the zone within which explicit limitations are placed on the location of facilities, special natural habitats and the movement of people to prevent their being exposed to excessive location-based risk, as defined by the risk acceptance criteria. All off-site risks depend on the population density, and other vulnerability criteria, within the *maximum consequence distance*. Beyond this, the risks are deemed to be tolerable. The relationship between location-based risk and maximum consequence distance is shown in Figure 13.7. It may be that, for especially hazardous sites, the population normally located within the safety distance, apart from plant operatives, may be nil. In this case, the societal risk is determined by the population density between the safety distance and the maximum consequence distance.

Conflicts between central government and community interests are common in land planning. One example is the November 1984 disaster at a liquid petroleum gas plant operated by the Mexican national oil corporation (PEMEX) in Mexico City. In this event a series of explosions caused an estimated 500 deaths and 2,500 injuries, together with the partial destruction of a nearby working-class district (Johnson, 1985). Within a few days of the event, the government decided not to rebuild this devastated area and ordered the creation of a 14 ha 'commemorative park'. However, this plan did not reflect the residents'

Box 13.3 The 2005 explosions and fires at the Buncefield fuel site, UK

On 11 December 2005, serious explosions and fires occurred at the Buncefield oil storage and transfer depot on the outskirts of Hemel Hempstead, Hertfordshire (Buncefield Major Incident Investigation Board, 2006a). This COMAH facility began operating in 1968, when few other activities or buildings existed nearby. All the evidence suggests that the main explosion occurred due to the ignition of a vapour cloud originating from Tank 912 in Bund A (see Figure 13.5). In turn, this was due to an escape from the tank of about 300 t of unleaded petrol, due to over-filling caused by the failure of safety systems designed to prevent such an accident. In total, over 20 large fuel-storage tanks were involved. The resulting fire, described as the largest in peace-time Europe, burned for 32 h and closed sections of the M1 motorway, despite the efforts of up to 1,000 fire-fighters.

No one died, but 43 people were injured and 2,000 persons were evacuated. Twenty business premises, employing 500 people, were destroyed and a further 60 businesses, employing 3,500 people, suffered damaged. At least 300 residential properties were damaged and fuel supplies to London and

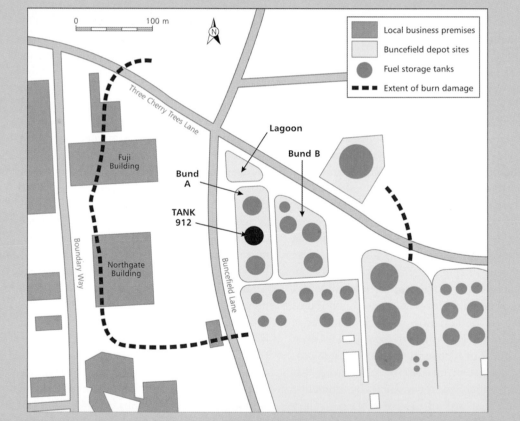

Figure 13.5 Map of the pre-incident layout of the Buncefield fuel depot site showing the extent of burn damage. After Buncefield Major Incident Investigation Board (2006b), Health and Safety Executive. Used with permission.

Plate 13.3 A view of the fire at the Buncefield oil storage terminal near Hemel Hempstead, England on 11 December 2005. Buncefield, the fifth-largest oil-products storage depot in the UK, was destroyed following a series of explosions. (Photo: Courtesy of Chiltern Air Support Unit and Buncefield Investigation at www.buncefield investigation.gov.uk/images/index.htm)

parts of south-east England, including Heathrow airport, were severely disrupted. There was a significant impact on health services when 244 people, mainly members of the emergency services with respiratory complaints, attended local hospitals (Hoek *et al.*, 2007). Economic losses were placed at £1 billion, mainly due to compensation claims against site operators, plus costs of the emergency response, losses in the aviation sector and the cost of the subsequent investigation. Criminal proceedings were brought against five defendants for neglect of health and safety procedures. After a complex corporate criminal trial, guilty verdicts were delivered against the defendants in June 2010.

This disaster raised many concerns about high-hazard sites. Apart from technical issues around the design and operation of these facilities, many recommendations were made for improving emergency preparedness and land use planning (Buncefield Major Incident Investigation Board, 2008a). The area surrounding Buncefield had been subject to incremental development during recent decades, placing more and more people and property at risk, and the main recommendation from the Investigation Board was for a thorough review of the existing land planning system for high-hazard sites. Specific matters suggested for review included moving away from expressing harm in terms of 'dangerous doses' towards a risk of fatality, better integration of land use planning with the COMAH regulations, taking account of societal risk by factoring in the size and distribution of the population around the site, proper cost-benefit analysis of restricting development near hazardous sites and making the entire process more transparent and accessible to the general public (Buncefield Major Incident Investigation Board, 2008b). It seems essential that, in a crowded island like Britain, more attention should be given to the population that is at risk from high-hazard facilities, especially those – like Buncefield – where off-site development already exists.

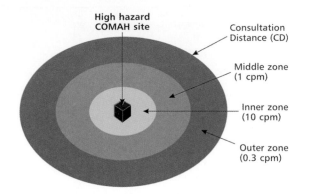

Figure 13.6 Idealized risk contours within the Consultation Distance around a typical high-hazard chemical site in the UK. The annual risk of any person receiving a 'dangerous dose' declines from 10 cpm in the Inner Zone to 1 cpm in the Middle Zone and to 0.3 cpm in the Outer Zone. After Buncefield Major Incident Investigation Board (2006a), Health and Safety Executive. Used with permission.

wishes for a complete re-siting of the PEMEX facility and involved the demolition of all the dwellings in the damaged zone and the resettlement of almost 200 families in other parts of the city. Such remotely taken decisions, made when the community was still recovering from the immediate aftermath of the disaster, can be seen as both arbitrary and insensitive, from a local perspective.

Acceptable land use control relies on a balanced appraisal of the probability of large escapes of toxic material or explosions from a site, the local consequences of a major accident and the need for particular types of hazardous activity in the regional or national interest. Economies of scale mean that large-scale sites often provide cheaper manufacturing and transport costs, but bigger operations also tend to create larger risks. In the USA the Environmental Protection Agency has used a GIS to map the releases of toxic chemicals in eight states to show that the largest releases have taken place near densely populated areas (Stockwell *et al.*, 1993). There is some evidence that hazardous industrial sites have

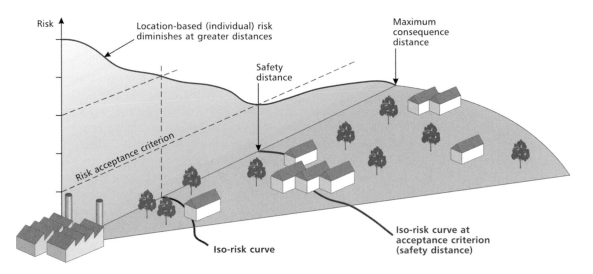

Figure 13.7 Example of risk acceptance and land planning around major hazard establishments. The safety distance is the point where the risk falls below the acceptance criteria (at greater distances the risk to individuals is deemed acceptable). The risk is zero beyond the maximum-consequence distance. Iso-risk curves show the pattern of individual risk. After Duijim (2009). Used with permission (personal communication).

proliferated in areas with lower-income and minority population groups. For example, in Phoenix, Arizona some disadvantaged communities face greater potential risks from proximity to releases of hazardous substances than do other social groups in the city (Bolin *et al.*, 2000).

The perceived risks of nuclear power have led to the location of these facilities in relatively remote or rural areas, but even greenfield industrial sites tend to attract later development and create planning tensions. The problem of allowing subsequent development around one existing high-hazard site in the UK is illustrated in Box 13.3. In the absence of strong planning controls, it is likely that unwanted hazardous facilities will continue to be placed by developers and governments where local resistance is less than at other candidate sites. Therefore, small rural communities with low income and high unemployment, which are remote from political influence, are most likely to have industrial hazards imposed upon them in the future.

FURTHER READING

Chapman, J. (2005) Predicting technological disasters: mission impossible? *Disaster Prevention and Management* 14: 343–52. An attempt to place these risks within a wider framework.

Cozzani, V., Campedel, M., Renni, E. and Krausman, E. (2010) Industrial accidents triggered by flood events: analysis of past accidents. *Journal of Hazardous Materials* 175: 501–9. This paper illustrates the current level of concern about na-tech disasters.

Duijm, N. J. (2009) *Acceptance Criteria in Denmark and the EU.* Environmental Project Report No. 1269, Danish Environmental Protection Agency, Ministry of the Environment, Copenhagen. One of the best explanations of the planning restrictions necessary for high-hazard industrial sites.

Mannan, M.S. *et al.* (2005) The legacy of Bhopal: the impact over the last 20 years and future direction. *Journal of Loss Prevention in the Process Industries* 18: 218–24. The worst-ever industrial disaster in Asia viewed with hindsight.

Phillips, B.D., Metz, W.C. and Nieves, L.A. (2005) Disaster threat: preparedness and potential response of the lowest income quartile. *Environmental Hazards* 6: 123–33. A reminder that technological disasters, like natural disasters, strike most severely at disadvantaged social groups.

Stanton, N.A. and Walker, G.H. (2011) Exploring the psychological factors in the Ladbroke Grove rail accident. *Accident Analysis and Prevention* 43: 1117–27. A specific example that reinforces the views of the Normal Accidents School.

WEB LINKS

Bhopal Disaster Information Centre www.bhopal.com

International Information Centre on the Chernobyl disaster www.chernobyl.inf/

International Atomic Energy Authority www.iaea.org/

List of recent technological disasters compiled through UNEP www.unepie.org/pc/apell/disasters/lists/technological

CHAPTER FOURTEEN

Environmental hazards in a changing world

14

A INTRODUCTION

Environmental hazards are centre stage, due to recent high-profile disasters, fresh perspectives on risk and concern about future trends. Safety issues of all kinds are prioritized by governments; public interest in risk is fuelled by the media and the spread of information on social networking websites. A shift in the 'landscapes of risk' has produced a comprehensive 'de-localization' of hazards and disasters. Social scientists increasingly recognize relationships between disasters and wider modernization processes, whilst physical scientists explore the global drivers behind extreme natural events. In brief, hazards and disasters are increasingly relevant to our understanding of many present-day societal and environmental issues, including the potential consequences of climate change and the prospects for sustainable human development.

Academic perspectives and policy implementation have changed. Official responses to disaster reduction and climate change adaptation have traditionally operated in separate spheres. Disaster planning and emergency operations were typically allocated to civil defence agencies, whilst longer-term climate change issues were man-aged by central government departments. Yet the shared priority is to reduce any possible adverse effects in the future that arise from interactions of physical and biological systems with human communities. Risk, human vulnerability and decision making under conditions of uncertainty are common themes, irrespective of time-scales and causal factors. Most tools required for 'natural disaster emergencies' (e.g. insurance, forecasting and warning, land planning) are also appropriate for 'adaptation to climate change'. Increasingly, these tools are employed on a community 'bottom-up' basis, in order to secure sustainable development into the future that goes beyond coping with a disaster emergency. The recognition of this complementary relationship has produced a move towards comprehensive *integrated disaster risk management*. When disaster risk reduction (DRR) interacts more positively with climate change adaptation (CCA) in order to lessen deaths, damages and other negative impacts from all environmental threats, greater synergy is likely to follow (Mercer, 2010; Romieu *et al.*, 2010).

This concluding chapter reviews hazards and disasters within this evolving framework. Following a brief account of globalization issues, the aim is to discuss aspects of disaster-related environmental change already recorded and to examine the evidence for anticipating further change in the future. The approach builds on what is already known about how human vulnerability influences disaster outcomes and takes a closer look at the key geophysical drivers, particularly those relating to the hydro-meteorological risks with links to the global climate.

B THE GLOBALIZATION OF HAZARD

Globalization is an umbrella term for many modern-day phenomena. Originally, the word was used to describe the upsurge in commercial exchanges following the decline of communism, although the geographical spread of economic power is not entirely without precedent (Pelling, 2003). However, Giddens (1990) saw new transport systems, including electronic communication and information flows, as key factors in shrinking the space and time constraints that limited trade and other interactions between spatially distant human societies. Since then, the term has widened to cover all types of global flows, including goods (trade), people (migration) and financial and cultural exchanges (international loans, development aid). These exchanges have grown, due to the rise of Western-style democracy and free-market ethics across the world. They can be measured by the *KOF Index of Globalization* (Dreher, 2006). This Index aggregates the separate roles of economic, social and political networks of global connection for each nation, to produce yearly snapshots of interconnectedness and mutual interdependence across the world (Figure 14.1). Unsurprisingly, the highest values are found in Western-style democratic nations; the lowest values are in poor, remote countries, sometimes with repressive governments. Most importantly, it is these regimes which are likely to be disaster prone.

Globalization is linked to modernization. It has already been shown that attitudes to risk in Western-style countries were reshaped by the work of Beck (1992) and his concept of the 'risk society' (see Chapter 4, p. 74.). Beck proposed that, in the early phases of modernization, risks were down-played because the overriding priority was material prosperity. As a result, side-effects of economic growth – such as a decline in environmental quality – were ignored. Towards the end of the twentieth century, the balance changed and safety issues assumed greater prominence. Risk assessment became a legal requirement for many activities in the public and private sectors. Such changes coincided with a growing public distrust of scientific experts and fears about 'dread' or 'new' hazards, such as complex technologies, some diseases and terrorism.

Beck's work was rooted in the European experience where prosperous residents were concerned about issues like nuclear power and food safety. These concerns have led to the enthusiastic health and safety culture found in most Western-style countries today. A growing aversion to risk in the MDCs is based on a perceived value of the future during a longer expected life-span. This view is less common in the LDCs, where more frequent hazard impacts occupy the thoughts of three-quarters of the people in the world – notably the poorest members of society living in urban slums or remote rural areas. Indeed, rather than improving safety and life expectancy, it is possible that modernization has brought added risk for some people.

Certain social scientists and political economists have returned to the roots of vulnerability in order to explain how globalization has redistributed and intensified risk for the poorest, most excluded people. This debate echoes a radical viewpoint of the 1980s, when capitalism was criticized for maximizing material growth to the extent that safety and prosperity for a favoured few were bought through increased vulnerability for many (Hewitt, 1983). Today, a similar concern is that increased economic activity and the related growth of greenhouse gas emissions contribute to climate changes with potential risks for everyone, but most of all, for those already disadvantaged.

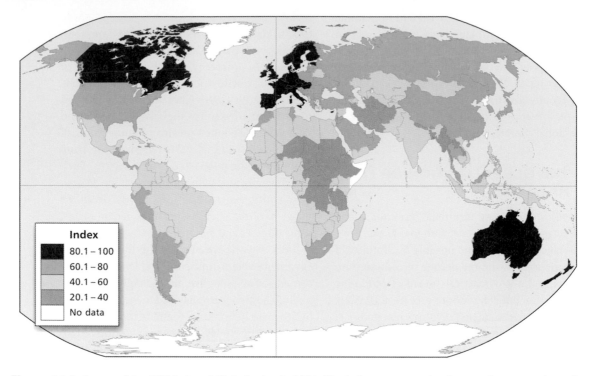

Figure 14.1 A map of the KOF Index of Globalization in 2011. The Index measures the degree of connectedness for individual nations based on economic, political and social criteria. The map has been compiled from data at http://www.kof.rthz.ch KOF Index of Globalisation 2008 (Dreher, Axel (2006): Does Globalisation Affect Growth? Evidence from a new Index of Globalisation, *Applied Economics* 38(10): 1091–1110.

Globalization has accelerated natural resource depletion and cemented large inequalities between rich and poor nations and between peoples. Development initiatives, designed to raise living standards for more of the world's population, have been found wanting; continuing on the current paths may threaten future sustainability. The high frequency of disaster strikes in recurring areas of the poorest countries has simply gone on for too long. Some researchers now believe that a new – more morally based – agenda is required that can lessen much of the existing disaster loss if greater political will and new thinking can be applied (Wisner, 2003).

On the other hand, there are few challenges to the goal of limitless growth based on capitalist ideals (Newell, 2011) and even to existing policy attitudes towards hazard and disaster. Several authors (Pelling, 2003; Adger and Brooks, 2003) have drawn attention to dangers inherent in the responses of UN agencies and other international bodies. These institutions have typically promoted global-scale action as a solution to local problems. At worst, such policies can impose inappropriate solutions, perhaps found successful in the MDCs, in preference to options based on indigenous, lower-cost solutions. An effort is required across all scales – local, national and global – so that billions of poor people can have greater ownership of their own futures and, where appropriate, can access disaster-reducing responses – like affordable insurance cover or hazard-resistant homes – that have proved effective elsewhere. Many of the limitations are already known.

Too many communities remain ill-equipped to deal with disasters today because of failures to apply existing knowledge and expertise.

The globalization of risk applies to geophysical hazards, especially climate-related threats. This is because of *teleconnections* which are the dynamic linkages between the oceans and the atmosphere in widely separated parts of the world. Some of these linkages are incorporated into routine hazard assessments; for example, the hemispheric-scale interactions between air and ocean masses that produce climatic variations like the *El Niño Southern Oscillation* (ENSO). Other worldwide 'surprise' threats exist. Extra-terrestrial comet impacts and so-called 'super' volcanic eruptions are part of the Earth's geological history. Similarly, the role of disease pandemics, like the Black Death of the fourteenth century, can be traced in written records. Such global catastrophes are unlikely to be confined to history. The non-linear nature of the climate system can lead to abrupt climate change, as in previous 'ice ages', and continued global warming could eventually produce 'tipping-points' with major consequences for the ENSO and monsoon systems as well as the Atlantic thermohaline circulation (Lenton *et al.*, 2008).

C ENVIRONMENTAL CHANGE

The future holds the prospect of environmental changes across the Earth on a scale unprecedented in historical times. These will add to the uncertainties that already surround hazards and disasters. Some possible outcomes from anthropogenic global warming are highly relevant for disaster risk assessment. Although many climate change scenarios carry low confidence levels, and the balance between negative and positive impacts will vary according to location, most changes will require major human adjustments and significant capital investments to fund adaptation measures. To this extent, *global environmental change* (GEC) itself can be seen as an environmental hazard.

Globalization has helped to make the world a more complex and vulnerable place. Major disasters damage economies far away through *interconnectedness* and 'echo disruption' to communications and supply chains reliant on modern technology. The $M_W = 7.6$ Chi-Chi earthquake that struck Taiwan in September 1999 damaged high-tech facilities making computer memory chips. Production was disrupted and wholesalers hoarded the chips; prices rose by up to five times and created a worldwide shortage (OECD, 2003a). The August 1999 earthquake in Turkey damaged telecom networks which normally handled some 3 million calls per day with a 99 per cent successful connection rate. Following the earthquake, the number of daily calls rose to more than 50 million; only 11 per cent were connected, causing serious disruption for the emergency services. On 26 December 2006, the $M_W = 7.1$ Hengchun seafloor earthquake off the south-west coast of Taiwan severed submarine communication cables serving much of East and South-East Asia and dislocated transactions on the international money markets. After hurricane 'Katrina' in 2005, the International Energy Agency reported a deliberate draw-down of European and Asian oil stocks by 60 million barrels that triggered a price rise in many parts of the world.

Environmental change includes entirely natural processes, human-modified natural processes and socio-economic processes. Turner *et al.* (1990) drew a useful distinction between systematic and cumulative changes.

- *Systematic changes* were defined as those directly affecting planetary-scale systems through human activity such as 'greenhouse' gas (GHG) emissions, industrial emissions of ozone-depleting gases and land-cover changes.
- *Cumulative changes* were exemplified by human-induced biodiversity loss, water pollution, deforestation and soil depletion – all factors operating on smaller scales but significant because they occurred pretty well everywhere.

Thus, systematic change is the risk of potential damage to an entire system, as compared with the cumulative losses to units within that system. Risk is amplified by linkages between the components and can be triggered by large-scale sudden events or a build-up of smaller changes over a period of time. This human-induced change is superimposed on the natural fluctuations in geophysical systems that create extreme events. Human actions add to disaster potential by the degradation of regional ecosystems, right through to the emissions of greenhouse gases that threaten irreversible global climate change. In other words, human-induced GEC places additional layers of uncertainty onto what has still to be learned about the behaviour of natural Earth systems. Some interactions are likely to compound the existing disaster potential.

Table 14.1 presents global change from a hazards-based perspective, using two equally important pathways that lead to disaster. *Socio-economic paths* increase vulnerability as a result of expanding population, urbanization and inequality (see Chapter 3, section G). Certain societal factors, such as biodiversity loss and air pollution, also have an ability to modify natural processes. On-going change from such sources builds slowly through 'creeping risks'. For example, climate change and environmental degradation may stimulate mass migrations of refugees (Warner *et al.*, 2010). *Geophysical paths* are the traditional 'natural hazard' routes to disaster. Some, like tectonic hazards, remain almost untouched by human activity. Others, like hydro-meteorological hazards and related near-surface threats – like landslides – are subject to unintended human influences. At present, it is difficult to predict the extent to which such changes will add to the disaster toll.

Table 14.1 Global change paths tending to increase environmental hazards

Type and pathway of change	Main drivers of hazardous variations and changes	Some hazard-related factors
Socio-economic path		
Widespread human-based trends and actions with cumulative effects that promote risk to more people	Modern globalization and economic development (includes contribution of GHG emissions to climate change)	Air pollution Water pollution Biodiversity loss Desertification
These factors raise human exposure to low-intensity hazards mainly at local or regional scales; cumulatively may attain global hazard significance	Population growth, socio-economic factors Industrialization, land pressure, technical innovation, urbanization	Soil erosion Land degradation
Geophysical path		
Variations and changes in the functioning of natural geo-spheric and bio-spheric systems, including modifications due to human actions.	Planetary-scale geophysical systems, mainly tectonic and hydro-climatic processes	Teleconnections and atmosphere-hydrosphere interactions
These factors influence the spatial and temporal variability of large-scale, often high-intensity hazards responsible for large losses of life and economic assets	El Niño-Southern Oscillation, North Atlantic Oscillation, Oceanic circulation Anthropogenic climate change Extra-terrestrial threats from asteroid and comet impact	Climate change Magnitude and frequency of floods and droughts Diseases Sea-level rise Extra-terrestrial threats

Source: Developed from concepts in Turner *et al.* (1990)

D AIR POLLUTION AND CLIMATE CHANGE

Regional-scale air pollution was detected in the mid-1960s when acid deposition was recorded over rural parts of Europe and North America, due to the transport of oxides of sulphur and nitrogen from industrial sources hundreds of kilometres away. Poor air quality now threatens to disturb regional climates elsewhere. Biomass burning of forests and agricultural wastes, combined with fossil fuel emissions from road vehicles, industry, power stations and inefficient domestic cookers are the chief sources of air pollution. Although *per capita* consumption of fossil fuel is relatively low in the LDCs, emissions of gaseous pollutants like carbon monoxide and particulate (aerosol) matter are growing rapidly and are linked to regional haze, ozone depletion and global warming (UNEP and C4, 2002). Individual particles remain in the atmosphere for only a few days but the resulting haze clouds absorb incoming solar radiation and exert a cooling effect directly opposed to the warming trend associated with greenhouse gas emissions. The aerosols also reduce precipitation by weakening the hydrological cycle (Ramanathan, *et al.*, 2001).

The problem is most marked in Asia, where an intermittent haze layer, known as the *Asian brown cloud*, overlies the region from Pakistan to China. During 1997–98, smoke pollution from biomass fires in Indonesia spread over South-East Asia and interacted with the seasonal monsoon circulation in various ways (Koe *et al.*, 2001). The pollution was associated with forest clearance but was mostly caused by fires within drained swamp vegetation. Previously, farmers used slash-and-burn methods during the dry season without creating disaster. But a government drive to raise production in the timber, palm-oil and rubber sectors led to forest clearance on an industrial scale and little has been done to control this practice (Harwell, 2000).

During 1997–98, over 31,500 fires destroyed over 9 million hectares of land (Stolle and Tomich 1999). Direct fire-related deaths were estimated at 1,000, over 20 million people were exposed to high levels of pollutants and about 40,000 people received hospital treatment for the effects of smoke inhalation. Total direct economic losses were estimated at around US$9.3 billion. In June 2002, an Agreement on Trans-boundary Haze Pollution was signed by some members of the Association of South East Asian Nations (ASEAN) to enhance fire-fighting ability and collaboration between neighbouring countries through the development of an early warning system using satellite imagery to detect local fires. Indonesia did not sign that agreement, and more than 40,000 fires broke out again in 2006.

It is estimated that anthropogenic sources are responsible for three-quarters of the Asian brown cloud. This haze layer, up to 3 km deep, reduces the amount of sunlight and solar energy received at the Earth's surface by 10–15 per cent during the winter monsoon season (December to April). There is also likely to be a reduction in evaporation, especially over the ocean surfaces, with effects on the regional climate and hydrology on a scale equivalent to that arising from global warming. Several scientists (Fu *et al.*, 1998; Xu, 2001; Bollasina *et al.*, 2011) have observed a downward trend in summer precipitation over south Asia linked to migration and weakening of the summer monsoon rain belt. The effects of aerosol pollution may extend beyond Asia and disturb precipitation patterns in Australia (Rotstayn, 2007). Ramanathan (2007) further suggested that warming in the Himalayas associated with the Asian brown cloud might be responsible for glacier retreat in the high mountains. There is concern about the effects of the haze layer on health hazards, especially related to respiratory diseases. Several Asian mega-cities (Beijing, Delhi, Jakarta, Calcutta and Mumbai) exceed WHO standards for suspended particulate matter and sulphur dioxide in the atmosphere and increases in these pollutants will create additional risk.

Plate 14.1 Firemen attempt to control a forest fire deliberately started in 1997 to clear land for development in East Kalimantan, Indonesia. Thousands of fires raged for several months and created a dense cloud of air pollution across much of South-East Asia. (Photo: Panos/Dermot Tatlow DTA00167INN)

1 Global-scale air pollution

Climate change is defined as a sustained shift in the average value of climatic elements (temperature, sunshine, precipitation, wind, etc.), either singly or in combination. The shift must be maintained for sufficient time to be measurable by *Climate Normals*, which are calculated over 30-year periods by the World Meteorological Organization and other bodies. In contrast, *climate variability* is expressed by differences in climatic elements from one year to the next. Short-term anomalies can group over a period of years, and may be temporarily mistaken for part of a trend, but they are not sustained and have little statistical effect on the Climate Normals.

Today, climate change due to *anthropogenic global warming* (AGW) is commonly perceived as the greatest environmental threat facing the world. Global warming has been detected by observed increases in the average surface temperature of the planet which could be caused by natural processes and/or human influences. Present shifts are widely linked to the growth in worldwide emissions of greenhouse gases (GHGs) and their ability to change climatic conditions by absorbing outgoing long-wave infrared radiation from the Earth's surface. The temperature rise over coming decades is expected to be greater and faster than at any time in the past 10,000 years. Possible effects range from more frequent inundations of low-lying coasts, especially where natural ecosystems such as salt marsh or mangroves have been removed, to increased river flow from snowmelt in alpine areas. Disease patterns will change. Serious consequences are predicted for countries highly dependent on natural resource use. The gap between LDCs and MDCs may well widen, due to the impacts on ecosystems already under stress in countries with few resources for adapting to, or mitigating, change.

Callendar (1938) was the first person to establish a clear link between fossil fuel consumption and global warming. He correctly estimated the nineteenth-century background level of atmospheric CO_2 at 290 ppm and noted that the recorded rise of 10 per cent between 1900 and 1935 closely matched the increase in fuel burned. The work of Keeling (1960), at the remote observatory of Mauna Loa, Hawaii, confirmed the overall rise in the concentration of

atmospheric CO_2 up to the present day (Figure 14.2). Regular fluctuations within the trend reflect the seasonal take-up and release of CO_2 by plants in the northern hemisphere, where most land-mass is located. Present concentrations of atmospheric CO_2 are over one-third higher than pre-industrial levels and account for about two-thirds of the enhanced greenhouse effect attributed to air pollution. Apart from CO_2, the presence of other greenhouse gases like CH_4, N_2O and O_3 has increased significantly in industrialized regions of the world. The rise in GHG concentrations is consistent with trends in human activity. For example, the burning of fossil fuels – coal, oil and natural gas – accounts for 70–90 per cent of all human emissions of CO_2. Unlike the aerosols within the Asian haze cloud, GHGs have a long life in the atmosphere and are evenly distributed across the globe.

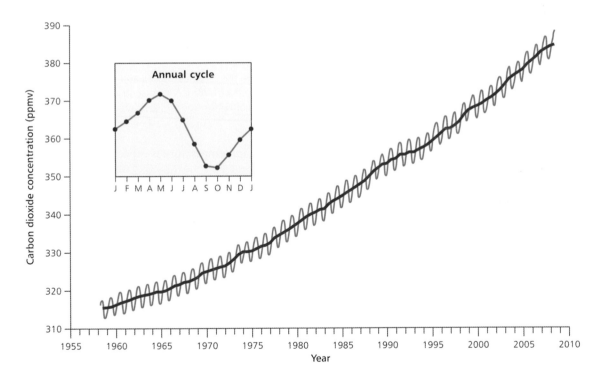

Figure 14.2 The increase in atmospheric carbon dioxide 1960–2010 measured at Mauna Loa, Hawaii. The inset shows the annual cycle of CO_2 within the overall annual trend. After http://en.wikipedia.org/wiki/File:Mauna_Loa_Carbon_Dioxide.png Reproduced under Creative Commons Attribution-Share Alike 3.0 Unported license.

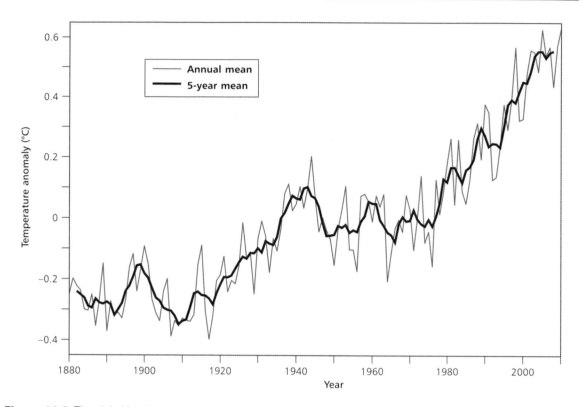

Figure 14.3 The global land–ocean temperature index 1880–2010. The mean annual values and the smoothed 5-year running mean illustrate a continuing rise that has accelerated since about 1970. After NASA Goddard Institute for Space Studies at http://www.giss.nasa.gove/gistemp/ Reproduced by courtesy of NASA/Goddard Institute for Space Studies.

Since the late nineteenth century, the average global temperature has risen by 0.8°C. The increase is greater than can be explained by natural variability but consistent with the increasing concentration of GHGs in the atmosphere. Figure 14.3 shows the combined global land and marine surface annual temperature record expressed as anomalies from the 1951–80 average (Hansen *et al.*, 2006). Over land areas more than 3,000 stations are used to compile this record, with the densest cover in the more populated regions. The much smaller marine dataset is derived from sea-surface temperature (SST) measurements taken by vessels at sea and from satellite observations. The warmest years have so far occurred in the 1990s and early 2000s. Without a reduction in greenhouse gas emissions, the global average

surface temperature is expected to rise between 1.4 and 5.8°C by the year 2100. Even if emissions stabilize, temperatures are expected to rise for centuries because of a lag in the response of the world's oceans. A progressive rise in global temperature is likely to change climatic extremes in complex and non-uniform ways (see section 14F).

E GEOPHYSICAL PATHS TO DISASTER

1 Time and space scales

Geophysical pathways vary greatly and it is helpful to distinguish between systematic hazards and rare hazards.

Systematic hazards are routine features of the geospheric and biospheric systems. Tectonic hazards tend to be more variable and less predictable than the atmosphere–ocean interactions that produce climatic anomalies and extremes. Short-term atmospheric variability may be overridden by longer-term changes that enhance hazard potential. Climate change is a trend that provides a frame within which shorter-term events can be placed. Future disaster impacts are hard to quantify, as when extrapolating links between global warming and patterns of severe storms. The link between sea-level rise and increased coastal flooding is clearer, in terms of both the science of sea-level rise and the likely socio-economic effects.

Rare hazards are unusual events capable of creating catastrophic global-scale losses (Coates, 2009). Evidence of these 'super' threats lies in the geological record rather than human experience. Attention is focused on marine hazards due to geophysical instability, like tsunamis, and collisions between planet Earth and extra-terrestrial objects. These are surprisingly new concerns. It was not until the late twentieth century that many crater-like depressions on the Earth's surface – previously attributed to volcanic origin – were recognized as the result of collisions with asteroids, comets and planetary debris. Possible future catastrophes include a major earthquake in the Pacific Northwest of the USA and the eruption of the Cumbre Vieja volcano on the Canary Island of La Palma, both capable of generating massive tsunami waves.

2 The El Niño Southern Oscillation (ENSO) and La Niña

The ENSO system produces important year-to-year climatic variations across the globe – in all ocean basins, on all seven continents and in the stratosphere (McPhaden *et al.*, 2006). The ENSO cycle involves a change in circulation within the tropical Pacific Ocean that normally happens every two to seven years around the Christmas season, hence the name 'El Niño' (The Christ Child). The term 'Southern Oscillation' refers to the cycle of varying strength in the atmospheric pressure gradient between the Indo-Australian low-pressure cell and the South Pacific high-pressure cell. The first sign of an ENSO event is a local incursion of abnormally warm surface water southward along the Peruvian coast, but, in association with changed air-flow patterns across the Pacific, an El Niño event can spread more widely and last for over a year.

Under non-ENSO climate conditions, shown in Figure 14.4a, the pressure gradient produces the *Walker cell circulation*, characterized by the flow of the south-east trade winds across the ocean and a convergence of low-level air in the western Pacific. The resulting vigorous uplift of moist air brings heavy seasonal rainfall to eastern Australia and much of south-east Asia. A westerly return flow aloft contributes to subsidence in the eastern Pacific, thereby completing the Walker cell. The offshore winds from Peru blow across up-welling cold water that is often at least 5°C colder than waters in the western Pacific. These cool sea-surface temperatures, and the stable descending air, maintain the dry conditions along the South American coast, while the up-welling cold water, rich in nutrients, supports an important fishing industry in Peru.

During El Niño events, the trade winds weaken along the equator as atmospheric pressure rises in the western Pacific and falls in the eastern Pacific. The changed gradient across the Pacific first weakens and then reverses the Walker cell (Figure 14.4b). Up-welling of cold water ceases along the Peruvian coast and anomalously warm water spreads throughout the Pacific Ocean basin. This contributes to a low-level onshore flow of moist, unstable air along the coast of South America that brings heavy rainfall and floods, sometimes accompanied by disease epidemics, to the normally arid areas of Peru and Ecuador. The fishing industry in Peru collapses. At the same time, negative anomalies in sea-surface temperature in the western Pacific Ocean lead to the displacement

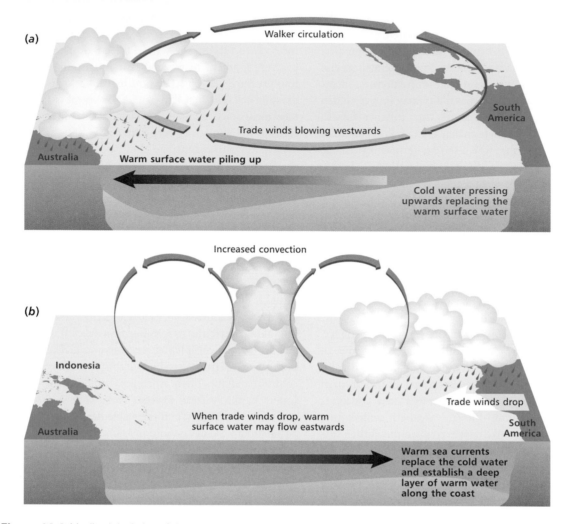

Figure 14.4 Idealized depiction of the two phases of the Walker circulation that form the Southern Oscillation pressure variation in the southern hemisphere: (a) normal cell pattern; (b) El Niño phase. When the normal Walker circulation becomes very strong, a La Niña phase of the cycle occurs.

of the convection zone usually centred over northern Australia and Indonesia.

ENSO has major hazard implications although it may not create higher-than-usual levels of activity in the atmosphere (Goddard and Dilley, 2005). ENSO cycles are driven by the amounts of heat stored in the upper layers of the tropical Pacific Ocean and offer the potential for forecasting and improved disaster preparation for associated climatic variations such as:

- rainfall patterns severely disrupted in many Pacific islands; drought in much of Micronesia and Hawaii, with failure of staple agricultural crops and forest fires
- temperature across New Zealand reduced; drier conditions in the north-east of the country
- warm and very wet summer (December–February) in Peru and coastal Ecuador, with floods; similar effects in southern Brazil and

northern Argentina, especially during spring and early summer; mild winter with heavy rain in central Chile
- hot and dry conditions in much of the Amazon basin, Colombia and Central America; similar effects in parts of south-east Asia and much of Australia, with bushfires and poor air quality
- in North America, winter warmer than normal in the upper Midwest states, the north-east, and Canada; central and southern California, north-west Mexico and the south-western US states cool and wet.

Strong El Niño episodes have long been associated with weather-related disasters, especially in years like 1982–83 and 1997–98. Floods, droughts, wildfires and diseases during the 1997–98 event claimed over 20,000 lives, with damages exceeding US$8 billion (IFRCRCS, 1999). The most severe impacts were in South America. Floods in Peru made 500,000 people homeless and destroyed US$2.6 billion-worth of public utilities, equivalent to nearly 5 per cent of national GDP. The Peruvian fishing industry declined by 96 per cent in the first three months of 1998, as compared to the same period the year before. ENSO-related hazards include:

Drought is significantly associated with ENSO events (Dilley and Heyman, 1995; Bouma *et al.*, 1997). During the 1877–78 event, drought and famine created 10 million deaths in northern China, 8 million in India, up to 1 million in Brazil and an unknown, but high number, of deaths in Africa. In India, severe drought is often El Niño-related but this is not always so (Kumar *et al.* 2006).

Flood events are linked to ENSO. Trenberth *et al.* (2002) found that average global temperatures are high during El Niño years and rainfall intensities rise, due to enhanced convective activity (Zhang *et al.*, 2007). For example, the 1993 river floods in the Midwest of the USA were linked to El Niño (Lott, 1994). Coastal flooding from tropical storms shows an ENSO signal, with a reduced number of cyclones in the Atlantic basin in ENSO years (Tang and Neelin, 2004) but increased risk during La Niña events. In late 1997, El Niño coastal storms in Peru carried a sea surge 15 km inland and flooded the main square in the coastal city of Trujillo. Along the Pacific coast of Colombia, short-term sea-level rises of about 30 cm have been recorded. Faced with marine erosion and flooding of the barrier islands, some villages have either moved further along the barrier islands or migrated to fossil beach ridges on the mainland (Correa and Gonzalez, 2000). There is also a correlation between El Niño events and damaging landslides and floods in Chile (Sepúlveda *et al.*, 2006).

Disease is firmly associated with El Niño events (Kovats *et al.*, 2003). This is because increased rainfall and humidity encourages the transmission of vector-borne infectious diseases in the tropics and subtropics. In the USA mild, humid summer weather in 1878 led to an epidemic of mosquito-borne yellow fever; in Memphis, Tennessee, 20,000 people died (McMichael, 2001). Epidemics of several mosquito-borne and rodent-borne diseases are triggered by ENSO phases (Bouma *et al.*, 1997). Specific examples include outbreaks of cholera in Bengal and dengue fever in Indonesia and northern South America (Bouma and Pascual, 2001; Gagnon *et al.*, 2001). In 1982–83, widespread floods tripled the cases of acute diarrhoeal diseases along the west coast of South America. Figure 14.5 shows the ENSO-related variability of epidemics of Ross River virus in south-east Australia. This is the most common mosquito-borne viral disease in Australia, with more than 5,000 cases annually

Wildfire hazard is increased during drought. Some of Australia's worst bushfires, such as the 1983 'Ash Wednesday' disaster, have been in El Niño years. During 1997–98, parts of south-east Asia suffered the worst drought for about 50 years. The rain forests were dry, uncontrolled forest clearance was taking place, and widespread fires broke out. Over 5,000,000 ha of forest, some containing several endangered species, was burned out in Kalimantan and Sumatra (Siegert *et al.*, 2001).

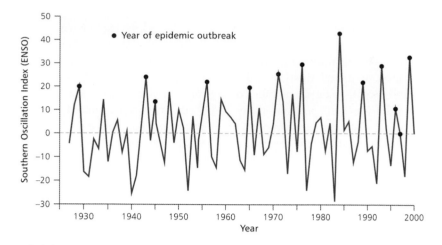

Figure 14.5 Relationships between strong El Niño events and epidemics of Ross River virus 1928–99 in south-east Australia. After McMichael (2001).

Atmospheric conditions opposite to El Niño are called *La Niña* (the girl) events. They occur when exceptionally strong Walker cell conditions exist. The cold surface water off the coast of South America spreads further north than usual to occupy a latitudinal band 1–2° wide around the equator, where it produces sea-surface temperatures as low as 20°C.

These colder-than-normal ocean temperatures inhibit the formation of rain-producing clouds over the equatorial Central Pacific. However, in northern Australia, Indonesia and Malaysia increased rainfall occurs in the northern-hemisphere winter months, whilst the Philippines and the Indian subcontinent experience increased precipitation during the northern-hemisphere summer. Increased rainfall is experienced over south-east Africa and northern Brazil during the northern-hemisphere winter. In North America, cold conditions are experienced across Alaska, western Canada, the northern Great Plains and the western United States. On the other hand, the south-eastern United States is generally warmer and drier than normal. In 2007, La Niña conditions coincided with an unusually strong summer monsoon in the Indian subcontinent, with several million people made homeless by floods and landslides, and very intense rainfall across central Africa.

3 The North Atlantic Oscillation (NAO)

The NAO is a rhythmic oscillation of air and ocean masses that imposes climatic variability on large parts of Europe and western Asia. The main alternation exists between polar influences to the north and subtropical influences to the south which control the strength and track of depression systems crossing the Atlantic Ocean from east to west. The most important hazard effects are in winter, due to the role played by the NAO in wind-storm hazards over north-western Europe (see Chapter 9).

The NAO is simpler than ENSO because the Atlantic Ocean is much smaller than the Pacific and exerts less climatic influence (Visbeck, 2002). In the North Atlantic, there are two semi-static pressure systems, the *Icelandic Low pressure system* and the *Azores High pressure system*. The *NAO Index* is a function of the relative strength of these two systems, usually measured between Lisbon (Portugal) and Reykjavik (Iceland). A *Positive Index* indicates a stronger-than-average pressure

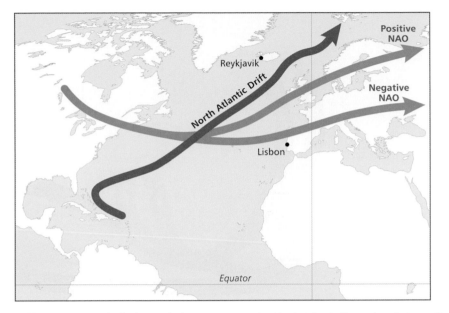

Figure 14.6 The average track of winter wind-storms across the North Atlantic Ocean in relation to the sign of the NAO and the warm water current of the North Atlantic Drift. Major hazard impacts are found in the United Kingdom and other countries in western Europe. Adapted from McPhaden *et al.* (2006) and other sources.

gradient, resulting in more intense winter storms that cross the Atlantic on a more northerly track (Figure 14.6). Winters in Europe are mild, with high rainfall totals, and summers tend to be cool. A *Negative Index* reflects a weak pressure gradient between the Azores and Iceland, leading to fewer, and weaker, winter Atlantic storms following a direct east–west track. Temperatures are then relatively high in summer and low in winter, with below-average rainfall. Some potential exists for NAO forecasting and a few precipitation models have been developed, but more refinement is necessary (Murphy *et al.*, 2001).

The NAO has been linked to several climate-related hazards. In the UK, Woodworth *et al.* (2007) showed how extreme sea level events and storm surges were related to the Index, although the sea level change was comparatively modest. Others have found relationships with river flows in the UK (Macklin and Rumsby, 2007) and the Iberian Peninsula (Trigo *et al.*, 2004). The NAO effect extends beyond the coastal parts of Europe.

Fagherazzi *et al.* (2005) noted links to flooding in Venice; Kaczmarek (2003) to floods in Poland; Cullen *et al.* (2002) to river floods in the Middle East; and Karabork (2007) to drought occurrence in Turkey. Some evidence supports a role for the NAO in landslide events in both Portugal (Zezere *et al.*, 2005) and Italy (Clarke and Rendell, 2006). The NAO may influence hazardous events elsewhere in the northern hemisphere. Correlations with drought have been found for both southern China (Xin *et al.*, 2006) and north-western China (Lee and Zhang, 2011). Further relationships may well be uncovered in the future.

4 Abrupt changes in ocean circulation

Over geological time-scales, sudden changes in global ocean currents have been associated with climatic disruptions during the Quaternary. In particular, global-scale cold periods have been associated with changes to ocean currents since

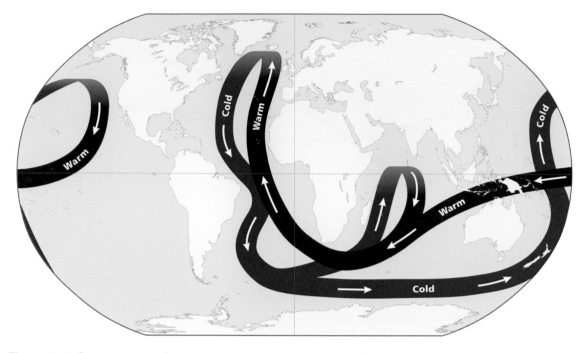

Figure 14.7 Conceptual view of the oceanic 'conveyor belt' associated with the sinking of cold, salty water in the North Atlantic Ocean. Variations in the strength of this circulation affect sea-surface temperatures and may influence the frequency of Atlantic hurricanes and the duration of drought in sub-Saharan Africa. After Broecker (1991). Reproduced with permission.

the Last Glacial Maximum, about 20,000 years ago (US National Academy of Sciences, 2002).

Most attention is on the *thermohaline ocean circulation* (THC) and its significance for the Gulf Stream in the North Atlantic Ocean (Broecker, 1991). The THC is the worldwide 'conveyor belt' that transports deep water around the world's oceans (Figure 14.7). A link with the North Atlantic Drift means that this surface current is driven by large-scale sinking of dense, saline water in the North Atlantic basin. The water here is highly saline because it consists of warm surface water (North Atlantic Drift/Gulf Stream) that has travelled far from the south, where relatively high rates of evaporation at the warm water surface increase salinity. After sinking, the deep water then flows southwards and eastwards through the Indian Ocean and wells up in the western Pacific. It completes the circuit by

returning as a surface flow westward through the Indian Ocean, before turning northwards to reach the North Atlantic again. This conveyor belt of water is of great importance in redistributing heat from the tropics to the subpolar regions. Changes to ocean currents would profoundly alter this distribution of heat, and thus the ways in which atmospheric systems operate.

It is considered that the greatest impacts would be in areas bordering the North Atlantic and in north-west Europe. This is because, due to the Gulf Stream, annual mean temperatures here are about 9°C above the global average for the latitude. Evidence exists for abrupt climatic deteriorations in the past. These range from the onset of major glacial episodes found in the geological record to the 200-year-long phase of the Little Ice Age in historic times; such downturns in climate may well be associated with changes to ocean current dynamics (Broecker, 2000; Alley, 2007).

It appears that the North Atlantic circulation has slowed down, perhaps by as much as 30 per cent, between 1957 and 2004 (Street-Perrott and Perrott, 1990; Bryden *et al.*, 2005). This is a significant change, and Lund *et al.* (2006) noted a strong negative correlation between the strength of Atlantic Ocean circulation and the occurrence of cool periods. The chances of a complete change in the state of the ocean circulation patterns, which is one doom-laden scenario occasionally promoted in the media, are limited, but smaller-scale changes may affect the climate (Broecker, 1997). For example, a reduction in the salinity and density of the North Atlantic Ocean could result from the increased rainfall expected due to the raised atmospheric greenhouse gas concentrations following AGW. In turn, this would weaken the Gulf Stream mechanism and reduce the pole-ward transfer of warm water and air.

5 Asteroid and comet impact

The rarest hazards are associated with 'super' volcanic events or the impact on Earth of a large object from space. The geological record certainly provides evidence of global catastrophes leading to 'mass extinctions' of life on Earth. The most extreme of these occurred 251 million years ago at the boundary between two of the great geological periods, the Permian and the Triassic. Fossil evidence suggests that this event killed 96 per cent of all marine species and 70 per cent of all land species (including plants, insects and vertebrate animals). The most-cited example is the *K/T extinction*, which occurred 65 million years ago, at the boundary between the Cretaceous and Tertiary periods, and which destroyed more than half the species on Earth. This event has been attributed to the impact of an extra-terrestrial object.

Although Planet Earth is constantly threatened by showers of debris from space, most approaching material burns up in the atmosphere so that only the very largest masses survive to reach the surface. These tend to be either *asteroids* (a range of solid objects varying in size from less than 1 km to about 1,000 km in diameter) or *comets* (diffuse bodies of gas and solid particles that orbit the Sun). Until recently, the chance of a large extra-terrestrial object striking Earth was deemed highly unlikely. This attitude was fostered by the limited surface evidence of previous impacts. Global geography decrees that any impactor from space has the greatest chance of hitting a marine surface, where it will leave little, if any, surface trace. Even if the object reached a land mass, the probability was – at least in the past – that it would strike an uninhabited region and the event would pass unnoticed.

Erosional processes and vegetation growth soon obscure the visible evidence of most impact craters. Indeed, it has been estimated that only 15 per cent of the Earth's surface – largely the arid areas – is suitable for the long-term retention of impact evidence. Even when impact craters were discovered, geologists tended to consider their origins in terms of more routine, earth-based processes, like volcanic activity, rather than extra-terrestrial forces. Isolated exceptions attracted only limited attention. In 1908, a stony impactor exploded in the lower atmosphere over Siberia and devastated the coniferous forest cover over a 2,000 km² area. This, like three similar events recorded later in the twentieth century, could have destroyed a major city, but there were few societal consequences and the event was largely ignored.

Asteroid and comet impact is a 'super hazard' with the greatest disaster potential of all known natural events. Risk assessments have changed, following the increased ability of telescopes to search space for *Near-Earth Objects* (NEOs), coupled with the recognition of additional fossil crater sites. According to McGuire *et al.* (2002), over 160 impact sites have now been identified and more are likely to be discovered. Only 13 per cent of these occur in a marine environment and most examples are in Scandinavia, Australia and North America, where the surface evidence is preserved in old and stable rocks. Table 14.2 lists a sample of these sites according to size and age. The most recent impactor with potential global importance formed the Zhamanshin crater about 900,000 years ago. However, scientific thinking changed with the proposal that an extra-terrestrial impact caused the massive extinction of life about 65 million years at the K/T geological boundary (Alvarez *et al.*, 1980). This theory was subsequently linked to the discovery of the Chicxulub Basin, Gulf of Mexico. Although buried beneath later sediments, this large impact crater was formed at the same time as the mass extinction, although its actual role in the extinction is controversial.

The hazards to human life and property resulting from an asteroid or meteor strike depend only partially on size. Other factors include the velocity of the body on impact, whether the strike is on land or sea and whether it occurs in a densely populated region. However, the scale of potential disaster can be estimated according to the approximate size and energy release of an impactor (Table 14.3). For collisions with asteroids between 200 m and 2 km in diameter, the most important

hazard would probably be a tsunami (Hills and Goda, 1998).

In order to create a *global-scale catastrophe* – defined by Chapman and Morrison (1994) as an event leading to the death of more than one-quarter of the world's population (>1.5 billion people) – an impactor would need to be between 0.5 and 5 km in diameter. At the upper end of this range, Toon *et al.* (1997) identified other processes which alter the composition of the atmosphere and bring about climate change: blast waves injecting dust and water into the atmosphere, soot production from burning forests, acid rain and ozone depletion. Under these conditions, most of the world's population would probably die within the ensuing months or years.

According to the UK Task Force (2000), a potential hazard exists if an NEO at least 150 m in diameter is on an orbit that will bring it within 7.5 million kilometres of the Earth. The risk of impact from comets is assessed at 10–30 per cent of that for asteroids. These hazards can be accommodated, at least partially, into conventional disaster-management strategies. Some forecasting and warning is possible. For example, a lead-time of 250–500 days between detection and impact has been estimated for long-period comets (Marsden and Steel, 1994), while the period for asteroids might extend to decades or more. NEOs could be detected earlier by the deployment of larger, wide-angled search telescopes dedicated to whole-sky observation. The NASA Spaceguard Survey, which has been operational since 1995, is concerned with NEOs greater than 1 km in size. To date, none has been found with an orbit that would lead to an impact with Earth. If such an object were located, appropriate measures would have to be taken.

There are enormous uncertainties about practical disaster reduction measures. For smaller-scale threats, it might be possible to evacuate people from the predicted impact area and from low-lying coastal zones at risk from tsunami waves. For regional and global-scale threats the only option would be to avert the collision. Some suggestions include the development of a defence

Table 14.2 Some known impact craters ranked by age – millions of years before the present (Ma)

Crater name	Country	Diameter (km)	Age (ma)
Barringer	United States	1.1	0.049
Zhamanshin	Kazakhstan	13.5	0.9
Ries	Germany	24	15
Popigai	Russia	100	35.7
Chicxulub	Mexico	170	64.98
Gosses Bluff	Australia	22	214
Manicouagan	Canada	100	290
West Clearwater	Canada	36	290
Acraman	Australia	90	>450
Kelly West	Australia	10	>550
Sudbury	Canada	250	1,850
Vredefort	South Africa	300	2,023

Source: After Grieve (1998)

Table 14.3 The likely energy release, environmental effects and possible fatality rates for different scales of extra-terrestrial impacts on Earth

Approximate scale of impact	Diameter of impactor	Energy (mt)	Frequency (years)	Environmental effects	Deaths
Tunguska-scale event	50–300 m	9–2000	250	Regional catastrophe – many deaths in devastated urban areas the size of Tokyo or New York, major tsunamis created if an ocean impact occurs	5×10^3
Large sub-global event	300–600 m	$2000–1.5 \times 10^4$	35×10^3	Regional catastrophe – land impact destroys an area the size of Estonia, large impact blast, earthquakes, regional fires over $10^4–10^5$ km², large tsunamis reach 1 km inland	3×10^5
Nominal global threshold	>1.5 km	2×10^5	5×10^5	Global catastrophe – land impact destroys an area the size of France, impact dust and soot from fires alter optical depth of atmosphere, prolonged cooling of atmosphere, possible loss of ozone shield	1.5×10^9
High global threshold	>5 km	10^7	6×10^6	Global catastrophe – land impact destroys an area the size of India, high concentrations of dust and sulphate levels reduce sunlight and photosynthesis ceases, vision becomes difficult, ecosystem destruction	1.5×10^9
Rare K/T-scale events	>10 km	10^8	10^8	Global catastrophe – land destruction approaches continental scale, major earthquakes, global fires on land, 100 m high tsunami waves reach 20 km inland, human vision ceases, all advanced life forms at risk	5×10^9

Source: Adapted from Chapman and Morrison (1994) and Toon *et al.*, (1997)

system which could target dangerous NEOs with nuclear explosives with a yield of more than 1 mt, in order to deflect, fracture or fragment the largest bodies (Simonenko *et al.*, 1994). However, fragmented particles could still pose a risk. An alternative strategy might be to fly a space vehicle alongside the object – perhaps for months or years – and use much smaller explosions, or other means, to steer the NEO into a new, safe orbit. At present, with concern for future risks dominated by anthropogenic climate change, there seems little prospect of a coordinated international effort to mitigate threats from outer space.

F CLIMATE CHANGE AND ENVIRONMENTAL HAZARDS

The Intergovernmental Panel on Climatic Change (IPCC) has shown that the Earth's surface temperature has risen over the past 150 years, albeit with some regional variations (Solomon *et al.*, 2007). The increase was in two main phases

– from 1910 to the 1940s (by 0.35°C) and, more strongly, from the 1970s to the present (by 0.55°C). Long-term trends in world precipitation were detected from 1900 onwards, with wetter conditions over North and South America, northern Europe and north and central Asia but drier conditions over the Sahel, southern Africa, the Mediterranean and southern Asia. Worldwide, there was an increase in intense precipitation events. Since 1900 the thermal expansion of sea-water, plus melt-water from land-based ice, has led to sea levels rising by about 3 mm yr^{-1}.

The IPCC has also concluded that anthropogenic global warming (AGW) influences, rather than natural climatic factors, are likely to be responsible for these trends. The concentration of CO_2 and other atmospheric greenhouse gases is higher than any other time in the last half-million years and the second half of the twentieth century was the warmest 50-year period in the northern hemisphere for some 1,300 years. Climate models using natural factors alone fail to reproduce the warming trend, but with the inclusion of rising GHG concentrations, the models replicate a spatial pattern of temperature change similar to that compiled from direct measurements. The AGW scenario is now accepted by most climate scientists, and global warming is expected to continue to the end of the present century. Some, like Trenberth (2011), believe that the evidence for AGW has become sufficiently strong to place the burden of proof on those who claim there is no human fingerprint on present climate change. Inevitably, pockets of scepticism linger in the public mind (Poortinga *et al.*, 2011; McCright and Dunlap, 2011).

There is concern that AGW will lead to more climate-related disasters. The IPCC documented observed increases in the frequency of certain extreme climatic events since about 1950 (Solomon *et al.*, 2007). For example, heat-waves became more common, as did intense precipitation events in the mid-latitudes. In some regions, drought incidence as measured by the Palmer Drought Severity Index (PDSI) also

increased. In the future, weather-related disasters may become more frequent because of changing atmospheric conditions. Higher global temperatures could create more physiological heat stress; more land evaporation might create more droughts; more atmospheric moisture could promote episodes of intense precipitation and floods; more atmospheric energy will be available for hurricanes, tornadoes and other severe storms; higher sea levels and storm surges might well lead to more coastal flooding (van Aalst, 2006).

In March 2012 the IPCC released the SREX Report, presenting evidence on possible links between extreme atmospheric events, climate change and disasters (Field *et al.*, 2012). This Special Report focused on weather and climate variables near the upper or lower statistical range of observed values, especially the adverse societal impact of such events and the strategies needed to manage potential risks from climate-related disasters in the future. The result was a much-improved understanding of climate-related hazards, not least because contributions from social scientists were included. The Report also contained important evidence on the confidence levels that can be attached to observed and projected changes to those atmospheric extremes with clear hazard implications.

However, the Report did acknowledge that many extreme atmospheric events are not, in themselves, severe enough to create disasters. Although poor people are more likely than others to be affected by some lesser extremes, the rather generalized definition of disaster adopted in the Report ('widespread adverse human, material, economic or environmental effects') falls short of the criteria and impact thresholds used in this book. The Report did not attempt to relate extreme events to conventional disaster thresholds for mortality or material loss, although much attention was given to economic impacts. As a result, only limited associations are likely to exist between many of the extreme events discussed in the Report and the content of humanitarian-based disaster archives such as EM-DAT. It is also important to remember that extreme climate events remain physical attributes, on whatever statistical scale is employed, whilst disasters are complex socio-economic responses to and consequences of the extremes.

1 Observed changes to extremes

An understanding of extreme climate events relies on the statistical interpretation of values that fall close to the upper or lower end of the observed range of variation. Confidence in the analysis, including the detection of trends, depends on the *quality of the data* (consistency of observation, reliability of reporting) and on the *quantity of the data* (sample size and areal coverage). By definition, extremes are rare events, so probability statistics are not well established. The nature of the data changes over the years, between seasons, across regions and will also differ for individual extremes. For example, air temperature measurements are more numerous than river flow data; there is better information on tropical cyclones than on droughts. Small-scale hazards like tornadoes and hail-storms have very poor sampling networks.

Therefore, the evidence is often localized in space, whilst short datasets invite confusion between variations and trends. For example, tropical cyclones – and their disaster impacts – are highly variable from year to year, largely depending on where and when a damaging storm makes landfall. The level of scientific understanding across different hazards inevitably differs. So far, few researchers have attempted a multi-decadal analysis of some severe weather extremes and the related disasters. Many recorded extremes will fail to breach the disaster thresholds adopted in this book. When more extreme events do occur in the future, they will remain infrequent and difficult to assess in terms of disaster significance, partly due to a lack of knowledge regarding physical processes but more likely because of on-going socio-economic change.

2 Projected changes to extremes

Many different climate simulation models are used to explain observed extreme events and to estimate future trends. These models can be grouped into either *Global Circulations Models* (GCMs) or the *Regional Climate Models* (RCMs) used to down-scale the GCM outputs. The resulting simulations depend on the specific inputs to the models, such as physical quantities, vertical atmospheric levels and the thresholds set. In addition, model confidence varies according to the spatial and temporal scales of resolution and the type of extreme phenomenon involved. The major deficiency in all models is an inability to represent fully some small-scale atmospheric processes such as clouds, convective systems and the interactions between the atmosphere and land or ocean surfaces, including key feedback processes. In general, GCMs struggle to represent hazardous phenomena like tropical cyclones and tornadoes, whilst all models tend to simulate changes in extreme temperatures better than trends in intense precipitation (Field *et al.*, 2012).

Unsurprisingly, differing results emerge. But most model outputs indicate an increase in maximum daily temperatures and in intense falls of daily precipitation by the end of twenty-first century. There is consensus, amounting to medium confidence, that droughts will intensify in some seasons and areas and that locations subject to high sea levels are very likely to experience a continuation of this problem. Complexity within the global climate system and lack of agreement between some scientists reduces the confidence levels for most other projections. In addition, all estimates of future extremes are to some extent dependent on assumptions regarding greenhouse gas emissions in coming years. It is important to state here that, given the present state of knowledge, it is not possible to attribute any extreme climatic event – or weather-related disaster – directly to greenhouse warming. Clearly, further progress in understanding depends on the repetition of past experiments as the data and the models improve over time.

The reliability of simulation by the models and the degree of model agreement are crucial for assessing the confidence to be placed on future scenarios. In a number of cases, projections of extreme events suggest trends which are small, compared to natural variability. For some extremes even the direction of change (increase or decrease) is unclear. For example, opinions vary about possible changes to ENSO due to global warming. The frequency and intensity of ENSO events has tended to increase; during the second half of the twentieth century, 12 El Niño events were recorded (1951, 1953, 1957–58, 1963, 1965, 1969, 1972, 1976–77, 1982–83, 1986–87, 1990–95, 1997–98). The four strongest, and also the four longest, happened since 1980. Yet Collins *et al.* (2005) found little evidence of a net change in El Niño dynamics as a result of global warming.

3 Extreme climate events and disasters

It is crucial to appreciate that extreme events alone may trigger, but do not create, disasters. Figure 14.8 shows how extreme events interact with, and can be modulated by, societal vulnerability and disaster-reduction measures. It is unlikely that these factors will remain constant; indeed, they may change rapidly over a few years for some locations. Without any change in climatic events, increasing human vulnerability in an area will amplify the risk of disaster; equally, the implementation of disaster reduction measures will restrict losses. Human intervention and societal change are most relevant for risk scenarios of hydro-meteorological extremes such as floods. Even where climate trends – either observed or projected – indicate more intense precipitation episodes, these are unlikely to increase flood disasters if better land planning or more flood protection engineering takes place. Equally, further invasion of floodplains or coastal

areas is likely to lead to greater losses from climatic extremes.

Accurate economic information and reliable damage estimates are not always available, but climate-related disasters appear to have become more expensive in recent decades. The reasons are controversial. For the most part, the data show how physical factors – like extreme climatic events – can be overridden by societal changes. EM-DAT information suggests that deaths due to natural disasters – including climate-related events – are either flat-lining or in decline – despite the occurrence of more frequent extreme events and increases in the world population. This is consistent with the introduction of disaster reduction measures, especially those that save human lives by means of early warning and emergency evacuation. In contrast, it is less easy to move infrastructure and other fixed assets out of harm's way.

It seems plausible that the growing invasion of hazard zones, combined with greater wealth and

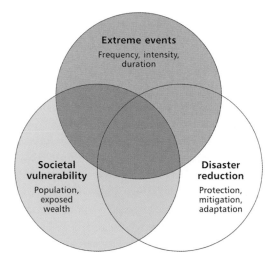

Figure 14.8 Diagrammatic view of the interactions between extreme events, societal vulnerability and disaster reduction measures. Extreme events alone do not determine disaster outcomes. Societal vulnerability tends to increase the hazard impact of extreme events; disaster-reduction measures tend to decrease the hazard impact of extreme events.

other assets, is the principal cause of any rise in disaster losses. Although most studies reveal a trend to higher monetary losses over time, no rise attributable to climate change has been found when the data are normalized for concurrent increases in the exposed population and the value of capital assets at risk (Bouwer, 2011). Of course, it is possible that clear links between extreme climate-related trends and disaster losses will eventually emerge, but it may take some time. For example, Crompton *et al.* (2011) concluded that it would take at least 120 years before a clear climate change signal emerged in the time-series of normalized US tropical cyclone losses. Similar time-scales might well apply to other weather-related disaster losses.

4 Climate-related disasters in the future

Table 14.4 lists the climate extremes of special concern due to their potential for increased disaster, based on the confidence levels attached by the IPCC to both post-1950 observations and projected future trends. Where world regions are specified, analysis is limited to the land areas with sufficient data. In general, trends with the higher confidence levels (such as heat-waves, coastal high water) are those which accord with expectations about climate change that results from global warming (AGW). Even so, no single extreme event can yet be directly attributed to AGW. The evidence for rises in other hazards is less

Table 14.4 Generalized changes to selected climatic extremes observed since 1950 and projected to 2100 that suggest increased potential for disaster

Extreme climate element or hazard	Type of change observed or expected	Main areas and regions affected	Confidence level for both observations and projections
Warm spells (heat waves)	Increase – changes in daily temperatures	Global	Very likely to increase
Intense precipitation events	Disproportionate increase – increased flash floods and major floods in mid/high latitudes	Mid-latitudes	Likely but with strong regional variations; limited to medium confidence in more floods
Drought (seasonal/year)	Increase in total areas affected	Many land regions	Medium confidence for some world regions
Tropical cyclones	Increase in storm lifetimes and Category 4–5 storms, impact exacerbated by sea level rise	Tropics	Likely increase in maximum wind speed but low confidence in changed long-term activity
Extreme extra-tropical storms	Increase in frequency and intensity; poleward shift in track	Northern hemisphere land areas	Likely poleward shift in main tracks in both hemispheres
Wildfires (links to drought)	Increase in frequency and intensity	Areas where drought occurs	Likely, but closely linked to drought episodes
High sea levels	Increase for most oceans, highest levels in coastal waters related to storm surge	Low-lying islands, estuaries and coasts	Very likely; high confidence that adverse effects will increase

Source: Adapted from data in Solomon *et al.* (2007) and Field *et al.* (2012)

Note: Confidence levels:

- Very likely – indicates a probability of 90–100 per cent
- Likely – indicates probability of 66–90 per cent
- Medium confidence – indicates direction of change without assessing the likelihood

secure. For example, low confidence is attached to trend identification for river floods, due to the scattered nature of the gauging network and the complicating effects of control works and land use changes in individual catchments. The data sample for very small-scale hazards (such as tornadoes and hail-storms) is too limited to permit analysis.

In brief, it remains difficult to harmonize climate change studies and detect reliable trends in extreme events, due to differences in investigation methods and the variety of datasets used. Further uncertainty is added when climate data is coupled with different *Storm Loss Models* used to estimate future economic damage. All modelling of future conditions is also dependent on the natural variability of the climate and assumptions about greenhouse gas emissions. Bearing these pitfalls in mind, the following sections provide some examples of how extreme weather events with disaster potential have changed over the last 50 years or so and how they may change in future. Where possible, assessments are made regarding the degree of confidence that can be attached to such estimates and the level of disaster risk involved.

5 Heat stress and other health risks

There is much interest in the effects of climate change on human health, notably the impact of extreme heat-waves (see Chapter 10, pp. 271–3). Existing heat-related morbidity and mortality is quite well documented; an estimated 1,000 deaths per year already occur in Australia, due to heat stress and related illness. The IPCC has high confidence that there will be higher maximum temperatures, with a much-increased frequency of heat-waves, by the end of the twenty-first century. Some researchers have warned of a doubling of mortality worldwide by 2020.

There is already evidence of a trend to more frequent heat-waves from several parts of the globe. Grundstein and Dowd (2011) used a thermal-based bio index to show a clear increase

between 1949 and 2010 across the USA in heat hazard days due to extreme temperatures. Following the 2003 heat-wave disaster in Europe, Schär *et al.* (2004) concluded that, despite an overall warming trend, this event was statistically unusual and therefore demonstrated more year-to-year variability in the European summer climate, introduced by greenhouse-gas forcing. Despite statistical complications arising from changes to temperature variability as well as mean values, there is agreement that the main health hazard comes from multi-day heat-waves and that these events can be reliably simulated by existing regional climate models (Fischer and Schär, 2010). The results indicate that the frequency of heat-wave days across Iberia and the Mediterranean region will increase from an average of 2 days per summer in 1961–90 to about 40 days per summer by 2071–2100.

Actual outcomes will be modified by societal factors. Risks are high for elderly people living in urban areas. This is not good news for Europe, which has the oldest population worldwide, with a median age expected to reach 47 years by 2050. Ballester *et al.* (2011) examined daily deaths in almost 200 areas of Europe and predicted that, by 2050, increased heat stress-related summer deaths would start to balance the reduction in existing deaths due to cold weather in the winter. For example, an exceptional summer heat-wave could cause up to 6,000 excess deaths (Kovats, 2008). Thus, total temperature-related mortality in Europe might well be lower at the end of the century if society adapted well to the changing climatic conditions. This would involve risk reduction for vulnerable sectors of the population and increasing access to air conditioning.

Rising temperatures will encourage the poleward spread of some important vector-borne diseases. Within the tropics, the extension of warm, humid seasons into higher altitudes could spread seasonal epidemics of malaria into new areas where the population has little or no immunity to the disease and where health care is limited (Martens *et al.*, 1998). The transfer of malaria from mosquitoes to humans is highly dependent on temperature and is most effective within the range 15–32°C with a relative humidity of 50–60 per cent (Weihe and Mertens, 1991). The largest impacts are expected at the margins of the present-day risk areas.

As in other cases, the accuracy of these estimates depends on the performance of the climate models, assumptions about future greenhouse gas emissions and societal change. Figure 14.9 shows the potential spread of *P. falciparum* malaria, which is clinically more dangerous than the more widely distributed *vivax* form of the disease, beyond the current limits of latitudes 50° north and south. Under the worst-case scenario, an additional 290 million people worldwide could be at risk by the 2080s, with the greatest increases in risk within China and central Asia as well as the eastern USA and Europe. Malaria is unlikely to take hold in the MDCs with effective health services but, as early as 2100, 60 per cent of the world population will be at risk unless precautionary health programmes are adopted.

6 Precipitation and floods

There is evidence of increased precipitation totals and more intense rainfall trends over certain regions during the twentieth century. Broadly speaking, precipitation totals have increased by 10–40 per cent in Europe and by about 10 per cent over the USA. Zhang *et al.* (2007) did attribute increased precipitation intensities to the effects of climate change following analysis of observed changes relative to the results from 14 different climate simulations. However, some regional patterns are not spatially uniform; a study of extreme rainfall trends in south-west India found large intra-regional differences (Pal and Al-Tabba, 2009).

It is not a straightforward task to translate trends of more precipitation into higher flood risks, although the incidence of floods has been related to past changes in the climate system. According to Knox (2000), geological evidence

suggests that the magnitude and frequency of floods shows a sensitivity to past climate changes smaller than those expected in the twenty-first century. Floods respond to changes in the atmospheric circulation, and large annual floods became more frequent after 1950 on the upper Mississippi river, USA (Figure 14.10). During these years, the upper westerly air circulation over the Midwest had a relatively strong meridional component leading to marked north–south exchanges of air masses and higher rainfall. Similarly, Petrow *et al.* (2009) indicated the extent to which a variation in air circulation patterns over Germany enhanced flood hazards during winter.

In the UK, climate scenarios envisage a five-fold increase in winter precipitation during the next 100 years and the insurance industry anticipates a doubling of the existing flood risk during this season alone (ABI, 2004). It is generally agreed that existing climate models lack the capability to link individual damaging floods explicitly to AGW, although Pall *et al.* (2011) deployed exceptional computing power to investigate widespread floods during 2000 in England and Wales. Simulations of the extreme river flows indicated that, in nine out of 10 cases, greenhouse gas emissions increased the risk of floods occurring in autumn 2000 by more than 20, and in two out of three cases by more than 90 per cent.

The tropics are vulnerable to future hydroclimatic changes mainly because of many large, unregulated rivers. For example, over 90 per cent of the drainage basin of the Ganges–Meghna–Brahmaputra river system lies upstream, beyond the control of Bangladesh. The flood regime may already be changing, since the annual area of Bangladesh flooded during 1980–99 was larger, and more variable, than during the period 1960–80. Climate change scenarios indicate substantial future increases in the peak discharges of the contributing rivers (Mirza, 2002).

Various studies anticipate increases in flood events and flood losses. Once again, the outcomes will be heavily influenced by societal change. For example, there is concern about future flood

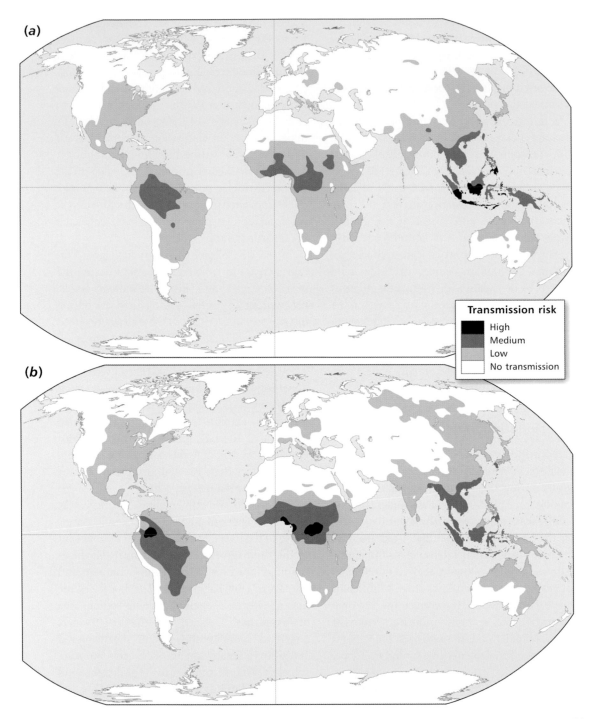

Figure 14.9 Potential spread of malaria (*P. falciparum*) due to climate change: (a) baseline conditions 1961–90; (b) scenario of risk areas estimated for the 2050s. Most of the increased risk is expected to take place near present-day margins. After Martens *et al.* (1998).

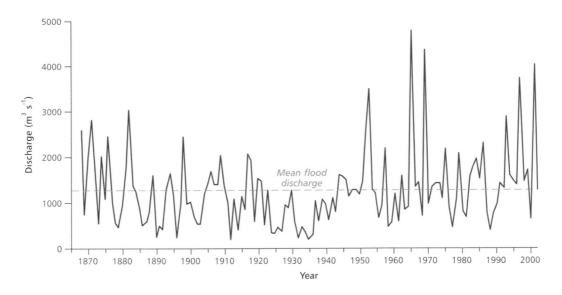

Figure 14.10 Annual maximum flood series for the Mississippi river at St Paul 1893–2002. Large floods were a feature during the late nineteenth century, and more especially since 1950. Adapted from US Geological Survey at http://www.waterdate.usgs.gov/mn/nwis//peak (accessed 1 August 2003).

risk in the deltaic areas of the Netherlands, due to a combination of population growth, economic development, land subsidence and sea-level rise. According to de Moel *et al.* (2011), the total amount of urban land at risk from sea or river floods increased six-fold during the twentieth century and may double again during the twenty-first century. In a polder area case study, Bouwer *et al.* (2010) estimated that, if no additional flood prevention measures were taken, annual economic losses arising from a combination of climate and societal change might well increase by as much as 96–719 per cent between 2000 and 2040. Most of the projected increase in costs was explained by expected growth in the asset values at risk.

7 Drought and wildfires

Trends in drought events and disasters are perhaps more difficult to identify. Droughts are tied to persistent anomalies in the atmospheric circulation, most likely triggered by changes in

tropical sea-surface temperatures (SSTs). There are strong links to El Niño–La Niña conditions, together with some association to the Asian monsoon system, and the interaction of these drivers is complex. Dai (2011) drew attention to increased global aridity since the 1970s; he concluded that recent warming has increased evaporation demands and probably altered atmospheric circulation patterns to produce an overall drying effect. For example, winter droughts in the Mediterranean region have become more common since the 1970s (Figure 14.11). Ten of the driest winters since 1902 have taken place in the last twenty years and climate modelling has indicated that roughly half of the switch to aridity could be attributed to climate change from greenhouse gases (Hoerling *et al.*, 2012).

Climate models also suggest increased aridity during the twenty-first century over much of Africa, southern Europe, the Middle East, most of the Americas, south-east Asia and Australia (Dai, 2011). Simulations for the south-western USA show dry events persisting for periods of 12 years

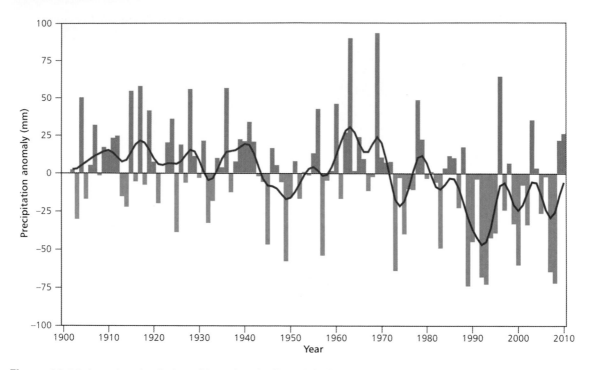

Figure 14.11 Annual totals of winter (November–April) precipitation in the Mediterranean region 1902–2010. The solid curve is the smoothed time-series using a 9-pt Gaussian filter. A distinct downturn in rainfall is evident since about 1970. After Hoerling *et al.* (2012). Reproduced from M. Hoerling *et al.*, On the increased frequency of Mediterranean drought. *Journal of Climate* 25 (2012): 2146–61. © American Meteorological Society. Reprinted with permission.

or more, a situation exacerbated by globally warmer temperatures that reduce spring snow-packs and soil-moisture levels later in the year (Cayan *et al.*, 2010). Climate change may also disturb conventional drought links such as the NAO. Folland *et al.* (2009) have predicted a more positive NAO Index in future, which is likely to increase the risk of summer drought in north-western Europe.

Increased aridity will have many hazard consequences, including those for wildfire. In the pre-industrial era, wildfires were driven mainly by precipitation patterns, before a shift to the current human-led regime. In turn, this regime is likely to be dominated by rising temperatures and very high fire risks in the twenty-first century (Pechony and Shindell, 2010). The largest burned-out areas are associated with temperatures above 28°C and

drought conditions, but South America is at particular risk because the extent of land clearance here outweighs the climatic drivers (Aldersley *et al.*, 2011). Overall, GCM simulations suggest that wildfire potential will grow significantly in many parts of the globe, including the USA, South America, central Asia, southern Europe, southern Africa and Australia (Liu *et al.*, 2010).

Other impacts are expected in the arid and semi-arid regions of the tropics, home to about 350 million people, many of whom live in poverty. For example, the number of people living in countries using more than 20 per cent of available water resources – a common indicator of water stress – will rise to over 5 billion by 2025, largely due to population growth. Adaptation to greater aridity will impose burdens on countries like Ethiopia which are already struggling to cope

(Conway and Schipper, 2011). Risk profiles for diseases will also change. For example, Lloyd *et al.* (2007) noted that the fatality rate from diarrhoea was highest during droughts, probably due to an increased use of unprotected water sources and poorer hygiene practices.

8 Tropical cyclones

A steep rise in the frequency and intensity of major (Category 3, 4 and 5) Atlantic hurricanes land-falling in the USA during the late twentieth and early twenty-first centuries has initiated debate about the possible effects of global warming. The 2005 Atlantic hurricane season was very active, with 26 named storms, five hurricanes and three tropical storms. Theoretically, rising sea-surface temperatures, together with increased water vapour in the lower troposphere, are likely to fuel greater storm activity, but the evidence is controversial. Indeed the subject is one of the most contested issues in climate science.

The debate began when Emanuel (1987 and 2000) theorized that a rise in global sea-surface temperatures could result in a 10–20 per cent increase in tropical cyclone wind speeds and a much larger rise in damage potential. An upward trend in hurricane activity since the mid-1970s was then identified (Emanuel, 2005). Webster *et al.* (2005) reported a decadal rise in Category 4 and 5 storms in most of the world's oceans and Hoyos *et al.* (2006) linked the activity to an upward trend in SSTs. The rise in storm activity is surprising, given the relatively small change in SSTs. To some extent, it also conflicts with previously held views that Atlantic hurricanes come in cyclical patterns of active and inactive spells linked to the El Niño Southern Oscillation (ENSO) and the North Atlantic Oscillation (NAO). As a result, Emanuel's hypothesis was rejected by other workers (Elsner *et al.*, 2000; Pielke *et al.*, 2005; Landsea, 2007).

More research is required because, although AGW may ultimately raise hurricane activity, the case is not yet proven (Shepherd and Knutson, 2007). Elsner (2003) and Trenberth (2005) have emphasized the large variability of hurricanes on multi-decadal timescales and suggested that any trends are likely to be small. Landsea (2005) also questioned the validity of the database and claimed that much longer time-series than those used so far did not display any trend related to global warming. This view was supported by Nyberg *et al.* (2007), who used proxy records to show that the frequency of major Atlantic hurricanes decreased progressively from the eighteenth century to reach anomalously low values in the 1970s and 1980s. In this context, the increased activity observed since 1995 can be interpreted as a return to normal hurricane patterns, rather than a product of anthropogenic climate change.

9 Extra-tropical storms

The winter storms that hit western Europe have also prompted questions about possible relationships with AGW. Storminess in the north-east North Atlantic region is closely tied to natural climate variability, often linked to the NAO pressure gradient. Wang *et al.* (2009) found substantial decadal-scale fluctuations between 1874 and 2007 in the North Sea region, with a maximum of winter storm activity in the early 1990s. This general pattern was confirmed for severe autumn and winter storms affecting Britain over an 85-year period and it was suggested that the high variability might obscure any long-term anthropogenic signal (Allan *et al.*, 2009).

However, there are some factors that may be relevant to future storm activity. Dong *et al.* (2011) demonstrated that NAO behaviour changed in the mid-1970s, when a pole-ward and east-ward shift in inter-annual variability took place, a shift that is consistent with the response from climate models attempting to simulate the increase in atmospheric concentrations of greenhouse gases (Figure 14.12). Despite a general pole-ward shift in storm tracks, the path in the north-east Atlantic is shifted south in winter, giving more storms and an increased frequency of strong winds over Britain, linked to a south-ward move of the jet stream in response to ocean warming in the central part of the North Atlantic Ocean (McDonald, 2011; Ulbrich *et al.*, 2009). Climate simulations project more extreme wind speeds over a large region stretching from the UK into northern France to Denmark, northern Germany and eastern Europe. Researchers have coupled wind speed outputs from GCM and RCM simulations to storm-loss models, in order to estimate future losses to the end of the twenty-first century. The results show a disproportionate increase in losses for the most extreme events and some geographical variability. For example, Donat *et al.* (2011) envisage a 25 per cent increase in losses for Germany. In a Europe-wide study, Schierz *et al.* (2010) claimed an average increase

in insurance losses of 104 per cent, with Denmark and Germany experiencing the largest increases (116 and 114 per cent, respectively).

10 Sea-level rise and coastal hazards

Further rises in sea level are one of the most certain outcomes of global warming. Mean global sea level has been rising since the late nineteenth century and the process is expected to accelerate during the twenty-first century. This will bring extra risks of marine flooding and coastal erosion to many areas. According to McGranahan *et al.* (2007), the coastal zone less than 10 m above sea level contains 10 per cent of the world's population. There is a higher share of exposure to

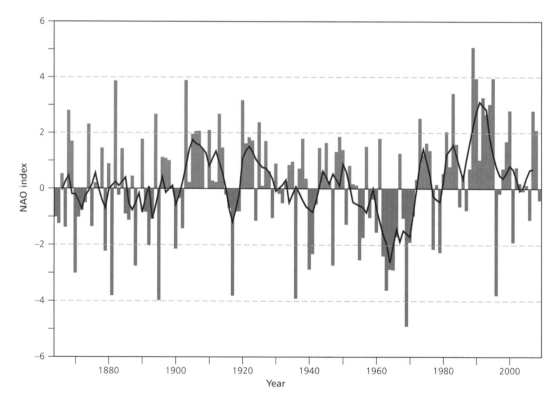

Figure 14.12 Annual variations in the NAO Index 1870–2010. Strongly positive Index values have been recorded since about 1970. After http://upload.wikimedia.org/wikipedia/commons/8/87/Winter-NAO Reproduced under Creative Commons Attribution-Share Alike 3.0 Unported license.

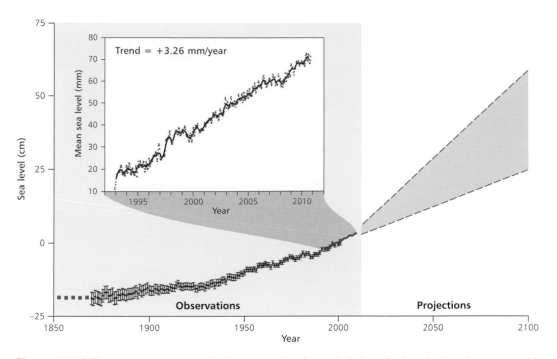

Figure 14.13 The progressive rise in global mean sea level recorded since the late nineteenth century, with the range of projected increases up to 2100. The red curve is based on tide-gauge data. The black curve is the altimetry record 1993–2009 with detailed inset trend. The shaded light blue zone represents the range of IPCC AR4 projections for the A1F1 greenhouse gas emission scenario. After Nicholls and Cazenave (2010). Reproduced from R. Nicholls and A. Cazenave, Sea-level rise and its impact on coastal zones. *Science* (June 2010): 1517–20. Reprinted with permission from AAAS.

flooding in the LDCs, not least in the coastal mega-cities, whilst the greatest relative risks are concentrated in small-island nations.

In the latter half of the twentieth century, the average rate of rate of sea-level rise (SLR) was calculated at 1.8 ± 0.5 mm yr^{-1}. More accurate measurements are available from high-precision altimeter satellites since the early 1990s. The data from 1993 to 2009 indicate a faster rate of SLR of 3.3 ± 0.4 mm yr^{-1}, suggesting that an accelerated rise is already taking place (Figure 14.13). IPCC (2007) projections were for increases above the 1990 level of about 0.2–0.6 m by the 2080s, but thermal expansion of sea-water coupled with a more rapid reduction of polar ice could produce future SLR of 1 m or more by 2100 (Nicholls and Cazenave, 2010).

Future sea-level estimates are subject to qualifications. For example, local land subsidence will increase risk in some areas, whereas further dam building along major rivers, holding back surface flows, could lessen the rate of rise. Disaster risk assessments depend not only on sea levels but on assumptions made about the coastal zones, such as the expected frequency of storm surges and future standards of coastal protection. If the incidence of storms is assumed to remain constant and the population at risk is assumed to grow at twice the national rate, in line with recent experience, most uncertainty lies in the assumptions made about sea defences. Without adaptive responses – such as improved sea defences or managed retreat from the shoreline – it appears that, by the 2080s, the number of people flooded by storm surge in a typical year could increase by five times over the 1990 total (Nicholls *et al.*, 1999).

Nicholls (1998) presented two scenarios:

- *constant protection*: where no changes are assumed from 1990 levels
- *evolving protection*: where sea defences are upgraded in line with projected increases in economic growth measured by GDP. This latter case mimics historical development but makes no extra allowance for future sea-level rise.

Table 14.5 lists the five world regions which contain more than 90 per cent of all potential flood victims. Given the above assumptions, it can be seen that, with *constant protection*, the number of people annually at risk increases from just over 7 million in 1990 to about 220 million in the 2080s; with *evolving protection* from floods the total number rises to less than 100 million.

The most serious impacts will be in low-lying coastal zones with high population densities, such as the delta areas of Egypt and Bangladesh. Egypt could lose 2 million ha of fertile land, displacing 8–10 million people; in Guyana, South America, a 1 m rise in sea level would displace 80 per cent of the population – about 600,000

people (IFRCRCS, 2002). These zones are expensive to protect because of the long shoreline and the need for on-going management of the waters held behind the coastal barriers. Major threats are faced by the small-island nations (the Maldives, Indian Ocean and Fiji, South Pacific Ocean), with villages located at sea level. Tuvalu is a country with 10,000 people, occupying a string of coral atolls only a few metres above sea level. People depend on rain-fed drinking water and a narrow range of primary products, but poor-quality agricultural land and shallow ground-water supplies are suffering salt contamination, due to higher sea levels. There is no money for engineered coastal defences, and managed retreat from the shoreline is impossible. Without an income stream to fund cyclone shelters and secure water storages, there appear to be no options other than relocation to other countries, such as Australia or New Zealand (Campbell *et al.*, 2005).

Even greater risks are possible in the future. Sea levels during previous interglacial periods have been up to 20 m higher than at present. If the Earth's climate continues to warm, polar ice sheets will decrease. For example, complete melting of the Greenland and West Antarctic ice sheets would result in a global sea-level rise of more than 10 m. Today, this would flood about one-quarter of the entire US population, mainly residents of the Gulf and East Coast states. Polar temperatures by the year 2100 may reach values similar to those 130,000–127,000 years before the present, when sea levels were several metres above present heights. A sea-level rise in excess of 5 m is possible within the next millennium (Overpeck *et al.*, 2006).

There is concern about the stability of the West Antarctic Ice Sheet (WAIS), which contains an estimated 25 million km^3 of water or about 70 per cent of the planet's freshwater reserves. Large parts of this ice mass rest on land below sea level, and slope inland, while its edges flow outwards to reach the coast and become floating ice shelves. In theory, this feature makes the ice unstable and prone to rapid disintegration. There is

Table 14.5 The world regions most vulnerable to coastal flooding due to future sea-level rise

Region	Average annual number of people flooded (millions)		
	1990	2080s	
		Constant protection	Evolving protection
South Mediterranean	0.2	13	6
West Africa	0.4	36	3
East Africa	0.6	33	5
South Asia	4.3	98	55
South-East Asia	1.7	43	21

Source: After Nicholls *et al.* (1999)

Plate 14.2 Male residents of Fogafale, Tuvalu in the Pacific Ocean reinforce their small protective wall with coral blocks. The wall is built on the ocean side of a coral atoll in an attempt to protect the property against high spring tides and storm surge. (Photo: Panos/Jocelyn Carlin JCR00072TUV)

considerable disagreement about how far the ice sheet has already thinned since the peak of the Earth's last major ice age 20,000 years ago and the pace of current melting. But most mass balance estimates suggest that ice loss is accelerating and that melt-water, rather than thermal expansion of oceanic water, will be the main contributor to sea-level rise in the twenty-first century. Model calculations suggest that partial disintegration of the WAIS could raise sea levels by about 3 m; complete melting would contribute 4.8 m to global sea level. As with so many environmental hazards, the future behaviour of the ice sheet is largely unknown and presents an uncertain threat.

FURTHER READING

Bouwer, L.M. (2011) Have disaster losses increased due to anthropogenic climate change? *Bulletin of the American Meteorological Society* 92: 39–46. A final cautionary note about the dangers of misinterpreting disaster data.

Huppert, H.E and Sparks, S.J. (2006) Extreme natural hazards: population growth, globalization and environmental change. *Philosophical Transactions of the Royal Society A: Mathematical, Physical and Engineering Sciences* 364: 1875–88. An impressive sweep through the key issues.

IPCC (2012) *Managing the Risk of Extreme Events and Disasters to Advance Climate Change Adaptation (SREX)* Special Report compiled by Working Groups I and II. The best indicator of future climate-related hazards currently available.

Knutson, T.R. *et al.* (2010) Tropical cyclones and climate change. *Nature Geoscience* 3: 152–63. This is a magisterial review of the science relating to this key hazard.

Trenberth, K.E. (2011) Attribution of climate variations and trends to human influences and natural variability. *WIRES Climate Change* 2: 925–30. A direct challenge to the climate-change doubters.

WEB LINKS

Intergovernmental Panel on Climate Change http://www.ipcc.*org*

World Meteorological Organization http://www.wmo.ch/pages/index_en.html

International Strategy for Disaster Reduction http://www.unisdr.org/

NOAA El Niño page http://www.elnino.noaa.gov/

NOAA Space Environment Center (SEC) www.sec.noaa.gov

National Oceanographic Data Center natural hazards page http://www.nodc.noaa.gov/General/Ocean themes/hazards.html

The Spaceguard Foundation http://spaceguard.esa.int/

The Earth Impact database http://www.unb.ca/passc/ImpactDatabase/index.html

Bibliography

Publications by more than five people are listed by lead author only.

Abersten, L. (1984) Diversion of a lava flow from its natural bed to an artificial channel with the aid of explosives: Etna, 1983. *Bulletin of Volcanology* 47: 1165–74.

ABI (Association of British Insurers) (2004) *Review of Planning Policy Guidance Note 25: Development and Flood Risk*. Consultation Response. ABI, London.

Abraham, J. *et al.* (2000) *Windstorms Lothar and Martin, December 26–28, 1999*. Risk Management Solutions, Newark, CA.

Abrahams, J. (2001) Disaster management in Australia: the national emergency management system. *Emergency Medicine* 13: 165–73.

Adams, J. (1995) *Risk*. UCL Press, London.

Adams, J., Maslin, M. and Thomas, E. (1999) Sudden climate transitions during the Quaternary. *Progress in Physical Geography* 23: 1–36.

Adams, W.C. (1986) Whose lives count? TV coverage of natural disasters. *Journal of Communication* 36: 113–22.

Adams, W.M. (1993) Indigenous use of wetlands and sustainable development in West Africa. *Geographical Journal* 159: 209–18.

Adger, W.N. (2006) Vulnerability. *Global Environmental Change* 16: 268–81.

Adger, W.N. and Brooks, N. (2003) Does global environmental change cause vulnerability to disaster? In Pelling, M. (ed.) *Natural Disasters and Development in a Globalizing World*, Routledge, London and New York, 19–42.

Adger, W.N. and Brown, K. (2010) Progress in global environmental change. *Global Environmental Change* 20: 547–9.

Adger, W.N., Hughes, T.P., Folke, C., Carpenter, S.R. and Rockström, J. (2005) Social-ecological resilience to coastal disasters. *Science* 309: 1036–9.

Ahern, M., Kovats, R.S., Wilkinson, P., Few, R. and Matthies, F. (2005) Global health impacts of floods: epidemiologic evidence. *Epidemiologic Reviews* 27: 36–46.

Ahmadizadeh, M. and Shakib, H. (2004) On the December 26 2003 South-Eastern Iran earthquake in Bam region. *Engineering Structures* 26: 1055–70.

Aini, M.S., Fakhru'l-Razi, A., Daud, M., Adam, N.M. and Kadir, R.A. (2005) Analysis of royal inquiry report on the collapse of a building in Kuala Lumpur. *Disaster Prevention and Management* 14: 55–79.

Akbari, M.E., Farshad, A.A. and Asadi-Lari, M. (2004) The devastation of Bam: an overview of health issues one month after the earthquake. *Public Health* 18: 403–8.

Aldersley, A., Murray, S.J. and Cornell, S.E. (2011) Global and regional analysis of climate and human drivers of wildfire. *Science of the Total Environment* 409: 3472–81.

Allan, R., Lindesay, J. and Parker, D. (1996) *The El Niño–Southern Oscillation and Climatic Vulnerability.* CSIRO Publishing, Collingwood, Victoria.

Allan, R., Tett, S. and Alexander, L. (2009) Fluctuations in autumn–winter severe storms over the British Isles 1920 to present. *International Journal of Climatology* 29: 357–71.

Alley, R.B. (2007) Wally was right: predictive ability of the North Atlantic 'conveyor belt' hypothesis for abrupt climate change. *Annual Review of Earth and Planetary Sciences* 35: 241–72.

Alley, W.M. (1984) The Palmer drought severity index: limitations and assumptions. *Journal of Climatology and Applied Meteorology* 23: 1100–9.

Alvarez, L., Alvarez, W., Asaro, F. and Michel, H.V. (1980) Extra-terrestrial cause for the Cretaceous–Tertiary extinction. *Science* 208: 1095–1108.

Angeli, M-G., Gasparetto, P., Menotti, R.M., Pasuto, A. and Silvano, S. (1994) A system of monitoring and warning in a complex landslide in North-Eastern Italy. *Landslide News* 8: 12–15.

Anonymous (2007a) *Natural Disaster Preparedness and Education for Sustainable Development.* UNESCO, Bangkok.

Anonymous (2007b) *The Role of Land-use Planning in Flood Management.* WMO/GWP Associated Programme on Flood Management, World Meteorological Organization, Geneva.

Anonymous (2009) *Flooding in England: A National Assessment of Flood Risk.* Environment Agency, Bristol.

Antonioni, G., Bonvinici, S., Spadoni, G. and Cozzani, V. (2009) Development of a framework for the risk assessment of NaTech accidental events. *Reliability Engineering and System Safety* 94: 1442–50.

Araña, V., Felpeto, A., Astiz, M., Ga_ia, A., Ortiz, R. and Abella, R. (2000) Zonation of the main volcanic hazards (lava flows and ashfall) in Tenerife, Canary Islands: a proposal for a surveillance network. *Journal of Volcanology and Geothermal Research* 103: 377–91.

Araya, A. and Stroosnijder, L. (2011) Assessing drought risk and irrigation need in northern Ethiopia. *Agricultural and Forest Meteorology* 151: 425–36.

Artunduaga, A.D.H. and Jiménez, G.P.C. (1997) Third version of the hazard map of Galeras volcano, Colombia. *Journal of Volcanology and Geothermal Research* 77: 89–100.

Associated Programme on Flood Management (2008) *Urban Flood Risk Management: A Tool for Integrated Flood Management.* AFPM Technical Document No, 11 – Flood Management Tools Series, World Meteorological Organization, Geneva.

Audru, J-C. *et al.* (2010) Major natural hazards in a tropical volcanic island: a review for Mayotte Island, Comoros Archipelago, Indian Ocean. *Engineering Geology* 114: 364–81.

Autier, P., D'Altilia, J-P., Delamalle, J-P. and Vercruysse, V. (1989) The food and nutrition surveillance systems of Chad and Mali: the 'SAP' after two years. *Disasters* 13: 9–32.

Autier, P. *et al.* (1990) Drugs supply in the aftermath of the 1988 Armenian earthquake. *The Lancet* (9 June): 1388–90.

Ayscue, J.K. (1996) Hurricane damage to residential structures: risk and mitigation. *Working Paper 94.* Natural Hazards Research and Applications Information Center, Boulder, CO.

Badia, A., Sauri, D., Cerdan, R. and Llurdes, J-C. (2002) Causality and management of forest fires in Mediterranean environments: an example from Catalonia. *Global Environmental Change B: Environmental Hazards* 4: 23–32.

Baker, E.J. (2000) Hurricane evacuation in the United States. In Pielke, R.A. Jr and Pielke, R.A. Sr (eds) *Storms,* vol. 1. Routledge, London and New York, 306–19.

Baker, E.J. (2011) Household preparedness for the aftermath of hurricanes in Florida. *Applied Geography* 31: 46–52.

Ballester, J. *et al.* (2011) Long-term projections and acclimatization scenarios of temperature-related mortality in Europe. *Nature Communications* 2: 1–8.

Bandyopadhyay, S., Kanji, S. and Wang, L. (2012) The impact of rainfall and temperature variation on diarrheal prevalence in sub-Saharan Africa. *Applied Geography* 33: 63–72.

Bankoff, G., Frerks, G. and Hilhorst, D. (2004) *Mapping Vulnerability: Disasters, Development and People.* Earthscan, London.

Baram, M.S. (1987) Chemical industry hazards: liability, insurance and the role of risk analysis. In Kleindorfer, P.R. and Kunreuther, H.C. (eds) *Insuring and Managing Hazardous Risks.* Springer-Verlag, New York, 415–42.

Barberi, F. and Carapezza, M.L. (1996) The problem of volcanic unrest: the Campí Flegrei case history. In Scarpa, R. and Tilling, R.I. (eds) *Monitoring and Mitigation of Volcano Hazards,* Springer-Verlag, Berlin, 771–86.

Barclay, J., Johnstone, J.E. and Matthews, A.J. (2006) Meteorological monitoring of an active volcano:

implications for eruption prediction. *Journal of Volcanology and Geothermal Research* 150: 339–58.

Bardsley, K.L., Fraser, A.S. and Heathcote, R.L. (1983) The second Ash Wednesday: 16 February 1983. *Australian Geographical Studies* 21: 129–41.

Barnett, A., Menighetti, J. and Prete, M. (1992) The market response to the Sioux City DC-10 crash. *Risk Analysis* 12: 45–52.

Barnett, A.G., Tong, S. and Clements, A.C.A. (2010) What measure of temperature is the best predictor of mortality? *Environmental Research* 110: 604–11.

Barnett, B.J. (1999) US government natural disaster assistance: historical analysis and a proposal for the future. *Disasters* 23: 139–55.

Barredo, J.I. (2010) No upward trend in normalized windstorm losses in Europe: 1970–2008. *Natural Hazards and Earth System Sciences* 10: 97–104.

Barry, R.G. and Chorley, R.J. (1987) *Atmosphere, Weather and Climate*. Methuen, London.

Basher, R. (2006) Global early warning systems for natural hazards:systematic and people-centred. *Philosophical Transactions of the Royal Society (A)* 364: 2167–82.

Batterbury, S. and Warren, A. (2001) The African Sahel 25 years after the great drought: assessing progress and moving towards new agendas and approaches. *Global Environmental Change* 11: 1–8.

Baum, R.L. and Godt, J.W. (2010) Early warning of rainfall-induced shallow landslides and debris flows in the USA. *Landslides* 7: 259–72.

Bechtol, V. and Laurian, L. (2005) Restoring straightened rivers for sustainable flood mitigation. *Disaster Prevention and Management* 14: 6–19.

Beck, U. (1992) *Risk Society: Towards a New Modernity*. Sage Publications, New Delhi.

Bedford, D. and Faust, L. (2011) Role of online communities in recent responses to disasters: tsunami, China, Katrina and Haiti. *Proceedings of the American Society for Information Science and Technology* 47: 1–3.

Beechley, R.W., van Bruggen. J. and Truppi, L.E. (1972) Heat island = death island? *Environmental Research* 5: 85–92.

Belletto de Pujo, J. (1985) Emergency planning: the case of Diablo Canyon nuclear power plant. *Natural Hazard Research Working Paper 51*. Institute of Behavioural Sciences, University of Colorado, Boulder, CO.

Below, R., Grover-Kopec, E. and Dilley, M. (2007) Documenting drought-related disasters. *The Journal of Environment and Development* 16: 328–44.

Below, R., Vos, F. and Guha-Sapir, D. (2010) *Moving towards Harmonization of Disaster Data: A Study of Six Asian Databases*. CRED Working Paper No. 272. Centre for Research on the Epidemiology of Disasters, Brussels.

Belt, C.B. Jr (1975) The 1973 flood and man's constriction of the Mississippi river. *Science* 189: 681–4.

Beringer, J. (2000) Community fire safety at the urban/rural interface: the bushfire risk. *Fire Safety Journal* 35: 1–23.

Bettencourt, S. *et al*. (2006) *Not If but When: Adapting to Natural Hazards in the Pacific Islands Region*. The World Bank, Washington, DC.

Bhowmik, N.G. (ed.) (1994) *The 1993 Flood on the Mississippi River in Illinois*. Miscellaneous Publication 151, Illinois State Water Survey, Champaign-Urbana, Il.

Binder, D. (1998) The duty to disclose geologic hazards in real estate transactions. *Chapman Law Review* 1: 13–56.

Bird, D.K., Gisladóttir, G. and Dominey-Howes, D. (2009) Resident perceptions of volcanic hazards and evacuation procedures. *Natural Hazards and Earth System Sciences* 9: 251–66.

Birkmann, J. (ed.) (2006) *Measuring Vulnerability to Natural Hazards: Towards Disaster Resilient Societies*. Teri Press, New Delhi.

Birkmann, J. (2011) First and second-order adaptation to natural hazards and extreme events in the context of climate change. *Natural Hazards* 58: 811–40.

Blaikie, P., Cannon, T., Davis, I. and Wisner, B. (1994) *At Risk: Natural Hazards, People's Vulnerability and Disasters*. Routledge, London and New York (new edn 2004).

Blanchard-Boehm, R.D., Berry, K.A. and Showalter, P.S. (2001) Should flood insurance be mandatory? Insights in the wake of the 1997 New Year's Day flood in Reno-Sparks, Nevada. *Applied Geography* 21: 199–221.

Bluestein, H.B. (1999) *Tornado Alley: Monster Storms of the Great Plains*. Oxford University Press, New York.

Bohonos, J.J. and Hogan, D.E. (1999) The medical impact of tornadoes in North America. *The Journal of Emergency Medicine* 17: 67–73.

Bolin, B. *et al*. (2000) Environmental equity in a sunbelt city: the spatial distribution of toxic hazards in Phoenix, Arizona. *Environmental Hazards* 2: 11–24.

Bollasina, M.A., Ming, Y. and Ramaswamy, V. (2011) Anthropogenic aerosols and the weakening of the South Asian summer monsoon. *Science* 334: 502–5.

Bollinger, G.A., Chapman, M.C. and Sibol, M.S. (1993) A comparison of earthquake damage areas as a function of magnitude across the United States. *Bulletin of the Seismological Society of America* 83: 1064–80.

Bolt, B.A. (1999) *Earthquakes*. W.H. Freeman, New York.

Bolt, B.A., Horn, W.L., Macdonald, G.A. and Scott, R.F. (1975) *Geological Hazards*. Springer-Verlag, Berlin.

Bommer, J.J. and Rodríguez, C.E. (2002) Earthquake-induced landslides in Central America. *Engineering Geology* 63: 189–220.

Born, P. And Viscusi, W.K. (2006) The catastrophic effects of natural disasters on insurance markets. *Journal of Risk and Uncertainty* 33: 55–72.

Bosher, L., Carrill, P., Dainty, A., Glass, J. and Price, A. (2007) Realising a resilient and sustainable built environment: towards a strategic agenda for the United Kingdom. *Disasters* 31: 236–55.

Botterill, L.C. (2003) Uncertain climate: the recent history of drought policy in Australia. *Australian Journal of Politics and History* 49: 61–74.

Bouchon, M., Hatzfeld, D., Jackson, J.A. and Haghshenas, E. (2006) Some insight on why Bam (Iran) was destroyed by an earthquake of relatively moderate size. *Geophysical Research Letters* 33: L09309.

Bouma, M.J. and van der Kaay, H.J. (1996) The El Niño–Southern Oscillation and the historic malaria epidemics on the Indian sub-continent and Sri Lanka: an early warning system for future epidemics? *Tropical Medicine and International Health* 1: 86–96.

Bouma, M.J. and Pascual, M. (2001) Seasonal and interannual cycles of epidemic cholera in Bengal 1891–1940 in relation to climate and geography. *Hydrobiologia* 460: 147–56.

Bouma, M.J., Kovat, R.S., Gobet, S.A., Cox, J.S.H. and Haines, A. (1997) Gobal assessment of El Niño's disaster burden. *The Lancet* 350: 1435–8.

Bouska, A. *et al.* (2005) *Hurricane Katrina: Analysis of the Impact on the Insurance Industry*. Towers Perrin, New York.

Bouwer, L.M. (2011) Have disaster losses increased due to anthropogenic climate change? *Bulletin of the American Meteorological Society* 92: 39–46.

Bouwer, L.M., Bubeck, P. and Aerts, J.C.J.H. (2010) Changes in future flood risk due to climate and development in a Dutch polder area. *Global Environmental Change* 20: 463–71.

Brabec, B., Meister, R., Stöckli, U., Stoffel, A. and Stucki, T. (2001) RAIFOS: Regional Avalanche Information and Forecasting System. *Cold Regions Science and Technology* 33: 303–11.

Brabolini, M. and Savi, F. (2001) Estimate of uncertainties in avalanche hazard mapping. *Annals of Glaciology* 32: 299–305.

Bradley, J.T. (1972) Hurricane Agnes: the most costly storm. *Weatherwise* 25: 174–84.

Bradstock, R.A., Gill, A.M., Kenny, B.J. and Scott, J. (1998) Bushfire risk at the urban interface estimated from historical weather records: consequences for the use of prescribed fire in the Sydney region of South-Eastern Australia. *Journal of Environmental Management* 52: 259–71.

Brambilla, S. and Manca, D. (2010) The Viarregio LPG railway accident: event reconstruction and modelling. *Journal of Hazardous Materials* 182: 346–57.

Brammer, H. (2000) Flood hazard vulnerability and flood disasters in Bangladesh. In Parker, D.J. (ed.) *Floods*, vol. 1. Routledge, London and New York, 100–15.

Bridger, C.A. and Helfand, L.A. (1968) Mortality from heat during July 1966 in Illinois. *International Journal of Biometeorology* 12: 51–70.

Briët, J.T. *et al.* (2008) Temporal correlation between malaria and rainfall in Sri Lanka. *Malaria Journal* 7: 1–14.

Brightmer, M.I. and Fantato, M.G. (1998) Human and environmental factors in the increasing incidence of dengue fever: a case study from Venezuela. *Geojournal* 44: 103–9.

Brodie, M.E., Weltzien, D., Altman, R.J. and Blendon, J.M. (2006) Experiences of Hurricane Katrina evacuees in Houston shelters: implications for future planning. *American Journal of Public Health* 96: 1402–8.

Broecker, W.S. (1991) The Great Ocean Conveyor. *Oceanography* 4: 79–89.

Broecker, W.S. (1997) Thermohaline circulation, the Achilles heel of our climate system: will man-made carbon dioxide upset the current balance? *Science* 278: 1582–88.

Broecker, W.S. (2000) Was a change in thermohaline circulation responsible for the Little Ice Age? *Proceedings of the National Academy of Science* 97: 1339–42.

Brooks, N. (2004) *Drought in the African Sahel: Long-term Perspectives and Future Prospects*. Working Paper 61, Tyndall Centre for Climate Change Research, University of East Anglia, Norwich.

Brotak, E.A. (1980) Comparison of the meteorological conditions associated with a major wildland fire in the United States and a major bushfire in Australia. *Journal of Applied Meteorology* 19: 474–6.

Brouwer, R. and Nhassengo, J. (2006) About bridges and bonds: community responses to the 2000 floods in Mabalane District, Mozambique. *Disasters* 30: 234–55.

Brown, J. and Muhsin, M. (1991) Case study: Sudan emergency flood reconstruction program. In Kreimer, A. and Munasinghe, M. (eds) *Managing Natural Disasters and the Environment*. Environment Department, World Bank, Washington, DC, 157–62.

Brugge, R. (1994) The blizzard of 12–15 March 1993 in the USA and Canada. *Weather* 49: 82–9.

Bründl, M., Romang, H.E., Bischif, N. and Rheinberger, C.M. (2009) The risk concept and its application in natural hazard risk management in Switzerland. *Natural Hazards and Earth System Sciences* 9: 801–13.

Brunetti, M.T., Peruccacci, S., Rossi, M., Luciani, S. and Guzetti, F. (2010) Rainfall thresholds for the possible occurrence of landslides in Italy. *Natural Hazards and Earth System Sciences* 10: 447–58.

Bryden, H.L., Longworth, H.R. and Cunningham, S.A. (2005) Slowing of the Atlantic meridional overturning circulation at 25 degrees N. *Nature* 438: 655–7.

Buchroithner, M.F. (1995) Problems of mountain hazard mapping using spaceborne remote sensing techniques. *Advances in Space Research* 15: 57–66.

Buller, P.S.J. (1986) *Gale Damage to Buildings in the UK: An Illustrated Review*. Building Research Establishment, Watford, Herts.

Buncefield Major Incident Investigation Board (2006a) *The Buncefield Investigation: Progress Report*. 21 February, Health and Safety Executive, London.

Buncefield Major Incident Investigation Board (2006b) *Buncefield Major Incident Investigation: Initial Report*. 13 July, Health and Safety Executive, London.

Buncefield Major Incident Investigation Board (2008a) *The Buncefield Incident 11 December 2005: Final Report*. Health and Safety Executive, London.

Buncefield Major Incident Investigation Board (2008b) *Recommendations on Land Use Planning and the Control of Societal Risk Around Major Hazard Sites*. July, Health and Safety Executive, London.

Burby, R.J. (2000) Land-use planning for flood hazard reduction. In Parker, D.J. (ed.) *Floods*, vol 2. Routledge, London and New York, 6–18.

Burby, R.J. (2001) Flood insurance and floodplain management: the US experience. *Environmental Hazards* 3: 111–22.

Burby, R.J. and Dalton, L.C. (1994) Plans can matter! The role of land use plans and state planning mandates in limiting the development of hazard-ous areas. *Public Administration Review* 54: 229–37.

Burby, R.J. *et al.* (1991) *Sharing Environmental Risks: How to Control Governments' Losses in Natural Disasters*. Westview Press, Boulder, CO.

Burton, I. and Kates, R.W. (1964a) The perception of natural hazards in resource management. *Natural Resources Journal* 3: 412–41.

Burton, I. and Kates, R.W. (1964b) The floodplain and the seashore: a comparative analysis of hazard-zone occupante. *Geographical Review* 54: 366–85.

Burton, I., Kates, R.W. and White, G.F. (1978) *The Environment as Hazard*, 2nd edn. Guildford Press, New York and London (rev. edn, 1993).

Bush, D.M., Webb, C.A., Young, R.S., Johnson, B.D. and Bates, G.M. (1996) Impact of hurricane 'Opal' on the Florida–Alabama coast. *Quick Response Report* 84. Natural Hazards Research and Applications Information Center, Boulder, CO.

Butler, I. (1997) Selected internet sites on natural hazards and disasters. *International Journal of Mass Emergencies and Disasters* 15: 197–215.

Buxton, M., Haynes, R., Mercer, D. and Butt, A. (2010) Vulnerability to bushfire risk at Melbourne's urban fringe: the failure of regulatory land use planning. *Geographical Research* 49: 1–12.

Cairncross, S., Hardoy, J.E. and Satterthwaite, D. (eds) (1990) *The Poor Die Young: Housing and Health in Third World Cities*. Earthscan, London.

Callendar, G.S. (1938) The artificial production of carbon dioxide and its influence on climate. *Quarterly Journal of the Royal Meteorological Society* 64: 223–40.

Campbell, D., Barker, D. and McGregor, D. (2011) Dealing with drought: small farmers and environmental hazards in southern St Elizabeth, Jamaica. *Applied Geography* 31: 146–58.

Campbell, G.L., Marfin, A.A., Lanciotti, R.S. and Gubler, D.J. (2002) West Nile virus. *The Lancet Infectious Diseases* 2: 519–29.

Campbell, J.R., Goldsmith, M. and Koshy, K. (2005) *Community Relocation as an Option for Adaptation to the Effects of Climate Change and Climate Variability in Pacific Island Countries (PICs)*. Final report to the Asia-Pacific Network, Bangkok.

Canuti, P., Casaglia, N., Canuti, F. and Fanti, R. (2000) Hydrogeological hazard and risk in archeological sites: some case studies in Italy. *Journal of Cultural Heritage* 1: 117–25.

Capra, L., Poblete, M.A. and Alvarado, R. (2004) The 1997 and 2001 lahars of Popocatépetl volcano (central Mexico): textural and sedimentological constraints on their origin and hazards.

Journal of Volcanology and Geothermal Research 131: 351–69.

Cardona, O.D. (1997) Management of the volcanic crises of Galeras volcano: social, economic and institutional aspects. *Journal of Volcanology and Geothermal Research* 77: 313–24.

Carn, S.A., Watts, R.B., Thompson, G. and Norton, G.E. (2004) Anatomy of a lava dome collapse: the 20 March 2000 event at Soufrière Hills volcano, Montserrat. *Journal of Volcanology and Geothermal Research* 131: 241–64.

Casagli, N., Catani, F., Ventisette, C.D. and Luzi, G. (2010) Monitoring, prediction and early warning using ground-based interferometry. *Landslides* 7: 291–301.

Cayan, D.R. *et al.* (2010) Future dryness in the southwest US and the hydrology of the early 21st century drought. *Proceedings of the National Academy of Sciences of the United States of America* 107: 21271–6.

Cekan, J. (1992) Seasonal coping strategies in Gental Mali: five villages during the 'Soudiere'. *Disasters* 16: 66–73.

Cenderelli, D.A. and Wohl, E.E. (2001) Peak discharge estimates of glacial-lake outburst floods and 'normal' climatic floods in the Mount Everest region, Nepal. *Geomorphology* 40: 57–90.

Chadambuka, A. *et al.* (2012) The need for innovative strategies to improve immunisation services in rural Zimbabwe. *Disasters* 36: 161–73.

Chakraborty, J., Tobin, G.A. and Montz, B.E. (2005) Population evacuation: assessing spatial variability in geophysical risk and social vulnerability to natural hazards. *Natural Hazards Review* 6: 23–33.

Chambers, R. and Conway, G.R. (1992) *Sustainable Rural Livelihoods: Practical Concepts for the 21st Century.* Discussion Paper 296, Institute for Development Studies, University of Sussex.

Chang, Y., Wilkinson, S., Brunsden, D., Seville, E. and Potangaroa, R. (2011) An integrated approach: managing resources for post-disaster reconstruction. *Disasters* 35: 739–65.

Changnon, S.A. (2000) Impacts of hail in the United States. In Pielke, R.A. Jr and Pielke, R.A. Sr (eds) *Storms*, vol. 2. Routledge, London and New York, 163–91.

Changnon, S.A. and Semonin, R.G. (1966) A great tornado disaster in retrospect. *Weatherwise* 19: 56–65.

Changnon, S.A. and Changnon, D. (2005) Snowstorm catastrophes in the United States. *Environmental Hazards* 6: 158–66.

Changnon, S.A., Pielke, R.A. Jr, Changnon, D., Sylves, R.T. and Pulwarty, R. (2000) Human factors explain the increased losses from weather and climate extremes. *Bulletin of the American Meteorological Society* 81: 437–42.

Chapman, C.R. and Morrison, D. (1994) Impacts on the Earth by asteroids and comets: assessing the hazard. *Nature* 367: 33–9.

Chapman, D. (1999) *Natural Hazards.* Oxford University Press, Melbourne.

Chapman, J. (2005) Predicting technological disasters: mission impossible? *Disaster Prevention and Management* 14: 343–52.

Chappelow, B.F. (1989) Repair and restoration of supplies in Jamaica in the wake of hurricane Gilbert. *Distribution Developments* (June): 10–14.

Chen, C.C. *et al.* (2001) Psychiatric morbidity and post-traumatic symptoms among survivors in the early stage following the 1999 earthquake in Taiwan. *Psychiatry Research* 105: 13–22.

Chester, D. (1993) *Volcanoes and Society.* E. Arnold, London.

Chester, D.K. and Duncan, A.M. (2010) The impact of Eyjafjallajökull volcanic eruption on air transport. *Geographical Journal (Commentary).* Online at http://geographicaljournal.rgs.org/index.php/home/93-the-impact-of-the-eyjafjallajoekull-volcanic-eruption-on-air-transport

Chester, D.K., Dibben, C.J.L. and Duncan, A.M. (2002) Volcanic hazard assessment in Western Europe. *Journal of Volcanology and Geothermal Research* 115: 411–35.

Chester, D.K., Degg, M., Duncan, A.M. and Guest, J.E. (2001) The increasing exposure of cities to the effects of volcanic eruptions: a global survey. *Global Environmental Change B: Environmental Hazards* 2: 89–103.

Childs, D.Z., Cattadori, I.M., Suwonkerd, W., Prajakwong, S. and Boots, M. (2006) Spatio-temporal patterns of malaria incidence in northern Thailand. *Transactions of the Royal Society of Tropical Medicine and Hygiene* 100: 623–31.

Chotard, S., Mason, J.B., Oliphant, N.P., Mebrahtu, S. and Hailey, P. (2010) Fluctuations in wasting in vulnerable child populations in the Greater Horn of Africa. *Food and Nutrition Bulletin* 31: S219–S233.

Choudhury, N.Y., Paul, A. and Paul, B.K. (2004) Impact of coastal embankment on the flash flood in Bangladesh; a case study. *Applied Geography* 24: 241–58.

Clapperton, C.M. (1986) Fire and water in the Andes. *Geographical Magazine* 58: 74–9.

Clark, K.M. (1997) Current and potential impact of hurricane variability on the insurance industry. In Diaz, H. and Pulwarty, R.S. (eds) *Hurricanes: Climate and Socioeconomic Impacts*. Springer-Verlag, Heidelberg.

Clarke, M.L. and Rendell, H.M. (2006) Hindcasting extreme events: the occurrence and expression of damaging floods and landslides in southern Italy. *Land Degradation and Development* 17: 365–80.

Clay, E. *et al.* (1999) *An Evaluation of HMG's Response to the Montserrat Volcanic Emergency. Vol. I.* Department for International Development, London.

Clermont, C., Sanderson, D., Sharma, A. and Spraos, H. (2011) *Urban Disasters – Lessons from Haiti*. Disasters Emergency Committee, London.

Coates, J.F. (2009) Risks and threats to civilization, humankind and the earth. *Futures* 41: 694–705.

Colby, J.D., Mulcahy, K.A. and Yong, W. (2000) Modelling flooding extent from hurricane Floyd in the coastal plain of North Carolina. *Global Environmental Change B: Environmental Hazards* 2: 157–68.

Collins, M. and CMIP Modelling Groups (2005) El Niño or La Niña-like climate change? *Climate Dynamics* 24: 89–104.

Collins, T.W. (2005) Households, forests and fire hazard vulnerability in the American West: a case study of a California community. *Environmental Hazards* 6: 23–37.

Colwell, R.R. (1996) Global climate and infectious disease: the cholera paradigm. *Science* 274: 2025–31.

Comfort, L.K. (1996) Self-organisation in disaster response: the Great Hanshin, Japan, earthquake of January 17, 1995. *Quick Response Report* 78. Natural Hazards Research and Applications Information Center, Boulder, CO.

Comfort, L.K. (1999) *Shared Risk: Complex Systems in Seismic Response*. Pergamon Press, Oxford.

Comfort, L.K. (2006) Cities at risk: Hurricane Katrina and the drowning of New Orleans. *Urban Affairs Review* 41: 501–16.

Conlon, K.C., Rajkovich, N.B., White-Newsome, J.L., Larson, L. and O'Neill, M.S. (2011) Preventing cold-related morbidity and mortality in a changing climate. *Maturitas* 69: 197–202.

Conway, D. and Schipper, E.L.F. (2011) Adaptation to climate change in Africa: challenges and opportunities identified from Ethiopia. *Global Environmental Change* 21: 227–37.

Cook, B.I., Seager, R. and Miller, R.L. (2011) Atmospheric circulation anomalies during two persistent North American droughts; 1932–1939 and 1948–1957. *Climate Dynamics* 36: 2339–55.

Cook, E.R., Seager, R., Cane, M.A. and Stahle, D.W. (2007) North American drought: reconstructions, causes and consequences. *Earth Sciences Review* 81: 93–134.

Cordes, J.J., Gatzlaff, D.H. and Yezer, A.M. (2001) To the water's edge and beyond: effects of shore protection projects on beach development. *Journal of Real Estate Finance and Economics* 22: 287–302.

Correa, I.D. and Gonzalez, J.L. (2000) Coastal erosion and village relocation: a Colombian case study. *Ocean and Coastal Management* 43: 51–64.

Council for Reducing Major Industrial Accidents (2002) *Risk Management Guide for Major Industrial Accidents*. Montreal, Canada.

Cova, T.J. (2005) Public safety in the urban–wildland interface: should fire-prone communities have a maximum occupancy? *Natural Hazards Review* 6: 99–107.

Cova, T.J., Drews, F.A., Siebeneck, L.K. and Musters, A. (2009) Protective actions in wildfires: evacuation or shelter in place? *Natural Hazards Review* 10: 151–62.

Covello, V.T. and Mumpower, J. (1985) Risk analysis and risk management: an historical perspective. *Risk Analysis* 5: 103–20.

Cox, D. *et al.* (1992) Estimation of risk from observation on humans. In *Risk*. Royal Society, London, 67–87.

Cozzani, V., Campedel, M., Renni, E. and Krausman, E. (2010) Industrial accidents triggered by flood events: analysis of past accidents. *Journal of Hazardous Materials* 175: 501–9.

Crandell, D.R., Mullineaux, D.R. and Miller, C.D. (1979) Volcanic-hazards studies in the Cascades Range of the western United States. In Sheets, P.D. and Grayson, D.K. (eds) *Volcanic Activity and Human Ecology*. Academic Press, London, 195–219.

CRED (2005) Are natural disasters increasing? *CRED Crunch 2*. Centre for Research on the Epidemiology of Disasters, Université Catholique de Louvain, Louvain-la-Neuve.

CRED (2012) Natural Disasters in 2011. *CRED Crunch 27*. Centre for Research on the Epidemiology of Disasters, Université Catholique de Louvain, Louvain-la-Neuve.

Crompton, R.P. and McAneney, K.J. (2008) Normalized Australian insured losses for meteorological hazards 1967–2006. *Environmental Science and Policy* 11: 371–8.

Crompton, R.P., Pielke, R.A. and McAneney, K.J. (2011) Emergence timescales for detection of anthropogenic climate change in US tropical cyclone loss data. *Environmental Research Letters* 6:014003, 4pp.

Cross, J.A. (1985) *Residents' Acceptance of Hurricane Hazard Mitigation Measures*. Final Summary Report, NSF Grant no. CEE-8211441, University of Wisconsin, Oshkosh.

Cross, J.A. (2001) Megacities and small towns: different perspectives on hazard vulnerability. *Environmental Hazards* 3: 63–80.

Cruden, D.M. and Krahn, J. (1978) Frank rockslide, Alberta, Canada. In Voight, B. (ed.) *Rockslides and Avalanches, vol. 1 Natural Phenomena*. Elsevier, Amsterdam, 97–112.

CSSC (California Seismic Safety Commission) (1995) *Northridge Earthquake: Turning Loss to Gain*. Report CSSC95–01, Governors Executive Order W-78–94. Sacramento, CA.

CSSC (California Seismic Safety Commission) (2003) *Status of the Unreinforced Masonry Building Law*. Report CSSC 2003–03, Sacramento, CA.

CSSC (California Seismic Safety Commission) (2006) *Status of the Unreinforced Masonry Building Law*. Progress Report SSC 2006–04 to the Legislature, Sacramento, CA.

CSSC (California Seismic Safety Commission) (2009) *The Study of Household Preparedness: Preparing California for Earthquakes*. Report CSSC 09–03, Sacramento, CA.

Cullen, H.M., Kaplan, A., Arkin, P.A. and Demenocal, P.B. (2002) Impact of the North Atlantic Oscillation on Middle Eastern climate and streamflow. *Climatic Change* 55: 315–38.

Cummins, J. and Mahul, O. (2009) *Catastrophe Risk Financing in Developing Countries: Principles for Public Intervention*. World Bank, Washington, DC.

Cunningham, C.J. (1984) Recurring natural fire hazards: a case study of the Blue Mountains, New South Wales, Australia. *Applied Geography* 4: 5–57.

Cutter, S.L. (1984) Emergency preparedness and planning for nuclear power plant accidents. *Applied Geography* 4: 235–45.

Cutter, S.L. (1993) *Living with Risk: The Geography of Technological Hazards*. E. Arnold, London and New York.

Cutter, S.L. (1996) Vulnerability to environmental hazards. *Progress in Human Geography* 20: 529–39.

Cutter, S.L. (2003) GI science, disasters and emergency management. *Transactions in GIS* 7: 439–45.

Cutter, S.L. and Ji, M. (1997) Trends in US hazardous materials transportation spills. *Professional Geographer* 49: 318–31.

Cutter, S.L., Boruff, B.J. and Shirley, W.L. (2003) Social vulnerability to environmental hazards. *Social Science Quarterly* 84: 242–61.

Cutter, S.L. *et al.* (2008) A place-based model for understanding community resilience to natural disasters. *Global Environmental Change* 18: 598–606.

Dai, A. (2011) Drought under global warming. *WIRES Climate Change* 2: 45–65.

Daley, W.R., Smith, A., Paz-Argandona, E., Malily, J. and McGeehin, M. (2000) An outbreak of carbon monoxide poisoning after a major ice storm in Maine. *The Journal of Emergency Medicine* 18: 87–93.

Daly, M., Poutasi, N., Nelson, F. and Kohlhase, J. (2010) Reducing the climate vulnerability of coastal communities in Samoa. *Journal of Sustainable Development* 22: 265–81.

Dando, W.A. (1980) *The Geography of Famine*. Edward Arnold, London.

Daniel, H. 2001. Replenishment versus retreat: the cost of maintaining Delaware's beaches. *Ocean and Coastal Management* 44: 87–104.

Darcy, J. and Hofmann, C-A. (2003) *According to Need? Needs Assessment and Decision-making in the Humanitarian Sector*. Report 15, Humanitarian Policy Group. Overseas Development Institute, London.

Dash, N. and Morrow, B.H. (2000) Return delays and evacuation order compliance: the case of Hurricane Georges and the Florida Keys. *Global Environmental Change B: Environmental Hazards* 2: 119–28.

da Silva Curiel, A., Wicks, A., Meerman, M., Boland, L. and Sweeting, M. (2002) Second generation disaster-monitoring microsatellite platform. *Acta Astronautica* 51: 191–7.

David, E. and Mayer, J. (1984) Comparing costs of alternative flood hazard mitigation plans. *Journal of the American Planning Association* 50: 22–35.

Davies, J.B., Sandström, S., Shorrocks, A. and Wolff, E. (2009) The global pattern of household wealth. *Journal of International Development* 21: 1111–24.

Davis, I. (1978) *Shelter after Disaster*. Oxford Polytechnic Press, Oxford.

Day, J.W. *et al.* (2007) Restoration of the Mississippi delta: lessons from hurricanes Katrina and Rita. *Science* 315: 1679–84.

Decker, R.W. (1986) Forecasting volcanic eruptions. *Annals and Review of Earth and Planetary Science* 14: 267–91.

de Loë, R. and Wojtanowski, D. (2001) Associated benefits and costs of the Canadian Flood Damage Reduction Program. *Applied Geography* 21: 1–21.

de Moel, H., Aerts, J.C.J.H. and Koomen, E. (2011) Development of flood exposure in the Netherlands during the 20th and 21st century. *Global Environmental Change* 21: 620–7.

Department of Humanitarian Affairs (1994) *Strategy and Action Plan for Mitigating Water Disasters in Vietnam.* United Nations Development Programme, New York and Geneva.

de Scally, F.A. and Gardner, J.S. (1994) Characteristics and mitigation of the snow avalanche hazard in Kaghan valley, Pakistan Himalaya. *Natural Hazards* 9: 197–213.

de Sherbinin, A., Schiller, A. and Pulsipher, A. (2007) The vulnerability of global cities to climate hazards. *Environment and Urbanization* 19: 39–64.

de Souza, A.B. Jr (2000) Emergency planning for hazardous industrial areas: a Brazilian case study. *Risk Analysis* 20: 483–93.

de Vries, J. (1985) Analysis of historical climate–society interaction. In Kates, R.W., Ausubel, J.H. and Berberian, M. (eds) *Climate Impact Assessment.* New York, John Wiley, 273–91.

de Waal, A. (1989) *Famine that Kills: Dorfur, Sudan, 1984–85.* Clarendon Press, Oxford.

Di, J. and Jian, L. (2011) Managing tsunamis through early warning systems: a multidisciplinary approach. *Ocean and Coastal Management* 54: 189–99.

Dilley, M. and Heyman, B.N. (1995) ENSO and disaster: droughts, floods and E1 Niño–Southern Oscillation warm events. *Disasters* 19: 181–93.

Dilley, M., Chen, R.S., Deichmann, U., Lerner-Lam, A. and Arnold, M. (2005) *Natural Disaster Hotspots: A Global Risk Analysis.* World Bank Publications, Washington, D.C.

Dinar, A. and Keck, A. (2000) Water supply variability and drought impact and mitigation in sub-Saharan Africa. In Wilhite, D.A. (ed.) *Drought: A Global Assessment*, vol. 2. Routledge, London and New York, 129–48.

Dohler, G.C. (1988) A general outline of the ITSU master plan for the tsunami warning system in the Pacific. *Natural Hazards* 1: 295–302.

Dolan, J.F. *et al.* (1995) Prospects for larger or more frequent earthquakes in the Los Angeles Metropolitan Region. *Science* 267: 199–205.

Dolan, M.A. and Krug, S.E. (2006) Paediatric disaster preparedness in the wake of Katrina: lessons to be learned. *Clinical Pediatric Emergency Medicine* 7: 59–66.

Dominey-Howes, D. and Minos-Minopoulos, D. (2004) Perceptions of hazard and risk on Santorini. *Journal of Volcanology and Geothermal Research* 137: 285–310.

Donald, J.R. (1988) Drought effects on crop production and the US economy. *The Drought of 1988 and Beyond.* Proceedings of a Strategic Planning Seminar, 18 October 1988, Rockville, MD: National Climate Program Office, 143–62.

Donat, M.G. *et al.* (2011) Future changes in European winter storm losses and extreme windspeeds inferred from GCM and RCM multi-model simulations. *Natural Hazards and Earth System Sciences* 11: 1351–70.

Dong, B., Sutton, R.T. and Woollings, T. (2011) Changes in inter-annual NAO variability in response to greenhouse gas forcing. *Climate Dynamics* 37: 1621–41.

Donovan, A.R. and Oppenheimer, C. (2011) The 2010 Eyjafjallajökull eruption and the reconstruction of geoegraphy. *The Geographical Journal* 177: 4–11.

Doocy, S., Gabriel, M., Collins, S., Robinson, C. and Stevenson, P. (2006) Implementing cash-for-work programmes in post-tsunami Aceh: experiences and lessons learned. *Disasters* 30: 277–96.

Douglas, M. and Wildavsky, A. (1982) *Risk and Culture.* University of California Press, Berkeley, CA.

Dow, K. and Cutter, S.L. (1997) Repeat Response to Hurricane Evacuation Orders. *Quick Response Report* 101. Natural Hazards Research and Applications Information Center, Boulder, CO.

Dow, K. and Cutter, S.L. (2000) Public orders and personal opinions: household strategies for hurricane risk assessment. *Global Environmental Change B: Environmental Hazards* 2: 143–55.

Drabek, T.E. (1991) *Microcomputers in Emergency Management.* Monograph no. 51. Institute of Behavioral Science, University of Colorado, Boulder, CO.

Dreher, A. (2006) Does globalization affect growth? Evidence from a new Index of Globalization. *Applied Economics* 38: 1091–110.

Drury, A.C., Olson, R.S. and Van Belle, D.A. (2005) The politics of humanitarian aid: US foreign disaster assistance 1964–1995. *The Journal of Politics* 67: 454–73.

Duijm, N.J. (2009) *Acceptance Criteria in Denmark and the EU.* Environmental Project Report No. 1269, Danish Environmental Protection Agency, Ministry of the Environment, Copenhagen.

Duijsens, R. (2010) Humanitarian challenges of urbanization. *International Review of the Red Cross* 92: 351–68.

Dumas, L.J. (1999) *Lethal Arrogance: Human Fallibility and Dangerous Technologies*. St. Martin's Press, New York.

Dunlap, R.E. and Scarce, R. (1991) The polls-poll trends: environmental problems and protection. *Public Opinion Quarterly* 55: 651–72.

Dunning, S.A., Rosser, N.J., Petley, D.N. and Massey, C.R. (2006) Formation and failure of the Tsatichhu landslide dam Bhutan. *Landslides* 3: 107–13.

Dymon, U.J. (1999) Effectiveness of geographic information systems (GIS) applications in flood management during and after hurricane 'Fran'. *Quick Response Report* 114. Natural Hazards Research and Applications Information Center, Boulder, CO.

Dynes, R. (2004) Expanding the horizons of disaster research. *Natural Hazards Observer* 28 (4): 1–2.

Earth Policy Institute (2005) *World Economic Outlook Database*. From International Monetary Fund at www.imf.org/external/pubs/ft/weo (updated April 2005).

EERI (Earthquake Engineering Research Institute) (2004) *Preliminary Observations on the Bam, Iran, Earthquake of December 26, 2003*. Earthquake Engineering Research Institute, Oakland, CA.

Eisensee, T. and Stromberg, D. (2007) News droughts, news floods and US disaster relief. *The Quarterly Journal of Economics* 122: 693–728.

Ellemor, H. (2005) Reconsidering emergency management and indigenous communities in Australia. *Environmental Hazards* 6: 1–7.

Elliott, J.R. and Pais, J. (2006) Race, class and Hurricane Katrina: social differences in human responses to disaster. *Social Science Research* 35: 295–321.

Elliott, J.R. and Pais, J. (2010) When nature pushes back: environmental impact and the spatial redistribution of socially vulnerable populations. *Social Science Quarterly* 91: 1187–202.

El-Masri, S. and Tipple, G. (2002) Natural disaster, mitigation and sustainability: the case of developing countries. *International Planning Studies* 7: 157–75.

Elsner, J.B. (2003) Tracking hurricanes. *Bulletin of the American Meteorological Society* 84: 353–6.

Elsner, J.B., Jagger, T. and Niu, X-F. (2000) Changes in the rates of North Atlantic major hurricane activity during the 20th century. *Geophysical Research Letters* 27: 1743–6.

Emanuel, K.A. (1987) The dependence of hurricane intensity on climate. *Nature* 326: 483–5.

Emanuel, K.A. (2000) A statistical analysis of tropical cyclone intensity. *Monthly Weather Review* 128: 1139–52.

Emanuel, K.A. (2005) Increasing destructiveness of tropical cyclones over the past 30 years. *Nature* 436: 686–8.

Emery, A.C. (2005) *Good Practice in Emergency Preparedness and Response*. UNEP and International Council on Mining and Metals, London.

Emmi, P.C. and Horton, C.A. (1993) A GIS-based assessment of earthquake property damage and casualty risk: Salt Lake City, Utah. *Earthquake Spectra* 9: 11–33.

Eraybar, K. *et al.* (2010) An exploratory study on perceptions of seismic risk and mitigation in two districts of Istanbul. *Disasters* 34: 71–92.

Ericksen, N.J. (1986) *Creating Flood Disasters? New Zealand's Need for a New Approach to Urban Flood Hazard*. National Water and Soil Conservation Authority, Wellington.

Fagherazzi, S., Fosser, G., D'Alpaos, L. and D'Odorico, P. (2005) Climatic oscillations influence the flooding of Venice. *Geophysical Research Letters* 32: L19710.

Falck, L.B. (1991) Disaster insurance in New Zealand. In Kreimer, A. and Munasinghe, M. (eds) *Managing Natural Disasters and Environment*. Environment Department, World Bank, Washington, DC, 120–5.

Fang, H., Han, D., Guojian, H. and Chen, M. (2012) Flood management selections for the Yangzte River midstream after the Three Gorges Project operation. *Journal of Hydrology* 433: 1–11.

FEMA (Federal Emergency Management Agency) (1989) *Alluvial Fans: Hazards and Management*. Federal Emergency Management Agency, Washington, DC.

Fell, R. (1994) Landslide risk assessment and acceptable risk. *Canadian Geotechnical Journal* 31: 261–72.

Federal Interagency Stream Restoration Working Group (2001) *Stream Corridor Restoration: Principles, Processes and Practices*. USDA–Natural Resources Conservation Service, http://www.nrcs.usda.gov/Internet/FSE_DOCUMENTS/stelprdb1044574.pdf (accessed 17 July 2012).

Ferris-Morris, M. (2003) *Planning for the Next Drought: Ethiopia Case Study*. United States Agency for International Development, Washington, DC.

Field, C.B. *et al.* (2012) *Managing the Risks of Extreme Events and Disasters to Advance Climate Change Adaptation (SREX)*. A Special Report of Working Groups I and II of the IPCC, Cambridge University Press, Cambridge.

Field, R.D., van der Werf, G.R. and Shen, S.S.P. (2009) Human amplification of drought-induced biomass burning in Indonesia since 1960. *Nature Geoscience* 2: 185–8.

Finch, C., Emrich, C.T. and Cutter, S.L. (2010) Disaster disparities and differential recovery in New Orleans. *Population and Environment* 31: 179–202.

Fischer, E.M. and Schär, C. (2010) Consistent geographical patterns of changes in high-impact European heatwaves. *Nature Geoscience* 3: 398–403.

Fischer, G.W., Morgan, M.G., Fischhoff, B., Nair, I. and Lave, L.B. (1991) What risks are people concerned about? *Risk Analysis* 11: 303–14.

Fischhoff, B., Lichtenstein, S., Slovíc, P., Derby, S.L. and Keeney, R.L. (1981) *Acceptable Risk*. Cambridge University Press, Cambridge.

Florida Department of Community Affairs (2005) *Protecting Florida's Communities: Land Use Planning Strategies and Best Development Practices for Minimizing Vulnerability to Flooding and Coastal Storms*. Florida Department of Community Affairs, Tallahassee, FL.

Flynn, J., Slovic, P. and Mertz, C.K. (1993a) Decidedly different: expert and public views of risks from a radioactive waste repository. *Risk Analysis* 13: 643–8.

Flynn, J., Slovic, P. and Mertz, C.K. (1993b) The Nevada initiative: a risk communication fiasco. *Risk Analysis* 13: 497–502.

Folland, C.K. *et al.* (2009) The summer North Atlantic Oscillation: past, present and future. *Journal of Climate* 22: 1082–103.

Fothergill, A. and Peek, L.A. (2004) Poverty and disasters in the United States: a review of recent sociological findings. *Natural Hazards* 32: 89–110.

Fouchier, R., Kuiken, T., Rimmelzwaan, G. and Osterhaus, A. (2005) Global task force for influenza. *Nature* 435 (26 May): 419–20.

Fouillet, A.R. *et al.* (2008) Has the impact of heat waves on mortality changed in France since the European heat wave of summer 2003? A study of the 2006 heat wave. *International Journal of Epidemiology* 37: 309–17.

Foxworthy, B.L. and Hill, M. (1982) *Volcanic Eruptions of 1980 at Mount St Helens: The First 100 Days*. Geological Survey Professional Paper 1249. Government Printing Office, Washington, DC.

Francis, P.W. (1989) Remote sensing of volcanoes. *Advances in Space Research* 9: 89–92.

Fratkin, E. (1992) Drought and development in Marsabit District, Kenya. *Disasters* 16: 119–30.

Friis, H. (2007) International nutrition and health. *Danish Medical Bulletin* 54: 55–7.

Fritzsche, A.F. (1992) Severe accidents: can they occur only in the nuclear production of electricity? *Risk Analysis* 12: 327–9.

Fu, C.B., Kim, J-W. and Zhao, Z.C. (1998) Preliminary assessment of impacts of global change on Asia. In Galloway, J.N. and Melillo, J.M. (eds) *Asian Change in the Context of Global Climate Change*. Cambridge University Press, Cambridge.

Fuchs, S. and McAlpin, M.C. (2005) The net benefit of public expenditures on avalanche defence structures in the municipality of Davos, Switzerland. *Natural Hazards and Earth System Sciences* 5: 319–30.

Fujita, T.T. (1973) Tornadoes around the world. *Weatherwise* 26: 56–62, 79–83.

Fukuchi, T. and Mitsuhashi, K. (1983) Tsunami countermeasures in fishing villages along the Sanriku coast, Japan. In Iida, K. and Iwasaki. T. (eds) *Tsunamis*. D. Reidel, Boston, MA, 389–96.

Gabriel, K.M.A. and Endlicher, W.R. (2011) Urban and rural mortality rates during heat-waves in Berlin and Brandenburg, Germany. *Environmental Pollution* 159: 2044–50.

Gadgil, S., Vinayachandran, P.N. and Francis, P.A. (2003) Droughts of the Indian summer monsoon: role of clouds over the Indian Ocean. *Current Science* 85: 1713–19.

Gagnon, A.S., Bush, A.B.G. and Smoyer-Tomic, K.E. (2001) Dengue epidemics and the El Niño Southern Oscillation. *Climate Research* 19: 35–43.

Gall, M., Borden, K.A. and Cutter, S.L. (2009) When do losses count? *Bulletin of the American Meteorological Society* 90: 799–809.

Galle, B. *et al.* (2003) A miniaturised ultraviolet spectrometer for remote sensing of SO_2 fluxes: a new tool for volcanic surveillance. *Journal of Volcanology and Geothermal Research* 119: 241–54.

Galloway, G.E., Boesch, D.F. and Twilley, R.K. (2009) Restoring and protecting coastal Louisiana. *Issues in Science and Technology* 25: unpaginated.

Gardner, G.T. and Gould, L.C. (1989) Public perceptions of the risks and benefits of technology. *Risk Analysis* 9: 225–42.

Gardoni, P. and Murphy, C. (2010) Gauging the societal impacts of natural disasters using a capability approach. *Disasters* 34: 619–36.

Garner, A.C. and Huff, W.A.K. (1997) The wreck of Amtrak's Sunset Limited: news coverage of a mass transport disaster. *Disasters* 21: 4–19.

Garside, R., Johnston, D., Saunders, W. and Leonard, G. (2009) Planning for tsunami evacuations: the case of the Marine Education Centre, Wellington, New Zealand. *The Australian Journal of Emergency Management* 24: 28–31.

Garvin, T. (2001) Analytical paradigms: The epistemiological distances between scientists, policy makers and the public. *Risk Analysis* 21: 443–55.

Gaume, E. *et al.* (2009) A compilation of data on European flash floods. *Journal of Hydrology* 367: 70–8.

Gayà, M. (2011) Tornadoes and severe storms in Spain. *Atmospheric Research* 100: 334–43.

Gehrels, T. (ed.) (1994) *Hazards due to Asteroids and Comets.* University of Arizona Press, Tucson.

Gemmell, I., McLoone, P., Boddy, F.A., Dickinson, G.J. and Watt, G.C.M. (2000) Seasonal variation in mortality in Scotland. *International Journal of Epidemiology* 29: 274–9.

Giannini, A., Saravanan, R. and Chang, P. (2003) Oceanic forcing of Sahel rainfall on inter-annual to inter-decadal time scales. *Science* 302: 1027–30.

Giddens, A. (1990) *The Consequences of Modernity.* Polity Press, Cambridge.

Gill, A.M. (2005) Landscape fires as social disasters: an overview of the bushfire problem. *Environmental Hazards* 6: 65–80.

Gillespie, T.W., Chu, J., Frankenburg, E. and Thomas, D. (2007) Assessment and prediction of natural hazards from satellite imagery. *Progress in Physical Geography* 31: 459–70.

Gislason, S.R. *et al.* (2011) Characterization of Eyjafjallajökull volcanic ash particles and a protocol for rapid risk assessment. *Proceedings of the National Academy of Sciences* 108: 7307–12.

Givri, J.R. (1995) Satellite remote sensing data on industrial hazards. *Advances in Space Research* 15: 87–90.

Glantz, M.H. (ed.) (1988) *Drought and Hunger in Africa: Denying Famine a Future.* Cambridge University Press, Cambridge.

Glickman, T.S., Golding, D. and Silverman, E.D. (1992) *Acts of God and Acts of Man: Recent Trends in Natural Disasters and Major Industrial Accidents.* Discussion Paper CRM 92–02. Resources for the Future, Washington, DC.

Goddard, L. and Dilley, M. (2005) El Niño: catastrophe or opportunity? *Journal of Climate* 18: 651–65.

Goff, F. *et al.* (2001) Passive infrared remote sensing evidence for large intermittent CO_2 emissions at Popocatépetl volcano, Mexico. *Chemical Geology* 177: 133–56.

Gokceoglu, C. and Sezer, E. (2009) A statistical assessment on international landslide literature (1945–2008). *Landslides* 6: 349–51.

Goldberg, M.S., Gasparrini, A., Armstrong, B. and Valois, M-F. (2011) The short-term influence of temperature on daily mortality in the temperate climate of Montreal, Canada. *Environmental Research* 111: 853–60.

Golding, B.W. (2009) Long lead-times for flood warnings: reality or fantasy? *Meteorological Applications* 16: 3–12.

González, F.I. (1999) Tsunami. *Scientific American* 280: 56–65.

Goosse, H., Barriat, P.Y., Lefebvre, W., Loutre, M.F. and Zunz, V. (2008) *Introduction to Climate Dynamics and Climate Modelling.* Online textbook at http://www.climate.be/textbook Université Catholique de Louvain, Belgium (accessed 4 September 2011).

Gouveia, C., Trigo, R.M. and Dacamara, C.C. (2009) Drought and vegetation stress monitoring using satellite data. *Natural Hazards and Earth System Sciences* 9: 185–95.

Govaerts, A. and Lauwerts, B. (2009) *Assessment of the Impact of Coastal Defence Structures.* OSPAR Commission, London

Govere, J.M., Durrheim, D.N., Coetzee, M. and Hunt, R.H. (2001) Malaria in Mpumalanga Province, South Africa, with special reference to the period 1987–1999. *South African Journal of Science* 97: 55–8.

Graham, N.E. (1977) Weather surrounding the Santa Barbara fire: 26 July 1977. *Weatherwise* 30 (4): 158–9.

Gray, W.M. and Landsea, C.W. (1992) African rainfall as a precursor of hurricane-related destruction on the US east coast. *Bulletin of the American Meteorological Society* 73: 1352–64.

Green, C.H., Parker, D.J. and Tunstall, S.M. (2000) *Assessment of Flood Control and Management Options.* Working Paper IV (4), prepared for World Commission on Dams, Cape Town.

Greenberg, M.R., Sachsman, D.B., Sandman, P.M. and Salomone, K.L. (1989) Network evening news coverage of environmental risk. *Risk Analysis* 9: 119–26.

Greene, J.P. (1994) Automated forest fire detection. *STOP Disasters* 18: 18–19.

Gregg, C.E., Houghton, B.F., Johnston, D.M., Paton, D. and Swanson, D.A. (2004) The perception of volcanic risk in Kona communities from Mauna Loa and Hualalai volcanoes, Hawaii. *Journal of Volcanology and Geothermal Research* 130: 179–96.

Grieve, R.A.F. (1998) Extra-terrestrial impacts on earth: the evidence and the consequences. In Grady, M.M., Hutchinson, R., McCall, G.J.H. and Rothery, D.A. (eds) *Meteorites: Flux with Time and Impact Effects.* Special Publication 140, Geological Society of London, London, 105–31.

Griffiths, J.S., Hutchinson, J.N., Brundsen, D., Petley, D. and Fookes, P.G. (2004) The reactivation of a landslide during the construction of the Ok Ma tailings dam, Papua New Guinea. *Quarterly Journal of Engineering Geology and Hydrogeology* 37: 173–86.

Gruber, U. and Haefner, H. (1995) Avalanche hazard mapping with satellite data and a digital elevation model. *Applied Geography* 15: 99–114.

Grundstein, A. and Dowd, J. (2011) Trends in extreme apparent temperatures over the United States 1949–2010. *Journal of Applied Meteorology and Climatology* 50: 1650–3.

Gruntfest, E. (1987) Warning dissemination and response with short lead times. In Handmer, J.W. (ed.) *Flood Hazard Management*. Geo Books, Norwich, 191–202.

Guffanti, M., Ewert, J.W., Gallina, G.M., Bluth, G.J.S. and Swanson, G.L. (2005) Volcanic ash hazard to aviation during the 2003–2004 eruptive activity of Anatahua volcano, Commonwealth of the Northern Marianas Islands. *Journal of Volcanology and Geothermal Research* 146: 241–55.

Guha-Sapir, D. and Below, R. (2002) *The Quality and Accuracy of Disaster Data: A Comparative Analysis of Three Global Data Sets*. Working Paper prepared for the Disaster Management Facility, World Bank, CRED, Brussels.

Guha-Sapir, D. and Below, R. (2006) Collecting data on disasters: easier said than done. *Asian Disaster Management News* 12: 9–10.

Guha-Sapir, D., Hargitt, D. and Hoyois, P. (2004) *Thirty Years of Natural Disasters 1974–2003*. Presses Universitaires de Louvain, Louvain-la-Neuve, France.

Guintran, J-O., Delacollette, L. and Trigg, P. (2006) *Systems for the Early Detection of Malaria Epidemics in Africa*. World Health Organization, Geneva.

Guo, H., Hu, Q., Zhang, Q. and Feng, S. (2012) Effects of the Three Gorges Dam on Yangtze River flow and river interaction with Poyang Lake, China: 2003–2008. *Journal of Hydrology* 416: 19–27.

Gupta, J.P. (2002) The Bhopal tragedy: could it have happened in a developed country? *Journal of Loss Prevention in the Process Industries* 15: 1–4.

Guttman, N.B. (1997) Comparing the Palmer drought index and the standardized precipitation index. *Journal of the American Water Resources Association* 34: 113–21.

Guzzetti, F. (2000) Landslide fatalities and the evaluation of landslide risk in Italy. *Engineering Geology* 58: 89–107.

Haile, M. (2005) Weather patterns, food security and humanitarian response in sub-Saharan Africa. *Philosophical Transactions of the Royal Society B – Biological Sciences* 360: 2169–82.

Haines, A., Kovacs, R.S., Cambell-Lendrum, D. and Corvalan, C. (2006) Climate change and human health: impacts, vulnerability and mitigation. *Lancet* 367: 2101–9.

Hall, J.W. *et al.* (2003) Quantified scenarios analysis of drivers and impacts of changing flood risk in England and Wales 2030–2100. *Environmental Hazards* 5: 51–65.

Hamilton, L.S. (1987) What are the impacts of Himalayan deforestation on the Ganges–Brahmaputra lowlands and delta? Assumptions and facts. *Mountain Research and Development* 7: 256–63.

Hammer, B.O. and Schmidlin, T.W. (2000) Vehicle-occupant deaths caused by tornadoes in the United States 1900–1998. *Global Environmental Change B: Environmental Hazards* 2: 105–18.

Hammer, B.O. and Schmidlin, T.W. (2002) Response to warnings during the 3 May 1999 Oklahoma City tornado: reasons and relative injury rates. *Weather and Forecasting* 17: 577–81.

Hammond, L. and Maxwell, D. (2002) The Ethiopian crisis of 1991–2000: lessons learned, questions unanswered. *Disasters* 26: 262–79.

Hancox, G.T. (2008) The 1979 Abbotsford landslide, Dunedin, New Zealand: a retrospective look at its nature and causes. *Landslides* 5: 177–88.

Handmer, J.W. (1987) Guidelines for floodplain acquisition. *Applied Geography* 7: 203–21.

Handmer, J.W. (1999) Natural and anthropogenic hazards in the Sydney sprawl: is the city sustainable? In Mitchell, J.K. (ed.) *Crucibles of Hazard*. United Nations University Press, Tokyo, 138–85.

Handmer, J.W. and Tibbits, A. (2005) Is staying at home the safest option during bushfires? Historical evidence for an Australian approach. *Environmental Hazards* 6: 81–91.

Hansen, J. *et al.* (2006) Global temperature change. *Proceedings of the National Academy of Sciences of the USA* 103: 14288–93.

Hanson, S. (2011) A global ranking of port cities with high exposure to climate extremes. *Climatic Change* 104: 89–111.

Hansson, S.O. (2010) Promoting inherent safety. *Process Safety and Environmental Protection* 88: 168–72.

Harlan, S.L., Brazel, A.J., Prashad, L., Stefanov, W.L. and Larsen, L. (2006) Neighbourhood microclimates and vulnerability to heat stress. *Social Science and Medicine* 63: 2847–63.

Harp, E.L., Reid, M.E., McKenna, J.P. and Michael, J.A. (2009) Mapping of hazard from rainfall-triggered landslides in developing countries:

examples from Honduras and Micronesia. *Engineering Geology* 104: 295–311.

Harwell, E.E. (2000) Remote sensibilities: discourses of technology and the making of Indonesia's natural disaster. *Development and Change* 31: 307–40.

Haskell, R.C. and Christiansen, J.R. (1985) Seismic bracing of equipment. *Journal of Environmental Sciences* 9: 67–70.

Haug, R. (2002) Forced migration, processes of return and livelihood construction among pastoralists in northern Sudan. *Disasters* 26: 70–84.

Hay, I. (1996) Neo-liberalism and criticisms of earthquake insurance arrangements in New Zealand. *Disasters* 20: 34–48.

Hay, S.I. *et al.* (2009) A world malaria map: *Plasmodium falciparum* endemicity in 2007. *PLoS Med* 6: 286–302.

Hayes, B.D. (2004) Interdisciplinary planning of non-structural flood hazard mitigation. *Journal of Water Resources Planning and Management* 130: 15–25.

Haylock, H.J.K. and Ericksen, N.J. (2000) From state dependency to self-reliance. In Wilhite, D.A. (ed.) *Drought*, vol. 2. Routledge, London and New York, 105–14.

Haynes, K., Handmer, J., McAneney, J., Tibbits, A. and Coates, L. (2010) Australian bushfire fatalities 1900–2008: exploring trends in relation to the 'Prepare, Stay and Defend or Leave Early' policy. *Environmental Science and Policy* 13: 185–94.

Hazard Mitigation Team (1994) *Southern California Firestorms*. FEMA-1005-DR-CA Report. Federal Emergency Management Agency, San Francisco, CA.

Hazarika, S. (1988) *Bhopal: The Lessons of a Tragedy*. Penguin Books, New Delhi.

Hazell, P.B.R. and Hess, U. (2010) Drought insurance for agricultural development and food security in dryland areas. *Food Security* 2: 395–405.

Healey, D.T., Jarrett, F.G. and McKay, J.M. (1985) *The Economics of Bushfires: The South Australian Experience*. Oxford University Press, Melbourne.

Health and Safety Executive (1978) *Canvey: An Investigation of Potential Hazards from Operations in the Canvey Island/Thurrock Area*. HMSO, London.

Health and Safety Executive (2004) *Thirty Years On and Looking Forward*. Health and Safety Executive, London.

Healy, J.D. (2003) Excess winter mortality in Europe: a cross-country analysis identifying key risk factors. *Journal of Epidemiology and Community Health* 57: 784–9.

Heim, R. (2002) A review of twentieth-century drought indices used in the United States. *Bulletin of the American Meteorological Society* 83: 1149–65.

Helm, P. (1996) Integrated risk management for natural and technological disasters. *Tephra* 15: 4–13.

Heltberg, R. (2007) Helping South Asia cope better with natural disasters: the role of social protection. *Development Policy Review* 25: 681–98.

Hemrich, G. (2005) Matching food security analysis to context: the experience of the Somalia Food Security Assessment Unit. *Disasters* 29 (Supplement 1): 567–91.

Hendrickson, D., Armon, J. and Mearns, R. (1998) The changing nature of conflict and famine vulnerability: the case of livestock raiding in Turkana District, Kenya. *Disasters* 22: 185–99.

Hewitt, K. (ed.) (1983) *Interpretations of Calamity*. Allen and Unwin, Boston, MA and London.

Hewitt, K. and Burton, I. (1971) *The Hazardousness of a Place: A Regional Ecology of Damaging Events*. Department of Geography, University of Toronto, Toronto.

Hilker, N., Badoux, A. and Hegg, C. (2009) The Swiss flood and landslide damage database 1972–2007. *Natural Hazards and Earth System Sciences 9*: 913–25.

Hills, J.G. and Goda, M.P. (1998) Tsunami from asteroid and comet impacts: the vulnerability of Europe. *Science of Tsunami Hazards* 16: 3–10.

Hinkel, J. (2011) Indicators of vulnerability and adaptive capacity: towards a clarification of the science–policy interface. *Global Environmental Change* 21: 198–208.

Hinman, G.W., Rosa, E.A., Kleinhesselink, R.R. and Lowinger, T.C. (1993) Perceptions of nuclear and other risks in Japan and the United States. *Risk Analysis* 14: 449–55.

Hochrainer-Stigler, S. *et al.* (2011) *The Costs and Benefits of Reducing Risk from Natural Hazards to Residential Structures in Developing Countries*. Working Paper 2011–01. Wharton Risk Management and Decision Processes Center, University of Pennsylvania, Philadelphia.

Hoek, M.R., Bracebridge, S. and Oliver, I. (2007) Health impacts of the Buncefield oil depot fire, December 2005: study of accident and emergency records. *Journal of Public Health* 29: 298–302.

Hoerling, M. *et al.* (2012) On the increased frequency of Mediterranean drought. *Journal of Climate* 25: 2146–61.

Hohenemser, C., Kates, R.W. and Slovic, P. (1983) The nature of technological hazard. *Science* 220: 378–84.

Hohl, R., Schiesser, H-H. and Knepper, I. (2002) The use of weather radars to estimate hail damage to automobiles: an exploratory study in Switzerland. *Atmospheric Research* 61: 215–38.

Holcombe, E. and Anderson, M. (2010) Tackling landslide risk: helping land use policy to reflect unplanned housing realities in the Eastern Caribbean. *Land Use Policy* 27: 798–800.

Holzer, T.L. (1994) Loma Prieta damage largely attributed to enhanced ground shaking. *EOS: Transactions of the American Geophysical Union* 75 (26): 299–301.

Hoque, B.A. *et al.* (1993) Environmental health and the 1993 Bangladesh cyclone. *Disasters* 17: 144–52.

Horikawa, K. and Shuto, N. (1983) Tsunami disasters and protection measures in Japan. In Iida, K. and Iwasaki, T. (eds) *Tsunamis*. D. Reidel, Boston, MA, 9–22.

Horney, J.A., MacDonald, P.D.M., Willigen, M.U., Berke, P.R. and Kaufman, J.S. (2010) Individual actual or perceived property flood risk: did it predict evacuation from Hurricane Isabel in North Carolina in 2003? *Risk Analysis* 30: 501–11.

Houze, R.A. Jr *et al.* (2007) Hurricane intensity and eyewall replacement. *Science* 315 (2 March): 1235–9.

Howe, P. and Devereux, S. (2004) Famine intensity and magnitude scales: a proposal for an instrumental definition of famine. *Disasters* 28: 353–72.

Howe, P.D. (2011) Hurricane preparedness as anticipatory adaptation: a case study of community businesses. *Global Environmental Change* 21: 711–20.

Hoyos, C.D., Agudelo, P.A., Webster, P.J. and Curry, J.A. (2006) Deconvolution of the factors contributing to the increase in global hurricane intensity. *Science* 312: 94–7.

Huang, Z., Rosowsky, D.V. and Sparks, P.R. (2001) Long-term hurricane risk assessment and expected damage to residential structures. *Reliability Engineering and System Safety* 74: 239–49.

Hughes, P. (1979) The great Galveston hurricane. *Weatherwise* 32: 148–56.

Hulme, M. (2001) Climatic perspectives on Saharan desiccation: 1973–1998. *Global Environmental Change* 11: 19–29.

Hung, J.J. (2000) Chi-Chi earthquake-induced landslides in Taiwan. *Earthquake Engineering and Engineering Seismology* 2: 25–33.

Hungr, O., Corominas, J. and Eberhardt, E. (2005) Estimating landslide motion mechanism, travel distance and velocity. In Hungr, O., Fell, R., Couture, R. and Eberhardt, E. *Landslide Risk Management*. Taylor and Francis, London, 99–128.

Hunt, E.D., Hubbard, K.G., Wilhite, D.A., Arkebauer, T.J. and Dutcher, A.L. (2009) The development and evaluation of a soil moisture index. *International Journal of Climatology* 29: 747–59.

Hupp, C.R., Osterkamp, W.R. and Thornton, J.L. (1987) *Dendrogeomorphic Evidence and Dating of Recent Debris Flows on Mount Shasta, Northern California*. US Geological Survey Professional Paper 1396-B, Washington, DC.

Hürlimann, M., Copons, R. and Altimir, J. (2006) Detailed debris flow hazard assessment in Andorra: a multidisciplinary approach. *Geomorphology* 78: 359–72.

Hurrell, J.W., Kushnir, Y. and Visbeck, M. (2001) The North Atlantic Oscillation. *Science* 291: 603–4.

Hwang, S., Xi, J., Cao, Y., Feng, X. and Qia, X. (2007) Anticipation of migration and psychological stress and the Three Gorges project, China. *Social Science and Medicine* 65: 1012–24.

Institution of Civil Engineers (1995) *Megacities: Reducing Vulnerability to Natural Disasters*. Thomas Telford, London.

Interagency Performance Evaluation Taskforce (2006) *Performance Evaluation of the New Orleans and Southeast Louisiana Hurricane Protection System*. Draft Final Report. Vol. I, Executive Summary and Overview. US Army Corps of Engineers.

Intergovernmental Oceanographic Commission (2008) *Tsunami Preparedness: Information Guide for Disaster Planners*. Manuals and Guides 49, UNESCO, Paris.

IFRCRCS (1994) *World Disasters Report 1994*. Martinus Nijhoff, Dordrecht.

IFRCRCS (1999) *World Disasters Report 1999*. International Federation of Red Cross and Red Crescent Societies, Geneva.

IFRCRCS (2000) *World Disasters Report 2000*. International Federation of Red Cross and Red Crescent Societies, Geneva.

IFRCRCS (2002) *World Disasters Report 2002*. International Federation of Red Cross and Red Crescent Societies, Geneva.

IFRCRCS (2004) *World Disasters Report 2004*. International Federation of Red Cross and Red Crescent Societies, Geneva.

IFRCRCS (2005) *World Disasters Report 2005*. International Federation of Red Cross and Red Crescent Societies, Geneva.

IFRCRCS (2006) *World Disasters Report 2006*. International Federation of Red Cross and Red Crescent Societies, Geneva.

IFRCRCS (2009) *World Disasters Report 2006. Geneva, International Federation of Red Cross and Red Crescent Societies.*

IFRCRCS (2010) *World Disasters Report 2010.* International Federation of Red Cross and Red Crescent Societies, Geneva.

IPCC (2007) Summary for policymakers. In Parry, M.L., Canziani, O.F., Palutikof, J.P., van der Linden, P.J. and Hanson, C.E. (eds) *Climate Change 2007: Impacts, Adaptation and Vulnerability.* Contribution of Working Group II to the Fourth Assessment Report, Intergovernmental Panel on Climate Change, Cambridge University Press, Cambridge, 7–22.

ISDR Secretariat (2003) *Living with Risk: Turning the Tide on Disasters towards Sustainable Development.* United Nations, Geneva.

ISDR Secretariat (2006) Support for Sri Lanka from the United Nations University. *Disaster Reduction in Asia Pacific–ISDR Informs* 2: 63–5.

Isler, P.L., Merson, J. and Roser, D. (2010) 'Drought proofing' Australian cities: implications for climate change adaptation and sustainability. *World Academy of Science, Engineering and Technology* 70: 352–60.

Ives, J.D. and Messerli, B. (1989) *The Himalayan Dilemma: Reconciling Development and Conservation.* Routledge, London.

Jackson, E.L. and Burton, I. (1978) The process of human adjustment to earthquake risk. In *The Assessment and Mitigation of Earthquake Risk.* UNESCO, Paris, pp. 241–60.

Javelle, P., Fouchier, C., Arnaud, P. and Lavabre, J. (2010) Flash flood warning at ungauged locations using radar rainfall and antecedent soil moisture estimations. *Journal of Hydrology* 394: 267–74.

Jayaraman, V., Chandrasekhar, M.G. and Rao, V.R. (1997) Managing the natural disasters from space technology inputs. *Acta Astronautica* 40: 291–325.

Jiang, T., Zhang, Q., Zhu, D. and Wu, Y. (2006) Yangtze floods and droughts (China) and teleconnections with ENSO activities (1470–2003). *Quaternary International* 144: 29–37.

Jibson, R.W. and Baum, R.L. (1999) *Assessment of Landslide Hazards in Kaluanui and Maakua Gulches, Oahu, Hawaii Following the 9 May Sacred Falls Landslide.* Open File Report 99–364, US Geological Survey, Reston, VA.

Jóhannesdóttir, G. and Gisladóttir, G. (2010) people living under threat of volcanic hazard in southern Iceland: vulnerability and risk perception. *Natural Hazards and Earth System Sciences* 10: 407–20.

Jóhannesson, T. (2001) Run-up of two avalanches on the deflecting dams at Flateyri, north-western Iceland. *Annals of Glaciology* 32: 350–4.

Johnson, D.P. and Wilson, J.S. (2009) The socio-spatial dynamics of extreme urban heat events: the case of heat-related deaths in Philadelphia. *Applied Geography* 29: 419–34.

Johnson, K. (1985) *State and Community during the Aftermath of Mexico City's November 19, 1984, Gas Explosion.* Special Publication 13, Institute of Behavioral Science, University of Colorado, Boulder, CO.

Jones, D.K.C. (1992) Landslide hazard assessment in the context of development. In McCall, G.J.H., Laming, J.C. and Scott, S.C. (eds) *Geohazards*, Chapman and Hall, London, 117–41.

Jones, D.K.C. (1995) The relevance of landslide hazard to the International Decade for Natural Disaster Reduction. In *Landslide Hazard Mitigation with Particular Reference to Developing Countries, Proceedings of a Conference.* Royal Academy of Engineering, London, 19–33.

Jones, D.K.C., Lee, E.M., Hearn, G.J. and Gene, S. (1989) The Catak landslide disaster, Trabzon province, Turkey. *Terra Nova* 1: 84–90.

Jones, F.O. (1973) *Landslides of Rio de Janeiro and the Serra das Araras escarpment.* Professional Paper 697, US Geological Survey, Washington, DC.

Jones, S.R. and Mangun, W.R. (2001) Beach nourishment and public policy after Hurricane Floyd: where do we go from here? *Ocean and Coastal Management* 44: 207–20.

Jones-Lee, M.W., Hammerton, M. and Philips, P.R. (1985) The value of safety: the results of a national survey. *Economic Journal* 95: 49–72.

Jonkman, S.N., Kok, M. and Vrijling, J.K. (2008) Flood risk assessment in the Netherlands: a case study for the Dike Ring, South Holland. *Risk Analysis* 28: 1357–74.

Jonkman, S.N., Maaskant, B., Boyd, E. and Levitan, M.L. (2009) Loss of life caused by the flooding of New Orleans after Hurricane Katrina: analysis of the relationship between flood characteristics and mortality. *Risk Analysis* 29: 676–98.

Jowett, A.J. (1989) China: the demographic disaster of 1958–1961. In Clarke, J.L, Curson, P., Kayastha, S.L. and Nag, P. (eds) *Population and Disaster*, Basil Blackwell, Oxford, 137–58.

Joyce, K.E., Belliss, S.E., Samsonov, S.V., McNeil, S.J. and Glassey, P.J. (2009) A review of the status of remote sensing and image processing techniques for mapping natural hazards and disasters. *Progress in Physical Geography* 33: 183–207.

Kaczmarek, Z. (2003) The impact of climate variability on flood risk in Poland. *Risk Analysis* 3: 559–66.

Kafali, C. (2011) *Regional Wind Vulnerability in Europe.* Report 04.2011, AIR Currents, AIR Worldwide, Boston.

Kahn, M. (2005) The death toll from natural disasters: the role of income, geography and institutions. *The Review of Economics and Statistics* 87: 271–84.

Kaiser, R., Spiegel, P.B., Henderson, A.K. and Gerber, M.L. (2003) The application of Geographic Information Systems and Global Positioning Systems in humanitarian emergencies: lessons learned, programme implications and future research. *Disasters* 27: 127–40.

Kajoba, G.M. (1992) Food security and the impact of the 1991–92 drought in Zambia. Unpublished text of lecture delivered at the University of Stirling, October.

Karabork, M.C. (2007) Trends in drought patterns of Turkey. *Journal of Environmental Engineering and Science* 6: 45–52.

Karanci, A.N. and Rüstemli, A. (1995) Psychological consequences of the 1992 Erzincan (Turkey) earthquake. *Disasters* 19: 8–18.

Karpisheh, L., Mirdamadi, M., Hosseini, J.F. and Chizari, M. (2010) Iranian farmers' attitudes and management strategies dealing with drought: a case study in Fars Province. *World Applied Sciences Journal* 10: 1122–8.

Karter, M.J. (1992) *Fire Loss in the United States during 1991.* Fire Analysis and Research Division, National Fire Protection Association, Quincy, MA.

Kaskaoutis, D.G. *et al.* (2011) Satellite monitoring of the biomass-burning aerosols during the wildfires of August 2007 in Greece: climate implications. *Atmospheric Environment* 45: 716–26.

Kasperson, R.E. *et al.* (1988) The social amplification of risk: a conceptual framework. *Risk Analysis* 8: 177–87.

Kassahun, A., Snyman, H.A. and Smit, G.N. (2008) The impact of rangeland degradation on the pastoral production systems, livelihoods and perceptions of the Somali pastoralists in Eastern Ethiopia. *Journal of Arid Environments* 72: 1265–81.

Kates, R.W. (1962) *Hazard and Choice Perception in Flood Plain Management.* Paper 78, Department of Geography, University of Chicago, Chicago, IL.

Kates, R.W. (1971) Natural hazard in human ecological perspective: hypotheses and models. *Economic Geography* 47: 438–51.

Kates, R.W. *et al.* (2001) Sustainability science. *Science* 292: 641–2.

Keating, B.H. and McGuire, W. (2000) Island edifice failures and associated tsunami hazards. *Pure and Applied Geophysics* 157: 899–955.

Keefer, D.K. (1984) Landslides caused by earthquakes. *Bulletin of the Geological Society of America* 95: 406–21.

Keefer, D.K. *et al.* (1987) Real-time landslide warning during heavy rainfall. *Science* 238: 921–5.

Keeling, C.D. (1960) The concentration and isotopic abundance of carbon dioxide in the atmosphere. *Tellus* 12: 200–3.

Keeney, R.L. (1995) Understanding life-threatening risks. *Risk Analysis* 15: 627–37.

Keeney, R.L. and von Winterfeldt, D. (1994) Managing nuclear waste from power plants. *Risk Analysis* 14: 107–8.

Keeves, A. and Douglas, D.R. (1983) Forest fires in South Australia on 16 February 1983 and consequent future forest management aims. *Australian Forestry* 46: 148–64.

Kellet, J. (2010) *The Pakistan Flooding: Three Months on and the Inequitable Response Remains.* Global Humanitarian Assistance at http://www.global humanitarianassistance.org (accessed 27 March 2011).

Kelly, M. (1992) Anthropometry as an indicator of access to food in populations prone to famine. *Food Policy* 17: 443–54.

Kelly, M. and Buchanan-Smith, M. (1994) Northern Sudan in 1991: food crisis and the international relief response. *Disasters* 18: 16–34.

Kerle, N. and Oppenheimer, C. (2002) Satellite remote sensing as a tool in lahar disaster management. *Disasters* 26: 140–60.

Kervyn, F. (2001) Modelling topography with SAR interferometry: illustrations of a favourable and less favourable environment. *Computers and Geosciences* 27: 1039–50.

Key, D. (ed.) (1995) *Structures to Withstand Disaster.* Institution of Civil Engineers and Thomas Telford, London.

Keyantash, J. and Dracup, J.A. (2002) The quantification of drought: an evaluation of drought indices. *Bulletin of the American Meteorological Society* 83: 1167–80.

Khan, F.I. and Abbasi, S.A. (2001) An assessment of the likelihood of occurrence and the damage potential of domino effect (chain of accidents) in a typical cluster of industries. *Journal of Loss Prevention in the Process Industries* 14: 283–306.

Kick, E.L., Fraser, J.C., Fulkerson, G.M., McKinney, L.A. and de Vries, D.H. (2011) Repetitive flood victims and acceptance of FEMA mitigation offers: an analysis with community-system policy implications. *Disasters* 35: 510–39.

Kidon, J., Fox, G., McKenney, D. and Rollins, K. (2002) An enterprise-level economic analysis of losses and financial assistance for eastern Ontario maple syrup producers from the 1998 ice storm. *Forest Policy and Economics* 4: 201–11.

Kilburn, C.R.J. and Petley, D.N. (2003) Forecasting giant catastrophic slope collapse: lessons from Vajont, northern Italy. *Geomorphology* 54: 21–32.

Kim, N. (2012) How much more exposed are the poor to natural disasters? Global and regional measurement. *Disasters* 36: 195–211.

Kintisch, E. (2005) Levees came up short, researchers tell Congress. *Science* 310: 953–5.

Kirchsteiger, C. (1999) Trends in accidents, disasters and risk sources in Europe. *Journal of Loss Prevention in the Process Industries* 12: 7–17.

Kitler, M.E., Gavinio, P. and Lavanchy, D. (2002) Influenza and the work of the World Health Organisation. *Vaccine (Supplement 2)* 20: 5–14.

Kiyono, J. and Kalantari, A. (2004) Collapse mechanism of adobe and masonry structures during the 2003 Iran Bam earthquake. *Bulletin of Earthquake Research Institute, University of Tokyo* 13: 157–61.

Klein, R.J.T., Nicholls, R.J. and Thomalla, F. (2003) Resilience to natural hazards: how useful is this concept? *Environmental Hazards* 5: 35–45.

Kleinberg, E. (2002) *Heat Wave: A Social Autopsy of Disaster in Chicago.* University of Chicago Press, Chicago and London.

Kling, G.W. *et al.* (2005) Degassing Lakes Nyos and Momoun: defusing certain disaster. *Proceedings of the National Academy of Sciences of the USA* 102: 1485–90.

Knox, J.C. (2000) Sensitivity of modern and Holocene floods to climate change. *Quaternary Science Reviews* 19: 439–57.

Knutson, T.R. *et al.* (2010) Tropical cyclones and climate change. *Nature Geoscience* 3: 152–63.

Kobayashi, Y. (1981) Causes of fatalities in recent earthquakes in Japan. *Journal of Disaster Science* 3: 15–22.

Kocin, P.J. and Uccellini, L.W. (2004) A snowfall impact scale derived from North-East storm snowfall distributions. *Bulletin of the American Meteorological Society* 85: 177–94.

Koe, L.C.C., Arellano, A.F. Jr and McGregor, J.L. (2001) Investigating the haze transport from 1997 biomass burning in south-east Asia: its impact upon Singapore. *Atmospheric Environment* 35: 2723–34.

Kuhn, K., Campbell-Lendrum, D., Haines, A. and Cox, J. (2005) *Using Climate to Predict Infectious Disease Epidemics.* World Health Organization, Geneva.

Kummu, M. and Varis, O. (2010) The world by latitudes: a global analysis of human population, development level and environment across the north–south axis over the past half century. *Applied Geography* 31: 495–507.

Korecha, D. and Barnston, A.G. (2007) Predictability of June–September rainfall in Ethiopia. *Monthly Weather Review* 135: 628–50.

Kovats, R., Bouma, M., Hajat, S., Worrall, E. and Haines, A. (2003) El Niño and health. *The Lancet* 362: 1481–9.

Kovats, S. (2008) *Health Effects of Climate Change in the UK.* Health Protection Agency, London.

Kreibich, H., Christenberger, S. and Schwarze, R. (2011) Economic motivation of households to undertake private precautionary measures against floods. *Natural Hazards and Earth System Sciences* 11: 309–21.

Krewski, D., Clayson, D. and McCullough, R.S. (1982) Identification and measurement of risk. In Burton, I., Fowle, C.D. and McCullough, R.S. (eds) *Living with Risk.* Environmental Monograph 3. Institute of Environmental Studies, University of Toronto, Toronto, 7–23.

Krishnamurthy, P.K., Fisher, J.B. and Johnson, C. (2011) Mainstreaming local perceptions of hurricane risk into policy-making: a case study of community GIS in Mexico. *Global Environmental Change* 21: 143–53.

Kuhn, K., Campbell-Lendrum, D., Haines, A. and Cox, J. (2005) *Using Climate to Predict Infectious Disease Epidemics.* World Health Organization, Geneva.

Kumar, K.K., Rajagopatan, B., Hoerling, M., Bates, G. and Cane, M. (2006) Unravelling the mystery of Indian monsoon failure during El Niño. *Science* 314: 115–19.

Kunii, O., Nakamura, S., Abdur, R. and Wakai, S. (2002) The impact on health and risk factors of the diarrhoea epidemics in the 1998 Bangladesh floods. *Public Health* 116: 68–74.

Kunreuther, H. (2008) Reducing losses from catastrophic risk through long-term insurance and mitigation. *Social Research* 75: 905–30.

Kunreuther, H. and Pauly, M. (2006) Rules rather than discretion: lessons from Hurricane 'Katrina'. *Journal of Risk and Uncertainty* 33: 101–16.

Kusler, J. and Larson, L. (1993) Beyond the ark: a new approach to US floodplain management. *Environment* 35: 7–34.

Lacoursiere, P.E.J-P. (2005) Bhopal and its effects on the Canadian regulatory framework. *Journal of Loss Prevention in the Process Industries* 18: 353–9.

Lagadec, P. (1982) *Major Technological Risk: An Assessment of Industrial Disasters.* Pergamon Press, Oxford.

Lagadec, P. (1987) From Seveso to Mexico and Bhopal: learning to cope with crises. In Kleindorfer,

P.R. and Kunreuther, H.C. (eds) *Insuring and Managing Hazardous Risks.* Springer-Verlag, New York, 13–27.

Lagadec, P. (2004) Understanding the French 2003 heat wave experience: beyond the heat, a multi-layered challenge. *Journal of Contingencies and Crisis Management* 12: 160–9.

Lambert, S.J. (1996) Intense extratropical northern hemisphere winter cyclone events: 1899–1991. *Journal of Geophysical Research* 101: 21319–25.

Lander, J.F., Whiteside, L.S. and Lockridge, P.A. (2003) Two decades of global tsunamis 1982–2002. *Science of Tsunami Hazards* 21: 3–88.

Landsea, C.W. (2000) Climate variability of tropical cyclones. In Pielke, R.A. Jr and Pielke, R.A. Sr (eds) *Storms*, vol. 1. Routledge, London and New York, 220–41.

Landsea, C.W. (2005) Meteorology: hurricanes and global warming. *Nature* 438: E11–E12.

Landsea, C.W. (2007) Counting Atlantic tropical cyclones back to 1900. *Eos* 88: 197–202.

Landsea, C.W., Harper, B.A., Hoarau, K. and Knaff, J.A. (2006) Can we detect trends in extreme tropical cyclones? *Science* 313: 452–4.

Lane, L.R., Tobin, G.A. and Whiteford, L.M. (2003) Volcanic hazard or economic destitution: hard choices in Baños, Ecuador. *Environmental Hazards* 5: 23–34.

Langenbach, R. (2005) Performance of the earthen Arg-e-Bam (Bam citadel) during the Bam, Iran, earthquake. *Earthquake Spectra* 21: S345–74.

Larsen, J.C., Dennison, P.E., Cova, T.J. and Jones, C. (2011) Evaluating dynamic wildfire evacuation triggers using the 2003 Cedar Fire. *Applied Geography* 31: 12–19.

Lavigne, F., Thouret, J.C., Voight, B., Suwa, H. and Sumaryono, A. (2000) Lahars at Merapi volcano, central Java: an overview. *Journal of Volcanology and Geothermal Research* 100: 423–56.

Lazo, J.K. and Waldman, D.M. (2011) Valuing improved hurricane forecasts. *Economics Letters* 111: 43–6.

Le, T.V.H., Nguyen, H.N., Wolanski, E., Tran, T.C. and Haruyama, S. (2006) The combined impact on the flooding of Vietnam's Mekong river delta of local man-made structures, sea level rise and dams upstream in the river catchment. *Estuarine, Coastal and Shelf Science* 65: 1–7.

Leader, N. (2000) *The Politics of Principle: The Principles of Humanitarian Action in Practice.* Report 2, Humanitarian Policy Group, Overseas Development Institute, London.

Ledoux, L., Cornell, S., O'Riordan, T., Harvey, R. and Banyard, L. (2005) Towards sustainable flood and coastal management: identifying drivers of, and obstacles to, managed realignment. *Land Use Policy* 22: 129–44.

Lee, H.F. and Zhang, D.D. (2011) Relationship between NAO and drought disasters in north-western China in the last millennium. *Journal of Arid Environments* 75: 1114–20.

Leimena, S.L. (1980) Traditional Balinese earthquake-proof housing structures. *Disasters* 4: 147–50.

Lein, J.K. and Stump, N.I. (2009) Assessing wildfire potential within the wildland–urban interface: a south-eastern Ohio example. *Applied Geography* 29: 21–34.

Lenton, T.M. *et al.* (2008) Tipping elements in the Earth's climate system. *Proceedings of the National Academy of Sciences* 105: 1786–93.

Leone, F. and Lesales, T. (2009) The interest of cartography for a better perception and management of volcanic risk: from scientific to social representations. The case of Mt Pelée volcano, Martinique (Lesser Antilles). *Journal of Volcanology and Geothermal Research* 186: 186–94.

Lewis, H.W. (1990) *Technological Risk.* W.W. Norton, New York.

Liao, Y-H. *et al.* (2005) Deaths related to housing in 1999 Chi-Chi, Taiwan, earthquake. *Safety Science* 43: 29–37.

Ligon, B.L. (2004) Dengue fever and dengue hemorrhagic fever; a review of the history, transmission, treatment and prevention. *Seminars in Paediatric Infectious Diseases*, 15, Elsevier, 60–5.

Ligon, B.L. (2006) Infectious diseases that pose specific challenges after natural disasters; a review. *Seminars in Paediatric Infectious Diseases*, 17, Elsevier, 36–45.

Lin, F.C., Zhu, W. and Sookhanaphibarn, K. (2011) Observation of tsunami radiation at Tōhoku by remote sensing. *Science of Tsunami Hazards* 30: 223–32.

Lindell, M.K. and Perry, R.W. (1996) Identifying and managing conjoint threats: earthquake-induced hazardous materials releases in the US. *Journal of Hazardous Materials* 50: 31–40.

Lindell, M.K. and Perry, R.W. (2000) Household adjustment to earthquake hazards: a review of research. *Environment and Behavior* 32: 461–501.

Lindell, M.K. and Whitney, D.J. (2000) Correlates of household seismic hazard adjustment adoption. *Risk Analysis* 20: 13–25.

Lindell, M.K., Lu, J-C. and Prater, C.S. (2005) Household decision-making and evacuation in response to Hurricane Lili. *Natural Hazards Review* 6: 171–9.

Linnerooth-Bayer, J., Mechler, R. and Pflug, G. (2005) Refocusing disaster aid. *Science* 309: 1044–6.

Litman, T. (2006) Lessons from Katrina and Rita: what major disasters can teach transportation planners. *Journal of Transportation Engineering* 132: 11–18.

Liu, W.T. and Kogan, F.N. (1996) Monitoring regional drought using the Vegetation Condition Index. *International Journal of Remote Sensing* 17: 2761–82.

Liu, Y., Stanturf, J. and Goodrick, S. (2010) Trends in global wildfire potential in a changing climate. *Forest Ecology and Management* 259: 685–97.

Lloyd, S.J., Kovats, R.S. and Armstrong, B.G. (2007) Global diarrhoea morbidity, weather and climate. *Climate Research* 34: 119–27.

Lott, J.N. (1994) The US summer of 1993: a sharp contrast in weather extremes. *Weather* 49: 370–83.

Loughnan, M., Nicholls, N. and Tapper, N. (2010) Mortality-temperature thresholds for ten major populations centres in rural Victoria, Australia. *Health and Place* 16: 1287–90.

Lounibos, L.P. (2002) Invasions by insect vectors of human disease. *Annual Review of Entomology* 47: 233–66.

Lucas-Smith, P. and McRae, R. (1993) Fire risk problems in Australia. *STOP Disasters* 11: 3–4.

Luke, R.H. and McArthur, A.G. (1978) *Bushfires in Australia*. Australian Government Publishing Service, Canberra.

Luna, E.M. (2001) Disaster mitigation and preparedness: the case of NGOs in the Philippines. *Disasters* 25: 216–26.

Lund, D.C., Lynch-Stieglitz, J. and Curry, W.B. (2006) Gulf Stream density structure and transport during the past millennium. *Nature* 444: 601–4.

Maantay, J. and Maroko, A. (2009) Mapping urban flood risk: flood hazards, race and environmental justice. *Applied Geography* 29: 111–24.

McAneney, J., Chen, K. and Pitman, A. (2009) 100-years of Australian bushfire losses: is the risk significant and is it increasing? *Journal of Environmental Management* 90: 2819–22.

McCabe, G.J., Palecki, M.A. and Betancourt, J.L. (2004) Pacific and Atlantic Ocean influences on multi-decadal drought frequency in the United States. *Proceedings of the National Academy of Sciences* 101: 4136–41.

McCright, A.M. and Dunlap, R.E. (2011) Cool dudes: the denial of climate change among conservative white males in the United States. *Global Environmental Change* 21: 1163–72.

McDaniels, T.L., Karalet, M.S. and Fischer, G.W. (1992) Risk perception and the value of safety. *Risk Analysis* 12: 495–503.

McDonald, R.E. (2011) Understanding the impact of climate change on Northern Hemisphere extra-tropical cyclones. *Climate Dynamics* 37: 1399–425.

McEntire, D.A. (2004) Development, disasters and vulnerability: a discussion of divergent theories and the need for their integration. *Disaster Prevention and Management* 13: 193–8.

McGee, T.K. (2005) Completion of recommended WUI fire mitigation measures within urban households in Edmonton, Canada. *Environmental Hazards* 6: 147–57.

McGee, T.K. and Russell, S. (2003) 'It's just a natural way of life…' An investigation of wildfire preparedness in rural Australia. *Environmental Hazards* 5: 1–12.

McGranahan, G., Balk, D. and Anderson, B. (2007) The rising tide: assessing the risks of climate change and human settlements in low elevation coastal zones. *Environment and Urbanization* 19: 17–37.

MacGregor, D. *et al.* (1994) Perceived risks of radioactive waste transport through Oregon: results of a state-wide survey. *Risk Analysis* 14: 5–14.

McGregor, A. (2010) Sovereignty and the responsibility to protect: the case of Cyclone Nargis. *Political Geography* 29: 3–4.

McGuire, B., Mason, I. and Kilburn, C. (2002) *Natural Hazards and Environmental Change*. E. Arnold, London.

McGuire, L.C., Ford, E.S. and Okoro, C.A. (2007) Natural disasters and older US adults with disabilities: implications for evacuation. *Disasters* 31: 49–56.

McKay, J.M. (1983) Newspaper reporting of bushfire disaster in south-eastern Australia: Ash Wednesday 1983. *Disasters* 7: 283–90.

McLennan, J. and Birch, A. (2005) A potential crisis in wildfire emergency response capability? Australia's volunteer firefighters. *Environmental Hazards* 6: 101–7.

McMichael, T. (2001) *Human Frontiers, Environments and Disease: Past Patterns, Uncertain Futures*. Cambridge University Press, Cambridge.

McNutt, S.R. (1996) Seismic monitoring and eruption forecasting of volcanoes: a review of the state of the art and case histories. In Scarpa, R. and Tilling, R.I. (eds) *Monitoring and Mitigation of Volcano Hazards*, Springer-Verlag, Berlin, 99–146.

McPhaden, M.J., Zebiak, S.E. and Glantz, M.H. (2006) ENSO as an integrating concept in earth science. *Science* 314: 1740–5.

Macrae, J. *et al.* (2002) *Uncertain Power: The Changing Role of Official Donors in Humanitarian Action*. Report 12, Humanitarian Policy Group, Overseas Development Institute, London.

McSaveney, M.J. (2002) Recent rockfalls and rock avalanches in Mount Cook National Park, New Zealand. In Evans, S.G. and DeGraff, J.V. (eds) *Catastrophic Landslides: Effects, Occurrence and Mechanisms.* Geological Society of America, Boulder, CO, 35–70.

Macklin, M.G. and Rumsby, B.T. (2007) Changing climate and extreme floods in the British uplands. *Transactions of the Institute of British Geographers* 32: 168–86.

Major, J.J., Schilling, S.P., Pullinger, C.R., Escobar, C.D. and Howell, M.M. (2001) *Volcano-Hazard Zonation for San Vicente Volcano, El Salvador.* Open-File Report 01–387, US Geological Survey, Vancouver, Washington.

Malheiro, A. (2006) Geological hazards in the Azores archipelago: volcanic terrain instability and human vulnerability. *Journal of Volcanology and Geothermal Research* 156: 158–71.

Malingreau, J.P. and Kasawanda, X. (1986) Monitoring volcanic eruptions in Indonesia using weather satellite data: the Colo eruption of July 28 1983. *Journal of Volcanology and Geothermal Research* 27: 179–94.

Malmquist, D.L. and Michaels, A.F. (2000) Severe storms and the insurance industry. In Pielke, R.A. Jr and Pielke, R.A. Sr (eds) *Storms,* vol.1. Routledge, London and New York, 54–69.

Malone, A.W. (2005) The story of quantified risk and its place in slope safety policy in Hong Kong. In Glade, T., Anderson, M.G. and Crozier, M. (eds) *Landslide Hazard and Risk.* John Wiley, Chichester, 643–74.

Maneyena, S.B. (2006) The concept of resilience revisited. *Disasters* 30 (4): 433–50.

Mann, M.E. (2007) Climate over the past two millenia. *Annual Review of Earth and Planetary Sciences* 35: 111–36.

Mannan, M.S. *et al.* (2005) The legacy of Bhopal: the impact over the last 20 years and future direction. *Journal of Loss Prevention in the Process Industries* 18: 218–24.

Marchi, L., Borga, M., Preciso, E. and Gaume, E. (2010) Characterization of selected extreme flash floods in Europe and implications for flood risk management. *Journal of Hydrology* 394: 118–33.

Marco, J.B. (1994) Flood risk mapping. In Rossi, G., Harmancioglu, N. and Yevjevich, V. (eds) *Coping with Floods.* Kluwer, Dordrecht, 353–73.

Margreth, S. and Funk, M. (1999) Hazard mapping for ice and combined snow/ice avalanches – two case studies from the Swiss and Italian Alps. *Cold Regions Science and Technology* 30: 159–73.

Marinos, P., Bouckovalas, G., Tsiambaos, G., Sabatakakis, N. and Antoniou, A. (2001) Ground zoning against seismic hazard in Athens, Greece. *Engineering Geology* 62: 343–56.

Marsden, B.G. and Steel, D.I. (1994) Warning times and impact probabilities for long-period comets. In Gehrels, T. (ed.) *Hazards due to Asteroids and Comets.* University of Arizona Press, Tucson, 221–39.

Marsh, G.P (1864) *Man and Nature.* Charles Scribner, New York.

Martens, P., McMichael, A., Kovats, S. and Livermore, M. (1998) Impacts of climate change on human health: malaria. In *Climate Change and Its Impacts.* Meteorological Office and Department of Energy, Transport and the Regions, Bracknell, Berks.

Martin, W.E., Martin, I.M. and Kent, B. (2009) The role of risk perception in the risk mitigation process: the case of wildfire in high risk communities. *Journal of Environmental Management* 91: 489–98.

Mason, J. and Cavalie, P. (1965) Malaria epidemic in Haiti following a hurricane. *American Journal of Tropical Medicine and Hygiene* 14: 533–9.

Mathur, K. and Jayal, N.G. (1992) Drought management in India: the long term perspective. *Disasters* 16: 60–5.

Mattinen, H. and Ogden, K. (2006) Cash-based interventions: lessons from southern Somalia. *Disasters* 30: 297–315.

Matzarakis, A., Muthers, S. and Koch, E. (2011) Human biometeorological evaluation of heat-related mortality in Vienna. *Theoretical and Applied Climatology* 105: 1–10.

Maxwell, D. (2007) Global factors shaping the future of food aid: the implications for WFP. *Disasters* 31 (Supplement 1): S25–S39.

Meehl, G.A., Karl, T., Easterling, D.R. *et al.* (2000) An introduction to trends in extreme weather and climate events: observations, socio-economic impacts, terrestrial ecological impacts and model projections. *Bulletin of the American Meteorological Society* 81: 413–16.

Mejía-Navarro, M. and Garcia, L.A. (1996) Natural hazard and risk assessment using decision-support systems: Glenwood Springs, Colorado. *Environmental and Engineering Geoscience* 1: 291–98.

Meltsner, A.J. (1978) Public support for seismic safety: where is it in California? *Mass Emergencies* 3: 167–84.

Menoni, S. (2001) Chains of damages and failures in a metropolitan environment: some observations on the Kobe earthquake in 1995. *Journal of Hazardous Materials* 86: 101–19.

Mercer, D. (1971) Scourge of an arid continent. *Geographical Magazine* 45: 563–7.

Mercer, J. (2010) Disaster risk reduction or climate change adaptation: are we reinventing the wheel? *Journal of International Development* 22: 247–64.

Merz, B., Kreibich, H., Schwarze, R. and Thieken, A. (2010) Assessment of economic flood damage. *Natural Hazards and Earth System Sciences* 10: 1697–724.

Messerli, B., Grosjean, M., Hofer, T., Lautaro, N. and Pfister, C. (2000) From nature-dominated to human-dominated environmental changes. *Quaternary Science Reviews* 19: 459–79.

Meyer, S.J. (1993) A crop specific drought index for corn: application in drought monitoring and assessment. *Agronomy Journal* 85: 396–9.

Meze-Hausken, E. (2004) Contrasting climate variability and meteorological drought with perceived drought and climate change in northern Ethiopia. *Climate Research* 27: 19–31.

Mileti, D.S. and Darlington, J.D. (1995) Societal responses to revised earthquake probabilities in the San Francisco Bay area. *International Journal of Mass Emergencies and Disasters* 13: 119–45.

Mileti, D.S. and Myers, M.F. (1997) A bolder course for disaster reduction: imagining a sustainable future. *Revista Geofisica* 47: 41–58.

Mileti, D.S., Darlington, J.D., Passerine, E., Forrest, B.C. and Myers, M.F. (1995) Toward an integration of natural hazards and sustainability. *Environmental Professional* 17: 117–26.

Mileti, D.S. *et al.* (1999) *Disasters by Design: A Reassessment of Natural Hazards in the United States.* Joseph Henry Press, Washington, DC.

Miller, A. and Goidel, R. (2009) News organizations and information gathering during a natural disaster; lessons from Hurricane Katrina. *Journal of Contingencies and Crisis Management* 17: 266–73.

Miller, C.D., Mullineaux, D.R. and Crandell, D.R. (1981) Hazards assessments at Mount St Helens. In Lipman, R.W. and Mullineaux, D.R. (eds) *The 1980 Eruption of Mount St Helens, Washington.* US Geological Survey Professional Paper 1250: 789–802.

Millist, N. and Abdalla, A. (2011) *Benefit-cost Analysis of Australian Locust Control Operations for 2010–11.* ABARES Report, Australian Plague Locust Commission, Canberra.

Mills, E. (2005) Insurance in a climate of change. *Science* 309: 1040–3.

Mills, J.W., Curtis, A., Kennedy, B., Kennedy, S.W. and Edwards, J.D. (2010) Geospatial video for field data collection. *Applied Geography* 30: 533–47.

Mirza, M.M. Q. (2002) Global warming and changes in the probability of occurrence of floods in Bangladesh and implications. *Global Environmental Change* 12: 127–38.

Mirza, M.M.Q., Warrick, R.A., Ericksen, N.J. and Kenny, G.J. (2001) Are floods getting worse in the Ganges, Brahmaputra and Meghna basins? *Environmental Hazards* 3: 37–48.

Mishra, A.K. and Singh, V.P. (2010) A review of drought concepts. *Journal of Hydrology* 391: 202–16.

Mitchell, J.K. (1999) Natural disasters in the context of mega-cities. In Mitchell, J.K. (ed.) *Crucibles of Hazard.* United Nations University Press, Tokyo, 15–55.

Mitchell, J.K. (2003) European river floods in a changing world. *Risk Analysis* 23: 567–71.

Mitsch, W.J. and Day, J.W. (2006) Restoration of wetlands in the Mississippi–Ohio–Missouri (MOM) River Basin: experience and needed research. *Ecological Engineering* 26: 55–69.

Mizutani, T. and Nakano, T. (1989) The impact of natural disasters on the population of Japan. In Clarke, J.I., Curson, P., Kayastha, S.L. and Nag, P. (eds) *Population and Disaster.* Oxford, Basil Blackwell, 24–33.

Molyneux, D.H. (1998) Vector-borne parasitic diseases – an overview of recent changes. *International Journal of Parasitology* 28: 927–34.

Montero, J.C., Mir_n, I.J., Criado-Álvarez, J.J., Linares, C. and Diaz, J. (2010) Mortality from cold waves in Castile-La Mancha, Spain. *Science of the Total Environment* 408: 5768–74.

Montz, B.E. and Gruntfest, E.C. (1986) Changes in American floodplain occupancy since 1958: the experiences of nine cities. *Applied Geography* 6: 325–38.

Montz, B.E. and Gruntfest, E.C. (2002) Flash flood mitigation: recommendations for research and applications. *Environmental Hazards* 4: 15–22.

Montz, B.E. and Tobin, G.A. (2011) Natural hazards: an evolving tradition in applied geography. *Applied Geography* 31: 1–4.

Moore, P.G. (1983) *The Business of Risk.* Cambridge University Press, Cambridge.

Morris, A. (2003) Understandings of catastrophe: the landslide at La Josefina, Ecuador. In Pelling, M. (ed.) *Natural Disasters and Development in a Globalizing World.* Routledge, London, 157–69.

Morris, J. *et al.* (1982) Cholera among refugees in Rangsil, Thailand. *Journal of Infectious Diseases* 1: 131–4.

Morris, S.S. *et al.* (2002) Hurricane 'Mitch' and the livelihoods of the rural poor in Honduras. *World Development* 30: 49–60.

Morrison, D. (2006) Asteroid and comet impacts: the ultimate environmental catastrophe. *Philosophical Transactions of the Royal Society A: Mathematical, Physical and Engineering Sciences* 364: 2041–54.

Morrow, B.H. (1999) Identifying and mapping community vulnerability. *Disasters* 23: 1–18.

Mortimore, M.J. and Adams, W.M. (2001) Farmer adaptation, change and 'crisis' in the Sahel. *Global Environmental Change* 11: 49–57.

Morton, A. (1998) Hong Kong: managing slope safety in urban systems. *STOP Disasters* 33: 8–9.

Morton, J. and Barton, D. (2002) Destocking as a drought-mitigation strategy: clarifying rationales and answering critiques. *Disasters* 26: 213–28.

Mosquera-Machado, S. and Dilley, M. (2009) A comparison of selected global disaster risk assessment results. *Natural Hazards* 48: 439–56.

Moszynski, P. (2004) Cold is the main health threat after the Bam earthquake. *British Medical Journal* 328: 66.

Mothes, P.A. (1992) Lahars of Cotopaxi volcano, Ecuador: hazard and risk evaluation. In McCall, G.J.H., Laming, D.J.C. and Scott, S.C. (eds) *Geohazards*. Chapman and Hall, London, 53–63.

Mozumber, P., Helton, R. and Berrens, R.P. (2009) Provision of a wildfire risk map: informing residents in the wildland–urban interface. *Risk Analysis* 29: 1588–600.

Msangi, J.P. (2004) Drought hazard and desertification management in the dry lands of Southern Africa. *Environmmental Monitoring and Assessment* 99: 75–87.

Mulady, J.J. (1994) Building codes: they're not just hot air. *Natural Hazards Observer* 18: 4–5.

Munich Re (1999) *Topics 2000*. Report of the Geoscience Research Group, Munich Reinsurance Company, Munich.

Munich Re (2001) *Topics 2001*. Report of the Geoscience Research Group, Munich Reinsurance Company, Munich.

Munich Re (2002a) *Topics 2002*. Report of the Geoscience Research Group, Munich Reinsurance Company, Munich.

Munich Re (2002b) *Winter Storms in Europe (II): Analysis of 1999 Losses and Loss Potentials*. Geo Risks Research Department, Munich Re Group, Munich.

Munich Re (2005) *Topics Geo Annual Review: Natural Catastrophes 2005*. Geo Risks Research. Munich Re Group, Munich.

Munich Re (2006) *Topics Geo Annual Review: Natural Catastrophes 2006*. Geo Risks Research. Munich Re Group, Munich.

Muntán, E. *et al.* (2009) Reconstructing snow avalanches in the South-Eastern Pyrenees. *Natural Hazards and Earth System Sciences* 9: 1599–612.

Murphy, F.A. and Nathanson, N. (1994) The emergence of new virus diseases: an overview. *Seminars in Virology* 5: 87–102.

Murphy, S.J. *et al.* (2001) Seasonal forecasting for climate hazards: prospects and responses. *Natural Hazards* 23: 171–96.

Mustafa, D. (2003) Reinforcing vulnerability? Disaster relief, recovery and response to the 2001 flood in Rawalpindi, Pakistan. *Environmental Hazards* 5: 71–82.

Mustafa, D., Ahmed, S., Saruch, E. and Bell, H. (2011) Pinning down vulnerability: from narratives to numbers. *Disasters* 35: 62–86.

Nash, J.R. (1976) *Darkest Hours: A Narrative Encyclopaedia of Worldwide Disasters from Ancient Times to the Present*. Nelson Hall, Chicago, IL.

Nasrabadi, A.A., Naji, H., Mirzabeigi, G. and Dadbakhs, M. (2007) Earthquake relief: Iranian nurses' responses in Bam, 2003, and lessons learned. *International Nursing Review* 54: 13–18.

National Weather Service (2006) *Hurricane Katrina August 23–31 2005: Service Assessment*. National Oceanic and Atmospheric Administration, US Department of Commerce, Silver Spring, MD.

Neal, C.A. and Guffanti, M.C. (2010) *Airborne Volcanic Ash – A Global Threat to Aviation*. Fact Sheet 3116, US Geological Survey, Anchorage, AK.

Neal, J. and Parker, D.J. (1988) *Floodplain Encroachment: A Case Study of Datchet, UK*. Geography and Planning Paper 22, Middlesex Polytechnic, Enfield, Middlesex.

Nelson, A.C. and French, S.P. (2002) Plan quality and mitigating damage from natural disasters. *Journal of the American Planning Association* 68: 194–207.

Neumayer, E. and Barthel, F. (2011) Normalizing economic loss from natural disasters: a global analysis. *Global Environmental Change* 21: 13–24.

Newell, P. (2011) The elephant in the room: capitalism and global environmental change. *Global Environmental Change* 21: 4–6.

Newhall, C., Hendley, J.W. and Stauffer, P.H. (1997) *Benefits of Volcanic Monitoring Far Outweigh Costs – The Case of Mount Pinatubo*. Fact Sheet 115–97, US Geological Survey, Vancouver, WA.

Newhall, C., Stauffer, P.H. and Hendley, J.W. (1997) *Lahars of Mount Pinatubo*. Fact Sheet 114–97, US Geological Survey, Vancouver, WA.

Newhall, C.G. and Self, S. (1982) The Volcanic Explosivity Index (VEI): an estimate of explosive magnitude for historical volcanism. *Journal of Geophysical Research* 87: 1231–8.

Newman, C.J. (1976) Children of disaster: clinical observations at Buffalo Creek. *American Journal of Psychiatry* 133: 306–12.

Nicholls, N. (2011) What caused the eastern Australian heavy rains and floods of 2010/11? *Bulletin of the Australian Meteorological and Oceanographic Society* 24: President's Column, Friday March 11.

Nicholls, R.J. (1998) Impacts on coastal communities. In *Climate Change and its Impacts*. Meteorological Office and Department of Energy Transport and the Regions, Bracknell, Berks.

Nicholls, R.J. and Cazenave, A. (2010) Sea-level rise and its impact on coastal zones. *Science* (18 June): 1517–20.

Nicholls, R.J., Hoozemans, F.M.J. and Marchand, M. (1999) Increasing flood risk and wetland losses due to global sea-level rise: regional and global analyses. *Global Environmental Change* 9: S69–S87.

Njome, M.S., Suh, C.E., Chuyong, G. and de Wit, M.J. (2010) Volcanic risk perception in rural communities along the slopes of Mount Cameroon, West-Central Africa. *Journal of African Earth Sciences* 58: 608–622.

Noji, E.K. (ed.) (1997) *The Public Health Consequences of Disaster*. Oxford University Press, New York.

Noji, E.K. (2001) The global resurgence of infectious diseases. *Journal of Contingencies and Crisis Management* 9: 223–32.

Noji, E.K., Armenian, H.K. and Oganessian, A. (1993) Issues of rescue and medical care following the 1988 Armenian earthquake. *International Journal of Epidemiology* 22: 1070–6.

Nordhaus, W.D. (2010) The economics of hurricanes and implications of global warming. *Climate Change Economics* 1: 1–20.

Nordstrom, K.F., Jackson, N.L., Bruno, M.S. and de Butts, H.A. (2002) Municipal initiatives for managing dunes in coastal residential areas: a case study of Avalon, New Jersey, USA. *Geomorphology* 47: 137–52.

Norris, F.H., Stevens, S.P., Pfefferbaum, B., Wych, K.F. and Pfefferbaum, R.L. (2008) Community resilience as a metaphor, theory, set of capacities and strategy for disaster readiness. *American Journal of Community Psychology* 41: 127–50.

Nowak, R. (2007) The continent that ran dry. *New Scientist* (16 June): 8–11.

Noy, I. and Vu, T. (2010) The economics of natural disasters in a developing country: The case of Vietnam. *Journal of Asian Economics* 21: 345–54.

Nyberg, J. *et al.* (2007) Low Atlantic hurricane activity in the 1970s and 1980s compared to the past 270 years. *Nature* 447 (7 June): 698–701.

OCHA (2010) *Pakistan: Initial Floods Emergency Response Plan*. UN Office for the Coordination of Humanitarian Affairs, New York and Geneva.

OECD (2003a) *Emerging Risks in the 21st Century: An Agenda for Action*. Organization for Economic Cooperation and Development, Publications Service, Paris.

OECD (2003b) *OECD Guiding Principles for Chemical Accident Prevention, Preparedness and Response*. Organization for Economic Cooperation and Development, Publications Service, Paris.

Oechsli, F.W. and Buechly, R.W. (1970) Excess mortality associated with three Los Angeles September hot spells. *Environmental Research* 3: 277–84.

OFDA (Office of US Foreign Disaster Assistance) (1994) *Annual Report Financial Year 1993*. Office of US Foreign Disaster Assistance, Agency for International Development, Washington, DC.

Ojaba, E., Leonardo, A.I. and Leonardo, M.I. (2002) Food aid in complex emergencies: lessons from Sudan. *Social Policy and Administration* 36: 664–84.

Okrent, D. (1980) Comment on societal risk. *Science* 208: 372–5.

Oliveira, S., Andrade, H. and Vaz, T. (2011) The cooling effect of green spaces as a contribution to the mitigation of urban heat: a case study of Lisbon. *Building and Environment* 46: 2186–94.

Olsen, G.R., Carstensen, N. and Høyen, K. (2003) Humanitarian crises: what determines the level of emergency assistance? Media coverage, donor interests and the aid business. *Disasters* 27: 109–26.

Olshansky, R.B. (2006) Planning after Hurricane Katrina. *Journal of the American Planning Association* 72: 147–54.

Olshansky, R.B. and Wu, Y. (2001) Earthquake risk analysis for Los Angeles County under present and planned land uses. *Environment and Planning B: Planning and Design* 28: 419–32.

Olurominiyi, O.I., Mushkatel, A. and Pijawka, K.D. (2004) Social and political amplification of technological hazards: the case of the PEPCON explosion. *Journal of Hazardous Materials* A114: 15–25.

Omachi, T. (1997) *Drought Conciliation and Water Rights: Japanese Experience*. Water Series 1, Infrastructure Development Institute, Tokyo.

O'Meagher, B., Stafford-Smith, M. and White, D.H. (2000) Approaches to integrated drought risk management. In Wilhite, D.A. (ed.) *Drought*, vol.2. Routledge, London and New York, 115–28.

Orindi, V.A., Nyong, A. and Herrero, M. (2007) *Pastoral Livelihood Adaptation to Drought and Institutional Interventions in Kenya*. Occasional Paper 54, Human Development Office, UNDP, Geneva.

O'Rourke, T.D., Hozer, T., Rojahn, T. and Tierney, K. (2008) *Contributions of Earthquake Engineering to Protecting Communities and Critical Infrastructure from Multihazards*. Earthquake Engineering Research Institute, Oakland, CA.

Ostro, B.D., Roth, L.A., Green, R.S. and Basu, R. (2009) Estimating the mortality effect of the July 2006 California heat wave. *Environmental Research* 109: 614–19.

Othman-Chande, M. (1987) The Cameroon volcanic gas disaster: an analysis of a makeshift response. *Disasters* 11: 96–101.

Overpeck, J.T. *et al.* (2006) Paleoclimatic evidence for future ice-sheet instability and rapid sea-level rise. *Science* 311: 1747–50.

Pal, I. and Al-Tabba, A. (2009) Trends in seasonal precipitation extremes: an indication of 'climate change' in Kerala, India. *Journal of Hydrology* 367: 62–9.

Pall, P. *et al.* (2011) Anthropogenic greenhouse gas contribution to flood risk in England and Wales in autumn 2000. *Nature* 470: 382–5.

Palmer, W.C. (1965) *Meteorological Drought*. Research Paper 45, US Weather Bureau, Department of Commerce, Washington, DC.

Parasuraman, S. (1995) The impact of the 1993 Latur-Osmanabad (Maharashtra) earthquake on lives, livelihoods and property. *Disasters* 19: 156–69.

Parise, M. (2001) Landslide mapping techniques and their use in the assessment of landslide hazard. *Solar, Terrestrial and Planetary Science* 26: 697–703.

Park, S.K., Dalrymple, W. and Larsen, J.C. (2007) The 2004 Parkfield earthquake: test of the electromagnetic precursor hypothesis. *Journal of Geophysical Research – Solid Earth* 112: B05302.

Parker, D.J. (ed.) (2000) *Floods*, vols 1 and 2. Routledge, London and New York.

Parker, D.J., Priest, S.J. and Tapsell, S.M. (2009) Understanding and enhancing the public's behavioural response to flood warning information. *Meterological Applications* 16: 103–14.

Parker, D.J., Priest, S.J. and McCarthy, S.S. (2011) Surface water flood warning requirements and potential in England and Wales. *Applied Geography* 31: 891–900.

Pascual, M., Cazelles, B., Bouma, M.J., Chaves, L.F. and Koelle, K. (2008) Shifting patterns: malaria dynamics and rainfall variability in an African highland. *Proceedings of the Royal Society (B)* 275: 123–32.

Paté-Cornell, M.E. (1993) Learning from the Piper Alpha accident: a postmortem analysis of technical and organizational factors. *Risk Analysis* 13: 215–32.

Patel, T. (1995) Satellite senses risk of forest fires. *New Scientist* (11 March): 12.

Paul, A. and Rahman, M. (2006) Cyclone mitigation perspective in the islands of Bangladesh: a case study of Sandwip and Hatia Islands. *Coastal Management* 34: 199–215.

Paul, B.K. (1995) Farmers' and public responses to the 1994–95 drought in Bangladesh: a case study. *Quick Response Report* 76. Natural Hazards Research and Applications Information Center, Boulder, CO.

Paul, B.K. (1997) Survival mechanisms to cope with the 1996 tornado in Tangail, Bangladesh: a case study. *Quick Response Report* 92, Natural Hazards Research and Applications Information Center, Boulder, CO.

Paul, B.K. (2003) Relief assistance to 1998 flood victims: a comparison of the performance of the government and NGOs. *The Geographical Journal* 169: 75–89.

Paul, B.K. and Bhuiyan, R.H. (2004) The April 2004 tornado in north-central Bangladesh: a case for introducing tornado forecasting and warning systems. *Quick Response Report* 169. Natural Hazards Research and Applications Information Center, Boulder, CO.

Paul, B.K. and Dutt, S. (2010) Hazard warnings and responses to evacuation orders: the case of Bangladesh's cyclone 'Sidr'. *Geographical Review* 100: 336–55.

Paul, B.K., Brock, V.Y., Csiki, S. and Emerson, L. (2003) Public response to tornado warnings: a comparative study of the May 4 2003 tornadoes in Kansas, Missouri and Tennessee. *Quick Response Report 165*, Natural Hazards Research and Applications Information Center, Boulder, CO.

Peacock, W.G., Morrow, B.H. and Gladwin, H. (eds) (1997) *Hurricane Andrew: Ethnicity, Gender and the Sociology of Disasters*. Routledge, London and New York.

Peacock, W.G., Brody, S.D. and Highfield, W. (2005) Hurricane risk perceptions among Florida's single family homeowners. *Landscape and Urban Planning* 73: 120–35.

Pechony, O. and Shindell, D.T. (2010) Driving forces of global wildfires over the past millennium and the forthcoming century. *Proceedings of the National Academy of Sciences of the United States of America* 107: 19167–70.

Peduzzi, P. (2006) The Disaster Risk Index: overview of a quantitative approach. In Birkmann, J. (ed.) *Measuring Vulnerability to Natural Hazards*, Teri Press, New Delhi, 265–89.

Peduzzi, P., Dao, H. and Herold, C. (2005) Mapping disastrous natural hazards using global datasets. *Natural Hazards* 35: 265–89.

Peduzzi, P., Dao, H., Herold, C. and Mouton, F. (2009) Assessing global exposure and vulnerability towards natural hazards: the Disaster Risk Index. *Natural Hazards and Earth System Sciences* 9: 1149–59.

Peduzzi, P. *et al.* (2011) *Preview Global Risk Data Platform.* UNEP/GRID and UN/ISDR, Geneva.

Peek-Asa, C., Ramirez, M., Shoaf, K. and Seligson, H. (2002) Population-based case-control study of injury risk factors in the Northridge earthquake. *Annals of Epidemiology* 12: 525–6.

Peel, M.C., McMahon, T.A. and Finlayson, B.L. (2002) Variability of annual precipitation and its relationship to the El Niño–Southern Oscillation. *Journal of Climate* 15: 545–51.

Peiris, L.M.N., Rossetto, T., Burton, P.W. and Mahmood, S. (2006) *EEFIT Mission: October 8 2005 Kashmir Earthquake.* Preliminary Reconnaisance Report, London.

Pelling, M. (2003) Introduction. In Pelling, M. (ed.) *Natural Disasters and Development in a Globalizing World.* Routledge, London and New York.

Pelling, M. (2007) Learning from others: the scope and challenges for participatory disaster risk assessment. *Disasters* 31: 373–85.

Pelling, M. and Uitto, J.I. (2001) Small island developing states: natural disaster vulnerability and global change. *Global Environmental Change B: Environmental Hazards* 3: 49–62.

Penning-Rowsell, E.C. and Wilson, T. (2006) Gauging the impact of natural hazards: the pattern and cost of emergency response during flood events. *Transactions of the Institute of British Geographers* 31: 99–115.

Perla, R.I. and Martinelli, M. Jr (1976) *Avalanche Handbook.* Agriculture Handbook 489: US Department of Agriculture (Forest Service), Washington, DC.

Perrow, C. (1999) *Natural Accidents: Living with High-risk Technologies* (2nd edn). Princeton University Press, Princeton, NJ.

Perry, A.H. and Symons, L.J. (eds) (1991) *Highway Meteorology.* E. and F.N. Spon, London.

Perry, C.A. (1994) *Effects of Reservoirs on Flood Discharges in the Kansas and the Missouri River Basins, 1993.* Circular 120E, US Geological Survey, Denver, CO.

Perry, R.W. and Lindell, M.K. (1990) Predicting long-term adjustment to volcano hazard. *International Journal of Mass Emergencies and Disasters* 8: 117–36.

Perry, R.W. and Godchaux, J.D. (2005) Volcano hazard management strategies: fitting policy to patterned human responses. *Disaster Prevention and Management* 14: 183–95.

Peterson, D.W. (1996) Mitigation measures and preparedness plans for volcanic emergencies. In Scarpa, R. and Tilling, R.I. (eds) *Monitoring and Mitigation of Volcano Hazards*, Springer-Verlag, Berlin, 701–18.

Petley, D.N. (2005) Tsunami – how an earthquake can cause destruction thousands of kilometres away. *Geography Review* 18: 2–5.

Petley, D.N. (2009) Contribution to *Environmental Hazards: Assessing Risk and Reducing Disaster* (5th edn), Routledge, London and New York.

Petley, D.N. (2010) On the impact of climate change and population growth on the occurrence of fatal landslides in South, East and SE Asia. *Quarterly Journal of Engineering Geology and Hydrogeology* 43(4): 487–96.

Petley, D.N., Hearn, G.J. and Hart, A. (2005) Towards the development of a landslide risk assessment for rural roads in Nepal. In Glade, T., Anderson, M. and Crozier, M.J. (eds) *Landslide Hazard and Risk*, John Wiley, Chichester, 597–620.

Petley, D.N., Dunning, S.A., Rosser, N.J. and Kausar, A.B. (2006) Incipient earthquakes in the Jhelum valley, Pakistan, following the 8th October 2005 earthquake. In Marui, H. (ed.) *Disaster Mitigation of Debris Flows, Slope Failures and Landslides.* Frontiers of Science Series 47, Universal Academy Press, Tokyo, Japan, 47–56.

Petley, D.N. *et al.* (2007) Trends in landslide occurrence in Nepal. *Natural Hazards* 43: 23–44.

Petrazzuoli, S.M. and Zuccaro, G. (2004) Structural resistance of reinforced concrete buildings under pyroclastic flows: a study of the Vesuvian area. *Journal of Volcanology and Geothermal Research* 133: 353–67.

Petrow, T., Zimmer, J. and Merz, B. (2009) Changes in the flood hazard in Germany through changing frequency and persistence of circulation patterns. *Natural Hazards and Earth System Sciences* 9: 1409–23.

Peyret, M. *et al.* (2007) The source motion of 2003 Bam (Iran) earthquake constrained by satellite and ground-based geodetic data. *Geophysical Journal International* 169: 849–65.

Phillips, B.D., Metz, W.C. and Nieves, L.A. (2005) Disaster threat: preparedness and potential response of the lowest income quartile. *Environmental Hazards* 6: 123–33.

Phukan, A.C., Borah, P.K., Biswas, D. and Mahanta, J. (2004) A cholera epidemic in a rural area of north-east India. *Transactions of the Royal Society of Tropical Medicine and Hygiene* 98: 563–6.

Pidgeon, N. and O'Leary, M. (2000) Man-made disasters: why technology and organisations sometimes fail. *Safety Science* 34: 15–30.

Pielke, R.A. Jr (1997) Reframing the US hurricane problem. *Society and Natural Resources* 10: 485–99.

Pielke, R.A. Jr and Pielke, R.A. Sr (1997) *Hurricanes: Their Nature and Impacts on Society*. John Wiley, Chichester and New York.

Pielke, R.A. Jr and Landsea, C.W. (1998) Normalised hurricane damages in the United States 1925–95. *Weather and Forecasting* 13: 621–31.

Pielke, R.A. Jr and Pielke, R.A. Sr (2000) *Storms*, vols 1 and 2. Routledge, London and New York.

Pielke, R.A. Jr and Klein, R. (2005) Distinguishing tropical cyclone-related flooding in US Presidential Disaster Declarations. *Natural Hazards Review* 6: 55–9.

Pielke, R.A. Jr., Landsea, C.W., Mayfield, M., Laver, J. and Pasch, R. (2005) Hurricanes and global warming. *Bulletin of the American Meteorological Society* 86: 1571–5.

Pigram, J.J. (1986) *Issues in the Management of Australia's Water Resources*. Longman Cheshire, Melbourne.

Pingali, P., Alinovi, L. and Sutton, J. (2005) Food security in complex emergencies: enhancing food system resilience. *Disasters* 29 (SI): S5–S24.

Pinter, N. (2005) One step forward, two steps back on US floodplains. *Science* 308: 207–8.

Plafker, G. and Eriksen, G.E. (1978) Nevados Huascaran avalanches, Peru. In Voight, B. (ed.) *Rockslides and Avalanches, vol. 1: Natural Disasters*, Elsevier, Amsterdam, 48–55.

Platt, R.H. (1999) Natural hazards of the San Francisco Bay mega-city: trial by earthquake, wind and fire.

In Mitchell, J.K. (ed.) *Crucibles of Hazard*, United Nations University Press, Tokyo, 335–74.

Platt, R.H., Salvesen, D. and Baldwin, G.H. (2002) Rebuilding the North Carolina coast after Hurricane Fran: did public regulations matter? *Coastal Management* 30: 249–69.

Ploughman, P. (1995) The American print news media 'construction' of five natural disasters. *Disasters* 19: 308–26.

Pomonis, A., Coburn, A.W. and Spence, R.J.S. (1993) Seismic vulnerability, mitigation of human casualties and guidelines for low-cost earthquake resistant housing. *STOP Disasters* 12 (March–April): 6–8.

Poortinga, W., Spence, A., Whitmarsh, L., Capstick, S. and Pidgeon, N.F. (2011) Uncertain climate: an investigation into public scepticism about anthropogenic climate change. *Global Environmental Change* 21: 1015–24.

Pottier, N., Penning-Rowsell, E., Tunstall, S. and Hubert, G. (2005) Land use and flood protection: contrasting approaches and outcomes in France and in England and Wales. *Applied Geography* 25: 1–27.

Poumadère, M., Mays, C., Le Mer, S. and Blong, R. (2005) The 2003 heat wave in France: dangerous climate change here and now. *Risk Analysis* 25: 1483–94.

Powell, M.D. (2000) Tropical cyclones during and after landfall. In Pielke, R. Jr and Pielke, R. Snr (eds) *Storms*, vol. 1. Routledge, London, 196–219.

Pradhan, E.K. *et al.* (2007) Risk of flood-related mortality in Nepal. *Disasters* 31: 57–70.

Preuss, J. (1983) Land management guidelines for tsunami hazard zones. In Tida, K. and Iwasaki, T. (eds) *Tsunamis*. Reidel, Boston, MA, 527–39.

Provention Consortium (2007) *Construction Design, Building Standards and Site Selection*. Guidance Note 12, Provention Consortium Secretariat, Geneva.

Pultar, E., Cova, T.J., Raubal, M. and Goodchild, M.F. (2009) Dynamic GIS case studies: wildfire evacuation and volunteered geographic information. *Transactions in GIS* 13: 85–104.

Pulwarty, R.S., Wilhite, D.A., Diodato, D.M. and Nelson, D.I. (2007) Drought in changing environments: creating a roadmap, vehicles and drivers. *Natural Hazards Observer* 31 (5): 10–12.

Purtill, A. (1983) A study of the drought. *Quarterly Review of the Rural Economy* 5: 3–11.

Pyle, A.S. (1992) The resilience of households to famine in El Fasher, Sudan, 1982–89. *Disasters* 16: 19–27.

Qi, R. *et al.* (2012) Challenges and needs for process safety in the new millennium. *Process Safety and Environmental Protection* 90: 91–100.

Quarantelli, E.L. (ed.) (1998) *What is a Disaster?* Routledge, London and New York.

Rahn, P.H. (1984) Floodplain management program in Rapid City, South Dakota. *Bulletin of the Geological Society of America* 95: 838–43.

Rahu, M. (2003) Health effects of the Chernobyl accident: fears, rumours and the truth. *European Journal of Cancer* 39: 295–9.

Ramabrahmam, B.V. and Swaminathan, G. (2000) Disaster management plan for chemical process industries. Case study: investigation of release of chlorine to atmosphere. *Journal of Loss Prevention in the Process Industries* 13: 57–62.

Ramachandran, R. and Thakur, S.C. (1974) India and the Ganga floodplains. In White, G.F. (ed.) *Natural Hazards*. Oxford University Press, New York, 36–43.

Ramanathan, V. (2007) Warming trends in Asia amplified by brown cloud solar absorption. *Nature* 448: 575–8.

Ramanathan, V., Crutzen, P.J., Kiehl, J.T. and Rosenfeld, D. (2001) Aerosols, climate and the hydrological cycle. *Science* 294: 2119–24.

Ramsey, M.S. and Flynn, L.P. (2004) Strategies, insights and the recent advances in volcanic monitoring and mapping with data from NASA's Earth Observing System. *Journal of Volcanology and Geothermal Research* 135: 1–11.

Ranger, N. *et al.* (2011) An assessment of the potential impact of climate change on flood risk in Mumbai. *Climatic Change* 104: 139–67.

Rapanos, D. *et al.* (1981) *Floodproofing New Residential Buildings in British Columbia*. Ministry of Environment, Province of British Columbia, Victoria, BC.

Raphael, B. *et al.* Factors associated with population risk perceptions of continuing drought in Australia. *Australian Journal of Rural Health* 17: 330–7.

Raschky, P.A. (2008) Institutions and the losses from natural disasters. *Natural Hazards and Earth System Sciences* 8: 627–34.

Rashid, H. (2011) Interpreting flood disasters and flood hazard perceptions from newspaper discourse: tale of two floods in the Red River Valley, Manitoba. *Applied Geography* 31: 35–45.

Rasmussen, K. (1995) Natural events and accidents with hazardous materials. *Journal of Hazardous Materials* 40: 43–54.

Rauch, E. (2006) *Climate Change – Potential Impacts on the Insurance Industry*. Climate Change Seminar Report, Geo Risks Research, Munich Re Group, Munich.

Rauhala, J. and Schultz, D.M. (2009) Severe thunderstorm and tornado warnings in Europe. *Atmospheric Research* 93: 369–80.

Reason, J.T. (1990) *Human Error*. Cambridge University Press, Cambridge.

Reddick, C. (2011) Information technology and emergency management: preparedness and planning in US states. *Disasters* 35: 45–61.

Reichhardt, T., Check, E. and Marris, E. (2005) After the flood. *Nature* 437 (8 September): 174–6.

Renne, J. (2006) Evacuation and equity: a post-Katrina New Orleans diary. *Planning* 72: 44–6.

Renni, E., Krausman, E. and Cossani, V. (2010) Industrial accidents triggered by lightning. *Journal of Hazardous Materials* 184: 42–8.

Reyes, P.J.D. (1992) Volunteer observers' program: a tool for monitoring volcanic and seismic events in the Philippines. In G.J.H. McCall, D.J.C. Laming and S.C. Scott (eds) *Geohazards*, Chapman and Hall, London, pp. 13–24.

Rheinberger, C.M., Bründl, M. and Trasformi, E. (2009) Dealing with the white death: avalanche risk management for traffic routes. *Risk Analysis* 29: 76–94.

Rice, R. Jr *et al.* (2002) Avalanche hazard reduction for transportation corridors using real-time detection and alarms. *Cold Regions Science and Technology* 34: 31–42.

Robertson, I.N., Riggs, H.R., Yim, S. and Young, Y.L. (2006) Lessons from Katrina. *Civil Engineering*, April, 56–63.

Robock, A. (2000) Volcanic eruptions and climate. *Reviews of Geophysics* 38: 191–219.

Robock, A. (2002) Pinatubo eruption: the climatic aftermath. *Science* 295: 1242–4.

Rockström, J. *et al.* (2009) A safe operating space for humanity. *Nature* 461: 472–5.

Rodrigue, C.M. and Rovai, E. (1995) The 'Northridge' earthquake: differential geographies of damage, media attention and recovery. *National Social Science Perspectives Journal* 7: 98–111.

Rohrmann, B. (1994) Risk perception of different societal groups: Australian findings and cross-national comparisons. *Australian Journal of Psychology* 46: 150–63.

Rojas, O., Vrieling, A. and Rembold, F. (2011) Assessing drought probability for agricultural areas in Africa with coarse resolution remote sensing imagery. *Remote Sensing of Environment* 115: 343–52.

Roll Back Malaria Partnership (2011) *Global Malaria Action Plan.* World Health Organization, Geneva.

Romieu, E., Welle, T., Schneiderbauer, S., Pelling, M. and Vinchon, C. (2010) Vulnerability assessment within climate change and natural hazards contexts: revealing gaps and synergies through coastal applications. *Sustainability Science* 5: 159–70.

Romme, W.H. and Despain, D.G. (1989) The Yellowstone fires. *Scientific American* 261: 37–46.

Ropelewski, C.F. and Folland, C.K. (2000) Prospects for the prediction of meteorological drought. In Wilhite, D.A. (ed.) *Drought*, vol. 1. Routledge, London and New York, 21–40.

Rorig, M.L. and Ferguson, S.A. (2002) The 2000 fire season: lightning-caused fires. *Journal of Applied Meteorology* 41: 786–91.

Rose, G.A. (ed.) (1994) *Fire Protection in Rural America: A Challenge for the Future.* Rural Fire Protection in America Steering Committee, Report to Congress sponsored by the National Association of State Foresters, Washington, DC.

Rose, W.I. and Chesner, C.A. (1987) Dispersal of ash in the great Toba eruption, 75 k.a. *Geology* 15: 913–17.

Ross, S. (2004) *Toward New Understandings: Journalists and Humanitarian Relief Coverage.* Fritz Institute, San Francisco, CA.

Ross, T. and Lott, N. (2003) *A Climatology of 1980–2003 Extreme Weather and Climate Events.* Technical Report No. 2003–01, NOAA/NESDIS, National Climate Data Center, Ashville, NC.

Rotstayn, L. (2007) Have Australian rainfall and cloudiness increased due to the remote effects of Asian anthropogenic aerosols? *Journal of Geophysical Research* 112: DO9202.

Royal Society (1992) *Risk: Analysis, Perception and Management.* Report of a Royal Society Study Group. Royal Society, London.

Runqiu, H. (2009) Some catastrophic landslides since the twentieth century in the south-west of China. *Landslides* 6: 69–81.

Rural Fire Protection in America (1994) *Fire Protection in Rural America: A Challenge for the Future.* Rural Fire Protection in America, Washington, DC.

Russell, L.A., Goltz, J.D. and Bourque, L.B. (1995) Preparedness and hazard mitigation actions before and after two earthquakes. *Environment and Behaviour* 27: 744–70.

Sagan, L.A. (1984) Problems in health measurements for the risk assessor. In Ricci, P.F., Sagan, L.A. and Whipple, C.G. (eds) *Technological Risk Assessment*, Martinus Nijhoff, The Hague, 1–9.

Sagan, S.D. (1993) *The Limits of Safety: Organisations, Accidents and Nuclear Weapons.* Princeton University Press, Princeton, NJ.

Salcioglu, E., Basoglu, M. and Livanov, M. (2007) Posttraumatic stress disorder and comorbid depression among survivors of the 1999 earthquake in Turkey. *Disasters* 31: 115–29.

Sanchez, C., Lee, T.S., Batts, D., Benjamin, J. and Malilay, J. (2009) Risk factors for mortality during the 2002 landslides in Chuuk, Federated States of Micronesia. *Disasters* 33: 705–20.

Sanders, J.F. and Gyakum, J.R. (1980) Synoptic-dynamic climatology of the bomb. *Monthly Weather Review* 108: 1598–606.

Santella, N., Stenberg, L.J. and Sengul, H. (2010) Petroleum and hazardous material releases from industrial facilities associated with Hurricane Katrina. *Risk Analysis* 30: 635–49.

Sapir, D.G. and Misson, C. (1992) The development of a database on disasters. *Disasters* 16: 74–80.

Sass, O. (2005) Temporal variability of rockfall in the Bavarian Alps, Germany. *Arctic, Antarctic and Alpine Research* 37: 564–73.

Sauchyn, D.J. and Trench, N.R. (1978) LANDSAT applied to landslide mapping. *Photogrammetric Engineering and Remote Sensing* 44: 735–41.

Saunders, M.A. and Lea, A.S. (2005) Seasonal predictions of hurricane activity reaching the coast of the United States. *Nature* 434, April 21, 1005–8.

Scarpa, R. and Gasparini, P. (1996) A review of volcano geophysics and volcano-monitoring methods. In Scarpa, R. and Tilling, R.I. (eds) *Monitoring and Mitigation of Volcano Hazards*, Springer-Verlag, Berlin, 3–22.

Schär, C. *et al.* (2004) The role of increasing temperature variability in European summer heatwaves. *Nature* 427: 332–6.

Schaerer, P.A. (1981) Avalanches. In Gray, D.M. and Male, D.H. (eds) *Handbook of Snow.* Pergamon, Toronto, 475–518.

Schierz, C. *et al.* (2010) Modelling European winter storm losses in current and future climate. *Climate Change* 101: 485–514.

Schmidlin, T.W., King, P.S., Hammer, B.O. and Ono, Y. (1998) Risk factors for death in the 22–23 February 1998 Florida tornadoes. *Quick Response Report* 106. Natural Hazards Research and Applications Information Center, Boulder, CO.

Schmidtlein, M.C., Finch, C. and Cutter, S.L. (2008) Disaster declarations and major hazard occurrences in the United States. *The Professional Geographer* 60: 1–14.

Schuster, R.L. and Highland, L.M. (2001) *Socio-economic and Environmental Costs of Landslides in the Western Hemisphere*. Open File Report 01–276, US Geological Survey, Reston, VA.

Schwartz, R.M. and Schmidlin, T.W. (2002) Climatology of blizzards in the coterminous United States 1959–2000. *Journal of Climate* 15: 1765–72.

Seaman, J., Leivesley, S. and Hogg, C. (1984) Epidemiology of natural disasters. In *Contributions to Epidemiology and Biostatistics*, vol. 5, S. Karger, Basel.

Selvaraju, R. (2003) Impact of El Niño–Southern Oscillation on Indian foodgrain production. *International Journal of Climatology* 23: 187–206.

Sepúlveda, S.A., Rebolledo, S. and Vargas, G. (2006) Recent catastrophic debris flows in Chile: geological hazards, climatic relationships and human response. *Quaternary International* 158: 83–95.

Shaluf, I.M., Ahmadun, F.R. and Shariff, A.R. (2003) Technological disaster factors. *Journal of Loss Prevention in the Process Industries* 16: 513–21.

Shanmugasundaram, J., Arunachalam, S., Gomathinayagam, S., Lakshmanan, N. and Harikrishna, P. (2000) Cyclone damage to buildings and structures – a case study. *Journal of Wind Engineering and Industrial Aerodynamics* 84: 369–80.

Shaw, B.E. (1995) Frictional weakening and slip complexity in earthquake faults. *Journal of Geophysical Research* 100: 18239–52.

Shepherd, J.M. and Knutson, T. (2007) The current debate on the linkage between global warming and hurricanes. *Geography Compass* 1: 1–24.

Sherrie, B.K., Norris, F. and Galea, S. (2010) Measuring capacities for community resilience. *Social Indicators Research* 99: 227–47.

Showalter, P.S. and Myers, M.F. (1994) Natural disasters in the United States as release agents of oil, chemicals or radiological materials between 1980–89: analysis and recommendations. *Risk Analysis* 14: 169–81.

Shrubsole, D. (2000) Flood management in Canada at the crossroads. *Environmental Hazards* 2: 63–75.

Siebert, L. (1992) Threats from debris avalanches. *Nature* 356: 658–59.

Siebert, L., Simkin, T. and Kimberly, P. (2011) *Volcanoes of the World*. Smithsonian Institution, University of California Press, Berkeley, Los Angeles, London.

Siegert, F., Ruecker, G., Hinrichs, A. and Hoffman, A.A. (2001) Increased damage from fires in logged forests during droughts caused by El Niño. *Nature* 414: 437–40.

Siegrist, M. and Gutscher, H. (2008) Natural hazards and motivation for mitigation behaviour: people cannot predict the effect evoked by a severe flood. *Risk Analysis* 28: 771–8.

Sigurdsson, H. (1988) Gas bursts from Cameroon crater lakes: a new natural hazard. *Disasters* 12: 131–46.

Sigurdsson, H. and Carey, S. (1986) Volcanic disasters in Latin America and the 13 November eruption of Nevado del Ruiz volcano in Colombia. *Disasters* 10: 205–16.

Simkin, T., Siebert, L. and Blong, R. (2001) Volcano fatalities – lessons from the historical record. *Science* 291: 255.

Simmons, K.M. and Sutter, D. (2005) Protection from Nature's fury: analysis of fatalities and injuries from F5 tornadoes. *Natural Hazards Review* 6: 82–6.

Simmons, K.M. and Sutter, D. (2006) Direct estimation of the cost effectiveness of tornado shelters. *Risk Analysis* 26: 945–54.

Simonenko, V.A., Nogin, V.N., Petrov, D.V., Shubin, O.N. and Solem, J.C. (1994) Defending the earth against impacts from large comets and asteroids. In Gehrels, T. (ed.) *Hazards Due to Asteroids and Comets*. University of Arizona Press, Tucson, 929–53.

Simpson, D.M. (2002) Earthquake drills and simulations in community-based training and preparedness programmes. *Disasters* 26: 55–69.

Singhroy, V. (1995) SAR integrated techniques for geohazard assessment. *Advances in Space Research* 15: 67–78.

Sjöberg, L. (2001) Limits of knowledge and the limited importance of trust. *Risk Analysis* 21: 189–98.

Sjöberg, L., Moen, B-L. and Rundmo, T. (2004) *Exploring Risk Perception: An Evaluation of the Psychometric Paradigm in Risk Perception Research*. Rotunde Publications No. 84, Trondheim, Norway.

Skogdalen, J.E., Khorsandi, J. and Vinnem, J.E. (2012) Evacuation, escape and rescue experiences from offshore accidents including the Deepwater Horizon. *Journal of Loss Prevention in the Process Industries* 25: 148–58.

Slovic, P. (1986) Informing and educating the public about risk. *Risk Analysis* 6: 280–5.

Slovic, P., Flynn, J.H. and Layman, M. (1991) Perceived risk, trust, and the politics of nuclear waste. *Science* 254: 1603–7.

Small, C. and Naumann, T. (2001) The global distribution of human population and recent volcanism. *Global Environmental Change B: Environmental Hazards* 3: 93–109.

Smets, H. (1987) Compensation for exceptional environmental damage caused by industrial activities.

In Kleindorfer, P.R. and Kunreuther, H.C. (eds) *Insuring and Managing Hazardous Risks*. Springer-Verlag, New York, 79–138.

Smith, B.G. (2010) Socially distributing public relations: Twitter, Haiti and interactivity in social media. *Public Relations Review* 36: 329–35.

Smith, B.J. and de Sanchez, B.A. (1992) Erosion hazards in a Brazilian suburb. *Geographical Review* 6: 37–41.

Smith, D.I. (1989) A dam disaster waiting to break. *New Scientist* (11 November): 42–6.

Smith, D.I. (2000) Floodplain management: problems, issues and opportunities. In Parker, D.J. (ed.) *Floods*, vol. 1. Routledge, London and New York, 254–67.

Smith, K. and Ward, R. (1998) *Floods: Physical Processes and Human Impacts*. John Wiley, Chichester and New York.

Smith, W.D. and Berryman, K.R. (1986) Earthquake hazard in New Zealand: inferences from seismology and geology. *Bulletin of the Royal Society of New Zealand* 24: 223–42.

Smyth, C.G. and Royle, S.A. (2000) Urban landslide hazards: incidence and causative factors in Niterói, Rio de Janeiro State, Brazil. *Applied Geography* 20: 95–117.

Snow, R.W., Guerra, C.A., Noor, A.M., Myint, H.Y. and Hay, S.I. (2005) The global distribution of clinical episodes of *Plasmodium falciparum* malaria. *Nature* 434 (10 March): 214–17.

Sokolowska, J. and Tyszka, T. (1995) Perception and acceptance of technological and environmental risks: why are poor countries less concerned? *Risk Analysis* 15: 733–43.

Solecki, W.D. *et al.* (2005) Mitigation of the heat island effect in urban New Jersey. *Environmental Hazards* 6: 39–49.

Solomon, S.D. *et al.* (2007) *Climate Change 2007: The Physical Science Basis*. Contribution of Working Group I to the Fourth Assessment Report of the IPCC, Cambridge University Press, Cambridge.

Solomon, T. and Mallewa, M. (2001) Dengue and other emerging flaviviruses. *Journal of Infection* 42: 104–15.

Sommer, A. and Mosely, W.H. (1972) East Bengal cyclone of November 1970: epidemiological approach to disaster assessment. *The Lancet* 1: 1029–36.

Sousa, P.M. *et al.* (2011) Trends and extremes of drought indices throughout the 20th century in the Mediterranean. *Natural Hazards and Earth System Sciences* 11: 35–51.

Southern, R.L. (2000) Tropical cyclone warning-response strategies. In Pielke, R.A. Jr and Pielke,

R.A. Sr (eds) *Storms*, vol. 1. Routledge, London and New York, 259–305.

Spangle, W. and Associates Inc. (1988) *California at Risk: Steps to Earthquake Safety for Local Government*. California Seismic Safety Commission, Sacramento, CA.

Spence, R.J.S., Baxter, P.J. and Zuccaro, G. (2004) Building vulnerability and human casualty estimation for a pyroclastic flow: a model and its application to Vesuvius. *Journal of Volcanology and Geothermal Research* 133: 321–43.

Stanton, N.A. and Walker, G.H. (2011) Exploring the psychological factors in the Ladbroke Grove rail accident. *Accident Analysis and Prevention* 43: 1117–27.

Stark, K.P. and Walker, G.R. (1979) Engineering for natural hazards with particular reference to tropical cyclones. In Heathcote, R.C. and Thoms, B.G. (eds) *Natural Hazards in Australia*. Australian Academy of Science, Canberra, 189–203.

Starosolszky, O. (1994) Flood control by levees. In Rossi, G., Harmancogliu, N. and Yevjevich, V. (eds) *Coping with Floods*. Kluwer, Dordrecht, 617–35.

Starr, C. (1969) Social benefit versus technological risk. *Science* 165: 1232–8.

Starr, C. and Whipple, C. (1980) Risk of risk decisions. *Science* 208: 1114–19.

Stephenson, R. and Anderson, P.S. (1997) Disasters and the information technology revolution. *Disasters* 21: 305–34.

Stockwell, J.R., Sorenson, J.W., Eckert, J.W. Jr and Carreras, E.M. (1993) The US EPA Geographic Information System for mapping environmental releases of toxic chemical releases. *Risk Analysis* 13: 155–64.

Stoddard, A. (2003) Humanitarian NGOs: Challenges and trends. In Macrae, J. and Harmer, A. (eds) *Humanitarian Action and the 'Global War on Terror': A Review of Trends and Issues*. Report 14, Humanitarian Policy Group, Overseas Development Institute, London.

Stojanovic, T.A. and Ballinger, R.C. (2009) Integrated coastal management: a comparative analysis of four UK initiatives. *Applied Geography* 29: 49–62.

Stolle, F. and Tomich, T.P. (1999) The 1997–1998 fire event. *Nature and Resources* 35: 22–30.

Stott, P.A., Stone, D.A. and Allen, M.R. (2004) Human contribution to the European heatwave of 2003. *Nature* 432 (2 December): 610–14.

Street-Perrott, F.A. and Perrott, R.A. (1990) Abrupt climate fluctuations in the tropics: the influence of the Atlantic circulation. *Nature* 343: 607–12.

Strobl, E. (2012) The economic growth impact of natural disasters in developing countries: evidence from hurricane strikes in the Central American and Caribbean regions. *Journal of Development Economics* 97: 130–41.

Strömberg, D. (2007) Natural disasters, economic development and humanitarian aid. *Journal of Economic Perspectives* 21: 199–222.

Strunz, G. *et al.* (2011) Tsunami risk assessment in Indonesia. *Natural Hazards and Earth System Sciences* 11: 67–82.

Sullivent, E.E. *et al.* (2006) Non-fatal injuries following Hurricane Katrina, New Orleans, Louisiana, 2005. *Journal of Safety Research* 37: 213–17.

Sur, D. *et al.* (2006) The malaria and typhoid fever burden in the slums of Kolkata, India: data from a prospective community-based study. *Transactions of the Royal Society of Tropical Medicine and Hygiene* 100: 725–33.

Suryo, I. and Clarke, M.C.G. (1985) The occurrence and mitigation of volcanic hazards in Indonesia as exemplified at the Mount Merapi, Mount Kelut and Mount Galunggung volcanoes. *Quarterly Journal of Engineering Geology* 18: 79–98.

Sylves, R. (1996) The politics and administration of presidential disaster declarations. *Quick Response Report* 86. Natural Hazards Research and Information Center, Boulder, CO.

Tadesse, T., Haile, M., Senay, G., Wardlow, B.D. and Knutson, C.L. (2008) The need for integration of drought monitoring tools for proactive food security management in sub-Saharan Africa. *Natural Resources Forum* 32: 265–79.

Tang, B.H. and Neelin, J.D. (2004) ENSO influence on Atlantic hurricanes via tropospheric warming. *Geophysical Research Letters* 31: L24204.

Tanner, T.M. (2010) Shifting the narrative: child-led responses to climate change and disasters in El Salvador and the Philippines. *Children and Society* 24: 339–51.

Tayag, J.C. and Punongbayan, R.S. (1994) Volcanic disaster mitigation in the Philippines: experience from Mt Pinatubo. *Disasters* 18: 1–15.

Teka, O. and Vogt, J. (2010) Social perception of natural risks by local residents in developing countries – the example of the coastal area of Benin. *The Social Science Journal* 47: 215–24.

Tekeli-Ye_il, S., Dedeo_lu, N., Braun-Fahrlaender, C. and Tanner, M. (2010) Factors motivating individuals to take precautionary action for an expected earthquake in Istanbul. *Risk Analysis* 30: 1181–95.

Telford, J. and Cosgrave, J. (2007) The international humanitarian system and the 2004 Indian Ocean earthquake and tsunamis. *Disasters* 31: 1–28.

Teng, W.L. (1990) AVHRR monitoring of US crops during the 1988 drought. *Photogrammetric Engineering and Remote Sensing* 56: 1143–6.

Thieken, A.H., Petrow, T., Kreibich, H. and Merz, B. (2006) Insurability and mitigation of flood losses in private households in Germany. *Risk Analysis* 26: 383–91.

Thomalla, F. and Schmuck, H. (2004) 'We all knew that a cyclone was coming': disaster preparedness and the cyclone of 1999 in Orissa, India. *Disasters* 28: 373–87.

Thomas, M.F. (1994) *Geomorphology in the Tropics*. John Wiley, Chichester and New York.

Thomson, M.C. *et al.* (2006) Malaria early warnings based on seasonal climate forecasts from multi-model ensembles. *Nature* 439 (2 February): 576–9.

Thorarinsson, S. (1979) On the damage caused by volcanic eruptions with special reference to tephra and gases. In Sheets, P.D. and Grayson, D.K. (eds) *Volcanic Activity and Human Ecology*, Academic Press, London, 125–59.

Tickner, J. and Gouveia-Vigeant, T. (2005) The 1991 cholera epidemic in Peru; not a case of precaution gone awry. *Risk Analysis* 25: 495–502.

Tierney, K., Bevc, C. and Kuligowski, E. (2006) Metaphors matter: disaster myths, media frames and their consequences in Hurricane Katrina. *Annals of the American Academy of Political and Social Science* 604: 57–81.

Timmerman, P. (1981) *Vulnerability, Resilience and the Collapse of Society*. Environmental Monograph 1. Institute for Environmental Studies, University of Toronto, Toronto.

Timmerman, P. and White, R. (1997) Megahydropolis: coastal cities in the context of global environmental change. *Global Environmental Change* 7: 205–34.

Tinsley, J.C., Youd, T., Perkins, D.M. and Chen, A.T.F. (1985) Evaluating liquefaction potential. In Ziony, J.I. (ed.) *Evaluating Earthquake Hazards in the Los Angeles Region*. Department of the Interior, Washington, DC, 263–315.

Tobin, G.A. and Montz, B.E. (1997) The impacts of a second catastrophic flood on property values in Linda and Olivehurst, California. *Quick Response Report* 95. Natural Hazards Research and Applications Information Center, Boulder, CO.

Tobin, G.A. and Whiteford, L.M. (2002) Community resilience and volcano hazard: the eruption of Tungurahua and evacuation of the Faldas in Ecuador. *Disasters* 26: 28–48.

Todhunter, P.E. (2011) *Caveant admonitus* (Let the forewarned beware): the 1997 Grand Forks

(USA) flood disaster. *Disaster Prevention and Management* 20: 125–39.

Tolhurst, K. (2010) *Report on Fire Danger Ratings and Public Warning*. University of Melbourne, Melbourne.

Toon, O.B., Zahnle, K., Morrison, D., Turco, R.P. and Covey, C. (1997) Environmental perturbations caused by the impacts of asteroids and comets. *Annals of the New York Academy of Sciences* 822: 403–31.

Toya, H. and Skidmore, M. (2007) Economic development and the impacts of natural disasters. *Economics Letters* 94: 20–5.

Tralli, D.M., Blom, R.G., Zlotnicki, V., Donnellan, A. and Evans, D.L. (2005) Satellite remote sensing of earthquakes, volcanoes, flood, landslide and coastal inundation hazards. *ISPRS Journal of Photogrammetry and Remote Sensing* 59: 185–98.

Treby, E.J., Clark, M.J. and Priest, S.J. (2006) Confronting flood risk: implications for insurance and risk transfer. *Journal of Environmental Management* 81: 351–9.

Trenberth, K. (2005) Uncertainty in hurricanes and global warming. *Science* 308: 1753–4.

Trenberth, K., Branstator, G.W. and Arkin, P.A. (1988) Origins of the 1988 North American drought. *Science* 242: 1640–5.

Trenberth, K., Caron, J.M., Stepaniak, D.P. and Worley, S. (2002) Evolution of El Niño–Southern Oscillation and global atmospheric surface temperatures. *Journal of Geophysical Research* 107: DO4065.

Trenberth, K.E. (2011) Attribution of climate variations and trends to human influences and natural variability. *WIRES Climate Change* 2: 925–30.

Trigo, R.M., Guveia, C.M. and Barriopedro, D. (2010) The intense 2007–2009 drought in the Fertile Crescent: impacts and associated atmospheric circulation. *Agricultural and Forest Meteorology* 150: 1245–57.

Trigo, R.M. *et al.* (2004) North Atlantic Oscillation influence on precipitation, river flow and water resources in the Iberian Peninsula. *International Journal of Climatology* 24: 925–44.

Tsai, F., Hwang, J-H., Chen, L-C. and Lin, T-M. (2010) Post-disaster assessment of landslides in southern Taiwan after 2009 typhoon Morakot using remote sensing and spatial analysis. *Natural Hazards and Earth System Sciences* 10: 2179–90.

Tschoegl, L., Below, R. and Guha-Sapir, D. (2006) *An Analytical Review of Selected Data Sets on Natural Disasters and Impacts*. Centre for Research on the Epidemiology of Disasters, Brussels.

Tuffen, H. (2010) How will melting of ice affect volcanic hazards in the twenty-first century? *Philosophical Transactions of the Royal Society* 368: 2535–58.

Turner, B.A. (1994) Causes of disaster: sloppy management. *British Journal of Management* 5: 215–19.

Turner, B.L. (2010) Vulnerability and resilience: coalescing or paralleling approaches for sustainability science? *Global Environmental Change* 20: 570–6.

Turner, B.L. *et al.* (1990) The types of global environmental change: definitional and spatial-scale issues in their human dimensions. *Global Environmental Change* 1: 14–22.

UK Task Force (2000) *Report on Potentially Hazardous Near-Earth Objects*. HMSO, London.

Ulbrich, U. and Christoph, M. (1999) A shift of the NAO and increasing storm track activity over Europe due to anthropogenic greenhouse gas forcing. *Climate Dynamics* 15: 551–9.

Ulbrich, U., Leckebush, G.C. and Pinto, J.G. (2009) Extra-tropical cyclones in the present and future climate: a review. *Theoretical and Applied Climatology* 96: 117–31.

Ummenhofer, C.M., Sen Gupta, A., Li, Y., Taschetto, A.S. and England, M.H. (2011) Multi-decadal modulation of the El Niño–Indian monsoon relationship by Indian Ocean variability. *Environmental Research Letters* 6: 1–8.

UN Department of Humanitarian Affairs (1994) *Strategy and Action Plan for Mitigating Water Disasters in Vietnam*. United Nations, New York and Geneva.

UNDP (1998) *Human Development Report 1998*. Oxford University Press, Oxford.

UNDP (United Nations Development Programme) (2004) *Reducing Disaster Risk: A Challenge for Development*. UN Bureau for Crisis Prevention and Recovery, New York.

UNDRO (United Nations Disaster Relief Organization) (1985) *Volcanic Emergency Management*. United Nations, New York.

UNDRO (United Nations Disaster Relief Organization) (1990) *Disaster Prevention and Preparedness Project for Ecuador and Neighbouring Countries*. Project Report, Office of the Disaster Relief Coordinator, Geneva.

UNEP and C_4 (2002) *The Asian Brown Cloud: Climate and Other Environmental Impacts*. United Nations Environment Programme, Nairobi.

UNESCO (2007) *Natural Disasters Preparedness and Education for Sustainable Development*. UNESCO, Bangkok.

UNFPA (2007) *State of World Population*. United Nations Population Fund, New York.

Unganai, L.S. and Kogan, F.N. (1998) Drought monitoring and corn yield estimation in southern Africa from AVHRR data. *Remote Sensing of Environment* 63: 219–32.

UN/ISDR (UN/International Strategy for Disaster Reduction) (2004) *Living with Risk: A Global Review of Diasaster Reduction Initiatives*. United Nations, Geneva.

UN/ISDR (UN/International Strategy for Disaster Reduction) (2007) *Building Disaster Resilient Communities: Good Practices and Lessons Learned*. United Nations, Geneva.

UN/ISDR (UN/International Strategy for Disaster Reduction) (2009) *UNISDR Terminology on Disaster Risk Reduction*. United Nations, Geneva.

United Nations in Pakistan (2011) *Pakistan Floods One Year On*. United Nations Flood Response Team, Islamabad.

Urbina, E. and Wolshon, B. (2003) National review of hurricane evacuation plans and policies: a comparison and contrast of state practices. *Transportation Research A* 37: 257–75.

Usbeck, T. et al. (2010) Increasing storm damage to forests in Switzerland from 1858 to 2007. *Agricultural and Forest Meteorology* 150: 47–55.

US Department of Commerce (1994) *The Great Flood of 1993*. Natural Disaster Survey Report, Department of Commerce, Silver Spring, MD.

US National Academy of Sciences (2002) *Abrupt Climate Change: Inevitable Surprises*. National Research Council committee on Abrupt Climate Change, National Academy Press, Washington, DC.

Valery, N. (1995) Earthquake engineering: a survey. *The Economist* (22 April): 18–20.

van Aalst, M.K. (2006) The impact of climate change on the risk of natural disasters. *Disasters* 30: 5–18.

van Asch, T.W.J., Malet, J.P., van Beek, L.P.H. and Amittrano, D. (2007) Techniques, issues and advances in numerical modelling of landslide hazard. *Bulletin de la Societe Geologique de France* 178: 65–88.

van Dorp, J.R., Merrick, J.R.W., Harrald, J.R., Mazzuchi, T.A. and Grabowski, M. (2001) A risk management procedure for the Washington state ferry. *Risk Analysis* 21: 127–42.

Varnes, D.J. (1978) Slope movements and types of processes. In *Landslides: Analysis and Control*. Special Report 176, Transportation Research Board, National Academy of Sciences, Washington, DC, 11–13.

Vicente-Serrano, S.M. and L_pez-Moreno, J.I. (2006) The influence on atmospheric circulation at different spatial scales on winter drought variability through a semi-arid climatic gradient in north-east Spain. *International Journal of Climatology* 26: 1427–53.

Visbeck, M. (2002) The ocean's role in Atlantic climate variability. *Science* 297: 2223–4.

Vogel, G. (2005) Will a pre-emptive strike against malaria payoff? *Science* 310 (9 December): 1606–7.

Voight, B. (1996) The management of volcano emergencies: Nevado del Ruiz. In Scarpa, R. and Tilling, R.I. (eds) *Monitoring and Mitigation of Volcano Hazards*. Springer-Verlag, Berlin, 719–69.

Vranes, K. and Pielke, R.A. (2009) Normalized earthquake damage and fatalities in the United States: 1900–2005. *Natural Hazards Review* 10: 84–101.

Wadge, G. (ed.) (1994) *Natural Hazards and Remote Sensing*. Royal Society and Royal Academy of Engineering, London.

Walker, G., Mooney, J. and Pratts, D. (2000) The people and the hazard: the spatial context of major accident hazard management in Britain. *Applied Geography* 20: 119–35.

Walmsley, L. (2010) *Humanitarian Aid in 2009: Headlines from the Latest DAC Release*. Global Humanitarian Assistance at http://www.globalhumanitarianassistance.org/ (accessed 27 March 2011).

Walmsley, L. (2011) *Data and Guides Workstream – Map of Aid Players Infographic*. Global Humanitarian Assistance at http://www.globalhumanitarianassistance.org/ (accessed 27 March 2011).

Waltham, T. (2005) The flooding of New Orleans. *Geology Today* 21: 225–31.

Wang, J-F. and Li, L-F. (2008) Improving tsunami warning systems with remote sensing and geographical information system input. *Risk Analysis* 28: 1653–68.

Wang, X.L. et al. (2009) Trends and variability of storminess in the Northeast Atlantic region 1874–2007. *Climate Dynamics* 33: 1179–95.

Ward, S.N. and Day, S. (2001) Cumbre Vieja volcano – potential collapse and tsunami at La Palma, Canary Islands. *Geophysical Research Letters* 28: 3397–400.

Waring, S.C. and Brown, B.J. (2005) The threat of communicable diseases following natural disasters: a public health response. *Disaster Management and Response* 3: 41–7.

Warner, K. et al. (2010) Climate change, environmental degradation and migration. *Natural Hazards* 55: 689–715.

Water and Rivers Commission (2000) *Water Facts 14: Floodplain Management*. Water and Rivers Commission, Government of State of Western Australia, Perth.

Watson, J.T., Gayer, M. and Connolly, M.A. (2007) Epidemics after natural disasters. *Emerging Infectious Diseases* 13: 1–5.

Webb, J.D.C., Elsom, D.M. and Meaden, G.T. (2009) Severe hailstorms in Britain and Ireland: a climatological survey and hazard assessment. *Atmospheric Research* 93: 587–606.

Webster, P.J., Holland, G.J., Curry, J.A. and Chang, H-R. (2005) Changes in tropical cyclone number, duration and intensity in a warming environment. *Science* 309: 1844–6.

Weihe, W.H. and Mertens, R. (1991) Human well-being, diseases and climate. In Jager, J. and Ferguson, H.L. (eds) *Climate Change*. Cambridge University Press, Cambridge, 345–59.

Wells, L. (2002) Avalanche hazard reduction for transportation corridors using real-time detection and alarms. *Cold Regions Science and Technology* 34: 31–42.

Welsh, S. (1994) CIMAH and the environment. *Disaster Prevention and Management* 3: 28–43.

Wenger, D. (2006) Hazards and disasters research: How would the past 40 years rate? *Natural Hazards Observer* 31 (1): 1–3.

Werner, M., Reggiani, P., de Roo, A., Bates, P. and Sprokkereef, E. (2005) Flood forecasting and warning at the river basin and at the European scale. *Natural Hazards* 36: 25–42.

Werner, M., Cranston, M., Harrison, T., Whitfield, D. and Schellekens, J. (2009) Recent developments in operational flood forecasting in England, Wales and Scotland. *Meteorological Applications* 16: 13–22.

Wesolek, E. and Mahieu, P. (2011) The F4 tornado of August 3, 2008, in northern France: case study of a tornadic storm in a low CAPE environment. *Atmospheric Research* 100: 644–56.

Westerling, A.L., Hidalgo, H.G., Cayan, D.R. and Swetnam, T.W. (2006) Warming and earlier spring increase in western US forest wildfire activity. *Science* 313 (18 August): 940–3.

White, G.F. (1936) The limit of economic justification for flood protection. *Journal of Land and Public Utility Economics* 12: 133–48.

White, G.F. (1945) *Human Adjustment to Floods: A Geographical Approach to the Flood Problem in the United States*. Research Paper 29. Department of Geography, University of Chicago, Chicago, IL.

White, G.F. (ed.) (1974) *Natural Hazards: Local, National, Global*. Oxford University Press, New York.

White, I. and Howe, J. (2002) Flooding and the role of planning in England and Wales: a critical review. *Journal of Environmental Planning and Management* 45: 735–45.

White, P. (2005) War and food security in Eritrea and Ethiopia 1998–2000. *Disasters* 29 (SI): S92–S113.

Whitehead, J.C. *et al.* (2000) Heading for higher ground: factors affecting real and hypothetical hurricane evacuation. *Global Environmental Change B Environmental Hazards* 2: 133–42.

Whittow, J. (1980) *Disasters*. Penguin Books, Harmondsworth, Middlesex.

Whyte, A.V. and Burton, I. (1982) Perception of risk in Canada. In Burton, I., Fowle, C.D. and McCullough, R.S. (eds) *Living with Risk*. Institute of Environmental Studies, University of Toronto, Toronto, 39–69.

Wieczorek, G.F., Larsen, M.C, Eaton, L.S., Morgan, B.A. and Blair, J.L. (2001) *Debris-flow and Flooding Hazards Associated with the December 1999 Storm in Coastal Venezuela and Strategies for Mitigation*. Open File Report 01–144, US Geological Survey, Reston, VA.

Wigley, T.M.L. (1985) Impact of extreme events. *Nature* 316: 106–7.

Wilhite, D.A. (1986) Drought policy in the US and Australia: a comparative analysis. *Water Resources Bulletin* 22: 425–38.

Wilhite, D.A. (2002) Combating drought through preparedness. *Natural Resources Forum* 26: 275–85.

Wilhite, D.A. and Easterling, W.E. (eds) (1987) *Planning for Drought: Toward a Reduction of Societal Vulnerability*. Westview Press, Boulder, CO. and London.

Wilhite, D.A. and Vanyarkho, O. (2000) Drought: pervasive impacts of a creeping phenomenon. In Wilhite, D.A. (ed.) *Drought*, vol.1. Routledge, London and New York, 245–55.

Wilkinson, P., Armstrong, B. and Landon, M. (2001) *Cold Comfort: The Social and Environmental Determinants of Excess Winter Deaths in England 1986–1996*. The Policy Press, Bristol.

Williams, R.S. Jr. and Moore, J.G. (1983) *Man Against Volcano: The Eruption on Heimaey, Vestmannaeyjar, Iceland* (2nd edn). USGS Information Services, Reston, VA.

Willis, M. (2005) Bushfires – how can we avoid the unavoidable? *Environmental Hazards* 6: 93–9.

Willoughby, H.E., Jorgensen, D.P., Black, R.A. and Rosenthal, S.L. (1985) Project STORMFURY: scientific chronicle 1962–1983. *Bulletin of the American Meteorological Society* 66: 505–14.

Wilson, S.G. and Fischetti, T.R. (2010) *Coastline Population Trends in the United States: 1960 to*

2008. Current Population reports. Census Bureau, US Department of Commerce, Washington, DC.

Wisner, B. (2003) Changes in capitalism and global shifts in the distribution of hazard and vulnerability. In Pelling, M. (ed.) *Natural Disasters and Development in a Globalizing World*. Routledge, London and New York, 43–56.

Wisner, B., Blaikie, P., Cannon, T. and Davis, I. (2004) *At Risk: Natural Hazards, People's Vulnerability and Disasters*. Routledge, London and New York.

Witham, C.S. (2005) Volcanic disasters and incidents; a new database. *Journal of Volcanology and Geothermal Research* 148: 191–233.

Witt, V.M. and Reiff, F.M. (1991) Environmental health conditions and cholera vulnerability in Latin America and the Caribbean. *Journal of Public Health Policy* 12: 450–64.

Wiwanitkit, V. (2006) Correlation between rainfall and the prevalence of malaria in Thailand. *Journal of Infection* 52: 227–30.

Wolshon, B., Urbina, E., Wilmot, C. and Levitan, M. (2005) Review of policies and practices for hurricane evacuation. I: Transportation planning, preparedness and response. *Natural Hazards Review* 6: 129–42.

Woodworth, P.L., Flather, R.A., Williams, J.A., Wakelin, S.L. and Jevrejeva, S. (2007) The dependence of UK extreme sea levels and storm surges on the North Alantic Oscillation. *Continental Shelf Research* 27: 935–46.

World Bank (2000) *Republic of Mozambique: A Preliminary Assessment of Damage from the Flood and Cyclone Emergency of Feb–March 2000*. World Bank, Washington, DC.

World Commission on Dams (2000) *Dams and Development: A New Framework for Decision-making*. United Nations Environment Programme, Nairobi.

World Economic Forum (2010) *Global Risks 2010: A Global Risk Network Report*. World Economic Forum, Geneva.

Wrathall, J.E. (1988) Natural hazard reporting in the UK press. *Disasters* 12: 177–82.

Wright, T.L. and Pierson, T.C. (1992) *Living with Volcanoes: The USGS Volcano Hazards Program*. Circular 1073, US Geological Survey, Reston, VA.

Wu, Q. (1989) The protection of China's ancient cities from flood damage. *Disasters* 13: 193–227.

Wu, Y.M. *et al*. (2004) Progress on earthquake rapid reporting and early warning systems in Taiwan. In Chen, T., Panza, G.F. and Wu, Z.L. (eds) *Earthquake Hazard, Risk and Strong Ground Motion*. Seismological Press, Beijing, 463–86.

Xin, X.G., Yu, R.C., Zhou, T.J. and Wang, B. (2006) Drought in late spring of China in recent decades. *Journal of Climate* 19: 3197–206.

Xu, Q. (2001) Abrupt change of the mid-summer climate in central east China by the influence of atmospheric pollution. *Atmospheric Environment* 35: 5029–40.

Yagar, S. (ed.) (1984) *Transport Risk Assessment*. University of Waterloo Press, Waterloo, Ontario.

Yates, D. and Paquette, S. (2010) Emergency knowledge management and social media technologies: a case study of the 2010 Haitian earthquake. *Proceedings of the American Society for Information Science and Technology* 47: 1–9.

Young, S., Balluz, L. and Malily, J. (2004) Natural and technologic hazardous materials releases during and after natural disasters: a review. *Science of the Total Environment* 322: 3–20.

Zaman, M.Q. (1991) The displaced poor and resettlement policies in Bangladesh. *Disasters* 15: 117–25.

Zeckhauser, R. and Shepard, D.S. (1984) Principles for saving and valuing lives. In Ricci, P.F., Sagan, L.A. and Whipple, C.G. (eds) *Technological Risk Assessment*, Martinus Nijhoff, The Hague, 133–68.

Zerger, A., Smith, D.I., Hunter, G.J. and Jones, S.D. (2002) Riding the storm: a comparison of uncertainty modelling techniques for storm surge risk management. *Applied Geography* 22: 307–30.

Zezere, J.L., Trigo, R.M. and Trigo, I.F. (2005) Shallow and deep landslides induced by rainfall in the Lisbon region (Portugal): assessment of relationships with the North Atlantic Oscillation. *Natural Hazards and Earth System Sciences* 5: 331–4.

Zhang, H-L. (2004) *China: Flood Management*. Case Study in Integrated Flood Management, WMO/GWP Associated Programme on Flood Management, Geneva.

Zhang, X. *et al*. (2007) Detection of human influence on twentieth-century precipitation trends. *Nature* 448: 461–5.

Zhang, Y., Prater, C.S. and Lindell, M.K. (2004) Risk area accuracy and evacuation from Hurricane Brett. *Natural Hazards Review* 5: 115–19.

Zimmermann, M., Pozzi, A. and Stoessel, F. (2005) *Hazard Maps and Related Instruments: The Swiss System and its Application Abroad*. National Platform for Natural Hazards, Swiss Agency for Development and Cooperation, Bern.

Zobin, V.M. (2001) Seismic hazard of volcanic activity. *Journal of Volcanology and Geothermal Research* 112: 1–14.

Index

Abbasi, S.A. 376
acoustic-flow monitoring (AFM)
    stations 198
Adams, W.C. 27
adaptation to hazards 98–9, 118–36,
    124–36, 164–75, 193–204, 226–7,
    260–6, 295–8, 329–36, 363–9,
    393–401
adobe construction 153, 158–9
agricultural drought 345–7
aid programmes: expenditure on
    111–13; following major
    earthquakes 162; internal
    government aid 109–11;
    international aid 111–14; players
    involved in 109; *see also* disaster
    aid
air pollution 407–10
air travel 386
ALARP principle 89–90
alluvial fans 308–9
Alquist-Priolo earthquake zone 134–5
Andorra 133
animal behaviour 168
Ano Liosia, Athens 172–3
anthropogenic global warming (AGW)
    408, 420
anticyclones 271
Antonioni, G. 377
aridity 428–9
Armenian earthquake (1988) 163

ash falls 183–4, 190, 195
'Asian brown cloud' phenomenon 407
Asian Disaster Preparedness Center
    121
asteroid impact 417–19
atmospheric hazards 235–6
Australia 288–94, 342–5, 362–3
avalanche-deflecting dams 102
avalanches 149, 207, 218–33; artificial
    release of 221–2; defensive
    structures against 222–4; mapping
    and modelling of 232–3; survival
    rates from 226; warning systems
    for 227, 229; *see also* snow
    avalanches
aviation 183–4

Baker, E.J. 261, 266
Ballester, J. 424
Bam earthquake (2003) 48–51
Bangkok 316
Bangladesh 304–6, 329–30
Bankoff, G. 38
Barnett, A. 386
Barnett, B.J. 110–11
Barrows, Harlan H. 14
Batterbury, S. 358
Beck, U. 74, 403
behavioural paradigm with regard to
    hazards 14–18
Below, R. 28–9

Bhopal disaster (1984) 376, 380–3,
    391–2
bias in measurement of disasters
    29–30
biodiversity 65
'bioshields' 173
Black Death pandemic 273–4, 405
blame, allocation of 391
Blanchard-Boehm, R.D. 329
Blithe river 358–9
Bouma, M.J. 286
Bouska, A. 260
Bouwer, L.M. 36, 427
Brotak, E.A. 292
brushfires 286
Buchanan-Smith, M. 361
building codes 155, 159, 255–7
building construction methods
    99–105; and earthquake damage
    153–9
Buller, P.S.J. 253
Buncefield explosions and fires (2005)
    398–9
Burby, R.J. 127, 328
Bush, D.M. 259
bushfires 286–93
Butler, I. 227

California: hazard zoning in 134–5;
    mapping of seismic hazards 129
California Earthquake Authority 163

California Seismic Safety Commission 120, 164
Callendar, G.S. 409
Cameroon volcanic gas disaster (1986) 193
Canada 226–7, 318
Canvey Island 379
Capra, L. 185
carbon dioxide (CO2) 185, 189–90, 409, 420
cash-for-work schemes 114
Centre for Research on the Epidemiology of Disasters (CRED) 28–32, 338, 373
Chakraborty, J. 119
Chang, Y. 108
Chapman, C.R. 418
check dams 369
Chemical Emergency Preparedness Program (CEPP), US 394
Chernobyl accident (1986) 389–91, 395
Chicxulub Basin 418
cholera 281–6
civil engineering techniques 159–60
climate change 20–1, 59, 80, 317, 336, 402–11, 417–34; definition of 408; and environmental hazards 419–34
climate dynamics 351–2
climate normals and climate variability 408
climate-related disasters 35–7, 423–4
climate-related disease 278–83
climatic conditions and flooding 309
coastal areas, protection of 306–8, 313–15, 321, 336, 430–2
coastal high-hazard areas (CCHAs) 131
cohesion in resistance to landslides 211
cold stress 270–1
Collins, M. 422
Collins, T.W. 296
comet impact 417–19
Comfort, L.K. 48
compensation schemes 382
complexity paradigm with regard to hazards 19–20
complexity science 46–8
consultation distance 400

Control of Industrial Major Accident Hazards (CIMAH) regulations (1984) 393
Control of Major Accident Hazards (COMAH) regulations (1999) 393, 397, 399
Cook, E.R. 355–6
corn yields 346
cost-benefit ratios 317
coupled human–environment system (CHES) 7, 35, 47
Cozzani, V. 377
crater-lake lahars 203
craters 417–18
creep 227
Crompton, R.P. 423
crop-specific drought index (CSDI) 345
Cullen, H.M. 415
Cunningham, C.J. 290
Cutter, S.L. 53–4, 61, 94, 396
Cyclone Preparedness Programme (CPP), Bangladesh 122–3, 261
cyclone shelters 261, 432
cyclones, tropical 236–47, 258–62; deadliest in twentieth century 242; definition of 236; development of 238–40; future prospects for 429; hazards associated with 237, 240–7; satellite sensing of 262; see also extra-tropical storms

Dadaab refugee camp 362
Dai, A. 427
Dalton, L.C. 127
dams: failure of 380–3, 395; for flood control 319–21; for water storage 318; unsafe or inadequate 308
Darfur 351, 364
DART (Deep-ocean Assessment and Reporting of Tsunamis) programme 151–3, 171
Datchet 316
Davies, J.B. 42
de Moel, H. 427
de Waal, A. 351, 361
deaths: causes of 13; in high- and low-income countries 14; in historical disasters 23–5; as reported in different disaster archives 28
debris flows 213
decision theory 71

Deepwater Horizon accident (2010) 373–4
deforestation 66, 313
de-icing agents 391
deltas 306
demographic trends 63
dengue fever 276–7, 285
determinism in risk perception 84
developmental paradigm with regard to hazards 14–18
Diablo Canyon nuclear power plant 396
disaster aid 13, 106–8, 160–3, 193, 224, 258–9, 295, 323–5, 361–3
Disaster Impact Index (DII) 54
disaster impact pyramid 25
disaster reduction initiatives 20–1, 96–136
Disaster Risk Index (DRI) 40–2, 54–5
disaster risk reduction (DRR) 402
disasters: absolute and relative impact of 31; costliest in recent years 38–40; death tolls 23–5, 28, 373–5; definition of 11–12; management of 42–5; measurement of 28–31; protection before and recovery after 43–4; recording of 24–5, 33–5; reporting of 25–7; spatial pattern of 37–42; time trends in 31–2; types of impact 26
Disasters Emergency Committee (DEC) 108, 161
disease epidemics 268; and climate 278–83; definition of 273; and disasters 275–6; hazard reduction for 283–6; linked with El Niño 413; nature of 273–8; socio-economic factors in 278
dissonance in risk perception 84–5
'DNA model' of complexity 47–8
Dolan, J.F. 166
domino accidents 376
Donat, M.G. 430
Dong, B. 429
Douglas, M. 73
Dowd, J. 424
drainage and drainage basins 220, 311
'dread risks' 386
DROP model of disaster resilience 61

drought 12–13, 337–69, 427; *absolute* and *partial* 339; adaptation to 363–9; in Australia 342–5, 362–3; causes of 351–8; in Ethiopia 349–50; forecasts and warnings of 365; human factors in 356–8; mitigation of 361–3; protection from 358–61; severity of 341, 345, 420; types of 339–51; in the US 345–6

drought hazard, definition of 337–8

Duijm, N.J. 90

Duijsens, R. 67

Dumas, L.J. 378

Durham Fatal Landslide Database 205

dust bowl conditions 355

dykes 318

Dymon, U.J. 94

Dynes, R. 18

Earth Observing System (EOS) satellites 196

earthquake behaviour 142–6

earthquake hazards 139–41; adaptation to 164–75; and building construction methods 153–9; micro-zonation for 134–5; mitigation of impacts 160–4; *primary* 146–8; protection from 103, 106, 153–60; safety checklist for 165; *secondary* 148–53

earthquake shaking and landslides 212, 216

earthquakes: activity near volcanoes 196–7; frequency of occurrence at different magnitudes 144; largest since 1900 140

Ebola hemorrhagic fever 274

economic cost of disasters 32, 34, 40

ecosystem services 10

education, role of 286

Eisensee, T. 25–7

El Niño 286, 352, 411–14, 422

Eldfell eruption (1973) 189–90

electromagnetic field variation 167

electronic distance meters (EDMs) 197

Elliott, J.R. 64

El-Masri, S. 127

Elsner, J.B. 429

Emanuel, K.A. 429

embankments 318–19

Emergency Events Database (EM-DAT) 28–32, 39, 205, 338, 375–6, 421–2

emergency period of a disaster 106–8

emergency planning zones (EPZs) 394

emergency relief 347, 362

Emmi, P.C. 94

engineering paradigm with regard to hazards 14

ENSO (El Niño and Southern Oscillation) cycle 411–13, 422, 429

Environment Agency, UK 317, 330–1

environmental change, *systematic* and *cumulative* 405–6

environmental degradation 65

environmental hazards: causes of 7–8; and climate change 419–34; definition and categorisation of 4–11; main features of 8; spectrum of 8; threats created by 11

Environmental Protection Agency, US 394, 401

epicentre of an earthquake 143

epidemics *see* disease epidemics

epidemiology, definition of 275

Estonia sinking 386

Ethiopia 349–50, 357

European Union 257, 384, 393–4

evacuation planning 195, 198, 265–6, 297, 395–6

extra-tropical storms 252–3, 429–30

extreme events 8–9, 77, 268, 421–3; analysis of 78–81; and disasters 422–3; projected changes in 421–3; temperature hazards from 270–3

extreme rainfall alerts (ERAs) 330

Exxon Valdez oil spill 377

'eyewall replacement' process 239–40

Eyjafjallajökull eruption (2010) 183–4

Fagherazzi, S. 415

famine: definition of 348; precautions against 368

famine drought 347–51, 364

Farming Early Warning System (FEWS NET) 366

Federal Emergency Management Agency (FEMA), US 118, 261, 327, 333–4, 394

'Felix' (hurricane, 2007) 18–19

Ferguson rockslide, California 210

fire *see* wildfires

fire-breaks 297

fire detection 297

fire fountains 179

fire-raising, deliberate 290

fire services and rural fire-fighting groups 285

fire-weather warnings 296

Fischer, G.W. 85

Fischhoff, B. 74

flash floods 308, 330

flaviviruses 276–8

Flood Disaster Protection Act (US, 1973) 327–8

flood pulse 302

floodline warnings direct (FWDs) 330

floodplains and floodplain development 101, 304, 314–18, 331, 336

flood-prone environments 301–9

flood-proofing of buildings 322–3

flood-risk maps 132, 327, 329, 331

floods 299–336, 413, 425; abatement of 321–3; causes of 310; in Engand and Wales 301, 326–7; nature of 309–18; recent disasters in Europe 300; threats posed by 299–300

floodwalls 318

flow failure from soil liquifaction 148

'fogging' 284

Folland, C.K. 365, 428

Food and Agriculture Organization (FAO) 347

food aid 113, 339, 361

forecasting and warning systems (FWS) 124–6, 166–71, 195–8, 227–30, 262–4, 286, 296–7, 330–1, 365, 395–6

Fothergill, A. 64

Framework Convention on Climate Change (FCCC) 20

friction in resistance to landslides 211

Friis, H. 61

Fritzsche, A.F. 373

Fujita, T.T. (and Fujita scale) 247–8

Fukushima disaster (2011) 377

Galeras volcano, Colombia 201

Gall, M. 29–30

Galle City, Sri Lanka 119–20

Galtür disaster (1999) 218–20

Galunggung emergency (1982) 193

Garcia, L.A. 94

Gardner, G.T. 386

Garner, A.C. 27

genetic diversity 274

geochemical changes 197

geographical information systems (GIS) 94

geological maps, use of 231

Giddens, A. 74, 403

Gini index 62–3

Glickman, T.S. 379

global circulation models (GCMs) 421, 430

global cities 67

global environmental change (GEC) 7, 405–6

Global Influenza Surveillance Network 278–80

Global Information and Early Warning System (GIEWS) 366

Global Outbreak and Response Network 286

global positioning systems (GPS) 197

global warming 271–2, 314; *see also* anthropogenic global warming

globalization of hazards 20, 403–5

Goidel, R. 27

Goldberg, M.S. 270

'golden hour' after an earthquake 160

Gould, L.C. 386

Gouveia, C. 366

Grantham, Australia 334–5

greehouse gas (GHG) emissions 20, 409–10, 420–2, 427

ground motion: from earthquakes 147, 157; from volcanic action 185, 197

ground oscillation 148

ground-water levels 167–8

Grundstein, A. 424

Guatemala City earthquake (1976) 107

Guha-Sapir, D. 28

Guintran, J.-O. 285

Gutscher, H. 329

hailstorms 250–1

Haiti earthquake (2010) 44, 141

Hamilton, L.S. 313

Handmer, J.W. 288, 297

Hangwan earthquake (2008) 107

Hawaii 200

Hawaiian type of volcanic eruption 179

hazard evaluation 125

hazard-resistant design 99, 106, 153–60, 190, 255–8, 358–61

hazards: awareness of 11; definition of 11; magnitude and duration of 9; in relation to risk 11–12; *systematic* and *rare* 411; *see also* adaptation to hazards; mitigation of hazard impacts; protection from hazards

haze pollution 407

head loading and slope failure 214

'headline disasters' 12–14

Health and Safety Executive 393, 396–7

health and safety regulations 383

heat stress 271–3, 424

heat-waves, frequency of 424

Hewitt, K. 16

high reliability school of thought on technological accidents 378–9

Hilo, Hawaii 174–5, 189

Hinman, G.W. 388

Hochrainer-Stigler, S. 105

Hong Kong 227, 229

Horton, C.A. 94

hot-spot volcanoes 177–8

Hoyos, C.D. 429

Huang, Z. 258

Huascaran disasters (1962) 149

Huff, W.A.K. 27

human capital approach to risk management 90

Human Development Index (HDI) 40–1, 54

human ecology 14–15

humanitarian assistance 324, 368; expenditure on 111–13

Hungr, O. 227

hurricane modification 256

hurricane warnings 262–4

hurricanes 18–19, 103–4, 236–41, 247, 256–66, 429; evacuation from 265–6

hydrographs 313, 320

hydrological drought 340–5

IBM plant, San José 159

ice storms 254

Indian Ocean tsunami (2004) 170

industrial accidents 376–7

influenza 278–80

insurance 114–18, 163–4, 224–5, 259–60, 325–9, 363, 392–3, 425; government schemes 118, 224–5, 325

Integrated Coastal Zone Management (ICZM) 131

integrated disaster risk management 402

integrated flood management, principle of 331

Intergovernmental Climate Prediction and Applications Centre (ICPAC) 365

Intergovernmental Panel on Climate Change (IPCC) 419–20, 424

International Search and Rescue Group (ISAR) 161

Irish Potato Famine (1865–69) 268–9, 347

Javelle, P. 330

jökulhlaups 193–5

Jones, F.O. 214

Kaczmarek, Z. 415

Kahn, F.I. 376

Kajoba, G.M. 347

Karabork, M.C. 415

Kashmir earthquake (2005) 164

'Katrina' (hurrcane, 2005) 27, 33, 40, 100, 235, 242–4, 260, 263, 300, 313, 319, 377

Keating, B.H. 188

Keeling, C.D. 409

Kelly, M. 361
Kelut volcano, Java 203
Key, D. 153
Kim, N. 38
Klein, R. 58, 235
Knox, J.C. 425
Kobe earthquake (1995) 40, 140, 144, 161, 164–5
Kocin, P.J. 254
KOF Index of Globalization 403–4
Kogan, F.N. 345
Krakatoa eruption (1883) 187–8
Krishnamurthy, P.K. 88
Kroo Bay, Sierra Leone 283
Kuhn, K. 273
Kummu, M. 40
Kunreuther, H. 106, 117
Kyoto Protocol 20

Ladbroke Grove railway accident 386
Lagadec, P. 272, 376
lahars 185–7, 197–8
land use planning 127–36, 171–5, 198–204, 230–3, 264–6, 297, 331–6, 368–9, 396–401
Landsea, C.W. 236, 245–7, 429
landslide dams 212–13
landslides 148–51, 187, 205–17, 220–1, 228–32; causes of 214; classification of 209; hazard reduction for 220–1, 231–2; mapping of 230–1; triggering of 216–17; types of terrain affected by 208–9; warning systems 228
Las Coinas landslide, El Salvador (2001) 216–17
lava flows 182–4; artificial barriers against 189
Lee County, Florida 131–2
levee effect 315–17, 321
Lewis, H.W. 386, 388
lightning 250, 296–7, 377–8
Lin, F.C. 171
Lindell, M.K. 377
Liu, W.T. 345
Lloyd, S.J. 429
locusts 269–70
Loma Prieta earthquake (1989) 172
loss acceptance 97
Loughnan, M. 273
Love waves (L-waves) from earthquakes 147–8
Lund, D.C. 417

McCabe, G.J. 354
McEntire, D.A. 18
McGee, T.K. 59
McGranahan, G. 431
MacGregor, D. 390
McGuire, B. 417
McGuire, W. 188
macro-protection from hazards 99–103
macro-zonation 128–31
magma 177–9
Mahrashtra earthquake (1993) 140
malaria 280–1, 284–5, 425–6
malnutrition 348–51
mangrove forests 255
March, George Perkins 51
Marco, J.B. 331
marine floods 309
Marmara earthquakes (1999) 164
MARS (Major Accident Reporting System) database 384
mass movement hazards 205–33; human and economic losses from 205–6
Mattinen, H. 114
maximum consequence distance 397
Maxwell, D. 113
Mejía-Navarro, M. 94
Meltsner, A.J. 84
Memphis 120–2
Merapi volcano, Java 188
meteorological drought 339–40
Meteorological Office 339
Meyer, S.J. 345
micro-protection from hazards 103–5
micro-zonation 131–6
migration 364
Mileti, D.S. 12, 19
Millennium Development Goals 20–1, 284
Miller, A. 27
Mirza, M.M.Q. 313
Mississippi River 319–21, 427
'Mitch' (hurricane, 1998) 18–19, 63, 224, 244–5, 259
mitigation of hazard impacts 98, 101–3, 106–8, 160–4, 193, 224–5, 232, 258–66, 285–6, 295, 323–9, 361–3, 391
modernization theory 403
Modified Mercali (MM) earthquake intensity scale 145–6

monsoon conditions 354–5
Montero, J.C. 271
Morrison, D. 418
Mozambique floods (2000) 329
mudflows 186
Munich Re Group 25, 31–3
Muntán, E. 232
Mustafa, D. 59
Myers, M.F. 19

Nash, J.R. 373
NATCAT (natural catastrophes) catalogue 29, 32
na-tech incidents 376–8
National Flood Insurance Program (NFIP), US 260–1, 264, 327–34
National Hurricane Center (NHC), US 262–3
natural disasters 29–33; death toll of 23; role of IT in management of 91–5
natural hazards 12–13; definition of 4–5; exposure to 37–8; and societal decisions 14; world map of 97
natural selection 274
Naumann, T. 177
Neal, J. 316
near-earth objects (NEOs) 417–19
Nepal 311
Nevado del Ruiz eruption (1985) 186–7
New Orleans 242–4, 319, 377
New Zealand 165–7, 225, 321–2, 363
Newhall, C.G. 177
Nicholls, R.J. 432
Ningxia earthquake (1920) 140
Noji, E.K. 161, 275
non-governmental organizations (NGOs) 108
Nordhaus, W.D. 247
normal accidents school of thought on technological hazards 378–9
Normal distribution 79
normalization of data 36
normalized difference vegetation index (NVDI) 366
Norris, F.H. 61
North Atlantic Drift 416–17
North Atlantic Oscillation (NAO) 414–15, 428–30; NAO Index 428
North Sea floods (1953) 314

Northeast Snowfall Impact Scale (NESIS) 254
Noy, I. 52
nuclear industry 387–91, 401
numerical weather prediction (NWP) 330
Nyberg, J. 429

Oakland–Berkeley Hills firestorm (1991) 296
ocean circulation 415–17
Office for the Coordination of Humanitarian Affairs (OCHA) 111, 121
Ogden, K. 114
Olsen, G.R. 112
Orissa 124
Ostro, B.D. 273

Pacific island nations 69
Pacific Tsunami Warning System (PTWS) 169–70
Pais, J. 64
Pakistan 310
Pall, P. 425
Palmer drought severity index (PDSI) 341, 345, 420
Pan Caribbean Disaster Preparedness and Prevention Project 261
paradigms for environmental hazards 14–20
Parker, D.J. 316
Parkfield, California 168
pastoralists 357–8, 363–4, 367
Paté-Cornell, M.E. 379
Paul, A. 262
peak ground acceleration (PGA) values 128, 171
Peduzzi, P. 24–5, 241
Peek, L.A. 64
Peel, M.C. 353
Peléan type of volcanic eruption 179
Pelée, Mount 130–1, 179
Pelling, M. 48
Perry, R.W. 377
Petley, D.N. 207
Petrow, T. 425
Phillips, B.D. 394
physical environment, changes in 35–6
Pielke, R.A. Jr 235, 245–7
Pigram, J.J. 369

Pinatubo eruptions (1991) 125, 178–9, 182, 185, 190, 193, 198
Piper Alpha disaster 379, 393
planning *see* land use planning
Plinian type of volcanic eruption 178
Ploughman, P. 27
political aspects of environmental issues 64–5
pollution 379; *see also* air pollution
population growth 63
pore water pressure 214
poverty 61–6, 356
precautionary principle 74
'Prepare, Act, Survive' policy 297
preparedness and preparedness planning 118–24, 164–6, 193, 199, 226–7, 261–2, 295–6, 329, 363–5, 393–5
Presidential disaster declarations (PDDs), US 109–10
primary waves (P-waves) from earthquakes 146–8
probabilism in risk perception 85
protection from hazards 98–106, 153–60, 189–92, 220–1, 221–4, 255–8, 294–5, 318–23, 358–61, 391
public health matters 273–4
public response to forecasts and warnings 125
Punongbayan, R.S. 186
pyroclastic flows 176, 179–80

Qi, R. 385
quantitative precipitation forecasting (QPF) 330
quasi-natural hazards 5

Rahman, M. 262
rainfall 237, 309, 311, 330, 425; reliability of 338–9
Ramanathan, V. 407
rapidly-developing depressions 252
Rasmussen, K. 377
Rauch, E. 236
Rayleigh waves from earthquakes 147
Reason, James 51, 385
recovery period of a disaster 106–8
regional climate models (RCMs) 421, 430
regional forcing 355

regrading to reduce landslide risk 220–1
reinsurance 115
reliability of systems against disaster 58
relief period of a disaster 106–8
relocation to avoid risk 333–4, 432
remote areas 67–8
remote sensing 91–3, 231
Renni, E. 377–8
resilience: concept and definition of 58–61; indicators of 60–1
retrofitting 105–6, 322
Richter scale 143
rift volcanoes 177–8
risk acceptance criteria 399
risk analysis 87
risk assessment 71–2, 75–82, 403; as distinct from risk perception 81–2; quantitative 71, 75–6; three-step process of 75
risk aversion 90
risk awareness 118
risk evaluation 87–8
risk management 71–2, 86–91; participatory approach to 88–91; sequential approach to 87
risk perception 72–4, 386–7; and communication 81–3; in practice 84–6
'risk society' concept 74, 403
risks: *acceptable* and *tolerable* 74, 89–90; definition of 11; magnitude–frequency relationships 75–8; nature of 71–4; *objective* and *subjective* 81; statistical analysis of 75–81; *voluntary* and *involuntary* 72–3
river channel enlargement 319
river floods 309, 315, 328
river flow duration curves 341–2
road travel and road accidents 386–7, 391
rock falls 209–10
Roll Back Malaria campaign 284–6
Ropelewski, C.F. 365
Ross, S. 27
Ross River virus (RRV) 283, 414
rotational landslides 213
Russell, S. 59

safety distance 397
Saffir–Simpson scale 75–7, 236–7

Sagan, L.A. 13
Sahel region of Africa 352–8
St Helens, Mt 185, 202
San Andreas fault 164, 166
San Francisco Bay area 121, 166, 168, 172, 227–8
San Francisco earthquake (1906) 163
Sanriku coast, Japan 159–60, 173
SARS (severe acute respiratory syndrome) 274
satellite sensing 91–3, 262
Schär, C. 424
Schmidlin, T.W. 254
Schwartz, R.M. 254
sea walls 318
sea-level rise (SLR) 314, 411, 415, 430–1
sea-surface temperatures (SSTs) and sea-surface temperature anomalies (SSTAs) 352–4, 427, 429
secondary waves (S-waves) from earthquakes 146–8
sediment traps 189
seismic risk, reduction of 88–9
seismicity patterns 167, 171
Self, S. 177
self-reliant mitigation of hazard impacts 101–3
Selvaraju, R. 365
sensitivity to environmental hazards 10
September 11th 2001 attacks 20
'set-back' ordinances 172–3
Sherrie, B.K. 61
shore-protection schemes 264–6; see also coastal areas
Siebert, L. 185
Siegrist, M. 329
Simmons, K.M. 250
Sioux City disaster (1989) 386
Sjöberg, L. 73
Skidmore, M. 65
slope angle 211, 214
slope-stability equations 231
SLOSH (Sea, Lake and Overland Surges from Hurricanes) model 264
Slovic, P. 73
Small, C. 177
small-island nations 68–9, 432
Smith, D.I. 395
Smith, K. 309

snow avalanches 207, 218–20; impact pressure of 220; starting of 219–20; see also avalanches
snow fences 222–3
snow melt 311
snow rakes 233
snowstorms 254
social amplification 85
social communication systems 94–5
social vulnerability index (SoVI) 54
socio-cultural hazards 5
socio-economic environment, changes in 36–7
soil liquifaction 148
Soldiers Grove, Wisconsin 333–4
Soufrière Hills volcano 180–1, 191–2, 195
Spence, R.J.S. 190
'spot' fires 290–1
SREX Report (2012) 420
stakeholder groups 119
standardized precipitation index (SPI) 340–1
Starr, C. 72
stationarity assumption 79
Stoddard, A. 108
storm detection 262
storm hazards 235–67; attitudes to 257–8
storm loss models 424
storm-proofed properties 257–8
storm suppression 256
storm surges 238, 259–60, 264, 313–14
storms, 'dry' and 'wet' 236
Stott, P.A. 272
Strömberg, D. 13, 25–7
subduction volcanoes 177–8
Sumatra earthquake (2004) 154
summer storms 247–51
sustainability science 51–2
sustainable development 74
'sustainable hazard mitigation' (Mileti and Myers) 19
sustainable livelihoods approach 59
Sutter, D. 250
'swarm' models 196–7
'Swiss cheese' disaster model 51, 385
Switzeland 233

'T' shelters 163
Tambora eruption (1815) 177

Tayag, J.C. 186
technological accidents 30, 33, 372–5, 393; theories about causes of 388–9
technological change 68–9
technological hazards 5–11, 74, 371–401; definition of 5–7, 373; factors improving safety 380–5; factors increasing risk 379–80; scale and nature of 373–8
tectonic hazards 176–204
tectonic plates 142, 151
Tegucigalpa 245
Teka, O. 88
teleconnections between climatic anomalies 352, 405
tephra 180–2
terrorism 371
Tessina landslide warning system 228
thermohaline ocean circulation (THC) 416
Thieken, A.H. 325
threat recognition 125
Three Mile Island accident (1979) 394, 396
Tibbits, A. 297
tiltmeters 197
Tipple, G. 127
T?hoku earthquake (2011) 6–7, 117, 140, 150, 152
Tone river 360–1
Tonoas Island landslides (2002) 230–1
Toon, O.B. 418
Tornado Alley 247
tornadoes 103–4, 247–50, 261, 264; deadliest in the US 249
total fire bans 294
Toya, H. 65
translational landslides 211
transport industry and transportation accidents 376–80, 386–7
tree-planting 273
Trenberth, K.E. 413, 420, 429
tropical depressions and tropial storms 236–8; see also cyclones
tsunamis 149–53, 187–8; deaths from 151; forecasts and warnings of 168–71; protection from 159–60
Tuffen, H. 186
Turner, B.A. 385
Turner, B.L. 52, 405
typhoons 236

Uccellini, L.W. 254
Uniform Building Code, US 159
uniformitarianism, assumption of 77
United Nations 20, 108, 347–8, 351
unreinforced masonry structures
    (URMs) 153–5, 159
urban heat-island effect 273
urbanization 67–8, 311–18

Vaiont landslide (1963) 215
Valery, N. 106
Valmeyer, Illinois 334
van der Kaay, H.J. 286
Vargas landslides (1999) 224–5
Varis, O. 40
vector organisms 273
vegetation: and avalanche risk 233;
    and fire risk 296; and landslide risk
    212–13, 221
vegation condition index (VCI) 345
Ventnor 213
Vesuvius 198–9
Vietnam 307
Village Disaster Risk Management
    Training (VDRMT) model 122–3
viral profiles 278
Vogt, J. 88
volcanic eruptions 35, 151; effects on
    weather and climate 182;
    precursory phenomena 196
volcanic explosivity index (VEI)
    177–9
volcanic gases 184–5, 197
volcanic hazards 176–204; deaths and
    other human impacts from 176–7;
    emergency planning for 194–5;
    forecasting and warning of 195–8;

influence of distance on 179–80;
    mitigation of impacts 193; *primary*
    179–85; protection against
    189–92; *secondary* 185–8
volcanic islands 188
volcanoes, nature of 177–9
Vu, T. 52
vulnerability and capacities index (VCI)
    54
vulnerability to environmental hazards
    9, 15, 52–8; definitions of 52–3;
    determinants of 54–7; *drivers* of
    61–9; measurement of 54; socio-
    economic aspects of 54, 57, 64;
    variations at household and family
    level in 54, 56

Walker cell circulation 411–14
Wang, X.L. 429
Ward, R. 309
warfare 371
Warren, A. 358
water flow velocity 299–300
water management 369
water rationing 365
water sprays on lava flows 189
water table, changes to 214
weather conditions, general warnings
    based on 227–30
weather-related disasters 35–7, 235
weathering 211, 214
Webster, P.J. 429
West Antarctic Ice Sheet (WAIS)
    432–4
West Nile virus (WNV) 277–8
White, Gilbert 14–16, 51–2
Wildavsky, A. 73

wildfires 268, 286–98, 413, 428;
    availability of fuel for 290–1;
    hazard reduction for 294–8;
    human influences on 288–9;
    ignition of 290; nature of 288–94;
    and weather conditions 291–4
wildland–urban interface 287
Wilhite, D.A. 362–3
willingness to pay: for improved
    forecasts 263; for safety
    improvements 90–1, 386
wind-induced building failure
    257–8
wind-storms 251–4
wind stress on buildings 103
winter storms 251–5; most severe in
    north-east USA 255
Witham, C.S. 176
WM-DAT database 299
Woodworth, P.L. 415
World Health Organization (WHO)
    273, 278–80, 286
World Meteorological Organization
    330

Yangtze River 303–4
yellow fever 277
Yokohama conference on reduction
    of natural disasters (1994) 20
York 324
Yucca Mountain project 390–1

Zeebrugge ferry accident (1987) 386,
    391
Zhamanshin crater 418
Zhang, X. 425
zoning 128–36, 198–202, 232–3